Deep Inelastic Scattering

Deep Inelastic Scattering

Robin Devenish
and
Amanda Cooper-Sarkar
Department of Physics
University of Oxford.

*This book has been printed digitally and produced in a standard specification
in order to ensure its continuing availability*

OXFORD
UNIVERSITY PRESS

Great Clarendon Street, Oxford OX2 6DP
Oxford University Press is a department of the University of Oxford.
It furthers the University's objective of excellence in research, scholarship,
and education by publishing worldwide in
Oxford New York
Auckland Cape Town Dar es Salaam Hong Kong Karachi
Kuala Lumpur Madrid Melbourne Mexico City Nairobi
New Delhi Shanghai Taipei Toronto
With offices in
Argentina Austria Brazil Chile Czech Republic France Greece
Guatemala Hungary Italy Japan South Korea Poland Portugal
Singapore Switzerland Thailand Turkey Ukraine Vietnam

Oxford is a registered trade mark of Oxford University Press
in the UK and in certain other countries

Published in the United States
by Oxford University Press Inc., New York

© Oxford University Press 2004

The moral rights of the author have been asserted

Database right Oxford University Press (maker)

Reprinted 2009

All rights reserved. No part of this publication may be reproduced,
stored in a retrieval system, or transmitted, in any form or by any means,
without the prior permission in writing of Oxford University Press,
or as expressly permitted by law, or under terms agreed with the appropriate
reprographics rights organization. Enquiries concerning reproduction
outside the scope of the above should be sent to the Rights Department,
Oxford University Press, at the address above

You must not circulate this book in any other binding or cover
And you must impose this same condition on any acquirer

ISBN 978-0-19-850671-3

Preface

The book gives an introduction to deep inelastic scattering in high energy particle physics, with particular reference to nucleon structure. As an experimental technique, deep inelastic scattering can trace its lineage back to the very discovery of the nucleus by Rutherford using the scattering data from Geiger and Marsden. In the 1970s and 1980s deep inelastic scattering was crucial to the emergence of the quark-parton model and the development of Quantum Chromodynamics (QCD) as the theory of the strong interaction. More recently the data from the HERA collider has led to advances in the understanding of QCD and its extension into the non-perturbative and high density regimes.

The book is intended for graduate students and others new to the field, it updates the overview of the subject given by Roberts *The Structure of the Proton*, CUP 1990, but from a more experimental perspective. It is intended to be self-contained, but does assume a general understanding of particle physics at the level of Perkins *Introduction to High Energy Physics*, CUP 2000 and relativistic quantum mechanics at the level of Aitchison and Hey, *Gauge Theories in Particle Physics*, Adam Hilger 1989 or Halzen and Martin, *Quarks and Leptons*, Wiley 1984.

The history of the subject is covered briefly. The theoretical formalism and the experimental techniques needed to study it in a modern context are explained in detail. Particular attention is paid to the extraction of parton momentum distributions and to the recent extension of the measurements and the formalism to very low values of parton momentum. Looking to the future, a clear exposition of the formalism required to analyse the high luminosity, high momentum transfer data from the HERA-II ep collider is given and prospects for electroweak physics at large momentum transfer are explored. The importance of accurate parton distribution functions for advances in hadron collider physics at the Tevatron and LHC is outlined. A brief account of fully polarized deep inelastic scattering is given. The exposition of detail in the text ia augmented by problems at the end of most chapters. A particular feature of the book is to illustrate the physics by plots drawn from published papers.

A major subject, revitalised by HERA, but not covered here is diffractive physics, it requires a text to itself. Another large subject, beyond the scope of this book, is deep inelastic scattering from nuclear targets.

Ackowledgements

We are indebted to the DESY Directorate and, in particular, the late Bjorn Wiik, for their determination to build HERA in the first place. We thank our colleagues in the 'deep inelastic community' and particularly those in the H1 and ZEUS collaborations for working so hard to produce many of the results discussed in this book. In more detail, the authors would like to express their thanks to Allen Caldwell, Vladimir Chekelian, Frank Close, John Dainton, Albert DeRoeck, Jeff Forshaw, Max Klein, Peter Landshoff, Alan Martin, Richard Roberts, Dan Stump, Robert Thorne, Wu-ki Tung, Andreas Vogt and Mike Whalley for many useful discussions. We are very grateful to four anonymous reviewers for a critical reading of an early version of the manuscript.We thank Irmgard Smith for drawing some of the diagrams, and Pete Gronbech and Stig Topp-Jorgensen for keeping the compters running. We thank our editor at OUP, Sonke Adlung, and his assistants for encouragement and patience. Finally, special thanks to Tony Doyle without whom this project would not have started.

Figures

We are grateful to the following publishers or organisations, who retain the original copyright, for permission to include figures. The details of the individual publication are given in the caption to the figure.

The American Physical Society. From *Physical Review D*: figures 4.9, 5.9, 5,14, 6.1, 6.2, 6.3, 6.4, 6.5, 6.7, 8.4, 10.14, 10.15, 10.16, 10.27 and 11.3. From *Physical Review Letters*: figures 5.13, 9.24, 10.5, 10.7, 10.8, 10.10, 10.13, 10.17, 10.18 and 10.20. From *Reviews of Modern Physics*: figures 5.11 and 9.7.

Annual Reviews. From *Annual Review of Nuclear and Particle Science*: figures 11.8 and 11.9.

Springer Verlag. From *Zeitschrift für Physik C*: figures 1.4, 5.7 and 9.4. From *European Physical Journal C*: figures 5.10, 5.16, 6.6, 6.8, 6.9, 7.1, 7.5, 7.6, 7.7, 7.10, 9.21, 10.25, 12.2, 12.3 and 12.5.

Elsevier. From *Nuclear Physics B*: figures 5.12 and 11.2. From *Physics Letters B*: figures 4.8, 5.15, 5.17, 5.18, 7.8, 7.9, 9.3, 10.21, 11.5, 11.6, 11.7, 12.1, 12.4, 12.6, 12.7 and 12.8.

The Polish Physical Society. From *Acta Physica Polonica*: figures 9.13, 9.17, 9.18, 9.25, 9.26 and 9.27.

World Scientific. From *International Journal of Modern Physics A*: figure 11.1. From *Conference Proceedings*: figures 9.16 and 9.20.

Cambridge Univesity Press, for figure 3.12.

DESY Hamburg. From *Conference Proceedings*: figures 5.8, 8.5 and 8.7.

The H1 and ZEUS collaborations for figures 8.1, 8.2 and 8.3.

Glossary/acronyms

BFKL	Balitsky Fadin Kuraev Lipatov
BGF	Boson–gluon fusion
CC	Charged current (i.e. $W^{\pm}$ mediated weak interactions)
CCFM	Ciafaloni Catani Fiorani Marchesini
CGC	Colour glass condensate
CMS	Centre-of-mass system (sometimes CM)
CTEQ	Coordinated theoretical–experimental project on QCD
DGLAP	Dokshitzer Gribov Lipatov Altarelli Parisi
DIS	Deep inelastic scattering
Evolution	dependence on Q^2 or equivalent variable that is calculable in pQCD
GRV	Glück Reya Vogt, PDF global fitting team
HERA	Hadronen–Elektronen Ring Anlage (Hadron-electron collider at DESY)
ISR	Initial state radiation
LEP	Large electron–positron collider (at CERN)
LHC	Large hadron collider (at CERN)
LLA	Leading logarithm approximation
LO	Leading order
MRST	Martin Roberts Stirling Thorne, PDF global fitting team
$\overline{\text{MS}}$	Modified minimal subtraction (scheme of renormalisation)
NC	Neutral current (i.e. γ^* or Z^0 mediated interactions)
NLLA	Next-to-leading logarithm approximation
NLO	Next-to-leading order
NNLO	Next-to-next-to-leading order
OPE	Operator Product Expansion
PDF	Parton (momentum) density function
pQCD	Perturbative QCD
Q^2	4-momentum transfer squared in DIS
QCD	Quantum chromodynamics
QCDC	QCD Compton scattering
QED	Quantum electrodynamics
QPM	Quark–parton model
RGE	Renormalization group equation
Scale	renormalization — Q^2 value at which QCD parameters are regularized
	factorization — Q^2 value at which parton densities are regularized
Twist	mass–dimension minus spin of an operator; 'higher twist' terms are inverse powers of Q^2 w.r.t. the leading term

Contents

1 Introduction — 1
  1.1 Rutherford scattering — 1
    1.1.1 Form factors — 3
  1.2 Inelastic scattering — 4
  1.3 Heroic age of DIS — 6
  1.4 Importance of weak probes — 7
  1.5 Partons and Quantum Chromodynamics — 8
  1.6 Extraction of parton density functions — 10
  1.7 DIS at colliders — 11
  1.8 Other topics — 12
  1.9 Outlook — 12
  1.10 Problems — 13

2 The quark–parton model — 14
  2.1 The essential idea of the parton model — 14
    2.1.1 Elastic electron–muon scattering — 16
    2.1.2 Elastic electron–quark scattering — 18
  2.2 Inelastic e– or μ–hadron scattering — 19
  2.3 Neutrino–induced DIS — 23
    2.3.1 Elastic neutrino–electron scattering — 23
    2.3.2 Elastic neutrino–quark scattering — 25
    2.3.3 Inelastic neutrino–hadron scattering — 26
    2.3.4 Neutrino scattering via the neutral current — 27
  2.4 Quark–parton model tests — 28
  2.5 Sum rules — 31
  2.6 Summary — 32
  2.7 Problems — 33

3 QCD and formal methods — 35
  3.1 Quantum Chromodynamics — 36
    3.1.1 The QCD Lagrangian — 36
    3.1.2 QCD colour factors — 38
  3.2 Some simple tree–level diagrams — 40
  3.3 Infrared and collinear singularities — 43
  3.4 Renormalization — 45
    3.4.1 Regularization and renormalization schemes — 46
    3.4.2 Charge renormalization in QED — 47

		3.4.3 The QCD running coupling at leading order	50
	3.5	Renormalization group methods	52
		3.5.1 Anomalous dimensions	54
	3.6	The QCD running coupling	56
	3.7	The hadronic tensor $\mathbf{W}_{\mu\nu}$	57
		3.7.1 Light cone dominance	59
	3.8	The operator product expansion	60
		3.8.1 OPE applied to DIS	61
		3.8.2 RGE calculation of moments	64
	3.9	Factorization	65
	3.10	Summary	66
	3.11	Problems	67
4	**QCD improved parton model**		**69**
	4.1	Improving the parton model	69
		4.1.1 Splitting Functions	75
		4.1.2 Running coupling and the OPE	77
	4.2	The DGLAP equations	79
		4.2.1 $\mathbf{F}_2$ and $\mathbf{F}_L$	81
		4.2.2 A useful approximation	83
	4.3	DGLAP evolution at LO	84
		4.3.1 Valence quark evolution	84
		4.3.2 Singlet and gluon evolution	86
	4.4	Higher twist	87
	4.5	Heavy quarks	89
		4.5.1 Matching prescriptions for α_s	89
		4.5.2 Heavy quark production in DIS	91
		4.5.3 Variable flavour number schemes	91
	4.6	Beyond NLO	95
	4.7	Summary	95
	4.8	Problems	96
	4.9	Appendix: LO splitting functions and anomalous dimensions	100
5	**DIS experiments and data**		**101**
	5.1	Some numbers	101
	5.2	Kinematics	102
		5.2.1 Fixed target	102
		5.2.2 HERA collider	104
	5.3	Detectors	107
		5.3.1 The E665 muon scattering experiment	108
		5.3.2 The CCFR neutrino scattering experiment	110
		5.3.3 Detectors for the ep collider HERA	111
	5.4	Measurement of the cross-section	113
		5.4.1 Radiative corrections	115
		5.4.2 Acceptance and other corrections	117

x   Contents

    5.5 Nuclear effects   120
    5.6 Structure function data   122
        5.6.1 $\mathbf{F_2}$ 123
        5.6.2 $\mathbf{F_2^{\nu N}}$ and $\mathbf{x}F_3^{\nu N}$ 127
        5.6.3 $\mathbf{F_L}$ 130
        5.6.4 $\mathbf{F_2^c}$ 131
    5.7 $\mathbf{F_2^p}$ at very low $\mathbf{Q}^2$ 135
    5.8 Summary   136
    5.9 Problems   137
    5.10 Appendix: cross-sections and luminosity   138
        5.10.1 The size of DIS cross-sections   138
        5.10.2 Beam flux and luminosity   138

6 Extraction of parton densities   140
    6.1 Determining parton distribution functions   140
    6.2 Treatment of data sets in global analyses   141
        6.2.1 Nuclear binding corrections   141
        6.2.2 Data consistency   142
    6.3 Global fits: the general formalism   142
        6.3.1 The form of the parameterization   144
        6.3.2 The flavour composition of the sea   145
        6.3.3 Global fits: the relationship of the measurements to the parton distributions   146
    6.4 Results on PDF extraction   147
        6.4.1 Results from the experimental collaborations   148
        6.4.2 Results from the theoretical groups   153
    6.5 Information on PDFs from non-DIS processes   156
        6.5.1 Quark distributions   157
        6.5.2 The gluon distribution   158
    6.6 Theoretical and 'model' uncertainties   159
        6.6.1 Model assumptions   159
        6.6.2 Heavy quark production schemes   161
        6.6.3 Higher twist contributions   162
        6.6.4 The need to go beyond NLO and scale uncertainty   163
        6.6.5 Alternatives to the DGLAP evolution equations?   165
    6.7 Treatment of correlated systematic uncertainties   165
        6.7.1 Offset methods   167
        6.7.2 Hessian methods   168
        6.7.3 Diagonalization and eigenvector PDF sets   169
        6.7.4 Normalizations   170
        6.7.5 Comparison of offset and Hessian methods   170
        6.7.6 χ^2 tolerance   171

Contents   xi

		6.7.7	Uncertainties in predicting high energy cross-sections	174

      6.7.7 Uncertainties in predicting high energy cross-sections   174
      6.7.8 Alternative statistical techniques   174
  6.8 Dynamically generated partons, the GRV approach   175
  6.9 Future prospects for information on the gluon   176
  6.10 Summary   178
  6.11 Appendix: Comparability of evolution programs   178

7 α_s from scaling violations and jets at high Q^2   181
  7.1 Methods of determining $\alpha_s(M_Z^2)$ from structure function data   182
      7.1.1 Determinations of $\alpha_s(M_Z^2)$ from GLS sum-rule   183
  7.2 Determinations of $\alpha_s(M_Z^2)$ from structure function data: DGLAP NLO QCD fits   183
      7.2.1 Theoretical and model uncertainties   183
      7.2.2 Results   185
  7.3 Determinations of $\alpha_s(M_Z^2)$ from structure function data: extending the theoretical framework   188
  7.4 Jet production in DIS   189
      7.4.1 Breit frame kinematics   190
      7.4.2 Cross-section calculations   192
  7.5 Jet measures   194
      7.5.1 The cone algorithm   195
      7.5.2 The $\mathbf{k}_T$ cluster algorithm   196
      7.5.3 Longitudinally invariant $\mathbf{k}_T$ algorithm   196
  7.6 Description of jet data at NLO   197
  7.7 α_S from DIS jets   200
  7.8 Combined analysis of NC jet and inclusive data   203
  7.9 Summary   204
  7.10 Problems   206
  7.11 Appendix: transformation to the Breit frame   207

8 DIS at high Q^2   209
  8.1 Cross-sections for unpolarized lepton beams   209
      8.1.1 Neutral Current   209
      8.1.2 Charged Current   211
  8.2 Unpolarized high–$\mathbf{Q}^2$ data   212
      8.2.1 PDF extraction from high–$\mathbf{Q}^2$ data   214
  8.3 Cross-sections for polarized lepton beams   217
      8.3.1 Neutral Current   217
      8.3.2 Charged Current   218
  8.4 Extraction of electroweak parameters   218
      8.4.1 M_W measurements   218
      8.4.2 Measurements of the quark weak neutral couplings   221

8.5	Summary	223
8.6	Problems	223
8.7	Appendix: formalism for NC high-Q^2 DIS	224

9 DIS at low x — 228

- 9.1 Approaches at low x — 229
  - 9.1.1 Summation schemes — 230
  - 9.1.2 Low x and Mellin moments — 232
- 9.2 DGLAP at low x — 232
  - 9.2.1 Double asymptotic scaling — 234
  - 9.2.2 Singular input distribution — 235
- 9.3 Regge Theory — 236
  - 9.3.1 Hard and soft Pomerons — 241
- 9.4 The BFKL equation — 244
  - 9.4.1 Multiple gluon emission at small x — 244
  - 9.4.2 The reggeized gluon and the LO BFKL equation — 245
  - 9.4.3 BFKL for hadronic processes — 249
  - 9.4.4 Beyond the LO BFKL with fixed α_s — 251
  - 9.4.5 BFKL — discussion — 254
- 9.5 Angular ordering and the CCFM equation — 256
- 9.6 Unitarity and saturation — 260
  - 9.6.1 High density gluon dynamics — 261
- 9.7 Dipole models — general formalism — 263
- 9.8 Dipole models — examples — 266
  - 9.8.1 The Forshaw–Kerley–Shaw model — 266
  - 9.8.2 The Golec-Biernat–Wüsthoff model — 266
  - 9.8.3 Geometrical scaling — 269
  - 9.8.4 Dipole models and Q^2 evolution — 269
  - 9.8.5 Dipole models and the colour glass condensate — 270
- 9.9 The description of low-x inclusive data — 273
- 9.10 Summary — 275
- 9.11 Problems — 275

10 Hadron induced DIS — 278

- 10.1 Rapidity — 278
- 10.2 The cross-section for a hard hadronic process — 279
- 10.3 The Drell–Yan process — 281
  - 10.3.1 Kinematics and LO cross-section — 282
  - 10.3.2 Transverse momentum and QCD corrections — 285
- 10.4 W & Z boson production — 289
  - 10.4.1 Z boson production — 289
  - 10.4.2 W boson production — 291
  - 10.4.3 W decay asymmetry — 292
- 10.5 High p_T jet production — 294

10.5.1 Parton–parton kinematics	295
10.5.2 Single jet inclusive cross-section	296
10.5.3 Dijet cross-sections	299
10.6 Isolated photon production	304
10.7 Hadronic DIS at the LHC	307
10.7.1 Jet physics at the LHC	308
10.7.2 Collider luminosity	311
10.7.3 Parton-parton luminosity	312
10.8 Summary	314
10.9 Problems	314
10.10 Appendix: $\sum |\mathcal{M}|^2$ for tree-level sub-processes	316

11 Polarized DIS — 318

11.1 Formalism	318
11.2 The 'spin crisis'	319
11.3 Experiments and data	321
11.4 Theoretical framework	323
11.4.1 Sum rules	324
11.4.2 Wandzura–Wilczek relation	324
11.4.3 Axial anomaly	325
11.5 QCD analysis	326
11.6 Semi-inclusive asymmetries	331
11.7 Δg	333
11.7.1 HERMES	334
11.7.2 COMPASS	334
11.7.3 RHIC $\vec{p}\vec{p}$	336
11.8 Transverse spin asymmetries	338
11.9 Summary	339
11.10 Appendix: formalism for polarized DIS	339

12 Beyond the Standard Model — 343

12.1 General remarks	343
12.2 Quark compositeness and form factors	344
12.3 Direct leptoquark searches	345
12.4 Contact interactions	350
12.4.1 Compositeness	351
12.4.2 Leptoquarks	353
12.5 W production and anomalous high p_T leptons	354
12.5.1 Single top production	359
12.6 Summary	361
12.7 Problems	361
12.8 Appendix: LQ classification & CI couplings	362

A Dirac equation and some other conventions — 364

B Phase space and cross-sections — 368

C	DIS cross-sections	372
D	Feynman rules	378
E	Monte Carlo codes	383
F	Data sources	387
G	Parton Parameterizations	389
Bibliography		394
Index		400

1
Introduction

The basic idea of scattering experiments as a probe of spatial structure and forces is introduced. The meaning of the term 'deep inelastic' is explained using elementary arguments. A broadly historical, but brief, route into the structure of the book starts with an outline of the early measurements at SLAC using high energy electrons and the complementary results from the first high energy neutrino beams. The chapters on the parton model and the development of quantum chromodynamics provide the basis for all that follows. Much of the remainder of the book describes and explains the physics of deep inelastic scattering at high energy colliders, particularly HERA but also the Tevatron and the LHC.

1.1 Rutherford scattering

The use of scattering experiments to study the structure of atoms and their constituents has its origin in the work of Rutherford and co-workers in the first decades of the twentieth century. By measuring the energy and angle of a probe of known input momentum that has been scattered by an object, one can study aspects of the force producing the scattering and the distribution of 'charge' in the object. If the nature of the force is well known, for example by using a point-like electromagnetic probe such as an electron or muon, then the focus will be on the study of structure.

Start with a very simple example, elastic scattering of a point-like mass m from a fixed scattering centre with a potential of the form

$$V(r) = Zg^2 \frac{e^{-\mu r}}{r},$$

where units with $\hbar = c = 1$ are assumed (see Appendix A for details). The force is of the Yukawa type for the interaction of particle with 'charge' or coupling constant g with the charge Zg of the fixed force centre, through the mediation of a quantum of mass μ. The incident and final momenta of the probe and the scattering angle, θ, are shown in Fig. 1.1. As the force has a limited range, the probe particle is described by a free particle wave function for $r \gg \mu^{-1}$. The matrix element for the interaction then has the form

$$\langle \mathbf{k}'|V(r)|\mathbf{k}\rangle = \int e^{-i\mathbf{k}'\cdot\mathbf{r}} V(r) e^{i\mathbf{k}\cdot\mathbf{r}} d^3\mathbf{r}$$

2   Introduction

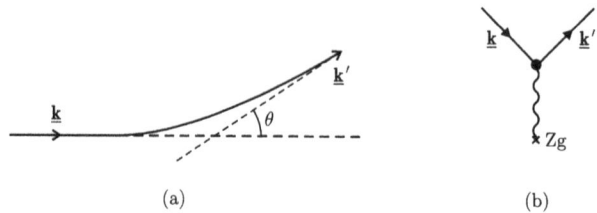

Fig. 1.1 Rutherford scattering. (a) Definition of the incident and final 3-momenta and the scattering angle. (b) Momentum space diagram for the scattering from a fixed force centre.

$$= \frac{4\pi Z g^2}{q^2 + \mu^2}, \tag{1.1}$$

where $q^2 = (\mathbf{k} - \mathbf{k}')^2$ is the squared momentum transfer. In this case, since there is no target recoil, $q^2 = 4|\mathbf{k}|^2 \sin^2(\theta/2)$. For more on the evaluation of the matrix element, see Problem 1 at the end of the chapter. Eq. (1.1) and Fig. 1.1(b) show the simplicity and essence of Feynman or momentum space diagrams — in momentum space the matrix element for the single scattering process ('lowest order') has the form of a coupling constant squared times a 'propagator' term, $(q^2 + \mu^2)^{-1}$. The spatial integral has been replaced by a product of factors. Feynman realised that one could derive 'rules' from a quantum field theory Lagrangian that would allow the contribution from a given diagram to be written down directly in momentum space. Another important feature of Feynman diagrams, which is not apparent for the present non-relativistic calculation, is that they are Lorentz invariant and give both time orderings for the relativistic propagator. The formal derivation of the Feynman rules for a relativistic field theory is a fairly elaborate process, but once established the rules are straightforward to use. There are many accessible introductions to field theory, for example, Aitchison and Hey (1989) or Peskin and Schroeder (1995). Simple Feynman diagrams will be used frequently, particularly in Chapters 2–4 and a summary of the rules for the standard model of particle physics is given in Appendix D.

Assuming that the interaction is sufficiently weak for first order perturbation to be adequate, the cross-section is given by application of Fermi's Golden Rule

$$\text{flux} \times \sigma = 2\pi |\langle \mathbf{k}'|V(r)|\mathbf{k}\rangle|^2 \rho_f,$$

where ρ_f is the density of final states. As target recoil is being ignored, ρ_f depends only on the scattered probe particle

$$\rho_f = \frac{k'^2 d\Omega \, dk'}{(2\pi)^3 dE},$$

where E is the total final state energy. Since the processes considered here involve beams of highly relativistic particles, one may ignore the mass of the beam particle and take $dk'/dE \to 1$. This also means that the beam particles are travelling with speed $\approx c$, so the flux factor in the above expression reduces to 1 and one obtains

$$\frac{d\sigma}{d\Omega} = \frac{k'^2}{(2\pi)^2} \left(\frac{4\pi Z g^2}{q^2 + \mu^2}\right)^2. \quad (1.2)$$

Taking the limit $\mu^2 \to 0$ and making the identification $g^2 = e^2/(4\pi\varepsilon_0)$, one finds the well-known expression for the Rutherford scattering differential cross-section (using $k'^2 = k^2 = |\mathbf{k}|^2$) and $q^2 = 4k^2 \sin^2(\theta/2)$)

$$\frac{d\sigma}{d\Omega} = \frac{\alpha^2 Z^2}{4k^2 \sin^4(\theta/2)}, \quad \text{or} \quad \frac{d\sigma}{dq^2} = 4\pi\alpha^2 \frac{Z^2}{q^4}. \quad (1.3)$$

The first form shows that the differential cross-section has a very strong dependence on the scattering angle and falls steeply as θ increases. This comes from the $1/q^4$ term and is a consequence of the exchange of a massless quantum — the photon. The cross-section for the Yukawa case, Eq. (1.2), shows that the strength of the angular behaviour will depend on the relative magnitudes of $|\mathbf{k}|^2$ and μ^2.[1] For $|\mathbf{k}|^2 \gg \mu^2$ the angular dependence will be strong and approach that of Rutherford scattering, whereas for $|\mathbf{k}|^2 \ll \mu^2$ the angular dependence will be almost negligible.

Although the full relativistic formulae are more complicated, these essential ideas will reappear many times in this book.

1.1.1 Form factors

The next step is to allow the charge on the fixed scattering centre to be distribution over a limited but finite region of space. Consider the electromagnetic case

$$V(r) = Zg^2 \int \frac{\rho(\mathbf{r}')}{|\mathbf{r} - \mathbf{r}'|} d^3\mathbf{r}' \quad \text{with} \quad \int \rho(\mathbf{r}') d^3\mathbf{r}' = 1,$$

where $\rho(\mathbf{r}')$ gives the relative spatial distribution of charge. The matrix element now involves a double integral, over $\mathbf{r}$ as before, and in addition, over $\mathbf{r}'$ covering the region in which the charge is non-zero.

$$\begin{aligned}\langle \mathbf{k}'|V(r)|\mathbf{k}\rangle &= \int \frac{\rho(\mathbf{r}')}{|\mathbf{r} - \mathbf{r}'|} e^{i\mathbf{q}\cdot\mathbf{r}} d^3\mathbf{r}' d^3\mathbf{r} \\ &= \frac{4\pi Z g^2}{q^2} F(q^2), \quad (1.4)\end{aligned}$$

---

[1] Strictly, one should consider q^2 rather than $|\mathbf{k}|^2$, but the magnitude of q^2 is limited by $|\mathbf{k}|^2$, essentially the incident energy available.

4 Introduction

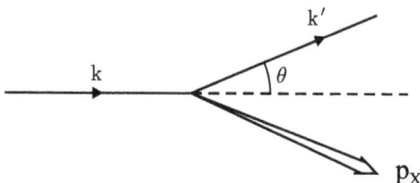

Fig. 1.2 Momenta and scattering angle for $\ell N \to \ell' X$ in the fixed–target frame.

where $\mathbf{q} = \mathbf{k} - \mathbf{k}'$ and $F(q^2)$ is the *form factor* of the charge distribution

$$F(q^2) = \int \rho(\mathbf{r}') e^{i\mathbf{q}\cdot\mathbf{r}'} \, d^3\mathbf{r}'.$$

The details of how to do the integrals are covered in Problem 2. The charge distribution ρ will involve one or more parameters related to the spatial extend over which the charge is distributed, giving rise to well-defined scales in the form factor. For $|qa| \ll 1$, where a gives the characteristic length scale of ρ, $F(q^2)$ may be approximated by

$$F(q^2) = 1 - \frac{q^2}{6}\langle r^2 \rangle, \tag{1.5}$$

where $\sqrt{\langle r^2 \rangle}$ is the root mean square radius of the charge distribution.

As the charge is now 'smeared out', the chance of a large-angle scatter is reduced compared to the point-like case and one expects the form factor to decrease quite rapidly as a function of q^2. As a simple example, it is found that the charge form factor of the proton may be parameterized by a dipole function of the form

$$F_p(q^2) = \frac{1}{(1 + q^2/m_0^2)^2},$$

where $m_0 \approx 0.84 \,\text{GeV}$ is a parameter fitted to the data. As Problem 3 shows, it is straightforward to relate m_0 to the root mean square charge radius of the proton and hence to get a measure of its size.

1.2 Inelastic scattering

So far the emphasis has been on elastic scattering and recoil has been ignored. Here, brief consideration is given to what can be learnt by measuring the energy of the scattered probe in both elastic and inelastic high energy collisions. The kinematics of the process $\ell(k) + N(p) \to \ell'(k') + X(p_X)$ are sketched in Fig. 1.2, for the fixed–target frame. Using 4-momentum conservation, $p^2 = m_N^2$, $p_X^2 = M_X^2$, ignoring the mass of the beam particle ℓ and defining the momentum transfer $q = k - k'$, one finds

$$k + p = k' + p_X, \quad \text{or} \quad p_X = q + p_X$$

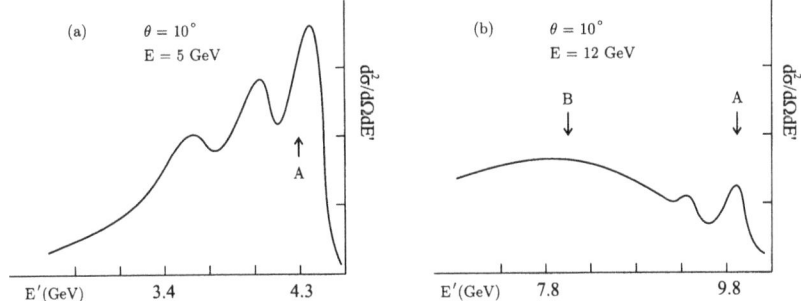

Fig. 1.3 The distributions of scattered energies, E', at a fixed scattering angle of $10°$ for $ep \to e'X$ in the fixed-target frame at two incident energies of 5 and 12 GeV. Note that the E' scale is decreasing right to left.

$$p_X^2 = p^2 + q^2 + 2p \cdot q, \text{ giving}$$
$$Q^2 \equiv -q^2 = 2p \cdot q + m_n^2 - m_X^2, \quad (1.6)$$

where the last equation defines Q^2. In the fixed-target frame in which the target is at rest initially, with E the incident energy and E' the scattered energy of the probe ℓ and scattering angle θ, the last equation becomes

$$2EE'(1 - \cos\theta) = 2m_N(E - E') + m_N^2 - m_X^2. \quad (1.7)$$

Also, allowing for recoil, $Q^2 = 4EE'\sin^2(\theta/2)$. Equation (1.6) or (1.7), shows that the scattering angle and scattered energy are not independent if the final state hadronic mass is specified ('exclusive process'). On the other hand, if one integrates over all hadronic final states ('inclusive process'), the two variables are independent and both are needed to describe the scattering. While the experimentalist will focus on how to measure E' and θ, it is best to use Lorentz invariant variables for more general discussions. The two most commonly used are Q^2 and $x = Q^2/(2p \cdot q)$, the latter is Bjorken x, which is dimensionless and has a convenient physical interpretation in the parton model. In the particular case of elastic scattering $m_X^2 = m_N^2$ giving $x = 1$ and

$$E' = \frac{E}{1 + (E/m_N)(1 - \cos\theta)}.$$

Figure 1.3(a) shows a sketch of a typical spectrum for an inclusive measurement of E' with an incident electron beam energy of 5 GeV at a fixed scattering angle θ of $10°$ on a proton target. The well defined peaks at values of E' close to the incident beam energy correspond to excitation of specific nucleon resonances with fixed M_X^2, such as the $\Delta(1232)$ indicate by the arrow at A. Plot (b) shows the same reaction at the same fixed scattering angle, but at the higher incident energy of 12 GeV. Now the peaks corresponding to individual resonances are much less prominent, but there is broad region marked at B where the cross-section is large. The scattering at large M_X and large momentum transfer is known as *deep inelastic scattering* or DIS.

One way to view the large cross-section is that it is made up of the many possible final states available at high energies. Another follows Rutherford's insight and attributes it to a single scattering of the probe from a constituent of the proton. It is the latter that will be explored in detail in this book. If the constituent were at rest and the scattering elastic, then E' ought to show a sharp peak at

$$E' = \frac{E}{1 + (E/m_c)(1 - \cos\theta)},$$

where m_c is the mass of the constituent. Of course for a particle confined to a linear dimension of order 1 fm, the uncertainty principle requires that the particle must have a minimum momentum of order 200 MeV and the position of the peak will be smeared out.

As will be elaborated in detail in later chapters, the double differential cross-section for high energy inelastic scattering mediated by virtual photon exchange has the form

$$\frac{d^2\sigma}{dQ^2\,dx} = \frac{4\pi\alpha^2}{Q^4 x}\left[(1-y)F_2(x,Q^2) + xy^2 F_1(x,Q^2)\right], \qquad (1.8)$$

where $y = Q^2/(sx)$, $s = (k+p)^2$ is the total ℓp centre-of-mass energy and $F_i(x, Q^2)$ are the *structure functions* of the proton. These latter are the extension to inclusive inelastic scattering of the form–factor for elastic scattering, note that they depend on both x and Q^2 in general. The variable y is related to the scattering angle in the lepton–parton centre-of-mass frame by $y = (1 - \cos\theta)/2$. The dominant Q^2 behaviour is the overall Q^{-4} from the virtual photon exchange. A similar expression may be written down for neutrino induced DIS, but without the steep Q^2 dependence and with a third structure function.

1.3 Heroic age of DIS

Electron scattering was used extensively as a tool by Hofstader and colleagues in the 1950s and 60s to determine the charge distribution and radii of nuclei. Precision measurements of the elastic form–factors of nucleons were also made at this time.

The heroic age of DIS opened in 1968 with the announcement of the first preliminary results of high energy inclusive inelastic scattering experiments with the (then) 20 GeV electron linear accelerator at Stanford (SLAC). The cross-section was large and it was found that at fixed values of x (in the range $0.1 < x < 0.8$) that the structure functions showed little or no variation with Q^2 (over a range roughly $1 < Q^2 < 10\,\text{GeV}^2$). This is in striking contrast to the rapid decrease, of over two orders of magnitude, found for the proton elastic form factors in the same range of Q^2.

Perhaps it is not so surprising that an inclusive inelastic structure function should show a much weaker dependence on Q^2 as it the sum over all possible final states, whereas for the elastic form factor one has to pay the additional penalty of keeping the proton intact. However this argument does not explain why the structure functions appear to depend only on the variable x.

The behaviour $F_i(x, Q^2) \to \tilde{F}_i(x)$ for $Q^2 \to \infty$ at fixed x had actually been predicted by Bjorken (1969) and is known as *Bjorken scaling*. Bjorken derived the result using Gell-Mann's quark model current algebra to give the behaviour of 'almost equal time' commutators of the proton's electromagnetic currents at infinite momentum and relating these to the structure functions using dispersion relations (some of the flavour of this rather formal approach will be covered in Chapter 3). In the same paper, Bjorken also predicted that the neutrino structure functions should scale and that the total νN cross-section, integrated over x and Q^2, should grow linearly with the laboratory energy of the incident neutrino beam.

Not long afterwards, Feynman gave a simple physical explanation for Bjorken scaling on the basis of the parton model which he had recently developed to explain the gross features of high energy inclusive hadron–hadron scattering (Feynman 1969). The parton model will be discussed in detail in Chap. 2, but the essence of the idea is that the electromagnetic or weak probe scatters *elastically* off a point-like constituent (parton) of the proton. There is no scale or dependence on Q^2 in F_i because the scattering centre is a point. The variable x is identified with the fraction of the proton's momentum carried by the struck parton, in a frame in which the proton has infinite longitudinal momentum. Feynman developed the parton model in more detail for DIS in (Feynman 1972). In addition to explaining Bjorken scaling, the parton model gives an explicit model for the structure functions, for example

$$F_2(x) = \sum_i e_i^2 x f_i(x),$$

where the sum is over partons with charge e_i and $f_i(x)$ are unknown, but universal functions for a given hadron, giving the probability that parton i is carrying a fraction x of the hadron's momentum. The f_i are known as parton momentum densities.

Full accounts of the development of SLAC and the first DIS measurements are given in the Nobel lectures for 1990 by Friedman, Kendall and Taylor and reproduced in Friedman *et al.* (1991).

1.4   Importance of weak probes

Having established that the parton model gives a natural explanation of Bjorken scaling, the next question is what are the partons and why are

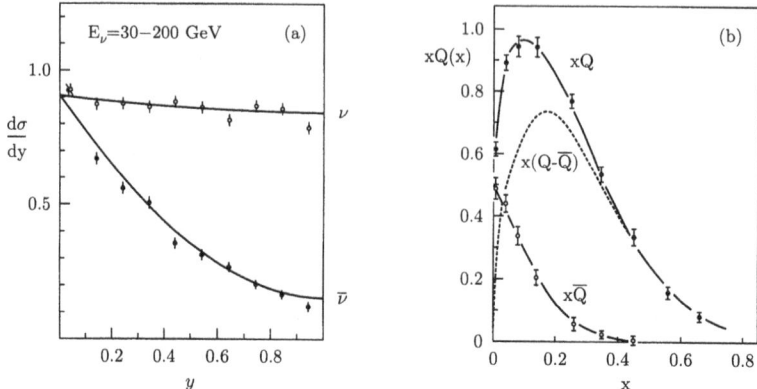

Fig. 1.4 (a) the $y = (1 - \cos\theta)/2$ distributions for ν and $\bar{\nu}$ scattering. (b) the summed quark, Q, and antiquark, $\bar{Q}$, momentum distributions extracted from them. (Adapted from de Groot et al. 1979).

they not seen in the hadronic final state debris? To be more precise, are partons quarks? Up to the discovery of Bjorken scaling, quarks had been used primarily as 'static' objects to explain the large number of meson and baryon resonances. Now there was an opportunity to get evidence of the quarks as dynamical objects.

If the partons have a well defined spin, the two electromagnetic structure functions F_1 and F_2 are related. It is possible to separate them by combining measurements of structure functions taken with different centre-of-mass energies. From this it was established that the partons have spin 1/2.

The virtual photon 'measures' the average charge squared on the partons. If partons could be isolated in the final state then exclusive measurements could pin down their properties. As they do not emerge, an alternative is to vary the probe. Neutrino scattering and the handedness of the weak interaction have played and continue to play a crucial role in disentangling the nature of partons and in establishing that they do indeed carry the charge and flavour quantum numbers of quarks. By exploiting the different angular dependence expected for ν and $\bar{\nu}$ scattering, it is possible to show that a $q\bar{q}$ sea exists at low x values, in addition to the 'valence' quarks associated with the hadron's static quantum numbers. Figure 1.4 shows an example of the neutrino cross-sections as a function of y — which is linearly related to the scattering angle — and the resulting Q and $\bar{Q}$ momentum distributions. Full details are given in Chapter 2.

1.5 Partons and Quantum Chromodynamics

The identification of partons with quarks and the universality of the parton momentum distributions led very quickly to many successful applications, but there were severe difficulties with the parton model. The first was clear evidence that there must be other constituents in the proton carrying momentum, but not coupling to either electromagnetic or weak probes. This came from the observation that the so-called momentum sum rule was not saturated by the measured parton distribution functions. If all constituents are summed over, then by definition

$$\sum_i \int_0^1 x f_i(x)\, dx = 1,$$

whereas the measured value for this quantity was only about 0.5. It was jokingly suggested that there must be 'glue' holding the proton together, and this would be carrying the missing fraction of the total momentum. The name gluon was retained to describe the field quantum of the strong interaction between quarks. The other major problem was that for the parton model to work, the partons had to be treated as essentially free objects, whereas to explain the non-appearance of partons (or quarks) in the final state it was necessary to assume that the force was strong enough to confine them in the hadron (at least up to the energy scales then explored). Both these problems were solved by the discovery and development of the non-Abelian gauge theory of the strong interaction — quantum chromodynamics (QCD) — in the early 1970s by Fritzsch, Gell-Mann, Leutwyler and others. The basics of QED and QCD, their similarities and differences and other more formal techniques are covered in Chapter 3. QED is an Abelian gauge theory with a single uncharged field quantum, the photon. QCD is a non-Abelian gauge theory based on an exact $SU(3)$ symmetry of 'colour' charges with eight massless, but coloured field quanta, the gluons. With its more complex symmetry structure and charged field quanta, QCD gives rise to a much richer physics than QED, much of which is still to be explored.

Two developments were crucial to the early success of QCD, the first was the proof by 't Hooft that it was a renormalizable theory and the second was the discovery of *asymptotic freedom* by Politzer, Gross and Wilczek. The latter is shorthand for the decrease of the strength of the effective coupling constant at small distances allowing perturbative methods to be used for quantitative predictions. In the other direction, it was assumed that the increase in the coupling strength at large distances could explain the confinement of quarks and gluons. QCD did not just provide a theoretical justification for the parton model, but went further and predicted that Bjorken scaling would be broken by terms depending on $\ln Q^2$ to first order. The experimental verification of the QCD predictions provided by muon induced DIS at the higher energies available at the proton accelerators of FNAL and CERN did much to establish QCD as *the* theory of the strong interaction and one of the pillars of the Standard Model of Particle Physics.

Early predictions of QCD were usually formulated in terms of the Operator Product Expansion, which though a powerful technique, does not lend

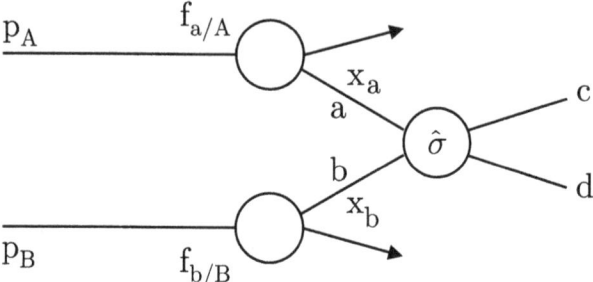

Fig. 1.5 The basic idea of the QCD improved parton model. Partons a, b from hadrons A, B undergo a hard scatter $ab \to cd$ described by the calculable cross-section $\hat{\sigma}$. The parton momentum densities are $f_{a/A}$ and $f_{b/B}$ respectively. For DIS, the parton density $f_{a/A}$ is replaced by the known coupling of the leptonic probe to virtual photon or weak boson.

itself to an easy physical interpretation. This changed with the development by Dokshitzer, Gribov, Lipatov, Altarelli and Parisi in the late 1970s of the more physically based formulation in terms of splitting functions for gluon bremsstrahlung by quarks ($q \to qg$) and $g \to q\bar{q}$ pair production, in close analogy to QED together with the pure non-Abelian gluon to gluon splitting, $g \to gg$. This led to the formulation of the 'QCD improved parton model' which provides the basis for all calculations involving strong hard-scattering processes. The key idea, taken over from the parton model, but now underpinned by QCD factorization theorems, is that the cross-section for a hard process may be written as a convolution of a calculable quark-gluon (parton–parton) process with the parton densities, as illustrated in Fig. 1.5. The latter ($f_{a/A}$, $f_{b/B}$ in the figure) cannot be calculated using perturbative techniques, but their evolution with scale, for example Q^2 in DIS, is calculable. Another view of factorization is that it separates the short–distance and rapid hard scatter from the long–distance and slower timescale of non-perturbative processes such as quark fragmentation. The QCD enhanced parton model is covered in detail in Chapter 4.

1.6 Extraction of parton density functions

Although the Q^2 evolution of parton momentum density functions is given by perturbative QCD, the forms of the density functions as functions of x are not determined. There are some overall constraints, such as sum rules to guarantee the correct quantum numbers of the hadron which must be respected. Eventually, lattice gauge calculations may provide estimates of some of the moments of parton densities. At present, the parton densities are parameterized at a starting scale using smooth functions of x with a limited number of parameters and evolved to the Q^2 of the data using next-to-leading order perturbative QCD. The parameters are then determined by a χ^2 fit. Chapter 5 outlines the main features of fixed target and collider experiments for lepton-induced DIS and gives a review of the data that forms the core for global fits to extract parton densities. The fitting procedure and results are discussed in in Chapter 6, including a discussion of error estimation. Accurate determination of parton momentum densities is crucial for many present and future experiments, with the uncertainty of the parton density often a limiting factor.

Apart from fermion masses, the only parameter in QCD to be determined from data is the 'strong fine structure constant', α_s. It is one of the parameters that may be varied in a global fit for parton densities, but in many fits it is set to the current best estimate from the Particle Data Group. One of the reasons for this is that at low x the behaviour of the gluon momentum density and α_s are strongly correlated. However DIS does offer at least two methods for determining α_s and its behaviour as a function of Q^2. One is through scaling violations of the nucleon structure function and the other is using jet rates at large Q^2. These topics are considered in Chapter 7, together with a brief account of jet production in DIS.

1.7 DIS at colliders

The HERA ep collider has extended the phase space for deep inelastic scattering at very large Q^2 and opened up a new domain for strong interaction physics at low x. High Q^2, which means $Q^2 \geq M_Z^2\,(M_W^2)$ is covered in Chapter 8. Now parity non-conserving effects and $\gamma - Z^0$ interference come into play, first measurements have been made at HERA-I and high–Q^2 physics is one of the main goals of the upgraded HERA-II programme. Another feature of the HERA-II programme is that the $e^\pm$ beams will be longitudinally polarized. Measurement of the neutral current (γ and Z^0 exchange) and charged current ($W^\pm$ exchange) processes at high Q^2 and large x, will provide an almost complete determination of the proton's parton density functions independently of fixed target data and allow study of fermion electroweak couplings at large space-like momentum transfers to complement the data from LEP at large time-like scales.

Possibly the most important results at HERA-I were the observation and precision measurement of the rise of the structure function F_2 at low x and the realisation of the importance of diffraction in high energy scattering. A sizeable fraction — at the 10% level — of deep inelastic scattering is diffractive, in which the proton either remains intact or is excited to a resonant state with the same quantum numbers. High energy diffraction is now a vast subject requiring a book to itself and is only touched on here.

HERA is able to reach x values of the order of nearly 10^{-6} at $Q^2 \sim 1\,\text{GeV}^2$ and this is the domain of gluon dominated physics. The standard perturbative QCD evolution of the structure functions may well be inadequate (though the standard fits to F_2 still describe almost all the data very well). At very small x, but quite possibly just outside the reach of HERA, the gluon density may be large enough that non-linear effects must also be considered. Low x physics is surveyed in Chapter 9. Although much seminal work was done before HERA started, the measured rise of F_2 revived interest in the subject. A lot has been learnt but the underlying reason for the rise is still the subject of vigorous debate.

Although lepton induced DIS is the main focus of this book, the QCD enhanced parton model allows hard scattering in hadron–hadron interactions to be considered as a complementary technique. Lepton pair and high E_T jet production are processes interesting in their own right, but they also provide input for determination of parton densities. These topics and others are considered in Chapter 10 mainly in the context of data from the Tevatron, but also looking ahead to the LHC. Accurate calculation and measurement of standard model processes at the LHC will be a vital first step in preparing for the expected new physics beyond the standard model. As the LHC is a hadron collider, this means that accurate knowledge of parton densities is a crucial input.

1.8 Other topics

One of the toughest challenges in DIS, both theoretically and experimentally is to understand the dynamics of nucleon spin. Much progress has been made since the 'spin crisis' of the late 1980s, the QCD framework is now better understood, but there are still open questions and the big challenge on the experimental side is to find ways of pinning down the contribution of the gluon to the nucleon spin. An outline of the problems and main results of polarized DIS are given in Chapter 11.

The final chapter of the book is a brief look at how effects beyond the standard model may be searched for using lepton induced DIS, with measurements from HERA as illustration. No attempt is made to give a complete account of BSM physics. The focus is on techniques such as 'contact interactions' which allow additions to the standard model Lagrangian from new physics at a much higher energy scale than accessible directly and direct searches for new particles like leptoquarks that could be formed from eq or νq states at HERA.

1.9 Outlook

For the immediate future, one looks forward to new results in DIS generally from HERA-II and the Tevatron Run-II. In fully polarized DIS, new results are expected from HERMES (polarized fixed target electron-nucleus scattering at HERA), COMPASS (polarized muon scattering at CERN) and polarized proton–proton scattering at RHIC (the relativistic heavy ion collider at Brookhaven. A few years of running at HERA after the high-Q^2 programme, devoted to the low x physics of ep and eA (electron–nucleus) interactions would add much to the understanding of high density gluon dynamics. In the slightly longer term there is the wealth of new data to come from the LHC at CERN. On a more speculative note are the possibilities for a very high energy ep collider formed by combining the TESLA

e^+e^- future linear collider with the proton beam at HERA — the THERA project — but maybe this is just a dream. What is clearly laid out in this book is the importance and vitality of the technique of deep inelastic scattering in high energy physics, nearly a century after the pioneering work by Rutherford, Geiger and Marsden.

1.10 Problems

1. Evaluate the matrix element for the Yukawa potential, Eq. (1.1). It is convenient to choose a coordinate system such that $\mathbf{q}$ defines the polar axis and $\mathbf{q} \cdot \mathbf{r} = qr \cos \theta$. Check the two forms for the Rutherford scattering cross-section.

2. Evaluate the matrix element for a charge distribution, Eq. ((1.4)). The integral is most easily evaluated by first making a change of variable $\mathbf{r}, \mathbf{r}' \rightarrow \mathbf{R}, \mathbf{r}'$ where $\mathbf{R} = \mathbf{r} - \mathbf{r}'$. Having done the angular integrals, the final integral over R is formally divergent, but may be performed by introducing a factor $\exp(-\mu R)$ and taking the limit $\mu \rightarrow 0$ after integration.

3. Verify the expansion of the form factor, Eq. (1.5). Using this expression and the proton dipole form–factor show that the root–mean–square charge radius of the proton is about 0.81 fm.
$$[\langle \mathbf{r}^2 \rangle = \int \mathbf{r}'^2 \rho(\mathbf{r}') d^3 \mathbf{r}'.]$$

2
The quark–parton model

This chapter introduces the quark–parton model (QPM), in which deep inelastic scattering is seen as the incoherent sum of point-like elastic scattering of spin-$\frac{1}{2}$ nucleon constituents (the partons). The underlying cross-section expressions for elastic electron–quark and quasi-elastic neutrino–quark scattering are developed. The full cross-section for lepton–nucleon deep inelastic scattering is constructed by folding the point-like elastic cross-sections with parton momentum densities. This then leads to expressions for the electromagnetic and neutrino nucleon structure functions in terms of parton density functions. The QPM structure functions exhibit Bjorken scaling, that is they are approximately independent of momentum transfer from the probe to the nucleon. This is a direct consequence of the underlying elastic scattering. An outline of other elementary tests of the QPM then follows and the chapter ends with a discussion of sum rules.

2.1 The essential idea of the parton model

The quark–parton model is developed by considering its application to the lepton-hadron scattering process. The commonly used Lorentz invariants will be defined for the process

$$\ell N \to \ell' X$$

where ℓ, ℓ' represent leptons (lepton is taken to include antileptons, unless it is necessary to distinguish them), N represents a nucleon and X represents the hadronic final state particles. The associated four vectors are k, k' for the incoming and outgoing leptons, respectively, and p for the target (or incoming) nucleon, see Fig. 2.1. The process is mediated by the exchange of a virtual vector boson, V^* (γ, W or Z), with 4-momentum given by

$$q = k - k'$$

and the 4-momentum P_X of the hadronic final state system X is given by

$$P_X = p + q.$$

Various Lorentz invariants are useful in the description of the kinematics of the process:

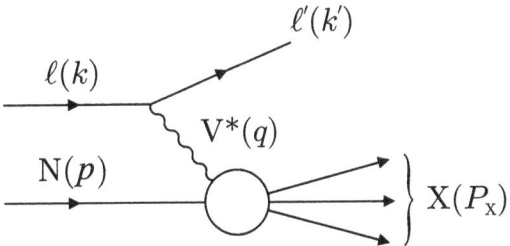

Fig. 2.1 Schematic diagram of lepton–hadron scattering via vector–boson exchange

$$s = (p+k)^2$$

is the centre-of-mass (CM) energy squared for the lp interaction;

$$Q^2 = -q^2$$

is the (negative of) the invariant mass squared of the virtual exchanged boson;

$$x = Q^2/2p \cdot q$$

is the Bjorken x variable, which will be interpreted in the QPM as the fraction of the momentum of the incoming nucleon taken by the struck quark; and

$$y = p \cdot q/p \cdot k,$$

which gives a measure of the amount of energy transferred between the lepton and the hadron systems. It is also related to the scattering angle in the lepton–quark centre-of-mass frame. Note that (ignoring masses),

$$Q^2 = sxy,$$

so that only two of these quantities are independent. Finally, the centre-of-mass energy of the γ^*p system (or equivalently the invariant mass of the final state hadronic system X) is often denoted by W, explicitly

$$W^2 = (q+p)^2 \quad (\equiv P_X^2).$$

The QPM grew out of the attempt by Feynman to provide a simple physical picture of the scaling that had been predicted by Bjorken and observed in the first high energy deep inelastic electron scattering experiments at SLAC, where F_2 was observed to be independent of Q^2 for x values around $x \sim 0.25$. The model states that the nucleon is full of point-like non-interacting scattering centres known as partons. The lepton–hadron reaction cross-section is then approximated by an incoherent sum of elastic lepton–parton scatters via an exchanged vector boson, see Fig. 2.2.

It turns out that only spin-$\frac{1}{2}$ partons (quarks or antiquarks) interact directly with the leptonic probe particles. For incident charged leptons at

16 The quark–parton model

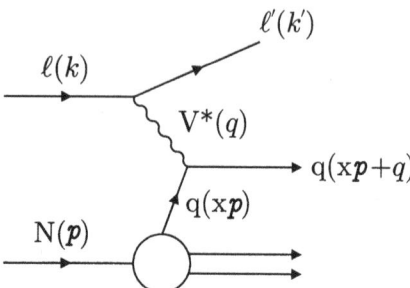

Fig. 2.2 Schematic diagram of lepton–hadron scattering in the quark-parton model

CM energies much less than M_W, M_Z, only electromagnetic interactions need be considered. Thus, the electromagnetic scattering of two spin-$\frac{1}{2}$ point-like particles is the underlying fundamental process.

2.1.1 Elastic electron–muon scattering

Electron–muon scattering provides an example of such a process which proceeds via the single photon exchange diagram of Fig. 2.3. The Feynman

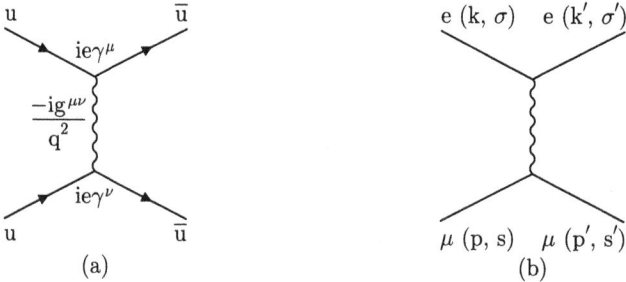

Fig. 2.3 Feynman diagram for electron–muon scattering

rules[1] for this diagram are elaborated in (a), the 4-momenta and spin labels are given in (b). Putting the external lines together with the vertex factors, the matrix element may be written in terms of the electron and muon currents

$$j_e^\nu = ie\bar{u}(k',\sigma')\gamma^\nu u(k,\sigma) \tag{2.1}$$

$$j_{\text{muon}}^\mu = ie'\bar{u}(p',s')\gamma^\mu u(p,s) \tag{2.2}$$

sandwiching the propagator, $-ig_{\mu\nu}/q^2$, to yield the matrix element

$$\mathcal{M} = i\frac{ee'}{q^2}\left[\bar{u}(k',\sigma')\gamma_\mu u(k,\sigma)\right]\left[\bar{u}(p',s')\gamma^\mu u(p,s)\right] \tag{2.3}$$

---

[1] The rules for QED are summarized in Appendix D.

Note that the charge on the muon current e' has been distinguished from that on the electron current to allow the formalism to be used for eq scattering, where the charge of the quark is not the same as that of the electron.

The cross-section for the process is then obtained from $|\mathcal{M}|^2$ by using the Fermi Golden Rule. For the unpolarized cross-section initial spin states must be averaged and the final spin states summed. This involves the contraction of two leptonic tensors

$$\frac{1}{4}\sum_{\text{spins}} |\mathcal{M}|^2 = \frac{e^2 e'^2}{q^4} L_e^{\mu\nu} L_{\mu\nu}^{\text{muon}} \quad (2.4)$$

The spin summations are performed using the well established trace techniques of Dirac algebra (see problem 1 at the end of the chapter) to obtain

$$L_e^{\mu\nu} = 2(k'^\mu k^\nu + k'^\nu k^\mu - (k'.k)g^{\mu\nu}) \quad (2.5)$$

where the mass of the electron has been neglected. Neglecting also the muon mass one finds

$$L_e \cdot L^{\text{muon}} = 8[(k' \cdot p')(k \cdot p) + (k' \cdot p)(k \cdot p')]$$

This may be written in terms of the Mandelstam variables

$$s = (k+p)^2 = (k'+p')^2; \quad t = (k-k')^2 = (p'-p)^2; \quad u = (k-p')^2 = (k'-p)^2$$

and ignoring masses

$$L_e \cdot L^{\text{muon}} = 2(s^2 + u^2)$$

finally identifying $u/s = y - 1$ one gets

$$\frac{1}{4}\sum_{\text{spins}} |\mathcal{M}|^2 = \frac{e^2 e'^2}{Q^4} 2s^2 [1 + (1-y)^2] \quad (2.6)$$

Since $p^2 = p'^2 = m_\mu^2$ and

$$(p+q)^2 = p'^2$$

one has

$$2p \cdot q + q^2 = 0$$

and consequently

$$x = \frac{Q^2}{2p \cdot q} = 1$$

and

$$Q^2 = sy.$$

Putting the matrix element together with the phase space and the flux factor (as detailed in Appendix B) the Fermi Golden Rule gives the differential cross-section

18 The quark–parton model

$$\frac{d\sigma}{dy} = \frac{e^2 e'^2}{8\pi Q^4}[1+(1-y)^2]s \qquad (2.7)$$

and since $e' = e$ this may be written in the form

$$\frac{d\sigma}{dy} = \frac{2\pi\alpha^2}{Q^4}[1+(1-y)^2]s \qquad (2.8)$$

This derivation and useful equivalent forms for the cross-section are collected in Appendix C.

It is illuminating to follow a more heuristic approach to the derivation of Eq. (2.8) by considering the spin directions of the electron and muon with respect to their directions of motion. There are two possibilities: the incoming electron and muon may have the same handedness[2] (RR or LL) or opposite handedness (RL or LR). If they have opposite handedness then there is net spin $J_z = +1(RL)$ or $J_z = -1(LR)$ along the beam direction. In the high energy limit handedness is conserved (see problem 2), hence $RL \to RL$ such that the amplitude for scattering at an angle θ is given by the rotation matrix[3] $d^1_{11}(\theta) = \frac{1}{2}(1 + \cos\theta)$. For $LR \to LR$, the amplitude is the same since $d^1_{-1-1} = d^1_{11}$. This case is illustrated in Fig. 2.4. The same result is obtained by considering the rotation of spin-$\frac{1}{2}$ through the angle θ for both the electron and the muon: $d_{++} = \cos(\theta/2)$ each, giving a total amplitude $\cos^2(\theta/2)$ for $RL \to RL$. Similarly for $LR \to LR$, since $d_{--} = d_{++}$.

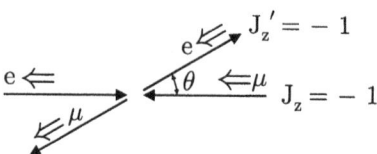

Fig. 2.4 Electron–muon scattering with opposite handedness, $LR \to LR$.

If the leptons have the same handedness, $LL \to LL$ or $RR \to RR$, there is no net spin along the beam direction and an isotropic distribution ensues. For an overall $J = 0$, $J_z = 0$ rotation, $d^0_{0,0} = 1$. Considering the rotations of the individual fermions, one needs to construct the $\uparrow\downarrow - \downarrow\uparrow$ spin state, giving $d_{++}d_{--} - d_{-+}d_{+-}$ or $\cos^2(\theta/2) + \sin^2(\theta/2) = 1$. The $LL \to LL$ case is illustrated in Fig. 2.5. To summarise, there are two distinguishable amplitudes: opposite handed $\cos^2(\theta/2) = (1-y)$ and same handed isotropic, giving a total cross-section with the dependence $1 + (1-y)^2$.

2.1.2 Elastic electron–quark scattering

The above formalism is now extended to electron–quark scattering in order to interpret lepton–hadron scattering as an incoherent sum of lepton–

---

[2] This is often known as chirality and in the high energy limit is the same as helicity.
[3] See the end of Appendix A for the definition of rotation matrices.

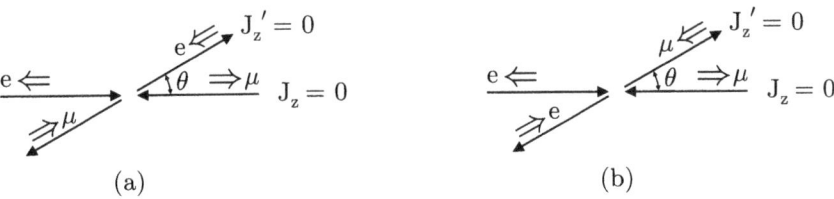

Fig. 2.5 Electron–muon scattering with same handedness, $LL \to LL$.

parton scatters. First, the charge of the quark may be $\tfrac{2}{3}e$ or $-\tfrac{1}{3}e$, which is written as $e' = e_i e$ in general. Second, the kinematic variables s, Q^2, y must be appropriate for electron–quark scattering, where the quark takes only a fraction x' of the momentum of the parent hadron. Clearly Q^2 and y are unaffected, but $p \to x'p$, implies $s \to x's$, if s is to be the centre-of-mass energy of the electron-hadron system. Thus using the formalism which led to Eq. (2.8), the cross-section for electron-quark scattering at a specific value of x' is given by

$$\frac{d\sigma}{dy} = \frac{2\pi\alpha^2}{Q^4}[1 + (1-y)^2]x's e_i^2. \qquad (2.9)$$

The relationship $Q^2 = sy$ becomes $Q^2 = x'sy$ and this gives x' an identification in terms of the external kinematic variables such that $x' = x = Q^2/2p.q$.

To obtain a cross-section for electron-hadron scattering from this result, a distribution function $q_i(x)$ which gives the probability that the struck quark i carries a fraction x of the hadron's momentum p must be defined. This gives the parton number, or density, distribution. It is also useful to define the momentum distribution $xq(x)$ (also called the momentum density), and this is what is meant by the term *parton distribution function* or PDF.

Thus, the double differential cross-section for incoherent scattering off all possible types of quark with a specific value of x is

$$\frac{d^2\sigma}{dx\,dy} = \frac{2\pi\alpha^2}{Q^4}[1+(1-y)^2]s\sum_i e_i^2 x q_i(x) \qquad (2.10)$$

or, equivalently,

$$\frac{d^2\sigma}{dx\,dQ^2} = \frac{2\pi\alpha^2}{xQ^4}[1+(1-y)^2]\sum_i e_i^2 x q_i(x) \qquad (2.11)$$

2.2 Inelastic e− or μ–hadron scattering

To appreciate these parton model results they must be compared with the general formulae for inelastic lepton–hadron scattering. In analogy to the leptonic tensor $L^e_{\mu\nu}$ a hadronic tensor is introduced[4]

---

[4]The formal definition of $W^{\mu\nu}$ is given in Appendix C.

$$d\sigma \sim L^e_{\mu\nu} W^{\mu\nu}$$

The hadronic tensor parameterizes our ignorance of the hadronic current, but it must have the general form

$$W^{\mu\nu} = -W_1 g^{\mu\nu} + \frac{W_2}{m^2} - i\epsilon^{\mu\nu\alpha\beta} p_\alpha q_\beta \frac{W_3}{2m^2} \qquad (2.12)$$
$$+ q^\mu q^\nu \frac{W_4}{m^2} + (p^\mu q^\nu + q^\mu p^\nu) \frac{W_5}{m^2} + i(p^\mu q^\nu - p^\nu q^\mu) \frac{W_6}{2m^2},$$

where m is the hadron mass and the W_i are real scalar functions (called structure functions) of two independent kinematic variables such as x, Q^2 or x, y. The expression for $W^{\mu\nu}$ can immediately be simplified since not all of these terms can contribute to the cross-section. The antisymmetric term W_6 is absent for unpolarized scattering since the lepton tensor is symmetric. For γ^* exchanges, considered so far, the W_3 term will not contribute since its contribution violates parity. Current conservation at the hadron vertex gives $q_\mu W^{\mu\nu} = q_\nu W^{\mu\nu} = 0$, so that

$$W_5 = -W_2 \frac{p \cdot q}{q^2}$$

and

$$W_4 = W_2 \left(\frac{p \cdot q}{q^2}\right)^2 + W_1 \frac{m^2}{q^2}$$

Thus, the hadronic tensor simplifies to

$$W^{\mu\nu} = W_1 \left(-g^{\mu\nu} + \frac{q^\mu q^\nu}{q^2}\right) + \frac{W_2}{m^2}\left(p^\mu - \frac{p \cdot q}{q^2} q^\mu\right)\left(p^\nu - \frac{p \cdot q}{q^2} q^\nu\right) \qquad (2.13)$$

The resulting differential cross-section expressed in terms of x, y, Q^2 is

$$\frac{d^2\sigma(\ell h)}{dx\,dy} = \frac{4\pi\alpha^2 s}{Q^4} \left[xy^2 F_1^{\ell h}(x,y) + (1-y) F_2^{\ell h}(x,y)\right], \qquad (2.14)$$

where mass terms have been ignored and the structure functions are redefined such that $F_1 = W_1$ and $F_2 = \nu W_2/m^2$, where $\nu = p \cdot q$. The longitudinal structure function is defined as $F_L = F_2 - 2x F_1$, so that the cross-section may also be written as

$$\frac{d^2\sigma(\ell h)}{dx\,dQ^2} = \frac{2\pi\alpha^2}{xQ^4} \left[Y_+ F_2^{\ell h}(x, Q^2) - y^2 F_L^{\ell h}(x, Q^2)\right], \qquad (2.15)$$

where $Y_+ = 1 + (1-y)^2$. The structure functions F_1, F_2, F_L in these equations are labelled with the superscript ℓh to indicate that they are appropriate to the charged lepton–hadron process.

Comparing Eqn. (2.15) with the parton model result of Eqn. (2.11) implies that the parton model predicts

$$F_2^{\ell h}(x, Q^2) = \sum_i e_i^2 x q_i(x) \qquad (2.16)$$

where the fractional momentum of the struck quark has again been identified with the kinematic variable $x = Q^2/(2p \cdot q)$. Thus the parton model predicts Bjorken scaling, whic is that F_2 depends only on x not on Q^2.

The parton model also predicts,

$$F_L^{\ell h}(x, Q^2) = 0,$$

also known as the Callan-Gross relationship, $2xF_1^{\ell h} = F_2^{\ell h}$, as a consequence of scattering from spin-$\frac{1}{2}$ partons.

The latter result may be understood by relating the structure functions to the cross-sections for the virtual photon–hadron subprocess. The detailed formalism is given in Appendix C. For Q^2 large enough to ignore mass terms, these cross-sections are given by

$$\sigma_T = \frac{4\pi^2 \alpha}{Q^2(1-x)} 2x F_1, \qquad \sigma_L = \frac{4\pi^2 \alpha}{Q^2(1-x)} F_L$$

and

$$\sigma_T + \sigma_L = \frac{4\pi^2 \alpha}{Q^2(1-x)} F_2$$

where σ_T, σ_L are the cross-sections for transverse and longitudinal virtual photon scattering. Real photons are transversely polarized, $\sigma_L = 0$, but this does not have to be true for virtual photons. It is the Callan-Gross relationship for spin-$\frac{1}{2}$ partons which implies that $\sigma_L = 0$ and measurements of $R = \sigma_L/\sigma_T$ from early SLAC data indicated that this is true (approximately). If partons were, for example, spin 0, a very different prediction would be obtained. In this case, $\sigma_T = 0$ and $R \to \infty$, since a spin-0 parton cannot absorb a transverse photon with helicity ± 1. The formal derivation of this result is left as a problem.

At this point it is worth reviewing the derivation of the parton model results to make some of its limitations clear. Cross-section formulae in the fixed-target frame will be used, as is frequently done in some older text books. Start from Eq. (C.1) of Appendix C

$$\frac{d^2\sigma}{d\Omega dE'} = \frac{\alpha^2}{q^4} \frac{E'}{E} \frac{L \cdot W}{M}$$

where E and E' are the energies of the incident and scattered electron and $d\Omega$ is the solid angle at scattering angle θ. M is the target mass from

22 The quark–parton model

the flux factor. Using the results given in the appendix one finds that for general deep inelastic scattering on a hadron of mass M

$$\left.\frac{L \cdot W}{M}\right|_{\text{hadron}} = \frac{4EE'}{M}\left[2W_1 \sin^2\frac{\theta}{2} + W_2 \cos^2\frac{\theta}{2}\right], \quad (2.17)$$

where W_1, W_2 are the structure functions of Eq. (2.13).

Using Eq. (C.8), one may write an equivalent expression for elastic scattering of an electron from a point-like spin-$\frac{1}{2}$ object with charge $e_i e$ and mass m

$$\left.\frac{L \cdot W}{m}\right|_{\text{point}} = 4m e_i^2 EE'\left[\cos^2\frac{\theta}{2} + \frac{Q^2}{2m^2}\sin^2\frac{\theta}{2}\right]\delta(\nu - Q^2/2), \quad (2.18)$$

where $\nu = p \cdot q$. The delta function arises from not integrating out the mass-shell condition $p'^2 = m^2$ in the final-state phase-space factor. Note also that with the normalization convention for the W_i used in this book it is necessary to keep the target mass explicitly on the left-hand sides of the above equations. By equating the coefficients of the two independent angular functions one finds

$$\frac{W_2}{M} \to m e_i^2 \,\delta(\nu - Q^2/2); \quad 2\frac{W_1}{M} \to \frac{Q^2}{2m} e_i^2 \,\delta(\nu - Q^2/2).$$

In the parton model, the target hadron is NOT pictured as point-like but the supposition is that, if the probing virtual photon has sufficiently large Q^2, it may be able to resolve point-like partons within the hadron. To make use of the results developed so far, it is necessary to be more specific about the relationship $p_{\text{parton}} = x' p_{\text{hadron}}$. The components of the parton 4-momentum must be specified in terms of those of the hadron 4-momentum, such that if p_{hadron} is given by $(E, p_L, \mathbf{p}_T = 0)$ then p_{parton} is given by $(x'E, x'p_L, \mathbf{p}_T = 0)$. This implies a mass for the parton $m = x'M$. With these definitions the parton model structure functions for a single parton are given by

$$\frac{W_2}{M} = x'M e_i^2 \,\delta(\nu' - Q^2/2); \quad 2\frac{W_1}{M} = \frac{Q^2}{2x'M} e_i^2 \,\delta(\nu' - Q^2/2),$$

where the variable

$$\nu' = p_{\text{parton}} \cdot q = x' p_{\text{hadron}} \cdot q = x'\nu$$

(Note that this relationship would not hold if the parton were not collinear with the hadron.) Next, using the property of the delta function that $\delta(ax) = \frac{1}{a}\delta(x)$ and remembering that $x = Q^2/(2\nu)$, one finds for the contribution of a single parton

$$2F_1 \equiv 2W_1 = \frac{x}{x'} e_i^2 \,\delta(x' - x)$$

$$F_2 \equiv \frac{\nu W_2}{M^2} = x' e_i^2 \delta(x' - x).$$

Finally, using the parton density functions $q_i(x')$ and summing the contributions for one parton incoherently over all the partons in the hadron gives

$$F_2(x, Q^2) = \sum_i e_i^2 \int_0^1 dx' \, q_i(x') x' \delta(x' - x)$$

and

$$2F_1(x, Q^2) = \sum_i e_i^2 \int_0^1 dx' \, q_i(x') \frac{x}{x'} \delta(x' - x),$$

which are the parton model results already derived. Note also that the fractional momentum of the struck parton, x', is identified with Bjorken x through the delta function arising from the kinematics of the underlying elastic scatter.

In this derivation the concept of a parton of mass $x'M$ has been introduced so that both the parton energy and longitudinal momentum may be fractions x' of those of the hadron. This collinear approximation is only really possible in the infinite momentum frame, where all masses and transverse momenta may be neglected. In this frame time dilation slows down the rate at which partons interact with each other, so that during the short time that the virtual photon interacts with the parton, the parton is essentially free. This justifies the incoherent addition of the *probabilities* for individual $\gamma*$–parton interactions, rather than first constructing the full amplitude. On a much longer time-scale, parton–parton interactions occur such that the struck parton together with other partons forms colourless hadrons in the final state.

Such a picture can only really be valid at large Q^2, but surprisingly both Bjorken scaling and the Callan-Gross relationship were substantiated by the early SLAC data at Q^2 values in the range $1 - 10\,\text{GeV}^2$. In the next two chapters the underlying assumptions of the parton model, and whether scaling can be exact, will be explored in the context of quantum chromodynamics. The remainder of this chapter develops the evidence for the parton model further by considering the results from neutrino and antineutrino scattering.

2.3 Neutrino–induced DIS

Similar methods to those just used for charged-lepton DIS will be used to build up the QPM results for neutrino–induced DIS. Here, because of parity non-conservation of the weak interaction, there are differences between particle and antiparticle interactions.

2.3.1 Elastic neutrino–electron scattering

The Feynman diagrams for elastic neutrino– and antineutrino–electron scattering through the exchange of $W^\pm$ weak bosons are illustrated in

24 The quark–parton model

Fig. 2.6. Instead of the the electromagnetic current of Eq. (2.1) is replaced

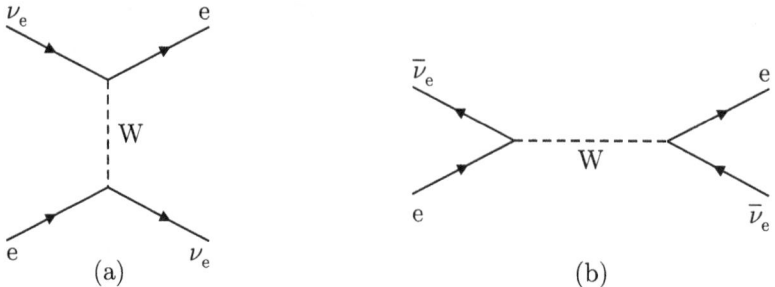

Fig. 2.6 (a) νe scattering. (b) $\bar{\nu} e$ scattering.

by

$$j^\nu = -i\frac{g_W}{\sqrt{2}} \bar{u}(k', \sigma')\frac{1}{2}\gamma^\nu(1 - \gamma^5)u(k, \sigma) \tag{2.19}$$

which has the $V - A$ form that projects out only left-handed particles and right-handed antiparticles. The weak coupling g_W has the value $g_W \sin(\theta_W) = e$ in electroweak unification. The propagator to be sandwiched between two currents of this form is now that of the massive W boson

$$-i\frac{(g_{\mu\nu} - p_\mu q_\nu/M_W^2)}{q^2 - M_W^2} \tag{2.20}$$

giving a matrix element

$$\begin{aligned} \mathcal{M} &= \frac{g_W^2}{2}\frac{1}{M_W^2 - q^2}\left[\bar{u}(k', \sigma')\frac{1}{2}\gamma_\mu(1 - \gamma^5)u(k, \sigma)\right] \\ &\times \left[\bar{u}(p', s')\frac{1}{2}\gamma^\mu(1 - \gamma^5)u(p, s)\right]. \end{aligned} \tag{2.21}$$

When $Q^2 = -q^2 \ll M_W^2$ this is usually written as

$$\mathcal{M} = \frac{G_F}{\sqrt{2}}\left[\bar{u}(k', \sigma')\gamma_\mu(1 - \gamma^5)u(k, \sigma)\right]\left[\bar{u}(p', s')\gamma^\mu(1 - \gamma^5)u(p, s)\right] \tag{2.22}$$

where $\frac{G_F}{\sqrt{2}} = \frac{g_W^2}{8M_W^2}$ defines Fermi's weak coupling constant G_F.

Evaluating the spin averaged squared matrix element and the flux and phase space factors gives

$$\frac{d\sigma(\nu e)}{dy} = \frac{G_F^2}{\pi}s \tag{2.23}$$

as the cross-section to Eq. (2.8) for neutrino–electron scattering. The antineutri electron cross-section is then obtained by crossing[5] as

---
[5] See Appendix B for an explanation of crossing symmetry.

$$\frac{d\sigma(\bar{\nu}e)}{dy} = \frac{G_F^2}{\pi}(1-y)^2 s \qquad (2.24)$$

The simple forms of these cross-sections can be readily understood by returning to the heuristic arguments based on handedness. Because of the $V - A$ form of the weak charged current, there are not many possibilities in spin averaging. The neutrino and electron are necessarily both left-handed and thus for νe scattering the isotropic distribution appropriate for $LL \to LL$ is obtained. The antineutrino is right-handed and thus for $\bar{\nu}e$ scattering the $(1-y)^2$ distribution appropriate for $RL \to RL$ is obtained.

2.3.2 Elastic neutrino–quark scattering

These results may now be applied to neutrino–quark scattering by defining quark and antiquark charged weak currents of the same $V - A$ form as the leptonic charged weak current, with exactly the same weak charge g_W. Hence, since this current projects out left-handed quarks and right-handed antiquarks, the scattering cross-sections at a specific value of x are given by

$$\frac{d\sigma(\nu q)}{dy} = \frac{G_F^2}{\pi} xs,$$

$$\frac{d\sigma(\bar{\nu}q)}{dy} = \frac{G_F^2}{\pi}(1-y)^2 xs,$$

$$\frac{d\sigma(\nu\bar{q})}{dy} = \frac{G_F^2}{\pi}(1-y)^2 xs,$$

$$\frac{d\sigma(\bar{\nu}\bar{q})}{dy} = \frac{G_F^2}{\pi} xs.$$

Thus, the parton model predictions for neutrino and antineutrino hadron scattering are,

$$\frac{d^2\sigma(\nu h)}{dx\,dy} = \frac{G_F^2 s}{\pi} \Sigma_i \left[x q_i(x) + (1-y)^2 x \bar{q}_i(x) \right] \qquad (2.25)$$

and

$$\frac{d^2\sigma(\bar{\nu} h)}{dx\,dy} = \frac{G_F^2 s}{\pi} \Sigma_i \left[(1-y)^2 x q_i(x) + x \bar{q}_i(x) \right] \qquad (2.26)$$

respectively, where the same parton distribution functions, introduced in Eqs (2.10) and (2.11), have been used. These cross-sections for neutrino and antineutrino DIS include antiquarks as well as quarks in the target nucleon as this is required to explain the experimental data. The reconciliation of the parton model for inelastic lepton–hadron scattering with the

26   *The quark–parton model*

static quark model, which pictures nucleons as made of three constituent quarks which give them their flavour properties, is made by identifying the constituent quarks as 'valence' quarks, and considering the nucleon to be made of three such quarks and a 'sea' of quark–antiquark pairs which have no overall flavour. Both the valence quarks and the sea quarks and antiquarks are identified as partons.

2.3.3 Inelastic neutrino–hadron scattering

To fully interpret the parton model predictions of Eqs (2.25) and (2.26) for neutrino and antineutrino hadron scattering they should be compared with the general formulae which may be derived from the hadronic tensor already given in Eq. (2.13). In the present case W_6 contributes in principle, and current conservation is not exact for the axial vector current such that there are residual terms from W_4 and W_5. However these contributions are all small terms of the order of the lepton mass. The only significant difference is that the parity violating contribution of W_3 cannot be omitted. The differential cross-sections which result are

$$\frac{d^2\sigma(\nu h)}{dx\,dy} = \frac{G_F^2 s}{2\pi}\left[xy^2 F_1^{\nu h}(x,y) + (1-y)F_2^{\nu h}(x,y) + y(1-\tfrac{y}{2})xF_3^{\nu h}(x,y)\right], \quad (2.27)$$

and

$$\frac{d^2\sigma(\bar{\nu} h)}{dx\,dy} = \frac{G_F^2 s}{2\pi}\left[xy^2 F_1^{\bar{\nu} h}(x,y) + (1-y)F_2^{\bar{\nu} h}(x,y) - y(1-\tfrac{y}{2})xF_3^{\bar{\nu} h}(x,y)\right], \quad (2.28)$$

where $\nu W_3/m^2 = F_3$. In terms of the longitudinal structure function F_L these become

$$\frac{d^2\sigma(\nu,\bar{\nu})}{dx\,dy} = \frac{G_F^2 s}{4\pi}\left[Y_+ F_2^{\nu,\bar{\nu}}(x,y) - y^2 F_L^{\nu,\bar{\nu}}(x,y) \pm Y_- xF_3^{\nu,\bar{\nu}}(x,y)\right], \quad (2.29)$$

where the plus sign applies for neutrino and the minus sign for antineutrino scattering, and $Y_- = 1 - (1-y)^2$. Comparing these with the quark–parton model formulae Eqs (2.25), (2.26) gives

$$F_L^{\nu,\bar{\nu}}(x,y) = 0$$

appropriate for scattering from only spin-$\tfrac{1}{2}$ partons, and

$$F_2^{\nu,\bar{\nu}}(x,y) = 2\sum_i x(q_i(x) + \bar{q}_i(x)),$$

$$xF_3^{\nu,\bar{\nu}}(x,y) = 2\sum_i x(q_i(x) - \bar{q}_i(x)),$$

so that Bjorken scaling is obtained. The parity violating structure function xF_3 is non-zero because the couplings of q and $\bar{q}$ are different due to their different handedness. The neutrino and antineutrino structure functions are distinguished from each other by the flavour of quarks and antiquarks which may enter the sums, as detailed in Section 2.4.

2.3.4 Neutrino scattering via the neutral current

To complete the formalism consider neutrino scattering via the exchange of the neutral weak boson, Z^0. For either neutrino–quark or neutrino–electron scattering only the Z^0 exchange diagram analogous to Fig. 2.6(a) is applicable, and the weak charged current of Eq (2.19) is replace by the weak neutral current

$$j^\nu = -ig_Z \bar{u}(k',\sigma')\frac{1}{2}\gamma^\nu(v_f - \gamma^5 a_f)u(k,\sigma) \tag{2.30}$$

where the vector, v_f, and axial-vector, a_f, electroweak couplings of the fermions are given by[6]

$$v_f = (T_{3f} - 2e_f \sin^2 \theta_W), \quad a_f = T_{3f} \tag{2.31}$$

This definition holds good for any fermion, whether lepton or quark, T_{3f}, is the 3rd-component of the weak isospin and θ_W is the Weinberg angle. Left-handed fermions form weak isospin doublets (ν_ℓ, ℓ^-), (u,d), (c,s), (t,b), with $T_{3f} = +1/2, -1/2$ for each pair, respectively. The weak coupling, g_Z, is given by $g_Z = g_W / \cos \theta_W$. The Z_0 propagator takes the same form as the W propagator (Eq. 2.20) so that when $Q^2 \ll M_Z^2$ one may write the scattering amplitudes in terms of the Fermi constant G_F, since $M_Z = M_W / \cos \theta_W$ and thus $g_Z/M_Z = g_W/M_W$.

The neutrino–electron differential cross-section is then given by

$$\frac{d\sigma(\nu e)}{dy} = \frac{G_F^2}{4\pi} s \left[(v_e + a_e)^2 + (1-y)^2(v_e - a_e)^2\right] \tag{2.32}$$

and the antineutrino–electron cross-section by

$$\frac{d\sigma(\bar{\nu} e)}{dy} = \frac{G_F^2}{4\pi} s \left[(v_e + a_e)^2(1-y)^2 + (v_e - a_e)^2\right]. \tag{2.33}$$

The neutrino–quark differential cross-section at a specific value of x is given by

$$\frac{d\sigma(\nu q)}{dy} = \frac{G_F^2}{4\pi} xs \left[(v_i + a_i)^2 + (1-y)^2(v_i - a_i)^2\right] \tag{2.34}$$

and the antineutrino–quark cross-section by

$$\frac{d\sigma(\bar{\nu} q)}{dy} = \frac{G_F^2}{4\pi} xs \left[(v_i + a_i)^2(1-y)^2 + (v_i - a_i)^2\right] \tag{2.35}$$

for quarks of flavour i. The neutrino–antiquark cross-section is given by

$$\frac{d\sigma(\nu \bar{q})}{dy} = \frac{G_F^2}{4\pi} xs \left[(v_i + a_i)^2(1-y)^2 + (v_i - a_i)^2\right] \tag{2.36}$$

[6] For a pedagogic account of weak NC couplings see Renton (1990).

and the antineutrino–antiquark cross-section by

$$\frac{d\sigma(\bar{\nu}\bar{q})}{dy} = \frac{G_F^2}{4\pi} xs \left[(v_i + a_i)^2 + (v_i - a_i)^2 (1-y)^2 \right]. \quad (2.37)$$

Putting these cross-sections together to form the ν, $\bar{\nu}$-hadron cross-sections, which take the usual form of Eq. (2.29), give the QPM predictions for the structure functions for neutrino–hadron scattering via the neutral current as

$$F_L^{\nu,\bar{\nu}}(x,y) = 0$$

as usual, and

$$F_2^{\nu,\bar{\nu}}(x,y) = \Sigma_i x (q_i(x) + \bar{q}_i(x))(v_i^2 + a_i^2), \quad (2.38)$$

$$xF_3^{\nu,\bar{\nu}}(x,y) = \Sigma_i x (q_i(x) - \bar{q}_i(x)) 2 v_i a_i. \quad (2.39)$$

The existence of neutral currents was originally established by the discovery of an unambiguous $\bar{\nu}-e$ elastic scattering event and the first measurements of the weak mixing angle θ_W were made using the ratios of the neutral to charged weak current cross-sections for neutrino and antineutrino scattering from isoscalar targets. A particularly advantageous combination is

$$\frac{\sigma_{NC}(\nu N) - \sigma_{NC}(\bar{\nu}N)}{\sigma_{CC}(\nu N) - \sigma_{CC}(\bar{\nu}N)} = \frac{1}{2} - \sin^2\theta_W. \quad (2.40)$$

2.4 Quark–parton model tests

A powerful test of the QPM can be made by comparing the structure functions obtained in charged lepton scattering with those obtained in neutrino and antineutrino scattering, as follows. Consider charged lepton scattering from a proton target. Clearly the result of Eqn. (2.16) should be generalized to give

$$F_2(x, Q^2) = \Sigma_i e_i^2 (x q_i(x) + x \bar{q}_i(x)), \quad (2.41)$$

and thus for a proton target

$$F_2^{\ell p}(x) = \frac{4}{9} x (u(x) + \bar{u}(x) + c(x) + \bar{c}(x)) + \frac{1}{9} x (d(x) + \bar{d}(x) + s(x) + \bar{s}(x)) \quad (2.42)$$

Nucleon parton distribution functions are *defined* to be those for the proton. For a neutron target strong isospin swapping ($u^n = d^p$, $d^n = u^p$) gives

$$F_2^{\ell n}(x) = \frac{4}{9} x (d(x) + \bar{d}(x) + c(x) + \bar{c}(x)) + \frac{1}{9} x (u(x) + \bar{u}(x) + s(x) + \bar{s}(x)) \quad (2.43)$$

and an isoscalar target is treated as an average of these two

$$F_2^{\ell N} = \frac{5}{18} x (u(x) + \bar{u}(x) + d(x) + \bar{d}(x)) + \frac{1}{9} x (s(x) + \bar{s}(x)) + \frac{4}{9} x (c(x) + \bar{c}(x)) \quad (2.44)$$

It is assumed that there is no significant bottom or top quark content in the nucleon. (Note that not all authors define a parton distribution for a heavy quark such as the charm quark. The treatment of heavy quarks is discussed in Section 4.5)

Now consider charged-current neutrino scattering from a proton target. Note that the W^+ exchange has to increase the charge of the parton by one unit. Hence, it can only hit quarks of charge $-\frac{1}{3}e$ or antiquarks of charge $-\frac{2}{3}e$, giving

$$\begin{align} F_2^{\nu p} &= 2x(d(x) + s(x) + \bar{u}(x) + \bar{c}(x)), \\ xF_3^{\nu p} &= 2x(d(x) + s(x) - \bar{u}(x) - \bar{c}(x)). \end{align} \quad (2.45)$$

For antineutrino scattering on a proton target the W^- exchange must decrease the charge of the struck parton by one unit. Hence, it can only hit quarks of charge $\frac{2}{3}e$ or antiquarks of charge $\frac{1}{3}e$, giving

$$\begin{align} F_2^{\bar{\nu} p} &= 2x(u(x) + c(x) + \bar{d}(x) + \bar{s}(x)), \\ xF_3^{\bar{\nu} p} &= 2x(u(x) + c(x) - \bar{d}(x). - \bar{s}(x)) \end{align} \quad (2.46)$$

Again it is assumed that there are no top or bottom quarks in the nucleon targets and that W^2 is above threshold for production of charmed quarks in the final state. (Below the charm threshold, d must be multiplied by $\cos^2\theta_c$ and s by $\sin^2\theta_c$ in Eq. 2.45 and $\bar{d}$ must be multiplied by $\cos^2\theta_c$ and $\bar{s}$ by $\sin^2\theta_c$ in Eq. (2.46), where θ_c is the Cabibbo mixing angle.)

Strong isospin swapping gives the corresponding results for a neutron target, and thus the results appropriate to an isoscalar target are

$$\begin{align} F_2^{\nu N}(x) &= x(u(x) + d(x) + \bar{u}(x) + \bar{d}(x) + 2s(x) + 2\bar{c}(x)), & (2.47) \\ F_2^{\bar{\nu} N}(x) &= x(u(x) + d(x) + \bar{u}(x) + \bar{d}(x) + 2\bar{s}(x) + 2c(x)), & (2.48) \end{align}$$

and

$$\begin{align} xF_3^{\nu N}(x) &= x(u(x) + d(x) - \bar{u}(x) - \bar{d}(x) + 2s(x) - 2\bar{c}(x)), & (2.49) \\ xF_3^{\bar{\nu} N}(x) &= x(u(x) + d(x) - \bar{u}(x) - \bar{d}(x) - 2\bar{s}(x) + 2c(x)). & (2.50) \end{align}$$

Hence, assuming $s(x) = \bar{s}(x), c(x) = \bar{c}(x)$, the structure functions F_2 for lepton and neutrino scattering on isoscalar targets are related as follows

$$F_2^{\ell N} = \frac{5}{18} F_2^{\nu N} - \frac{1}{3} x s(x) + \frac{1}{3} x c(x) \quad (2.51)$$

Since the contributions of strange and charmed sea quarks are small (see below), this implies

$$F_2^{\ell N} \approx \frac{5}{18} F_2^{\nu N}.$$

The observation that the ℓN and the νN structure functions were indeed related by the approximate factor of 5/18 established the relationship of the parton scatterers to the quarks of the constituent quark model and thus established the quark-parton model.

The early neutrino and antineutrino data were also used to develop the model further by determining the approximate shapes of the sea and valence quark momentum distributions in the nucleon. Since antiquark distributions are purely sea distributions, whereas quark distributions have both valence and sea contributions,

$$xq(x) = xq_v(x) + xq_{\text{sea}}(x), \quad x\bar{q}(x) = x\bar{q}_{\text{sea}}(x),$$

and it is assumed that

$$xq_{\text{sea}}(x) = x\bar{q}_{\text{sea}}(x).$$

Applying these relationships to the structure functions for (anti)neutrino scattering on isoscalar targets gives

$$xF_3^{\nu N} \approx xF_3^{\bar{\nu} N} \approx \frac{xF_3^{\nu N}(x) + xF_3^{\bar{\nu} N}(x)}{2} = x(u_v(x) + d_v(x)) = xV(x) \quad (2.52)$$

where $xV(x)$ represents a purely valence, or non-singlet,[7] distribution consisting only of u and d valence quarks, as suggested by the constituent quark model. Similarly,

$$F_2^{\nu N}(x) = F_2^{\bar{\nu} N}(x) = x(u_v(x) + d_v(x) + 2\bar{u}(x) + 2\bar{d}(x) + 2\bar{s}(x) + 2\bar{c}(x)) \quad (2.53)$$

so that

$$F_2^{\nu N}(x) = xV(x) + xS(x) \quad (2.54)$$

represents a combination of valence and sea, which is a singlet distribution.

Hence, the shapes of the valence and sea distributions may be extracted from the data on $\nu, \bar{\nu}$ structure functions. For large x, it is observed that $F_2^{\nu N} = xF_3^{\nu N}$ and so both distributions must be purely valence. Thus, the sea quarks are only significant at low momemtum fraction. At small x, $xF_3^{\nu N} \to 0$ and hence the valence distribution is now negligible. The valence and sea distributions extracted from the data are illustrated schematically in Fig. 1.4. These results can be interpreted as follows. If the target hadron consisted of just three non-interacting valence quarks, then there would be no sea distribution and the valence distribution would be a delta function at $x = \frac{1}{3}$. Of course the three quarks are confined in a volume the size of the proton and thus there must be a sizeable uncertainty in their momentum (called Fermi motion) which smears the delta function to a more gaussian-like distribution. This begs the question of what is confining the quarks. Since they must be interacting they can redistribute momenta amongst themselves. In the theory of QCD (to be developed in Chapter 3) quarks interact via the exchange of gluons, and these virtual gluons may also produce quark–antiquark pairs, which deplete the momentum of the valence quarks and produce a low momentum sea, as observed.

---

[7]The terminology singlet and non-singlet will be explained in context in Section 4.2.

At moderate Q^2, the sea is responsible for about 10% of the hadron's momentum, as may be deduced from the cross-section ratio, $\sigma(\nu N)/\sigma(\bar{\nu}N) \sim 0.43$ (see problem 4). Since the sea is considered as made up from $q\bar{q}$ pairs emitted by gluons, light $q\bar{q}$ pairs are more readily produced than than heavy $q\bar{q}$ pairs. This is why the strange and charmed quark contributions to the structure functions are small.

The charged lepton and ν, $\bar{\nu}$ DIS scattering data can be used to gain detailed information on the flavour decomposition of the valence and sea distributions. A lot of early phenomenology was done on the assumption of an SU(3) symmetric sea: $\bar{u} = u_{\text{sea}} = \bar{d} = d_{\text{sea}} = \bar{s} = s = K$, but $\bar{c} = c = 0$. This would give

$$\frac{F_2^{ep}}{F_2^{en}} = \frac{u_v + 4d_v + 4/3K}{4u_v + d_v + 4/3K} \tag{2.55}$$

The assumption of an SU(3) symmetric sea has been modified in the light of experience as explained in Chapter 6, however, the above relationship remains approximately true for moderate Q^2 values. Experimental results on this quantity indicate that $F_2^{en}/F_2^{ep} \to 1$ at small x, corresponding to dominance of the sea, whereas $F_2^{en}/F_2^{ep} \ll 1$ at large x indicating a dominance of u valence quarks. Since there is no sea at large x, the ratio $u_v/d_v = 2$, which is expected from the constituent quark model, would give $F_2^{en}/F_2^{ep} \to 2/3$. However, the experimental value is closer to $F_2^{en}/F_2^{ep} \to 1/4$, suggesting that, $d_v \to 0$ as $x \to 1$. Results on the ratio $F_2^{\nu p}/F_2^{\bar{\nu} p}$ confirm this (see problem 4). These results indicate that the d_v quark momentum distribution is softer than that of the u_v quark, at high x.

Further investigation of the use of DIS data to extract the detailed shapes of the parton distribution functions of different flavours is postponed until Chapter 6, because modern DIS data are so accurate that the corrections to the parton model embodied in QCD must first be considered.

2.5 Sum rules

To conclude this chapter the further support for the QPM which comes from sum rules is considered. Since $q(x)$ gives the quark number distribution the number sum rules

$$\int_0^1 dx u_v(x) = 2, \qquad \int_0^1 dx d_v(x) = 1 \tag{2.56}$$

are implied. Of course these relations cannot be directly verified, but their consequences can be, for example, the Gross Llewellyn-Smith sum rule

$$\int_0^1 dx F_3^{\nu N} \approx \int_0^1 dx (u_v(x) + d_v(x)) = 3 \tag{2.57}$$

was verified in early neutrino data on isoscalar targets. Similarly, the Adler sum rule

$$\int_0^1 \frac{dx}{x} \left(F_2^{\bar{\nu}p} - F_2^{\nu p} \right) = 2 \int_0^1 dx ((u_v(x) - d_v(x)) = 2 \qquad (2.58)$$

was verified using neutrino data on hydrogen targets.

Sum rules are more general than this simple QPM picture implies. They may be rigorously derived in QCD using the operator product expansion and current algebra to give results including QCD corrections where appropriate.

A very important sum rule, both historically and in present applications, is the momentum sum rule. Define the sum over all types of quarks and antiquarks in the nucleon as the singlet

$$\Sigma(x) = u(x) + \bar{u}(x) + d(x) + \bar{d}(x) + s(x) + \bar{s}(x) + c(x) + \bar{c}(x) \qquad (2.59)$$

then the momentum sum rule is expressed as

$$\int_0^1 dx \, x \Sigma(x) = 1 \qquad (2.60)$$

if quarks and antiquarks carry all of the momentum of the nucleon. This may be checked experimentally, since

$$x\Sigma(x) = F_2^{\nu N}$$

and the result

$$\int_0^1 dx \, F_2^{\nu N} \sim 0.5 \qquad (2.61)$$

implies that there is more momentum in the nucleon than that carried by the quarks and antiquarks. This gave impetus to the development of the theory of QCD in which the deficit in momentum is carried by the gluons.

2.6 Summary

This chapter has described the basic idea of the quark–parton model and provided the details for constructing the full deep inelastic cross-sections for electron– and neutrino–nucleon scattering. Crucial to the model are the parton momentum densities, which are assumed to be universal, although they cannot be calculated within the model. The various test of the model described in the chapter show it to be successful. However, a number of paradoxes remain:

- the partons are assumed to be essentially free particles, but at the same time are confined inside the hadron;
- although Bjorken scaling is approximately satisfied, there is clear evidence for scale breaking at large and small x;
- the failure to saturate the momentum sum rule with the contributions from charged partons alone shows that there are other nucleon constituents.

These puzzles led to the development of QCD as the theory of the strong interactions. How it in turn resolves the difficulties of the simple QPM will be taken up in Chapter 4 after the elements of QCD have been outlined in the next chapter.

2.7 Problems

1. The leptonic tensor $L_{\mu\nu}$. Using Eq. (2.3), show that when $\mathcal{M}^\dagger \mathcal{M}$ is written out in full and summed over all spin states one can identify the following expression (and an equivalent one for $L_{\mu\nu}^{\text{muon}}$)

$$L^e_{\mu\nu} = \frac{1}{2}\text{Tr}\left\{(\slashed{k}' + m)\gamma_\mu(\slashed{k} + m)\gamma_\nu\right\},$$

where the completeness relations for $u(k)$ and $u(k')$ have been used. Then using standard trace theorems for Dirac matrices show that

$$L^e_{\mu\nu} = 2(k'^\mu k^\nu + k'^\nu k^\mu - g^{\mu\nu}(k' \cdot k - m^2)).$$

Appendix A gives a summary of the Dirac equation.

2. Handedness, chirality and helicity. For a spin-$\frac{1}{2}$ particle, two-component helicity eigenstates satisfy $\sigma \cdot \mathbf{p}\chi_\pm/|\mathbf{p}| = \pm\chi_\pm$. Show that in the high energy limit $(m/E \to 0)$ the 4-component Dirac eigenstates of the chiral (handedness) operators $P_R = (1 + \gamma_5)/2$, $P_L = (1 - \gamma_5)/2$ are constructed from the spin-$\frac{1}{2}$ helicity eigenstates. By considering terms of the form $\bar{u}\Gamma u$, where $\Gamma = \gamma_\mu, \gamma_\mu\gamma_5, I$ and writing $u = u_R + u_L$ ($u_R = P_R u$, $u_L = P_L u$), show that vector and axial vector couplings do not flip helicity, but a scalar one does (all in the high energy limit).

3. Derive the equivalent equation to Eq. (2.18) for spin-0 quarks, by using the vertex factor $-ie(k+k')_\mu$ for the coupling of two spin-0 particles with momenta k, k' to the photon. Hence, deduce that $\sigma_T = 0$ in this case. Alternatively attempt a helicity-based argument using the results of the previous question.

4. Use Eqs (2.25), (2.26) to calculate the total cross-sections $\sigma(\nu N)$ and $\sigma(\bar{\nu}N)$ in terms of the integrals

$$\int_0^1 dx\, xq(x) = Q \quad \text{and} \quad \int_0^1 dx\, x\bar{q}(x) = \bar{Q}.$$

Hence, prove that

$$\frac{\bar{Q}}{Q} = \frac{3R-1}{3-R}, \quad \text{where} \quad R = \frac{\sigma(\nu N)}{\sigma(\bar{\nu}N)}.$$

5. Use Eqs (2.38), (2.39) to establish Eq. (2.40)
6. Consider Eqs. (2.45), (2.46) for large x. Hence, express the ratio $F_2^{\nu p}/F_2^{\bar{\nu}p}$ in terms of a ratio of valence quark distributions.

34 The quark–parton model

7. Experimental information on valence and sea distributions may also be obtained from charged lepton–nucleon scattering. Express $F_2^{\ell N}$ for an isoscalar target as a combination of valence and sea distributions. The difference $F_2^{\ell p} - F_2^{\ell n}$ (which may be deduced experimentally from data on proton and deuterium targets) can be expresses purely in terms of valence distributions. Justify this statement and hence derive the Gottfried sum rule

$$\int_0^1 \frac{dx}{x}(F_2^{\ell p} - F_2^{\ell n}) = \frac{1}{3}.$$

   By examining the assumptions you made in the derivation, suggest why this sum rule is violated.

8. Neutrino and antineutrino scattering data on proton and isoscalar targets potentially give a lot of information on the different parton flavour distributions. Suggest how the four cross-sections for the processes νp, $\bar{\nu} p$, νn, $\bar{\nu} n$, might be combined to extract the distributions u_v, d_v, $\bar{u}$, $\bar{d}$, assuming that $s = \bar{s} = 0$.

9. The momentum sum rule can also be checked from charged lepton hadron data. Define

$$\epsilon_u = \int_0^1 dx\, x(u+\bar{u}), \quad \epsilon_d = \int_0^1 dx\, x(d+\bar{d})$$

   and hence express $\int_0^1 dx\, F_2^{ep}$ and $\int_0^1 dx\, F_2^{en}$ in terms of these integrals (again assuming $s = \bar{s} = 0$). Experimental results yield

$$\int_0^1 dx\, F_2^{ep} \sim 0.18 \quad \text{and} \quad \int_0^1 dx\, F_2^{en} \sim 0.12.$$

   Deduce the fraction of nucleon momentum taken by gluons, ϵ_g.

3
QCD and formal methods

The chapter begins with a brief introduction to Quantum Chromodynamics, highlighting the similarities and differences with QED, particularly the more complicated interactions that follow from the fact that gluons carry the colour charge (Section 3.1). The crucial outcome is that QCD reconciles quarks completely confined in hadrons at large distances with the assumption of the parton model that at short distances the quarks interact almost freely in a hard lepton–hadron interaction.

Full QCD calculations are not attempted, instead essential tree level diagrams are evaluated using the corresponding QED processes and QCD colour factors (Section 3.2). The mathematical problems of infrared and ultraviolet divergences in both QED and QCD are touched upon in Sections 3.3, 3.4 with emphasis on the physical ideas underlying their resolution. For the practical applications considered in this book the most important consequence of renormalization is the 'running coupling' and this is covered in Sections 3.4.2, 3.4.3 and more formally in Section 3.6 after the technique of the renormalization group has been introduced in Section 3.5.

The rest of the chapter is concerned with an outline of the operator product expansion (Section 3.8), which is the technical tool used in much of the mathematical underpinning of the QCD-enhanced parton model. This first requires the connection between the deep inelastic structure functions in the Bjorken limit and the forward scattering amplitude for virtual-photon Compton scattering evaluated on the light cone (Section 3.7). Finally, the crucial idea of factorization is introduced. None of the topics in this chapter are treated with any rigour, the aim is to give an outline of key ideas and connections to the formal calculations. For the remainder of the book, particularly the next chapter, the important sections are those on the tree-level diagrams and the running coupling. There are many excellent textbooks on quantum field theory, for example, Peskin and Schroeder (1995) or Sterman (1993), and on QCD, for example, Muta (1998) or Ynduráin (1999), for those who want the details.

3.1 Quantum Chromodynamics

Before outlining the mathematical structure of QCD, it is worth recalling its important features. The idea that quarks carry a three–fold 'colour' charge was introduced to allow baryon wavefunctions (e.g. the Δ^{++}) to have simultaneously the correct permutational symmetry and satisfy Fermi–Dirac statistics. The word 'colour' and the designation of the three varieties of charge as red (r), green (g), blue (b) are convenient shorthands but one should consider colour charge as very closely analogous to the one–fold electric charge, as the mathematical structure of QCD will make clear. The theory also provides the basis for why only the 'colourless' qqq and q$\bar{\text{q}}$ combinations form 'confined' hadronic bound states.[1] The strong force quanta — the spin-1 massless gluons — also carry colour charges and this is the major difference between QCD and QED. Consider a q$\bar{\text{q}}$ meson and all the possible combinations of colour that a gluon exchanged between the q and $\bar{\text{q}}$ might carry: naively (r,g,b)⊗($\bar{\text{r}}$, $\bar{\text{g}}$, $\bar{\text{b}}$) gives nine combinations, however the r$\bar{\text{r}}$, g$\bar{\text{g}}$, b$\bar{\text{b}}$ 'neutral' combinations may be linearly combined to give three new states so that the nine combinations are split into one totally symmetric colour state (r$\bar{\text{r}}$ + g$\bar{\text{g}}$ + b$\bar{\text{b}}$)/$\sqrt{3}$ and eight others: r$\bar{\text{g}}$, r$\bar{\text{b}}$, g$\bar{\text{r}}$, g$\bar{\text{b}}$, b$\bar{\text{r}}$, b$\bar{\text{g}}$, (r$\bar{\text{r}}$ − g$\bar{\text{g}}$)/$\sqrt{2}$, (r$\bar{\text{r}}$ + g$\bar{\text{g}}$ − 2b$\bar{\text{b}}$)/$\sqrt{6}$. In group theoretic language, this is the outcome of combining the 3 and $\bar{3}$ representations of SU(3): $3 \otimes \bar{3} = 8 \oplus 1$. It explains why there are only eight gluons, the totally symmetric colour singlet gluon would be colourless and unconfined. It would couple equally to all colour charges – as the photon does to both charges in QED. The other two colour 'neutral' combinations are analogous to the electrically neutral members of strong isospin multiplets — they are not 'colourless'. The SU(3) symmetry of QCD is assumed to be exact.

3.1.1 The QCD Lagrangian

In QCD the strong force at short distances is assumed to have a similar space-time structure to QED. Both QED and QCD are examples of locally gauge (or phase) invariant quantum field theories. This is a very powerful symmetry that fixes the form of the interaction between the fundamental matter particles (the fermions) and the field quanta. QED is an Abelian gauge theory with the fermion wave-functions transforming as

$$\psi(x) \to \psi'(x) = e^{iq\theta(x)}\psi(x),$$

where q is the electric charge and $\theta(x)$ is a single space–time dependent phase factor. The QED Lagrangian density is given by

$$\mathcal{L}_{QED} = \sum_f \bar{\psi}_f(i\gamma_\mu D^\mu - m_f)\psi_f - \frac{1}{4}F^{\mu\nu}F_{\mu\nu},$$

---

[1] There is a nice account of this, together with other QCD–based results on the static properties of hadrons in Bowler (1990), Chapter 12.

where the sum runs over the charged fermions, $F^{\mu\nu} = \partial^\mu A^\nu - \partial^\nu A^\mu$ is the electromagnetic field strength tensor and $D^\mu = \partial^\mu + iqA^\mu$ is the *covariant derivative*. Note that the interaction term between the fermions and the EM field enters through the covariant derivative.

QCD is a non-Abelian gauge theory with a more complicated quark–gluon interaction generated by the SU(3) group structure of the colour charges. The fermion wave-functions transform as

$$\psi(x) \to \psi'(x) = e^{igt \cdot \theta(x)} \psi(x)$$

where g is the strong coupling constant and $t \cdot \theta$ represents the product of the colour group generators with a vector of space–time phase functions in colour space. The group generators t^a satisfy

$$[t^a, t^b] = if^{abc} t^c \qquad (3.1)$$

where f^{abc} are the SU(3) structure constants. The gluon field strength tensor is

$$F_a^{\mu\nu} = \partial^\mu A_a^\nu - \partial^\nu A_a^\mu + gf^{abc} A_b^\mu A_c^\nu, \qquad (3.2)$$

where A_a ($a = 1, \ldots, 8$) are the gluon fields and the final term represents the interaction of the gluons amongst themselves as they also carry colour charges. The quark spinor fields ψ_i transform as triplets under SU(3) with $i = 1, 2, 3$ running over the three colour indices. The Lagrangian density is given by

$$\mathcal{L}_{\text{QCD}} = \sum_f \bar{\psi}_f^i (i\gamma_\mu D^\mu - m_f)_{ij} \psi_f^j - \frac{1}{4} F_a^{\mu\nu} F_{\mu\nu}^a \qquad (3.3)$$

where the covariant derivative is

$$D_{ij}^\mu = \delta_{ij} \partial^\mu + ig(t^a)_{ij} A_a^\mu \qquad (3.4)$$

and $(t^a)_{ij}$ are 3×3 hermitian matrices, which for the fundamental triplet representation of SU(3) are $(\lambda^a)_{ij}/2$, where λ^a are the Gell-Mann matrices (the λ^a are listed along with the f^{abc} towards the end of Appendix D). The mass parameters, m_f, in the Lagrangian are only non-zero for the heavy flavours c, b, t. For the light quarks, mass is assumed to be generated by spontaneous breaking of chiral symmetry. In the absence of the mass term, the quark sector terms of the Lagrangian may be written as

$$\mathcal{L}_{\text{quark}} = \bar{\psi}_L i\gamma_\mu D^\mu \psi_L + \bar{\psi}_R i\gamma_\mu D^\mu \psi_R, \qquad (3.5)$$

where $\psi_{L,R} = \frac{1}{2}(1 \mp \gamma_5)\psi$ are the left- and right-handed chiral states, which in the massless limit are the same as left- and right-handed helicity states.[2] As there is no term connecting the left- and right-handed states in the massless Lagrangian, one may rotate the two helicity states independently.

---

[2] See Appendix A for details of Dirac spinors and Problem 1 at the end of chapter.

38 *QCD and formal methods*

The symmetry is not exact (otherwise every hadron would have a partner of opposite parity but the same mass), but strong isospin may be regarded as a consequence of the underlying chiral symmetry and other predictions based on spontaneously broken chiral symmetry for the light quarks are in agreement with strong interaction physics at scales small compared to 1 GeV.

Note that the same coupling constant g couples the gluon fields to themselves (in $F_a^{\mu\nu}$) and the gluon to the quark fields through the covariant derivative (D_{ij}^{μ}). The derivation of the Feynman rules from $\mathcal{L}_{\text{QCD}}$ and their use, even at the tree-level, is non-trivial as non-Abelian gauge theories give rise to greater complications in handling quantization and gauge invariance than QED. Briefly, in addition to the choice of gauge and gauge fixing term in the Lagrangian, to calculate with gauge invariance manifest requires the introduction of 'ghost' terms. The alternative is to give up manifest gauge invariance and ghost terms and work in a 'physical' or axial gauge in which the gluon fields satisfy $n_\mu A_a^\mu = 0$ where n_μ is a fixed space-like or null 4-vector. These complications will not be pursued here. Fortunately, most of the tree-level results that are required later can be derived from the more familiar QED calculation with the addition of colour factors. For completeness, the Feynman rules are given in Appendix D.

3.1.2 QCD colour factors

In QED, the strength of the coupling of the photon to two point-like particles with charges e_1 and e_2 is given by $e_1 e_2 \alpha_{\text{qed}}$. In QCD, the equivalent coupling strength for a gluon and two quarks with colour charges c_1 and c_2 is $\frac{1}{2} c_1 c_2 \alpha_s$ where $\alpha_s = g^2/4\pi$ and the $\frac{1}{2}$ is conventional. Usually, one is interested in expressions for physical processes like cross-sections in which case powers of the factor $\frac{1}{2} c_1 c_2$ have to be summed or averaged over the states involved to give the *colour factor*. For quark-gluon tree-level diagrams, the colour factor is the only difference between the QED and QCD expressions (apart from the change $\alpha_{\text{qed}} \rightarrow \alpha_s$).

Consider the colour factor for gluon exchange between a q and a q̄ in an overall colour singlet state $(r\bar{r} + g\bar{g} + b\bar{b})/\sqrt{3}$ — as would be the case in a meson.[3] The generic diagram is shown in Fig. 3.1, start with the r$\bar{\text{r}}$ interactions, there are four gluons that can contribute: (a) rb̄ for r$\bar{\text{r}}$ → b$\bar{\text{b}}$ (as shown) for which $c_1 c_2 = (+1)(-1) = -1$ (the sign of the colour charge is reversed for the q̄); (b) similarly rḡ for r$\bar{\text{r}}$ → g$\bar{\text{g}}$ gives -1; (c) gluons $(r\bar{r} + g\bar{g} - 2b\bar{b})/\sqrt{6}$ and $(r\bar{r} - g\bar{g})/\sqrt{2}$ can both contribute to r$\bar{\text{r}}$ → r$\bar{\text{r}}$, the first gives $c_1 c_2 = (\frac{1}{\sqrt{6}})(-\frac{1}{\sqrt{6}}) = -\frac{1}{6}$ and the second $(\frac{1}{\sqrt{2}})(-\frac{1}{\sqrt{2}}) = -\frac{1}{2}$. Summing over the final states (b$\bar{\text{b}}$, g$\bar{\text{g}}$, r$\bar{\text{r}}$) gives $-1-1-\frac{1}{6}-\frac{1}{2} = -\frac{8}{3}$. There are two more such factors to account for the b$\bar{\text{b}}$ and g$\bar{\text{g}}$ interactions and each is normalized by $\frac{1}{\sqrt{3}}\frac{1}{\sqrt{3}}$ from the totally symmetrised colour wavefunctions

---

[3]The treatment here follows Halzen and Martin (1984), Section 2.15, see also Renton (1990), Section 2.5.5.

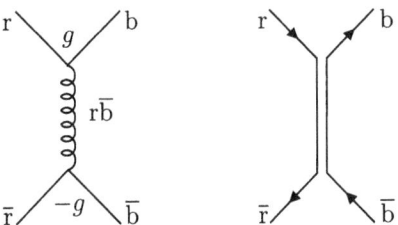

Fig. 3.1 The generic diagram for single gluon exchange between a qq̄ pair and the corresponding colour flow. The labelling is for the exchange of an rb̄ gluon producing a rr̄ → bb̄ interaction.

of the initial and final qq̄ state, so the overall result remains $-\frac{8}{3}$ and the colour factor in this case is $\frac{4}{3}$ (remembering the conventional factor of $1/2$).

An alternative method of calculating colour factors is to use the SU(3) group properties. The relationships are written for SU(N) as this is often done to make clear which are the colour factors before putting $N = 3$. The

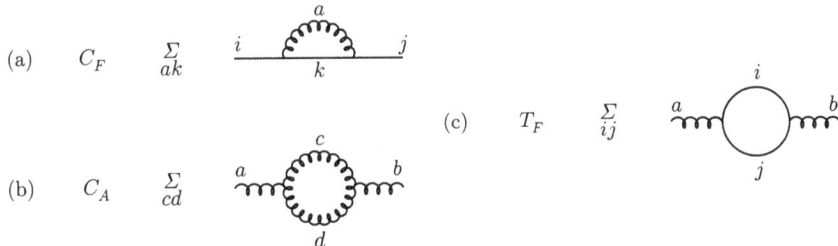

Fig. 3.2 Quark and gluon loop insertions associated with the colour factors: (a) C_F for the gluon loop on a quark line; (b) C_A for a gluon loop in a gluon line and (c) T_F for a quark loop in a gluon line.

relevant properties of the matrix representations of SU(N), t^a (fundamental, for quarks) and T^a (adjoint, for gluons), are:

$$[t^a, t^b] = if^{abc}t^c, \qquad [T^a, T^b] = if^{abc}T^c, \qquad (T^a)_{bc} = -if^{abc}. \qquad (3.6)$$

The t^a are normalized such that $\text{Tr}\left[t^a t^b\right] = T_F \delta^{ab}$, where $T_F = \frac{1}{2}$. The colour matrices then satisfy the following relations which define the constants C_F and C_A

$$\sum_a (t^a)_{ik}(t^a)_{kj} = C_F \delta_{ij}, \qquad C_F = \frac{N^2 - 1}{2N} \qquad (3.7)$$

$$\text{Tr}\,(T^a T^b) = \sum_{c,d} f^{acd} f^{bcd} = C_A \delta^{ab}, \qquad C_A = N. \tag{3.8}$$

C_F, C_A are the *Casimir* operators and, as they are group invariants, are used to label SU(3) representations. C_F, C_A and T_F are the colour factors associated with the quark and gluon loop insertions as shown in Fig. 3.2. For SU(3) $C_F = \frac{4}{3}$ and $C_A = 3$.

3.2 Some simple tree–level diagrams

Gluon bremsstrahlung and $q\bar{q}$ pair production processes will be considered. Both are important for $O(\alpha_s)$ improvements to the QPM and for large p_T jet production in DIS. The route followed here is to start with the equivalent QED calculation and then modify the result by the appropriate colour factor.

The QCD Compton process (QCDC) $\gamma^* q \to gq$ is one of the most useful for understanding the essential physics of how QCD modifies the quark–parton model. The related QED process is $\gamma^* e \to \gamma e$ which is Compton scattering with one photon virtual. The tree-level diagrams are shown in Fig. 3.3 which also defines the 4-momenta of the particles involved. The

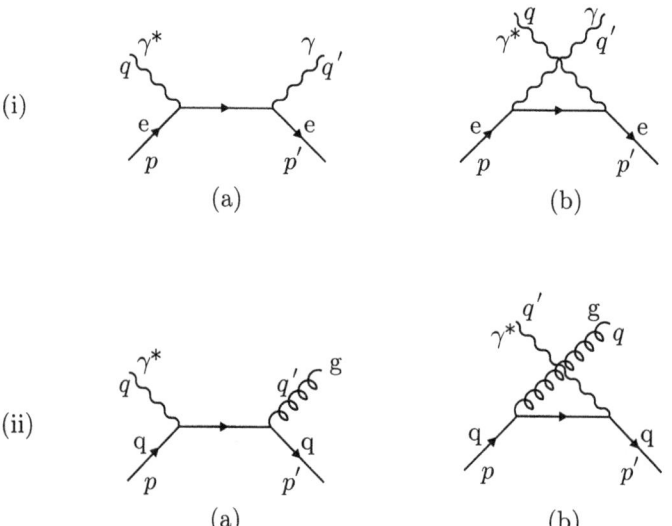

Fig. 3.3 (i) QED tree-level diagrams for $\gamma^* e \to \gamma e$, (a) s-channel, (b) u-channel; (ii) the equivalent diagrams for the QCD Compton process $\gamma^* q \to gq$.

usual kinematic Lorentz invariants are:

$$s \;=\; (q+p)^2 = (q'+p')^2 = 2q\cdot p - Q^2$$

$$t = (q-q')^2 = (p'-p)^2 = -2p \cdot p' \tag{3.9}$$
$$u = (q-p')^2 = (q'-p)^2 = -2q' \cdot p$$

where all masses have been neglected except for the virtual photon with $q^2 = -Q^2$. Both of the diagrams shown in the figure are required for gauge invariance. Using the Feynman rules for QED,[4] the amplitudes for the s and u channel processes are $\mathcal{M}_a$, $\mathcal{M}_b$ where:

$$\mathcal{M}_a = -ie^2 \varepsilon'^*_\nu \varepsilon_\mu \bar{u}(p') \gamma^\nu (\slashed{q}+\slashed{p}) \gamma^\mu u(p)/s$$
$$\mathcal{M}_b = -ie^2 \varepsilon'^*_\nu \varepsilon_\mu \bar{u}(p') \gamma^\mu (\slashed{p}-\slashed{q}') \gamma^\nu u(p)/u. \tag{3.10}$$

For the unpolarised cross-section one needs $|\mathcal{M}_a + \mathcal{M}_b|^2$ summed over the final spin states and averaged over the initial spin states:

$$\frac{1}{4} \sum_{\text{spins}} |\mathcal{M}_a + \mathcal{M}_b|^2 = 2e^4 \left[-\frac{u}{s} - \frac{s}{u} + \frac{2tQ^2}{su} \right]. \tag{3.11}$$

The three terms on the RHS come from $|\mathcal{M}_a|^2$, $|\mathcal{M}_b|^2$ and the cross term $\mathcal{M}_a \mathcal{M}_b^*$ respectively. The calculations are explained in many texts (for example Aitchison and Hey (1989) or Halzen and Martin (1984)) and some details are discussed in the problems at the end of this chapter. To get from this result to the equivalent one for $\gamma^* q \to gq$, one needs to replace e^4 and introduce the relevant colour factor. For future reference, e^2 is written in terms of α_{qed} (usually just written as α), so for a quark with charge $e_i e$

$$e^4 \to e^2 e_i^2 g^2 \to (4\pi)^2 \alpha \alpha_s e_i^2.$$

By looking at $|\mathcal{M}_a|^2$ diagrammatically, Fig. 3.3(ii)a, it is clear that the colour factor comes from the gluon loop insertion on the quark line $C_F = \frac{4}{3}$ (as in Fig. 3.2a). The expression for the QCDC spin summed and averaged squared matrix element is

$$\overline{|\mathcal{M}_{\text{QCDC}}|^2} = \frac{1}{4} \sum_{\text{spins}} |\mathcal{M}_a + \mathcal{M}_b|^2 = \frac{8}{3} (4\pi)^2 e_i^2 \alpha \alpha_s \left[-\frac{u}{s} - \frac{s}{u} + \frac{2tQ^2}{su} \right]. \tag{3.12}$$

$\overline{|\mathcal{M}|^2}$ is related to the cross-section by[5]

$$\frac{d\sigma}{d\Omega} = \frac{1}{(8\pi)^2 s} \frac{k}{k'} \overline{|\mathcal{M}|^2}, \tag{3.13}$$

where k and k' are the magnitudes of the initial and final state 3-momenta, respectively. Since QCDC involves a virtual photon in the initial state,

---
[4] The Feynman rules are given in Appendix D.
[5] See Appendix B for a summary of cross-section and phase space relations.

42 QCD and formal methods

a choice in the definition of k in the flux factor, $4k\sqrt{s}$, is required. This book follows the Hand convention, in which the initial state momentum is calculated as though the photon were real. Since the quarks are taken to be massless that means that $k = k' = \sqrt{s}/2$. Using this, the expression for the QCDC cross-section is

$$\left.\frac{d\sigma}{d\Omega}\right|_{\text{QCDC}} = \frac{2\,e_i^2\alpha\alpha_s}{3}\left[-\frac{u}{s}-\frac{s}{u}+\frac{2tQ^2}{su}\right]. \qquad (3.14)$$

In the γ^*q CMS, the invariant u is given by $u = -2kk'(1-\cos\theta)$ where θ is the angle between the incoming γ^* and the outgoing scattered quark. Because the partons are taken to be massless $\left.\frac{d\sigma}{d\Omega}\right|_{\text{QCDC}}$ is divergent as $\theta \to 0$. How this divergence is to be regulated will be discussed shortly.

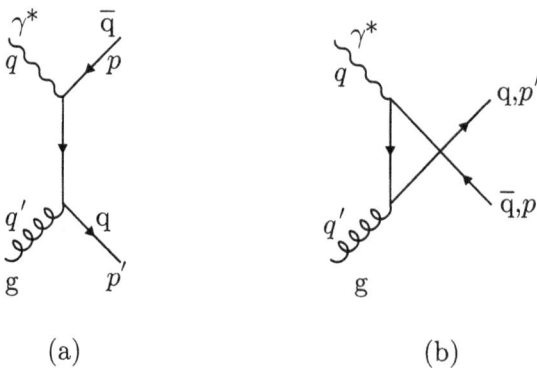

Fig. 3.4 The tree-level diagrams for the photon (or boson) gluon fusion process.

The second process that is important, particularly at low x and for heavy quark production, is the photon (or boson) gluon fusion process (BGF) for producing a q$\bar{\text{q}}$ pair. The tree-level diagrams are shown in Fig. 3.4. The corresponding QED process is the inverse of pair production ($e^+e^- \to \gamma^*\gamma$) in which one photon is virtual. The $\sum_{\text{spins}}|\mathcal{M}(ee \to \gamma^*\gamma)|^2$ may be calculated directly or 'crossing symmetry' may be used to write down the result immediately from that for Compton scattering.[6] The latter gives

$$\sum_{\text{spins}}|\mathcal{M}(ee \to \gamma^*\gamma)|^2 = (-1)\sum_{\text{spins}}|\mathcal{M}(\gamma^*e \to \gamma e)|^2 (s \leftrightarrow t, u \text{ unchanged})$$

Allowing for the different colour factor, in this case $\frac{1}{2}$ (T_F from the quark loop in the gluon line in the diagrams in $\overline{|\mathcal{M}_{\text{BGF}}|^2}$), one then gets

---
[6] See Appendix B for details.

$$|\mathcal{M}_{BGF}|^2 = (4\pi)^2 e_i^2 \alpha\alpha_s \frac{1}{2}\left[\frac{u}{t} + \frac{t}{u} - \frac{2sQ^2}{tu}\right],$$

where the kinematic invariants are now: $s = (q+q')^2$; $t = (q-p)^2$; $u = (q-p')^2$. Using the Hand convention for the virtual photon flux, the expression for the BGF cross-section is

$$\left.\frac{d\sigma}{d\Omega}\right|_{BGF} = \frac{1}{4}\frac{e_i^2 \alpha\alpha_s}{s}\left[\frac{u}{t} + \frac{t}{u} - \frac{2sQ^2}{tu}\right]. \quad (3.15)$$

In the γ^*g CMS, $t = -2kk'(1-\cos\theta)$ and $u = -2kk'(1+\cos\theta)$ where k and k' are the magnitudes of the initial and final state CM momenta and θ is the angle between the incoming γ^* and the outgoing $\bar{q}$. $\left.\frac{d\sigma}{d\Omega}\right|_{BGF}$ is divergent both when $\theta \to 0$ and $\theta \to \pi$.

3.3 Infrared and collinear singularities

A theory with massless particles appears to give rise to divergent cross-sections from the multiple emission of very low energy particles (infrared radiation) and from a singular matrix element when two particles are collinear. Consider the radiation of a soft photon from an electron emerging from some scattering process $e(p+k) \to \gamma(k) + e(p)$, where the 4-momenta are indicated in the brackets. Ignoring masses and inessential details, the cross-section for this final state radiation is

$$d\sigma \sim \left|\frac{\alpha}{(p+k)^2}\right|^2 \frac{d^3k}{E_k} \sim \frac{\alpha^2}{E_p^2(1-\cos\theta_{e\gamma})^2}\frac{dE_k}{E_k}d\Omega,$$

where $E_p, E_k, \theta_{e\gamma}$ are the energies of the final electron and photon and the angle between them. The collinear singularity arises as $\theta_{e\gamma} \to 0$. Integrating the characteristic dE_k/E_k spectrum of soft radiation down to zero energy gives the 'infrared' logarithmic singularity. There are two theorems arising from the structure of gauge field theories that overcome these problems by the inclusion of higher order processes ('radiative corrections'). The divergent terms are cancelled leaving a finite cross-section.

The first is the Bloch–Nordsieck Theorem. To give a flavour of how it works consider the scattering of an electron from a 'heavy muon' so that only radiation from the electron need be considered. The Born diagram, which is $O(e^2)$ in the electric charge, is shown in Fig. 3.5(a). The radiation of real photons from the initial and final state electrons are shown in diagrams (c). Collinear and infrared divergences of the sort just discussed are regulated by introducing a fictitious photon mass λ^2. This is a mathematical device to allow the integrals to be performed and the singularities isolated. The crucial point is that any measurement of a physical process will involve detectors with finite resolution in energy and angle and a minimum threshold energy E_0. Thus it will not be able to register a photon with $k < E_0$ or distinguish between an event in which the final state contains a single electron of energy E and one in which E is shared between an electron and a photon emitted at an angle $\theta_{e\gamma}$ less that the angular resolution of the detector. When observable quantities are calculated taking these constraints into account the results are finite.

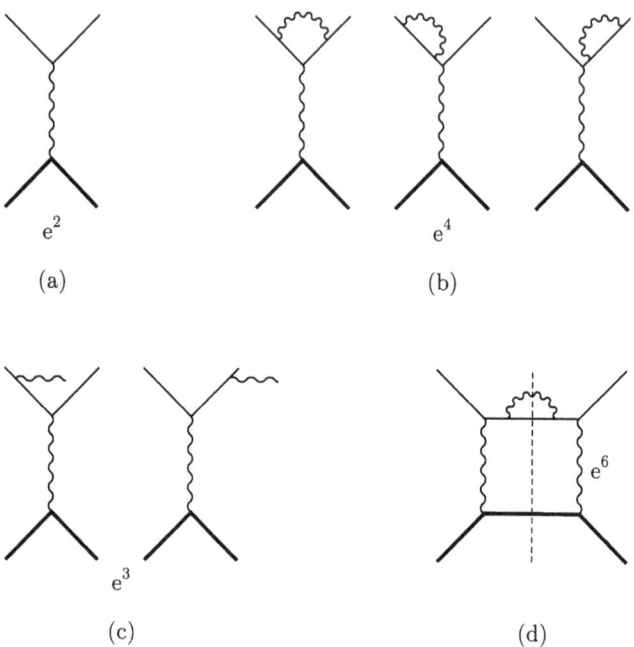

Fig. 3.5 Diagrams for $e\mu$ scattering: (a) $O(e^2)$ Born diagram; (b) $O(e^4)$ vertex and external line corrections; (c) $O(e^3)$ initial and final state radiation terms; (d) $O(e^6)$ cut diagram.

Rather than attempting to calculate a final state with an electron and a specific number of radiated photons, one calculates the cross-section for scattering into a final state with a single electron plus any number of photons with summed energy less than $E_{\min}$ or $\theta_{e\gamma} < \theta_{\min}$. The result is finite but depends on $E_{\min}$ and $\theta_{\min}$ and it would also appear still to depend on the photon mass λ^2. Consider now the terms shown in Fig. 3.5(b), these are $O(e^4)$ but the interference of these terms with the born diagram (a) gives terms of $O(e^6)$ in the cross-section — which is just the same order as the squared bremsstrahlung diagrams (c). The loops in diagrams (b) also give rise to logarithmic infrared divergences which can be regulated by the photon mass λ^2 and unitarity demands that the cutoffs be the same. Gauge invariance requires that the $\ln \lambda^2$ terms from the interference of (a) and (b) have exactly the same magnitude but opposite sign to those from $|(c)|^2$.

One is then left with terms like $\ln(E/E_{\min})$ or $\ln(E/m_e)$.

Unitarity may be used to verify the cancellation, it gives

$$2\mathrm{Im}\, A_{ab} = \sum_c (2\pi)^4 \delta^4(p_a - p_c) A_{ac} A^\dagger_{cb}$$

for the imaginary part of the scattering amplitude for initial state a to final state b, $\sum_c$ is a shorthand for the sum over a complete set of real intermediate states c. The $\delta^4()$ guarantees momentum conservation. In a perturbative calculation, the equation must be respected order by order. Fig. 3.5(d) shows the 'cut diagram' for the imaginary part of an $O(e^6)$ box. The particles in the intermediate state cut by the dashed vertical line are taken on the mass shell. Diagram (d) is $AA^\dagger$ from the RH of the $O(e^3)$ radiation diagrams (c). By attaching the photon loop in all possible ways to the electron lines in (d) and cutting the intermediate states in all possible ways, one accounts for all the terms at this order in the unitarity sum: (a)(b)† interference and $|(c)|^2$ terms. The divergent terms in the individual diagrams cancel in the unitarity sum which builds the complete $O(e^6)$ amplitude. Unitarity provides a very useful tool to ensure that the 'book keeping' is done correctly. This technique works for quantities like total cross-sections and structure functions that can be related to the imaginary parts of amplitudes. It is both more difficult and requires more care for less exclusive quantities like jet rates.

Often in QCD, one will be assuming quarks to be essentially massless thus infrared divergences arising from $\ln(E/m)$ terms as $m \to 0$ will remain even after the cancellation of the infrared gluon divergences. Another problem is that one may not be able to distinguish between a single coloured particle and an almost collinear jet of quarks and gluons with the same total colour charge. A theorem due to Kinoshita, Lee and Nauenberg helps here. They showed that in such circumstances, the divergences would cancel if all degenerate states (i.e. with the same mass) are summed. In particular, if the limit $m \to 0$ is taken, then states with gluons and massless $q\bar{q}$ pairs should be included. Generally, the more inclusive a final state is, the easier it is to apply the theorem and avoid divergences. Mass singularities from the initial state particles may remain and they will have to be circumvented in different ways.

3.4 Renormalization

Contributions to the perturbative expansion of scattering amplitudes beyond the leading order are usually formally divergent. For example, in QED quantum fluctuations introduce fermionic loops as higher order corrections to the photon propagator — see Fig. 3.6. The loop graphs and other higher order diagrams are divergent because of the unrestricted integration over the momentum flowing around loops. The way in which these divergences are regulated is known as renormalization. Not all field theories are renormalizable, for example the Fermi theory of weak processes is not, but the gauge theories of the standard model are. A renormalizable field theory is one in which the renormalization of a finite number of parameters ensures finite results for calculations to all orders of perturbation theory. Renormalization also has the effect of making what appear to be constants in the Lagrangian, e.g. the coupling g, become dependent on the scale of the process under study. A technical discussion of renormalization is not appropriate here, but some understanding of the idea is essential. Whether renormalization is really a fundamental requirement of field theory, or a consistent way of constructing an "effective theory" without knowledge of interactions at much higher energy scales, is still very much an open question. However, it is clear that the powerful constraints it imposes and the stunning success of the standard model make it a technique of lasting value. As the 'running coupling constant' is crucial to how QCD both underpins and extends the parton model, that will be the focus of discussion here. First, a simplified outline based on leading order charge renormalization in QED and QCD is given, followed by a brief account of the powerful methods of the 'renormalization group'.

3.4.1 Regularization and renormalization schemes

Renormalization involves two steps: firstly the sources of divergence must be identified and methods introduced to regularize them; secondly a finite number of parameters in the Lagrangian (e.g. masses, couplings) and wave function normalizations are renormalized to absorb the divergences. Counter terms may also have to be added to the Lagrangian. While the actual singularities are well defined, the finite parts 'left over' after the removal of the divergence will depend on the details of the regularization scheme. Consider the divergence arising from a Feynman loop integral (e.g. as shown in the second term of Fig. 3.6)

$$\int \frac{d^4k}{k^2 - m^2 + i\varepsilon}.$$

The integral diverges quadratically. An early method of regularization was that of Pauli and Villars in which the denominator is replaced by

$$\frac{1}{k^2 - m^2} \to \frac{1}{k^2 - m^2} - \frac{1}{k^2 - \Lambda^2}$$

where Λ is a large fixed mass — much larger than the mass or energy scales of interest in the problem at hand. The resulting integral is convergent, but depends explicitly on Λ. This method is conceptually simple and something similar will be used in Section 3.4.2 for charge renormalization. Ensuring that the regularization scheme respects gauge invariance is important, as the consequences of gauge invariance are necessary for the cancellation of divergences. A rather different approach, but one which is now standard in gauge theories, is the dimensional regularization of 't Hooft and Veltman.[7]

---

[7] For a pedagogic introduction, see Veltman (1994).

The essence of the idea is to evaluate the integrals in an n-dimensional space ($n \neq 4$) in which they converge, so the integral above becomes

$$\int \frac{d^n k}{k^2 - m^2 + i\varepsilon} \to i\pi^{\frac{n}{2}} \Gamma(1 - \frac{n}{2})(-m^2)^{\frac{n}{2}-1}.$$

Now

$$\Gamma(1 - \frac{n}{2}) = -\frac{2}{4-n} - 1 + \gamma_E + O(4-n),$$

where γ_E is the Euler constant and the divergence occurs as the pole at $n = 4$. The minimal subtraction scheme (MS) of 't Hooft and Veltman involves the introduction of counter terms to cancel only the singular terms at $n = 4$. However, a number of authors have argued that it is more convenient to remove the terms involving γ_E, which always appear in dimensional regularization, as well as the singular terms. This scheme is known as the modified minimal subtraction scheme ($\overline{\text{MS}}$). This latter is the one that is now preferred for most perturbative QCD calculations. Other schemes may be appropriate for particular calculations and will be mentioned when necessary.

3.4.2 Charge renormalization in QED

To get a feeling for the consequences of renormalization, consider the following simplified account of charge renormalization in QED.[8] Figure 3.6

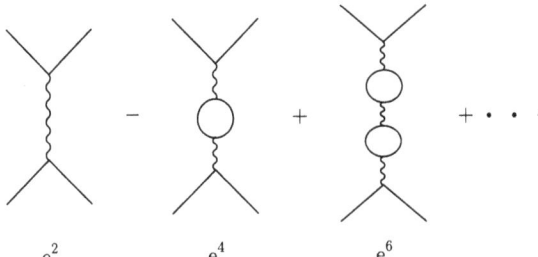

Fig. 3.6 Higher order loop corrections to the photon propagator for scattering between point like particles in QED.

shows the first few higher order loop diagrams for scattering between unlike point–like fermions at fixed CM energy (note the minus sign which arises from from each fermion loops). The amplitude depends only on the square of the 4-momentum transfer, q^2, and the modification of the photon propagator by a single fermion loop is

$$\frac{-ig_{\mu\nu}}{q^2} \to \frac{-ig_{\mu\nu}}{q^2} + \left(\frac{-ig_{\mu\alpha}}{q^2}\right) \Pi^{\alpha\beta} \left(\frac{-ig_{\beta\nu}}{q^2}\right).$$

[8] The approach taken here is close to that in either Halzen and Martin (1984) or Aitchison and Hey (1989).

$\Pi^{\alpha\beta}$ is the fermion loop term

$$\Pi^{\alpha\beta} = (-1)\int \frac{d^4k}{(2\pi)^4} \text{Tr}\left[(ie\gamma^\alpha)\frac{i(\slashed{k}+m)}{k^2-m^2}(ie\gamma^\beta)\frac{i(\slashed{q}-\slashed{k}+m)}{(q-k)^2-m^2}\right],$$

where k is the internal loop momentum and m the fermion mass. $\Pi^{\alpha\beta}$ may be reduced to the form

$$\Pi_{\alpha\beta} = -ig_{\alpha\beta}q^2\Pi(q^2) + \cdots, \tag{3.16}$$

where the missing terms are proportional to $q_\alpha q_\beta$ and give zero when the propagator is coupled to external charges (since $q_\alpha \Pi^{\alpha\beta} = 0$). The photon propagator now has the form

$$\frac{-ig_{\mu\nu}}{q^2}\left[1 - \Pi(q^2)\right].$$

The integral in $\Pi(q^2)$ diverges logarithmically and is controlled by the introduction of a high momentum cut-off Λ^2. At large $Q^2 = -q^2$, the leading term is

$$\Pi(Q^2) \approx \frac{\alpha_0}{3\pi}\ln\left(\frac{\Lambda^2}{Q^2}\right), \tag{3.17}$$

where $\alpha_0 = e_0^2/4\pi$ and e_0 is the 'bare' coupling or charge appearing in the Lagrangian. Note that dependence on the mass of the fermion in the loop has dropped out. The form of the amplitude for the scattering process, including the first loop correction, may be written

$$A_1(Q^2) = \alpha_0 A(Q^2)[1 - \alpha_0 \tilde{\Pi}(Q^2)], \tag{3.18}$$

where $\alpha_0 A(Q^2)$ is the tree-level amplitude and Π is rewritten as $\alpha_0 \tilde{\Pi}$ to display all factors of the coupling explicitly. This process is repeated for multiple fermion loop insertions giving a geometric series that can be summed

$$\begin{aligned}A_{\text{all}}(Q^2) &= \alpha_0 A(Q^2)[1 - \alpha_0\tilde{\Pi}(Q^2) + \alpha_0^2\tilde{\Pi}^2(Q^2) + \cdots]\\ &= \frac{\alpha_0 A_0(Q^2)}{1+\alpha_0\tilde{\Pi}(Q^2)}.\end{aligned}$$

The sum does not include all possible corrections to the photon propagator, but a class giving the largest corrections known as the 'leading logs'. The form of the above equation shows that the effect of the summed leading logs may be obtained from the lowest order matrix element by a simple redefinition of the coupling

$$\alpha(Q^2) = \frac{\alpha_0}{1+\alpha_0\tilde{\Pi}(Q^2)}.$$

Rearranging this and substituting for $\tilde{\Pi}$ gives

$$\frac{1}{\alpha(Q^2)} = \frac{1}{\alpha_0} + \frac{1}{3\pi} \ln\left(\frac{\Lambda^2}{Q^2}\right) \qquad (3.19)$$

which displays the dependence on α_0 and the cutoff parameter. Crudely, the idea of charge renormalization is to assume that α_0 is also infinite as $\Lambda^2 \to \infty$ but in such a way that the redefined or *renormalized* coupling α remains finite. It appears that the renormalization of the coupling does not depend on the quantum numbers of the external particles (other than their charges), but this is not obvious. Consider some other higher order

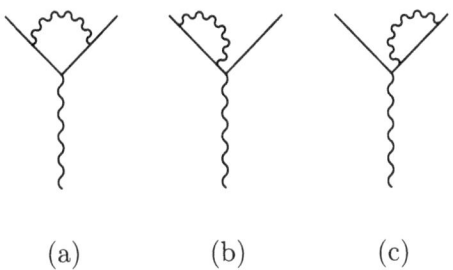

(a)　　　　(b)　　　　(c)

Fig. 3.7 Other order e^4 corrections to the coupling of the photon to a fermion.

corrections shown in Fig. 3.7, all give rise to infinities that must be renormalized and the result may well affect the renormalized charge in a process dependent way. The diagrams shown in the figure do modify the interaction in a manner that depends on the external particles. However the renormalization of charge is unaffected because the contributions that depend on the external particles cancel exactly between diagrams (a) and (b) plus (c). This is a beautiful result of gauge theories and is guaranteed by what are known as Ward–Takashi identities for QED and extended to non-Abelian gauge theories by Slavnov and Taylor. A simple example of gauge invariance and Ward identities in QED is discussed in Problem 3 at the end of the chapter.

As the cutoff Λ^2 is arbitrary and α_0 is not observable, one must remove dependence on both of them. This is done by *defining* the renormalized coupling at some scale μ^2, then using Eq. (3.19)

$$\frac{1}{\alpha(Q^2)} - \frac{1}{\alpha(\mu^2)} = \tilde{\Pi}(Q^2) - \tilde{\Pi}(\mu^2) = -\frac{1}{3\pi} \ln\left(\frac{Q^2}{\mu^2}\right)$$

or

$$\alpha(Q^2) = \frac{\alpha(\mu^2)}{1 - \frac{\alpha(\mu^2)}{3\pi} \ln\left(\frac{Q^2}{\mu^2}\right)}. \qquad (3.20)$$

If α is measured from some process at $Q^2 = \mu^2$, then the log correction term in the denominator of the above equation is zero. If the measurement is made at another scale, say $Q^2 > \mu^2$, then the effect of the renormalization is to introduce a finite correction term, which increases the effective coupling at that scale. The effect of renormalization is to make the coupling constant 'run' with the scale Q^2 of the process. For example, $\alpha_{qed} = 1/137$ from measurements at very low energies (e.g. Thomson scattering at the scale of the electron rest mass energy). Take this as the reference scale μ^2, then at the scale of M_Z the value of α has increased to $1/125$. An approximation to this result is discussed in Problem 5(i).

3.4.3 The QCD running coupling at leading order

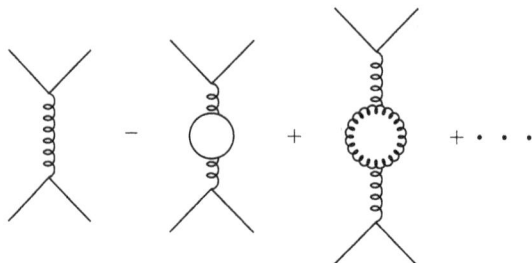

Fig. 3.8 Loop corrections to the gluon propagator.

Since the space-time structures of QED and QCD are similar, much of what has just been discussed may be taken over essentially unchanged for QCD. The QCD analog of the (bare) fine structure constant $(\alpha_S)_0 = g^2/4\pi$ may be defined in terms of the bare coupling constant g appearing in $\mathcal{L}_{QCD}$. The loop insertions for QCD involve two types of contribution, one from fermions (quarks) and one from gluons, as in Fig. 3.8. The gluon loop appears because gluons carry colour charges. Both contributions diverge logarithmically but with coefficients of opposite sign. Writing the overall coefficient of $\ln(Q^2/\Lambda^2)$ in either theory as αb_0, then for QED $b_0 = -1/3\pi$ (for electron loops only) and for QCD at a scale with n_f 'active' quark flavours (i.e. those with masses smaller than the scale) $b_0 = (33 - 2n_f)/12\pi$. In the latter case, the fermion loops contribute $-1/6\pi$ for each quark flavour[9] and the gluon self-coupling loops give a positive contribution of $11/4\pi$.

In terms of b_0 the generic leading order equation for the running coupling (Eq. 3.20) becomes

$$\alpha(Q^2) = \frac{\alpha(\mu^2)}{1 + \alpha(\mu^2) b_0 \ln\left(\frac{Q^2}{\mu^2}\right)}. \tag{3.21}$$

---

[9]This is the same as in QED when the factor 2 difference between QED and QCD couplings mentioned in Section 3.1.2 is allowed for. Many authors use $\alpha\beta_0/4\pi$ where $\beta_0 = 4\pi b_0$, here the the conventions of Ellis et al. (1996), Chapter 2 are followed.

For QCD, one defines the strong running coupling $\alpha_s(Q^2)$, satisfying Eq. (3.21), as $g^2(Q^2)/4\pi$ where $g(Q^2)$ is the renormalized coupling constant at scale Q^2. Because b_0 for QCD is positive, $\alpha_s(Q^2)$ decreases as Q^2 increases and formally $\alpha_s(Q^2) \to 0$ as $Q^2 \to \infty$. This remarkable result, known as *asymptotic freedom* is at the core of how QCD can reconcile confined quarks in hadrons on the scale of femtometres (small Q^2) with almost free quarks in 'hard processes' such as deep inelastic scattering at large Q^2. Figure 3.9

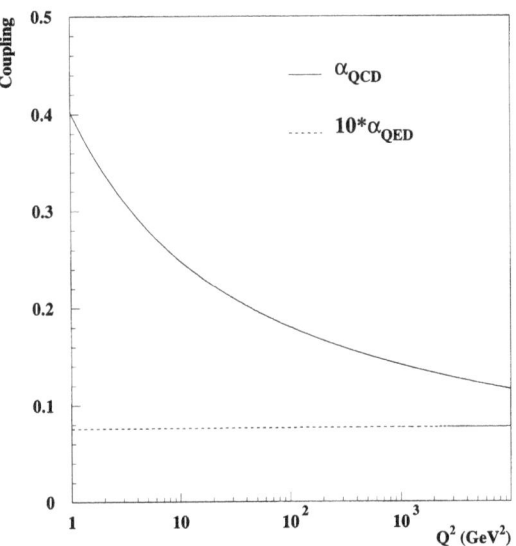

Fig. 3.9 The running couplings for QCD and QED as a function of Q^2. The calculations are at leading order and $n_f = 4$ for QCD and the contributions of e, μ and τ leptons for QED. The values of α_{qed} are multiplied by a factor of 10 to make the slow increase visible.

shows the running couplings for QED and QCD as functions of Q^2. For QCD, a reference value of $\alpha_s = 0.118$ at $Q^2 = M_Z^2$ has been assumed and four massless quark flavours used to calculate b_0. For QED, the reference scale for $\alpha = 1/137$ is taken to be $m_e c^2$ and three lepton flavours have been included.

There is no simple intuitive explanation for asymptotic freedom in QCD. In the case of QED, the virtual e^+e^- pairs surrounding a charge polarize the vacuum leading to a decrease in strength at larger distances and an increase as the bare charge is approached, a behaviour familiar from a classical charge placed in a polarizable medium. The fermion–antifermion loops in QCD act similarly, but the purely gluon loops act differently and

52 QCD and formal methods

in a sense 'spread out' the original colour at small distance scales through gluon splitting thus decreasing the effective strength of a hard interaction. The reason for the crucial difference in sign of the gluon contribution depends also on its spin-1 nature. Fermionic spin-1/2 operators anti-commute (bosonic spin-1 operators commute) and this leads to sign differences, the (-1) factor for fermionic loops being an example.

Another difference between QED and QCD is how to set the reference scale for defining α_s. In QED, it is natural to define α_{qed} using a long range (small Q^2) physical process. In QCD, there is no such equivalent because of confinement. It is sometimes convenient to define a new scale parameter Λ_{QCD},

$$\ln \Lambda_{QCD}^2 = \ln \mu^2 - \frac{1}{b_0 \alpha_s(\mu^2)}$$

in terms of which Eq. (3.21) may be rewritten to give

$$\alpha_s(Q^2) = \left[b_0 \ln \left(\frac{Q^2}{\Lambda_{QCD}^2} \right) \right]^{-1}. \qquad (3.22)$$

Either $\alpha_s(\mu^2)$ or Λ_{QCD} has to be determined from experimental data. Note that $\alpha_s(Q^2)$ and hence Λ_{QCD} depend on the number of active flavours through the occurrence of n_f in b_0. How to ensure that α_s behaves correctly at a threshold will be considered later. The value of Λ_{QCD} is expected to be of the order of a few 100 MeV, corresponding to the femtometre size of hadrons. At, or indeed somewhat above, this energy scale $\alpha_s(Q^2)$ becomes large and perturbative methods are no longer applicable.

There is further complication in setting the reference scale, within the leading order (LO) framework described so far the scale Λ_{QCD} cannot be defined uniquely as the following argument shows. Make the change of scale $\Lambda_{QCD} \to \Lambda'_{QCD} = \Lambda_{QCD}/\kappa$ in Eq. (3.22) to give

$$\alpha'_s = \frac{1}{b_0 \ln \left(\frac{Q^2 \kappa^2}{\Lambda_{QCD}^2} \right)} = \alpha_s \left[1 + \alpha_s b_0 \ln \kappa^2 \right]^{-1}$$

next expand in powers of α_s

$$\alpha'_s = \alpha_s - \left(b_0 \ln \kappa^2 \right) \alpha_s^2 + O(\alpha_s^3)$$

to show that the change in α_s may be absorbed in the $O(\alpha_s^2)$ term. To estimate the scale dependence of a calculation, one needs the next-to-leading order (NLO) contributions. It is appropriate at this point to consider a more powerful method of analysing the consequences of renormalization, before returning to question of reference scale.

3.5 Renormalization group methods

The renormalization group method exploits the requirement that a complete perturbative calculation of a physical quantity, Γ, should not depend on the particular choice of the scale μ^2 at which the renormalized constants are defined. Γ is invariant with respect to the 'group' of all possible scales. The approach was developed by Stückelberg, Peterman, Gell-Mann and Low in the 1950s for studying the high energy behaviour of perturbation theory and later by Callan and Symanzik in the 1970s. For simplicity consider Γ to be a physical quantity at a large energy scale $\sqrt{s}$ (such as a cross-section at high energy or a decay rate for a very massive state) so that effects of finite mass are $O(m^2/s)$ and may be ignored. The unrenormalized calculation in terms of bare quantities gives $\Gamma_B(s, \alpha_0, \Lambda^2)$ where α_0 is the bare coupling and Λ is the high energy cut-off to regularize the perturbative calculation. Renormalized parameters are defined at scale μ^2, particularly the coupling α and the normalization scale factor Z such that[10]

$$\Gamma_B(s, \alpha_0, \Lambda^2) = Z(\mu^2)^{-1}\Gamma(s, \alpha(\mu^2), \mu^2),$$

where dependence on other variables such as angles is suppressed. From the earlier discussion about renormalization, Γ does not depend on the cut-off Λ^2 and equivalently Γ_B does not depend on μ^2. Using $d\Gamma_B/d\mu^2 = 0$, or more conveniently the logarithmic derivative, one can immediately write down a differential equation for Γ

$$\mu^2 \frac{\partial \Gamma}{\partial \mu^2} + \mu^2 \frac{\partial \alpha}{\partial \mu^2} \frac{\partial \Gamma}{\partial \alpha} - \mu^2 \frac{\Gamma}{Z} \frac{\partial Z}{\partial \mu^2} = 0.$$

Defining $\beta(\alpha) = \mu^2 \dfrac{\partial \alpha}{\partial \mu^2}$ and $\gamma(\alpha) = -\dfrac{\mu^2}{Z} \dfrac{\partial Z}{\partial \mu^2}$, one gets the renormalization group equation (RGE),

$$\left[\mu^2 \frac{\partial}{\partial \mu^2} + \beta(\alpha)\frac{\partial}{\partial \alpha} + \gamma(\alpha)\right] \Gamma(s, \alpha(\mu^2), \mu^2) = 0. \qquad (3.23)$$

Naively, it might be expected that if Γ has energy dimension D (for example a cross-section has $D = -2$ and a decay rate $D = -1$)[11], then it will scale as

$$\Gamma(s, \alpha(\mu^2), \mu^2) = s^{D/2}\Gamma(1, \alpha(\mu^2), \mu^2/s). \qquad (3.24)$$

This does not happen because of the appearance of $\ln(s/\mu^2)$ terms from the renormalization. The renormalization group approach enables a change in energy scale to be 'traded' for a change of scale of the renormalized coupling, thus eliminating or reducing the size of the logarithmic corrections.

First, consider a dimensionless quantity (such as the ratio of cross-sections in e^+e^- annihilation $R_{e^+e^-}$) for which $\gamma(\alpha) = 0$. A change in the scale at which the renormalized constants are defined from $\mu^2 \to \mu_2^2$ will

---

[10] $Z(\mu^2)$ is defined so that matrix elements of the renormalised operator $\hat{A}$ (corresponding to Γ) at scale μ^2 are finite, $\hat{A}(\mu^2) = Z(\mu^2)\hat{A}_B$, where the bare operator is constructed from the fundamental fields of the theory.

[11] See the start of Appendix A for a comment on dimensions.

be compensated for by a change in the coupling from $\alpha \to \alpha_2$ (using the notation $\alpha = \alpha(\mu^2)$, $\alpha_2 = \alpha(\mu_2^2)$) such that

$$\Gamma(s, \alpha, \mu^2) = \Gamma(s, \alpha_2, \mu_2^2), \quad \text{where} \quad \beta(\alpha_2) = \mu_2^2 \frac{\partial \alpha_2}{\partial \mu_2^2}.$$

Take $\mu_2^2 = \lambda \mu^2$ where $\lambda = e^t$ then $t = \ln(\mu_2^2/\mu^2)$ and $\partial t/\partial \mu_2^2 = 1/\mu_2^2$, giving

$$\beta(\alpha_2) = \mu_2^2 \frac{\partial \alpha_2}{\partial \mu_2^2} = \mu_2^2 \frac{\partial \alpha_2}{\partial t} \frac{\partial t}{\partial \mu_2^2} = \frac{\partial \alpha_2}{\partial t}. \tag{3.25}$$

The differential equation for α_2 may be integrated to give an implicit relation between α and α_2 — the running coupling constant

$$t \equiv \ln\left(\frac{\mu_2^2}{\mu^2}\right) = \int_{\alpha(\mu^2)}^{\alpha(\mu_2^2)} \frac{dx}{\beta(x)} \tag{3.26}$$

provided $\beta(x)$ does not vanish in the range of integration. The function β depends only on the coupling and may be calculated perturbatively. Finally, consider the case when the scale $\mu_2^2 = s$, using the above relations one has

$$\Gamma(s, \alpha, \mu^2) = \Gamma(s, \alpha(s), s) = \Gamma(1, \alpha(s), 1), \tag{3.27}$$

where the last equality follows in this special case (since Γ is dimensionless it must be a function of s/μ^2 and α only). This result shows that Γ at the large scale s may be calculated simply by running α to the scale s using the running coupling constant defined by Eq. 3.26.

3.5.1 Anomalous dimensions

If the physical quantity has non-zero energy dimension D and non-zero $\gamma(\alpha)$, the solution of the renormalization group equation is more complicated. An outline of the method is as follows. Start from Eq. (3.23) for Γ at a scale $s\lambda$

$$\left[\mu^2 \frac{\partial}{\partial \mu^2} + \beta(\alpha) \frac{\partial}{\partial \alpha} + \gamma(\alpha)\right] \Gamma(s\lambda, \alpha, \mu^2) = 0,$$

where $\alpha = \alpha(\mu^2)$ as before. Now Γ is a homogeneous function of s and μ^2 with dimension $D/2$, thus it satisfies

$$s\lambda \frac{\partial \Gamma}{\partial(s\lambda)} + \mu^2 \frac{\partial \Gamma}{\partial \mu^2} = \frac{D}{2}\Gamma(s\lambda, \alpha, \mu^2).$$

Subtracting these two equations gives (s is a constant here)

$$\left[-\lambda \frac{\partial}{\partial \lambda} + \beta(\alpha) \frac{\partial}{\partial \alpha} + \omega(\alpha)\right] \Gamma(s\lambda, \alpha, \mu^2) = 0, \tag{3.28}$$

where $\omega(\alpha) = D/2 + \gamma(\alpha)$. Now define t such that

$$\lambda = e^{-t}, \quad \text{and} \quad -\lambda\frac{\partial}{\partial \lambda} = \frac{\partial}{\partial t}$$

and at the same time redefine the above equation with $\bar{\alpha}$ at scale μ^2/λ (or $\mu^2 e^t$). The coupling $\bar{\alpha}$ will satisfy an equation like (3.25), so with μ^2 fixed and λ as a function of t

$$\left(\frac{\mu^2}{\lambda}\right)\frac{\partial \bar{\alpha}}{\partial(\mu^2/\lambda)} = \beta(\bar{\alpha}), \quad \text{but} \quad \frac{\partial(\mu^2/\lambda)}{\partial t} = \frac{\mu^2}{\lambda}, \quad \text{so} \quad \frac{\partial \bar{\alpha}}{\partial t} = \beta(\bar{\alpha}).$$

Taken together these changes give

$$-\lambda\frac{\partial}{\partial \lambda} + \beta(\bar{\alpha})\frac{\partial}{\partial \bar{\alpha}} \to \frac{\partial}{\partial t} + \frac{\partial \bar{\alpha}}{\partial t}\frac{\partial}{\partial \bar{\alpha}} \to \frac{d}{dt}$$

and Eq. (3.28) becomes a simple first order differential equation

$$\left[\frac{d}{dt} + \omega(\bar{\alpha})\right]\Gamma(se^{-t}, \bar{\alpha}, \mu^2) = 0.$$

The solution takes the standard form

$$\Gamma(se^{-t}, \bar{\alpha}, \mu^2)\exp\left(\int_0^t \omega(\bar{\alpha})dt'\right) = \Gamma(s, \alpha, \mu^2),$$

where the 'constant' on the right hand side is fixed by the initial conditions at $t = 0$, i.e. $\lambda = 1$, $\alpha = \alpha(\mu^2)$. The above equation shows how scaling s down is compensated by running α up to a correspondingly higher scale, together with the exponential factor, which is the new feature. It is more convenient to have an equation for an upward scaling of s and this is accomplished by the simple expedient of setting $se^{-t} = \mu^2$ in the above equation, so that $t = \ln(s/\mu^2)$ and $\bar{\alpha} = \alpha(s)$. Finally, rewriting the equation from right to left gives

$$\Gamma(s, \alpha, \mu^2) = \Gamma(\mu^2, \alpha(s), \mu^2)\exp\left(\frac{D}{2}t + \int_0^t \gamma(t')dt'\right). \tag{3.29}$$

The function γ is known as the *anomalous dimension*, as it determines how differently Γ scales from the 'naive' expectation given by the energy dimension D. It can be calculated perturbatively. The importance of this result is that even if $\Gamma(\mu^2, \alpha(s), \mu^2)$ cannot be calculated perturbatively (as is often the case), how Γ *changes* as a function of energy scale s can be.

What has been sketched above concentrates on the key result of the renormalization group approach, but is grossly simplified. The anomalous dimension is not a universal function and depends on the fields involved in the interaction. For QCD, there will be separate γ functions for quark and gluon fields. The details are to be found in the references noted in the introduction to this chapter.

3.6 The QCD running coupling

Return now to Eq. (3.26) for the running coupling constant

$$\ln\left(\frac{\mu^2}{\mu_0^2}\right) = \int_{\alpha(\mu_0^2)}^{\alpha(\mu^2)} \frac{dx}{\beta(x)}$$

In QCD β has been calculated as an expansion in α_s (fixed) to the third significant term (full references are given in Ellis et al. (1996), Chapter 2)

$$\beta(\alpha_s) = -b_0\alpha_s^2 - b_1\alpha_s^3 - b_2\alpha_s^4 + O(\alpha_s^5) \tag{3.30}$$

and with n_f the number of active flavours,

$$\begin{aligned} b_0 &= \frac{33 - 2n_f}{12\pi}, \\ b_1 &= \frac{153 - 19n_f}{24\pi^2}, \\ b_2 &= \frac{77139 - 15099n_f + 325n_f^2}{3456\pi^3}. \end{aligned} \tag{3.31}$$

If the equation for $\alpha(\mu^2)$ is integrated keeping only the first term in the expansion of $\beta(\alpha_s)$, the result is

$$\frac{1}{\alpha_s(\mu^2)} - \frac{1}{\alpha_s(\mu_0^2)} = b_0 \ln(\mu^2/\mu_0^2)$$

or

$$\alpha_s(\mu^2) = \frac{\alpha_s(\mu_0^2)}{1 + \alpha_s(\mu_0^2) b_0 \ln(\mu^2/\mu_0^2)},$$

which is the leading order result obtained in Eq. (3.21). Keeping the first two terms in the expansion (3.30) the result, now to NLO, may be written

$$\frac{1}{\alpha_s(\mu^2)} - b_0 \ln \mu^2 - \frac{b_1}{b_0} \ln\left(\frac{1}{\alpha_s(\mu^2)} + \frac{b_1}{b_0}\right) = \\ \frac{1}{\alpha_s(\mu_0^2)} - b_0 \ln \mu_0^2 - \frac{b_1}{b_0} \ln\left(\frac{1}{\alpha_s(\mu_0^2)} + \frac{b_1}{b_0}\right). \tag{3.32}$$

This an implicit equation for $\alpha_s(\mu^2)$ in terms of $\alpha_s(\mu_0^2)$ and $\ln(\mu^2/\mu_0^2)$ which can be solved numerically. To get the NLO equivalent of Λ_{QCD} replace the constant righthand side of the above equation by $-\beta_0 \ln \Lambda^2 - b_1 (\ln b_0)/b_0$ and replace μ^2 by Q^2 to give

$$\ln\left(\frac{Q^2}{\Lambda^2}\right) = \frac{1}{b_0 \alpha_s(Q^2)} - \frac{b_1}{b_0^2} \ln\left(\frac{1}{b_0 \alpha_s(Q^2)} + \frac{b_1}{b_0^2}\right), \tag{3.33}$$

where Λ is $\Lambda_{\mathrm{QCD}}(\mathrm{NLO})$. Note that there is still an ambiguity in the definition of Λ which can be absorbed in the higher order terms in α_s and Λ will depend on n_f. For $\alpha_s(Q^2)$ small Eq. (3.33) can be inverted approximately to give

$$\alpha_s(Q^2) = \frac{1}{b_0 \ln(Q^2/\Lambda^2)} \left[1 - \frac{b_1}{b_0^2} \frac{\ln[\ln(Q^2/\Lambda^2)]}{\ln(Q^2/\Lambda^2)} + \cdots \right]. \quad (3.34)$$

Because of the ambiguities in the definition of Λ at both LO and NLO, it has become conventional to quote determinations of the strong coupling in terms of α_s directly at some convenient scale, instead of $\Lambda_{\rm QCD}$. Often the scale is chosen to be M_Z^2.

The importance of the running coupling and the renormalization group method is that they provide a systematic way to sum potentially large logs. Consider for simplicity a case in which Γ is dimensionless, for example, the quantity $R_{e^+e^-}$ and return to Eq. (3.27)

$$\Gamma(s, \alpha, \mu^2) = \Gamma(1, \alpha(s), 1).$$

The equation says that the effect of a change in energy scale is compensated for by the change in the running coupling $\alpha(s)$. This means that in the calculation of Γ as a perturbation expansion in $\alpha(s)$ certain series of $\ln(s/\mu^2)$ terms get summed. Expand the righthand side of the above equation in powers of $\alpha(s)$ and working to LO substitute $\alpha(s) = \alpha_\mu/(1+\alpha_\mu b_0 t)$, where $t = \ln(s/\mu^2)$ and $\alpha_\mu = \alpha(\mu^2)$, to give

$$\begin{aligned}
\Gamma(1, \alpha(s), 0, 1) &= \Gamma_0 + \Gamma_1 \alpha_\mu [1 + \alpha_\mu b_0 t]^{-1} + \cdots \\
&= \Gamma_0 + \Gamma_1 \alpha_\mu \sum_{n=0}^{\infty} (-\alpha_\mu b_0 t)^n \\
&= \Gamma_0 + \Gamma_1 \alpha_\mu [1 - \alpha_\mu b_0 t + \alpha_\mu^2 b_0^2 t^2 + \cdots]. \quad (3.35)
\end{aligned}$$

Thus in the renormalization group result on the first line the logs in the Γ_1 term, known as the 'leading logs', are summed 'automatically'. The last line shows the expansion in terms of α_μ that would be given without the renormalization group improvement. For t large, the renormalization group controls a potential breakdown of the perturbative expansion by the simple substitution of the fixed coupling by the running coupling $\alpha_\mu \to \alpha(s)$. If the next term in $\alpha(s)$ is included, $\Gamma_2 \alpha(s)^2$, and expanded in terms of α_μ and t, one gets

$$\Gamma_2 \alpha_s^2 \to \Gamma_2 \alpha_\mu^2 [1 - 2\alpha_\mu b_0 t + \cdots]$$

so at each power of α_μ, the log term is down by one power. These are the next-to-leading logs.

3.7 The hadronic tensor $W_{\mu\nu}$

The cross-section and structure functions for deep inelastic scattering were introduced in the context of the quark–parton model. This gives a good physical intuition, but to apply some of the more powerful mathematical

58 QCD and formal methods

techniques a more general approach is needed. In coordinate space, the structure of the deep inelastic scattering matrix element for $e(k) + p(p) \to e(k') + X$ is

$$S_{fi} = i \int d^4x\, d^4y\, J^e_\mu(x) D^{\mu\nu}(x-y) J^h_\nu(y),$$

where $J^e_\mu(x)$ is the leptonic EM current acting at x, $D^{\mu\nu}(x-y)$ is the photon propagator and $J^h_\nu(y)$ is the hadronic current acting at point y. Inserting expressions for the space–time structure of the currents and propagator and performing the spatial integrations gives

$$S_{fi} = i(2\pi)^4 \delta^4(p+k-p_X-k')\, \bar{u}(k')\gamma_\mu u(k) \frac{e^2}{q^2} \langle X|j^\mu(0)|p\rangle,$$

where j^μ is the hadronic EM current operator and normalisation factors have been ignored.[12] Taking the modulus squared of S_{fi}, summed and averaged over spins, gives the familiar expression for the cross-section in terms of a contraction between the leptonic and hadronic tensors. In the fixed-target frame, this takes the form

$$\frac{d^2\sigma}{d\Omega dE'} = \frac{\alpha^2}{q^4}\frac{E'}{mE} L_{\mu\nu} W^{\mu\nu},$$

where $q = k - k'$ (more details on cross-section formulae are given in Appendix C). For photon exchange the leptonic tensor is given by QED, ignoring masses (see also Section 2.1.1)

$$L_{\mu\nu} = \frac{1}{2}\mathrm{Tr}[\slashed{k}'\gamma_\mu \slashed{k}\gamma_\nu] = 2(k_\mu k'_\nu + k_\nu k'_\mu - k\cdot k' g_{\mu\nu}) \quad (3.36)$$

and the hadronic tensor is written in terms of the hadronic current

$$W_{\mu\nu} = \frac{1}{4\pi}\sum_X (2\pi)^4 \delta^4(q+p-p_X)\langle p|j^\dagger_\mu(0)|X\rangle\langle X|j_\nu(0)|p\rangle, \quad (3.37)$$

where the spin labels have been suppressed and the '$\sum_X |X\rangle\langle X|$' is a shorthand for the full sum and integral over the intermediate state phase space (see Eq. (C.2)). For inclusive leptoproduction, this expression may be reduced to the form of the imaginary part of the virtual-photon–proton forward scattering amplitude (virtual Compton scattering)

$$W_{\mu\nu}(p,q) = \frac{1}{4\pi}\int d^4z\, e^{iq\cdot z}\,\langle p|[j^\dagger_\mu(z), j_\nu(0)]|p\rangle. \quad (3.38)$$

The manipulation, which is essentially an application of unitarity, is shown diagrammatically in Fig. 3.10 and followed algebraically in Problem 7 at the end of the chapter. The full forward virtual Compton scattering amplitude

---

[12]It is instructive to compare this equation with that for electron–quark scattering in Chapter 2, Eq. (2.3).

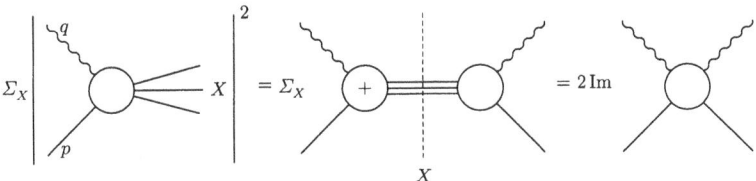

Fig. 3.10 The reduction of the DIS hadronic tensor $W_{\mu\nu}$ to the imaginary part of the forward γ^*p amplitude.

has a similar form to Eq. (3.38), but is related to the time-ordered product of the currents

$$T_{\mu\nu}(p,q) = i \int d^4z \, e^{iq\cdot z} \langle p|T(j_\mu^\dagger(z)j_\nu(0))|p\rangle. \qquad (3.39)$$

T and W are related by

$$W_{\mu\nu} = \frac{1}{2\pi} \mathrm{Im}\, T_{\mu\nu},$$

and both have the same Lorentz decomposition in terms of momentum tensors. The study of the mathematical structure of models for DIS thus becomes that of virtual Compton scattering.

The manipulations summarized in Fig. 3.10 are also those leading to the optical theorem for the total γ^*p total cross-section, which takes the form

$$\sigma_{\gamma^*p}^{\mathrm{tot}}(\lambda) = \frac{8\pi^2\alpha}{W^2 - m^2} \varepsilon_\mu^\dagger(\lambda) W^{\mu\nu} \varepsilon_\nu(\lambda),$$

where $\varepsilon_\mu(\lambda)$ is the polarisation 4-vector for the γ^* polarisation states $\lambda = \pm 1, 0$ (more details are give in Appendix C).

3.7.1 Light cone dominance

To see which part of phase space dominates the integral for $W_{\mu\nu}$, it useful to introduce *light cone variables*. Consider deep inelastic scattering in the rest frame of the target proton, with the z-axis along $\mathbf{q}$, then

$$p = (m,0,0,0), \qquad q = (\nu/m, 0, 0, \sqrt{(\nu/m)^2 + Q^2}),$$

where $\nu \equiv mq_0 = m(E - E')$ and $x = Q^2/(2\nu)$. The light cone variables are defined by

$$a^\pm = (a_0 \pm a_3)/\sqrt{2} \quad \text{and} \quad a \cdot b = a^+b^- + a^-b^+ - \mathbf{a_T} \cdot \mathbf{b_T},$$

where $\mathbf{a_T}$ is the two dimensional spatial vector representing the components of a perpendicular to the z-axis (and similarly for $\mathbf{b_T}$). In the Bjorken limit ($Q^2 \to \infty$, x fixed), $q_3 \to \nu/m + mx$, so

60 QCD and formal methods

$$q^+ = (2\nu/m + mx)/\sqrt{2} \to \sqrt{2}\nu/m, \qquad q^- = -mx/\sqrt{2},$$

and

$$q \cdot z = q^+ z^- + q^- z^+ \to \sqrt{2}\nu z^-/m - mxz^+/\sqrt{2}.$$

The integration over the exponential in Eq. (3.38) may be written as

$$\int d^4z\, e^{iq\cdot z} \ldots \to \int dz^-\, e^{iq^+z^-} \int dz^+ d^2z_T\, e^{iq^-z^+} \ldots$$

Causality requires the commutator to vanish for $z^2 < 0$. If either q^+z^- or q^-z^+ are large, the rapid oscillations of the exponential factor will suppress contributions to the integral. So the integral will be given by contributions for which $|z^-| < m/(\sqrt{2}\nu)$ and $|z^+| < \sqrt{2}/(mx)$ or $|z_0^2 - z_3^2| < 1/Q^2$. Thus in the Bjorken limit contributions with $z^2 \to 0$ dominate, which means that the current commutator on the light cone is required. The timescale on which the virtual photon interacts with the target nucleon is also very short as $|z_0| \approx |z_3| \approx O(1/\sqrt{Q^2})$ — the partons are almost 'frozen' in position. The physical picture of the parton model is an impulse approximation. Mathematically, light cone dominance is also crucial as it allows use of of the operator product expansion for evaluating the integrand.

Bjorken used this approach in his seminal work on the behaviour of the inelastic structure functions at large Q^2 and ν. He argued that on the light cone the commutator of currents in Eq. (3.38), $[j_\mu^\dagger(z), j_\nu(0)]$, could be related to that of free quarks in Gell-Mann's current algebra. This then allowed its evaluation in the Bjorken limit and the demonstration that 'scaling' resulted, that is, that the structure functions depend on the ratio $Q^2/(2\nu)$ but not on Q^2 and ν independently.

3.8 The operator product expansion

The operator product expansion (OPE) is method for studying the structure of field theories in the light-cone limit. It also provides the framework for the all-important factorization theorem which underpins one of the basic assumptions of the QCD improved parton model.

Consider a very simple example of the time-ordered product of two free particle scalar fields $T(\phi(x)\phi^\dagger(y))$. Since a quantum is created by field operator $\phi^\dagger(y)$ at y and annihilated by $\phi(x)$ at x, this is closely related to the propagator $\Delta(x-y)$. Formally

$$T(\phi(x)\phi^\dagger(y)) = \Delta(x-y)\,\mathbf{1} + :\phi(x)\phi^\dagger(y):,$$

where

$$\Delta(x-y) = -\frac{2}{(2\pi)^4}\frac{1}{(x-y)^2 + i\epsilon}$$

and the second term is the normal ordered product of the fields which is non-singular as $x \to y$. The singularity of the operator product as $x \to y$

has been isolated as a product of a known function and an operator (here the unit operator). Knowing that the mass dimension[13] of a scalar field ϕ is 1, the power of the leading singularity in $(x-y)$ must be $0-1-1 = -2$ on dimensional grounds.

The OPE of Wilson is a generalization of the above example. The singular part of the product of operators $A(x)B(y)$ is given by

$$A(x)B(y) \to \sum_j C_j(x-y) N_j(y) \quad \text{as} \quad (x-y)^2 \to 0,$$

where C_j are singular coefficient functions and N_j are non-singular local operators. Dimensional counting suggests that

$$C_i(x-y) \sim (x-y)^{d_{N_i} - d_A - d_B} \quad \text{as} \quad (x-y)^2 \to 0,$$

where d_O is the mass dimension of operator O. This would be the case in a free field theory but it gets modified by renormalization. There are two properties of the OPE that make it a powerful method: first the dimension of an operator increases as more fields or derivatives are added; secondly because the relationship is one between *operators* the coefficient functions are universal for that product, even though the matrix elements of the operators will be process dependent. The first point implies that in a Taylor expansion of an operator product about the light cone, successively higher order terms have increasing d_{N_i} and thus *less* singular coefficient functions. As the full exposition of the application of the OPE to DIS is rather technical only the essential steps will be outlined here, details are given in Muta (1998) or Ynduráin (1999).

3.8.1 OPE applied to DIS

The starting point is Eq. (3.39), which shows that the virtual Compton amplitude is related to the product of two EM current operators. To keep things simple the flavour and tensor structure will be ignored and only a generic structure function considered (close to the case for non-singlet F_1 or F_2/x). Apply the operator product expansion to $T(j^\dagger(z)j(0))$

$$iT(j^\dagger(z)j(0)) = \sum_{j,n} C_{j,n}(z^2, \mu^2) z^{\mu_1} \cdots z^{\mu_n} N_{j,n}^{\mu_1 \cdots \mu_n}(\mu^2), \quad (3.40)$$

where μ^2 is the renormalization scale and the non-singular operators $N_j(z,\mu^2)$ have been expanded in powers of z

$$N_j(z,\mu^2) = \sum_n z^{\mu_1} \cdots z^{\mu_n} N_{j,n}^{\mu_1 \cdots \mu_n}(\mu^2),$$

---

[13] With units $\hbar = c = 1$, mass or energy has dimension $[\text{length}]^{-1}$, thus the mass or energy dimension of a field operator may be deduced from the fact that $\int d^4 x L$ must be dimensionless. Here $L = m^2 \phi \phi^\dagger - \partial^\mu \phi \partial_\mu \phi^\dagger$ and it follows immediately that ϕ has mass dimension 1.

Table 3.1 Dimension, spin and twist of some operators.

Operator	Mass dimension	Spin	Twist
Scalar field $\phi(x)$	1	0	1
$\phi^\dagger \partial_{\mu_1} \cdots \partial_{\mu_n} \phi$	$n+2$	n	2
Vector field $A_\mu(x)$	1	1	0
$\bar{\psi}\gamma_\mu\psi$	3	1	2

where $N_{j,n}^{\cdots}$ are referred to as 'spin n' operators. In a free field theory $C_{j,n}$ would behave like $(z^2)^{\tau/2-d_J}$ for $z^2 \sim 0$, where $\tau = d_{j,n} - n =$ (dimension−spin) is the *twist*[14] of the operator $N_{j,n}$ with mass dimension $d_{j,n}$ and d_J is the dimension of the EM current. In a theory with interactions like QED or QCD, the singular behaviour will be modified by logarithms introduced by renormalization, which is also why the scale μ^2 appears. For DIS the operators are products of currents and are at least bilinear in the fields, so that the smallest twist is 2. This then gives the leading most singular contribution to the OPE. The dimension, spin and twist for some simple operators are shown in Table 3.1. Redefining $C_{j,n}$ relative to the leading twist by $C_{j,n}(z^2,\mu^2) = C'_{j,n}(z^2,\mu^2)(z^2)^{\tau/2-1}$, the OPE expansion (3.40) is inserted into Eq. (3.39) to give

$$T(q^2,\nu) = \sum_{j,n} \langle p|N_{j,n}^{\mu_1\cdots\mu_n}(\mu^2)|p\rangle \int d^4z\, e^{iq\cdot z}\, z^{\mu_1}\cdots z^{\mu_n} C'_{j,n}(z^2,\mu^2)(z^2)^{\tau/2-1}.$$

The integral is a Fourier transform which will be proportional to $q_{\mu_1}\cdots q_{\mu_n}$ and defines $\tilde{C}_{j,n}(q^2,\mu^2)$

$$\int d^4z\, e^{iq\cdot z}\, z^{\mu_1}\cdots z^{\mu_n} C'_{j,n}(z^2,\mu^2)(z^2)^{\tau/2-1}$$
$$= q^{\mu_1}\cdots q^{\mu_n} \tilde{C}_{j,n}(q^2,\mu^2)(-q^2)^{1-\tau/2}.$$

The matrix element of the operator $N_{j,n}^{\mu_1\cdots\mu_n}(\mu^2)$ cannot be evaluated completely, but as it can only depend on p, the Lorentz structure (up to contractions) must be proportional to $p_{\mu_1}\cdots p_{\mu_n}$ and this defines a reduced matrix element

$$\langle p|N_{j,n}^{\mu_1\cdots\mu_n}(\mu^2)|p\rangle = p_{\mu_1}\cdots p_{\mu_n} \bar{N}_{j,n}(\mu^2).$$

When these two expressions are taken together to form $T(q^2,\nu)$, the 4-momenta contract to give $(q\cdot p)^n$ and using $q\cdot p = Q^2/(2x)$ gives

$$T(q^2,\nu) = \sum_{j,n} C_{j,n}(Q^2,\mu^2)\bar{N}_{j,n}(\mu^2)\left(\frac{1}{x}\right)^n\left(\frac{1}{Q^2}\right)^{\tau/2-1}, \quad (3.41)$$

where $C_{j,n}(Q^2,\mu^2) = \tilde{C}_{j,n}(q^2,\mu^2)(Q^2/2)^n$.

[14] A more physical interpretation of twist will become apparent towards the end of this section.

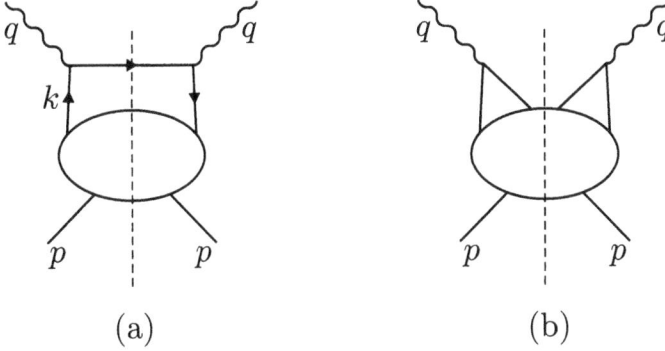

Fig. 3.11 Virtual Compton scattering diagrams for the covariant parton model approach: (a) the leading QPM diagram, (b) a suppressed higher order contribution (higher 'twist').

Having established the OPE expression for $T(q^2, \nu)$, the next step is to relate it to the structure function. This is done by using a dispersion relation for T. Since T is an analytic crossing symmetric function of ν ($T(q^2, \nu) = T(q^2, -\nu)$), it satisfies[15]

$$T(q^2, \nu) = \frac{1}{\pi} \int_{Q^2/2}^{\infty} d\nu' \, \mathrm{Im}\, T(q^2, \nu') \left[\frac{1}{\nu' - \nu} + \frac{1}{\nu' + \nu} \right].$$

Now $\mathrm{Im}\, T = 2\pi W$, and using $\nu = Q^2/(2x)$ to change variables one finds

$$\begin{aligned} T(q^2, x) &= 4 \int_0^1 \frac{dx'}{x'(1 - x'^2/x^2)} W(q^2, x') \\ &= 4 \sum_2^{\infty} \left(\frac{1}{x}\right)^n \int_0^1 dx' \, x'^{n-1} W(q^2, x'), \end{aligned} \quad (3.42)$$

where the sum starts at $n = 2$ and runs over even integers only (from the symmetry of T). By comparing Eqs (3.41) and (3.42) one can read off the OPE expression for the even-n moments of the structure function

$$\int_0^1 dx' \, x'^{n-1} W(q^2, x') = \frac{1}{4} \sum_j C_{j,n}(Q^2, \mu^2) \bar{N}_{j,n}(\mu^2) \left(\frac{1}{Q^2}\right)^{\tau/2 - 1}. \quad (3.43)$$

The OPE shows that the moments of structure functions or parton densities can be expanded as a power series in $1/Q^2$. At large Q^2, the moments will be given to a good approximation by the leading twist 2 terms for which the Q^2 dependence is contained in the coefficient functions $C_{j,n}$.

---

[15] Masses have been ignored in determining the lower limit of the integral.

These may calculated perturbatively, but the matrix elements $\bar{N}_{j,n}$ are non-perturbative objects. The higher twist terms correspond to a power series expansion in $1/Q^2$. For leading twist, the above equation may be simplified to read

$$\int_0^1 dx'\, x'^{n-1} W(q^2, x') = C_n(Q^2, \mu^2) \bar{N}_n(\mu^2)/4 + O(1/Q^2). \tag{3.44}$$

Figures 3.11(a) and (b) show examples of leading twist 2 and higher twist contributions to the virtual Compton amplitude, respectively. For a crossing anti-symmetric structure function, the OPE method gives the odd-n moments. For both symmetric and antisymmetric cases, the missing moments have to be found by analytic continuation in n.

3.8.2 RGE calculation of moments

The final step in the OPE approach is to use the renormalization group equation to calculate the coefficients $C_n(Q^2, \mu^2)$, or to be more precise to calculate their Q^2 dependence. This involves a knowledge of their anomalous dimensions, which can be calculated perturbatively. Suppose $\gamma(n, \alpha)$ is anomalous dimension corresponding to C_n then

$$C_n(Q^2, \mu^2) = C_n(\mu^2, \mu^2) \exp\left(\int_\alpha^{\alpha(Q^2)} \frac{\gamma(n, \alpha')}{\beta(\alpha')} d\alpha' \right), \tag{3.45}$$

where Eq. (3.25) has been used to replace the integral over t in Eq. (3.29) by one over α. Next expand γ as a power series in α, $\gamma(n, \alpha) = \gamma^0(n)\frac{\alpha}{2\pi} + O(\alpha^2)$ and use the expansion of $\beta(\alpha) = -b_0\alpha^2 + O(\alpha^3)$ to give

$$C_n(Q^2, \mu^2) = C_n(\mu^2) \exp\left(-\frac{\gamma^0(n)}{2\pi b_0} \int_\alpha^{\alpha(Q^2)} \frac{d\alpha'}{\alpha'} \right) = C_n(\mu^2) \left(\frac{\alpha(Q^2)}{\alpha} \right)^{d_n},$$

where $d_n = -\gamma^0(n)/(2\pi b_0)$. Finally, taking this result and Eq. (3.44) together, one has the leading order expression for the Q^2 dependence of the nth moment of the non-singlet structure function

$$M_n^{NS}(Q^2, \mu^2) = A_n(\mu^2) \left(\alpha(Q^2)\right)^{d_n} \quad n \geq 2 \text{ (n even)}, \tag{3.46}$$

where $A_n(\mu^2)$ collects together all the non-perturbatively calculable factors at a single scale and the Q^2 dependence is given through that of $\alpha(Q^2)$ to a known power. Since the powers are positive and increase slowly with increasing n[16], the structure function will tend to decrease at large x and increase at small x as Q^2 increases. In the early days of the application of pQCD to deep inelastic scattering moments played an important role in experimental tests.

---

[16]Expressions for $\gamma^0(n)$ are given in an appendix at the end of Chapter 4.

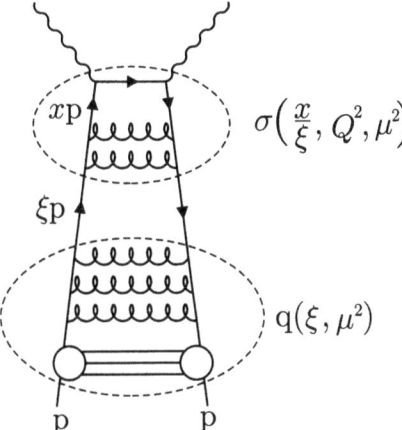

Fig. 3.12 Factorised form of the virtual Compton scattering amplitude from the QCD OPE approach (adapted from Roberts (1990), Fig.4.5).

3.9 Factorization

Factorization is the statement that the cross-section for DIS may be written as the convolution of two terms: a calculable hard scattering cross-section and a non-perturbative parton density. It is the basic assumption of the parton model and is contained in the form of the OPE expression given in Eq. (3.44) for the leading twist terms. This last and the (leading order) Eq. (3.46) show precisely how the moments of a structure function may be written as the product of a calculable term containing the Q^2 dependence and a non-perturbative matrix element. Returning to Eq. (3.44), define functions $q(x, \mu^2)$, $\sigma(z, Q^2, \mu^2)$ such that

$$\int d\xi \, \xi^{n-1} q(\xi, \mu^2) = \tilde{N}_n(\mu^2)/4$$

$$\int dz \, z^{n-1} \sigma(z, Q^2, \mu^2) = C_n(Q^2, \mu^2)$$

then Eq. (3.44) is equivalent to the convolution

$$W(q^2, x) = \int_x^1 \frac{d\xi}{\xi} q(\xi, \mu^2) \sigma\left(\frac{x}{\xi}, Q^2, \mu^2\right),$$

which is shown diagrammatically in Fig. 3.12. The function $q(\xi, \mu^2)$ is the non-perturbative piece that is identified with the parton density function (i.e. the probability to find a quark in the proton with momentum fraction ξ) and $\sigma(z, Q^2, \mu^2)$ is the cross-section for γ^*q scattering. The figure shows a 'gluon ladder' attached to blobs representing the proton (remember this is a diagram for the forward γ^*p scattering amplitude) which has been divided into two regions, one labelled by the hard cross-section σ and the other by the parton density function q. This should be compared with Fig. 3.11(a) which has only the very top (quark) rung of the ladder and no explicit gluon exchanges. As will be covered in great detail in the next chapter, the gluon rungs represent gluon radiation and the division shown in Fig. 3.12 is between those gluons corresponding to hard radiative corrections explicitly included in $\sigma(\gamma^* q)$ and the remaining softer gluons that are absorbed into the parton density $q(\xi, \mu^2)$. A further detail that needs explanation is that the scale at which the ladder is divided is known as the *factorization scale* μ_F^2 and it is not necessarily the same as the renormalization scale μ_R^2. It is through gluon radiation that Q^2 dependence, and thus scaling violations, enter the QCD improved parton model. The proof of factorization is a technically demanding task. Relevant classes of Feynman diagrams have to be analysed to ensure that all infrared singularities can be absorbed in the non-perturbative matrix element and that 'interference' diagrams such as Figure 3.11(b) only contribute at higher twist. Higher twist diagrams may correspond to more complicated classes of diagram than the simple ladder shown here — this point will be considered again towards the end of Chapter 9 on low-x physics. A good introduction to how factorization is established for DIS is to be found in Collins *et al.* (1989).

3.10 Summary

This chapter has outlined the essential features of QCD that are important for the QCD-enhanced parton model and deep inelastic scattering in the Bjorken limit. The basic idea of renormalization has been illustrated in QED using the example of loop insertions in the photon propagator. In QCD, renormalization leads to a running coupling that decreases at small distances, which is the crucial result that allows perturbative methods to be applied to hard scattering processes, and underpins the assumptions and successes of the parton model. Various formal methods such as the renormalization group and the operator product expansion have been sketched. Applying the OPE method to DIS gives the mathematical framework for verifying the assumptions of the QPM and for showing how gluon radiation and other higher order processes give rise to scaling violations that are calculable quantitatively. Although the OPE was crucial to the early development of the subject and is still an important mathematical tool, the more physically direct approach of splitting functions (to be introduced in the next chapter) is the one that is now the preferred way of performing calculations in the QCD-enhanced parton model. This is not only because of its physical interpretation, but because it enables calculation of a wide variety of processes in a way that is much easier to apply to actual experimental data, often only measured in limited regions of phase space.

3.11 Problems

1. Using the definitions in Appendix A, verify that $\mathcal{L}_{\text{quark}}$ of Eq. (3.5) does indeed reduce to $\bar{\psi}i\gamma_\mu D^\mu \psi$ and that $\bar{\psi}_L i\gamma_\mu D^\mu \psi_R = 0$.

2. Check the QED calculation of the Compton scattering cross-section outlined in Section 3.2. In particular, show that the interference term in the sum over spins is:

$$\sum_{spins} \mathcal{M}_a \mathcal{M}_b^* = \frac{e^4}{su} \text{Tr}[\not{p}'\gamma^\nu(\not{p}+\not{k})\gamma^\mu \not{p}\gamma_\nu(\not{p}-\not{k}')\gamma_\mu]. \qquad (3.47)$$

   Evaluate the trace for the case where the incident photon is virtual with mass $k^2 = -Q^2$ and other masses are neglected, hence verify the result given in Eq. (3.12).

3. Check the calculation of the BGF cross-section (Eq. (3.15)), either directly or by the application of crossing symmetry and suitable adjustments to the colour factors.

4. Gauge invariance and Ward identities. Consider a real photon state with momentum k and polarisation vector ε, $A_\mu = \varepsilon_\mu e^{-ik\cdot x}$, where $k \cdot \varepsilon = 0$ follows from the Lorentz condition $\partial_\mu A^\mu = 0$. EM gauge invariance states that physical results are invariant under the gauge transformation $A_\mu \to A_\mu - \partial_\mu f$, where f is scalar function satisfying $\partial_\mu \partial^\mu f = 0$. Taking $f = \lambda e^{-ik\cdot x}$ show that the gauge transformation is equivalent to the replacement $\varepsilon_\mu \to \varepsilon_\mu + \lambda k_\mu$. The amplitude for $\gamma e \to \gamma e$ (Compton scattering of real photons) has the form $\varepsilon_\mu^\dagger M^{\mu\nu} \varepsilon_\nu$, hence show that gauge invariance requires that $k_\mu M^{\mu\nu} = 0$, $M^{\mu\nu} k_\nu = 0$. Two tree level diagrams contribute to $M^{\mu\nu}$ (refer to Section 3.2), show that both are required to satisfy gauge invariance, i.e. $k_\mu M_a^{\mu\nu} \neq 0$, $k_\mu M_b^{\mu\nu} \neq 0$ but $k_\mu (M_a^{\mu\nu} + M_b^{\mu\nu}) = 0$. This is an example of a Ward identity.

5. (i) Using Eq. (3.21) for α_{qed} with $\alpha = 1/137$ at $\mu = 0.51\,\text{MeV}$ but including all relevant fermions (leptons and quarks) in the calculation of b_0 for QED, estimate the value of α_{qed} at the mass of the Z^0 boson (91.2 GeV).
   (ii) Using Eq. (3.22) plot α_s as a function of Q^2 between $1\,\text{GeV}^2$ and $10^5\,\text{GeV}^2$ for $n_f = 3$ and $\Lambda_{QCD} = 200\,\text{MeV}$.
   (iii) Convince yourself that you understand the NLO renormalization group expressions for α_s, Eqs (3.32) to (3.34). Estimate the size of the NLO correction in the last equation with Λ of 200 MeV and Q^2 values of 1, 10, 1000 GeV^2.

6. Show that the tensor:

$$L_{\mu\nu} = \frac{1}{2}\text{Tr}\left\{(\not{k}' + m)\gamma_\mu(\not{k} + m)\gamma_\nu\right\}$$

   is given by

$$L_{\mu\nu} = 2[k'_\mu k_\nu + k'_\nu k_\mu + \frac{q^2}{2}g_{\mu\nu}]$$

where $k^2 = m^2$, $k'^2 = m^2$ and $q^2 = (k - k')^2$. Show also that $q^\mu L_{\mu\nu} = 0$ and $L_{\mu\nu}q^\nu = 0$. $L_{\mu\nu}$ is known as the leptonic tensor and turns up in the calculation of cross-sections for ep scattering and e^+e^- annihilation.

7. The reduction of Eq. (3.37) to the form (3.38) is most easily demonstrated in the reverse direction. One needs: $j(z) = U(z)j(0)U^{-1}(z)$ and $U(z)|p\rangle = \exp(ip \cdot z)|p\rangle$ for translation by z; the representation

$$(2\pi)^4 \delta^4(q) = \int d^4z \exp(iq \cdot z),$$

and the use of momentum conservation to show that the second term in the commutator in Eq. (3.38) gives zero.

4
QCD improved parton model

The QCD improved parton model provides the foundation on which all the remaining chapters are based. The main ideas are introduced in Section 4.1, the tree level QCD Compton and boson–gluon fusion processes give rise to characteristic gluonic 'radiative corrections' to the zeroth order parton model. Infrared singularities are absorbed by renormalizing the parton densities, which thus become Q^2 dependent and no longer satisfy exact Bjorken scaling. However the Q^2 dependence can be calculated perturbatively using 'splitting functions'. The resulting formalism, known as the DGLAP evolution equations, is described in Section 4.2. To give an insight into the effect of these equations, some simple numerical examples at leading order are given in Section 4.3. At various points in the chapter, the relationship between the more formal Operator Product Expansion method outlined in the previous chapter and the DGLAP approach is indicated. The last three sections comment on a number of more technical issues, the one considered in greatest depth is the extension the DGLAP formalism to accommodate heavy quarks (Section 4.5).

4.1 Improving the parton model

The approach to the QCD improved parton model described here is an alternative to the formal methods outlined in Chapter 3. It was developed over a number of years by Dokshitzer, Gribov, Lipatov, Altarelli and Parisi — collectively referred to as DGLAP. An accessible account is given by Altarelli (1982), where references to the original papers may be found. The basic idea is summarised in Fig. 4.1 for $\gamma^* p$ interactions: (a) shows the zeroth order QPM process in which the γ^* is absorbed directly by a quark in the proton carrying momentum xp; (b) shows the generic QCD process in which the γ^* undergoes a more general hard scattering with a quark or gluon from the proton. In (b) there will be one or more gluonic interaction, either before or after the γ^* is absorbed by a quark.

To be more precise, a parton with momentum $p_i = \xi p$ interacts with the γ^* of virtuality Q^2 and 4-momentum q. Define $z = Q^2/(2p_i \cdot q) = x/\xi$, the analogue of Bjorken x for the γ^*-parton scattering. The masses of all

70 QCD improved parton model

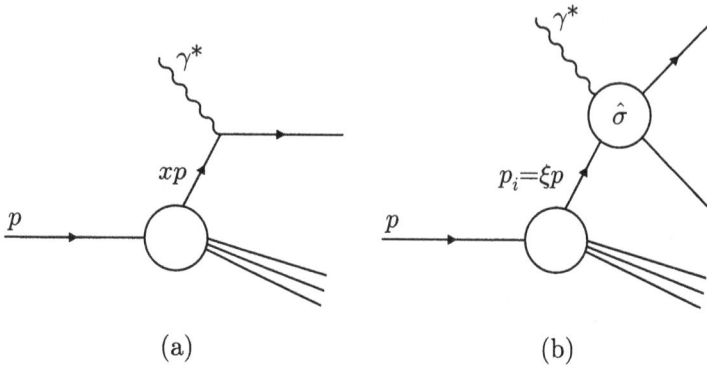

Fig. 4.1 The basic idea of the QCD improved parton model for γ^*p interactions: (a) shows the QPM approach; (b) the QCD improved version.

particles except the γ^* are taken to be zero. The cross-section $\hat{\sigma}$ for the hard interaction is calculable in pQCD and the overall γ^*p cross-section is then given by a convolution of $\hat{\sigma}$ with $f_i(\xi)$, the probability of finding the parton with momentum fraction ξ in the proton:

$$\sigma(x, Q^2) = \sum_i \int_0^1 dz \int_0^1 d\xi f_i(\xi) \delta(x - z\xi) \hat{\sigma}(z, Q^2). \qquad (4.1)$$

Note that the key feature of the parton model remains in the QCD approach, namely that the cross-section can be factored into a probability density for finding a parton and the probability of a hard scattering. This in turns relies on the time scale of the hard scatter being very much shorter that that of the subsequent hadronization of the final state. Mathematically, this picture is supported by QCD factorization theorems (see Section 3.9).

The physical processes that need to be 'added' to the QPM are the radiation of gluons by quarks and q$\bar{\text{q}}$ pair production by gluons. At high energies the structure functions are related to the cross-sections for transverse and longitudinal virtual photon scattering by[1]

$$\begin{aligned} \sigma_T &= \frac{4\pi^2 \alpha}{s} 2F_1 \\ \sigma_L &= \frac{4\pi^2 \alpha}{s} \left[\frac{F_2}{x} - 2F_1 \right], \end{aligned} \qquad (4.2)$$

where the Hand convention[2] has been used for the γ^* flux. It is convenient to define $\sigma_0 = \dfrac{4\pi^2 \alpha}{s}$, where in these expressions s is the γ^*p CM energy

---
[1] See Eqs (C.13) and (C.14) of Appendix C.
[2] See the discussion following Eq. (3.13) of Section 3.2.

squared. To build up the expression for a structure function, the γ^*-parton level cross-sections, $\hat{\sigma}_{T,L}$, are needed. To begin with consider the calculation of F_2 in terms of $(\hat{\sigma}_T + \hat{\sigma}_L)/\hat{\sigma}_0$ (with an obvious notation). As outlined in Problem 1 at the end of this chapter, it is straightforward to show that for a massless quark of charge e_i the γ^*-quark interaction of Fig. 4.1(a) gives

$$\hat{\sigma}_T = e_i^2 \hat{\sigma}_0 \delta(1-z),$$
$$\hat{\sigma}_L = 0. \qquad (4.3)$$

Substituting this in Eq. 4.1 gives

$$\frac{F_2(x, Q^2)}{x} = \int_0^1 dz \int_0^1 d\xi\, q_i(\xi) \delta(x - z\xi) e_i^2 \delta(1-z)$$
$$= \int_x^1 \frac{d\xi}{\xi} q_i(\xi) e_i^2 \delta(1 - x/\xi)$$
$$= e_i^2 q_i(x),$$

which agrees with the QPM result for a single quark flavour.

The first additional hard process beyond the QPM to be considered is gluon radiation from quark lines, there are two diagrams as shown in Fig. 4.2. These are essentially the diagrams of the QCD Compton process

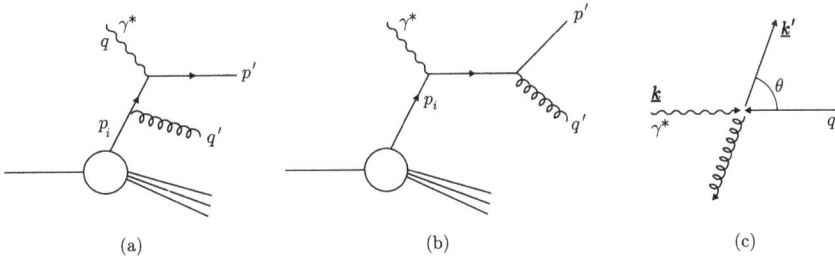

Fig. 4.2 Diagrams for gluon radiation from quarks in DIS via QCDC, together with the parton CM frame.

shown in Fig. 3.3. Using Eq. (3.14) with $t \to u$ and $u \to t$ and in terms of the kinematics of the hard scattering subprocess shown here, the parton level cross-section is

$$\hat{\sigma} = \int d\Omega \frac{2}{3} \frac{e_i^2 \alpha \alpha_s}{\hat{s}} \left[-\frac{\hat{t}}{\hat{s}} - \frac{\hat{s}}{\hat{t}} + \frac{2\hat{u}Q^2}{\hat{s}\hat{t}} \right], \qquad (4.4)$$

where $\hat{s} = (p_i + q)^2$, $\hat{t} = (q - p')^2$, $\hat{u} = (q - q')^2$. In the γ^*parton CMS θ is the scattering angle and k, k' the magnitudes of the initial and final state 3-momenta. Then

$$\hat{s} = 4k'^2 = 2p_i \cdot q + q^2 = Q^2(1-z)/z$$

72 QCD improved parton model

$$k = (\hat{s} + Q^2)/(2\sqrt{\hat{s}}), \qquad k' = \sqrt{\hat{s}}/2, \quad \text{so} \quad k'/k = 1 - z$$
$$\hat{t} = -2kk'(1 - \cos\theta), \qquad \hat{u} = -2kk'(1 + \cos\theta)$$

and using these, the terms in the [ ] brackets in Eq. (4.4) become

$$[\,] \rightarrow \left[\frac{1}{1-z}\left(\frac{1-c}{2}\right) + (1-z)\left(\frac{2}{1-c}\right) + \frac{2z}{1-z}\left(\frac{1+c}{1-c}\right) \right], \qquad (4.5)$$

where $c \equiv \cos\theta$. As it stands there are singularities for $z \to 1$ and $c \to 1$. The former corresponds to infrared gluon radiation and will be dealt with later as this singularity is cancelled by other terms. The angular integral will be dominated by the contribution from the the last two terms and the $1/(1-c)$ singularity will have to be regulated. The singularities occur as $\theta \to 0$ which corresponds to the gluon being emitted collinear with the initial parton direction. The residue at the $c = 1$ pole is $2(1 + z^2)/(1 - z)$. Rather than proceed with an angular integration directly, it is instructive to rewrite it in terms of the gluon (or quark) transverse momentum in the parton CMS,

$$p_t^2 = k'^2 \sin^2\theta \simeq \frac{\hat{s}}{4}\theta^2 \quad \text{for small } \theta.$$

Similarly for small angles $d\Omega \simeq 4\pi dp_t^2/\hat{s}$. Using these expressions and keeping only the dominant $1/(1-c)$ terms

$$\hat{\sigma} \simeq \hat{\sigma}_0 e_i^2 \frac{\alpha_s}{2\pi} \frac{4}{3}\left[\frac{1+z^2}{1-z}\right] \int_0^{p_t^2(\text{max})} \frac{dp_t^2}{p_t^2}, \qquad (4.6)$$

where $p_t^2(\text{max}) = \hat{s}/4 = Q^2(1-z)/4z$. The p_t^2 integral is regulated at the lower limit by the introduction of a cut-off, κ^2, and it gives

$$\int_{\kappa^2}^{p_t^2(\text{max})} \frac{dp_t^2}{p_t^2} = \ln\left(\frac{Q^2}{\kappa^2}\right) + \ln\left(\frac{1-z}{4z}\right).$$

This result shows explicity where the large logs come from, gluon radiation introduces non-zero p_t and integrating over the characteristic dp_t^2/p_t^2 radiation spectrum will introduce a $\ln(Q^2/\kappa^2)$ scaling violation in the structure function. Define the function

$$P_{qq}(z) = \frac{4}{3}\left[\frac{1+z^2}{1-z}\right], \qquad (4.7)$$

which is the probability distribution for $q \to q(z)g(1-z)$ splitting (where the ( ) indicate the fractions of the initial quark momentum), then $\hat{\sigma}$ may be written

$$\hat{\sigma}(z) = \hat{\sigma}_0 e_i^2 \frac{\alpha_s}{2\pi}\left[P_{qq}(z)\ln\left(\frac{Q^2}{\kappa^2}\right) + C(z)\right], \qquad (4.8)$$

where $C(z)$ includes the terms left over after the identification of the leading $\alpha_s \ln(Q^2/\kappa^2)$ term. $\hat{\sigma}(z)$ can now be added to the zeroth order QPM parton cross-section to give

$$\frac{F_2(x,Q^2)}{x} = \int_0^1 dz \int_0^1 d\xi\, q_i(\xi)\delta(x-z\xi)[e_i^2\delta(1-z) + \hat{\sigma}(z)/\hat{\sigma}_0]$$

$$= e_i^2 \int_x^1 \frac{d\xi}{\xi} q_i(\xi) \left[\delta(1-\frac{x}{\xi}) + \frac{\alpha_s}{2\pi}P_{qq}\left(\frac{x}{\xi}\right)\ln\left(\frac{Q^2}{\kappa^2}\right)\right],$$

keeping only the leading $\alpha_s \ln(Q^2/\kappa^2)$ term.

How is this result to be interpreted? Could one simply redefine the parton density by

$$q_i(x) \to q_i(x) + \Delta q_i(x, Q^2),$$

where

$$\Delta q_i(x, Q^2) = \frac{\alpha_s}{2\pi} \int_x^1 \frac{d\xi}{\xi} q_i(\xi) P_{qq}\left(\frac{x}{\xi}\right) \ln\left(\frac{Q^2}{\kappa^2}\right). \tag{4.9}$$

This would not be satisfactory as the parton density would then depend on the arbitrary cut-off, κ^2. The solution is similar to renormalization, the $q_i(x)$ used so far must be considered as unmeasurable 'bare' distributions. By introducing a new scale $\mu^2 \gg \kappa^2$, the soft non-perturbative physics is included in the renormalized parton density $q_i(x, \mu^2)$ which is now scale dependent. This is the (collinear) factorization scale and it is the scale at which the collinear singularity is absorbed into the parton density, $\ln(Q^2/\kappa^2) \to \ln(Q^2/\mu^2) + \ln(\mu^2/\kappa^2)$, and

$$q_i(x, \mu^2) = q_i^0(x) + \frac{\alpha_s}{2\pi} \int_x^1 \frac{d\xi}{\xi} q_i^0(\xi) P_{qq}\left(\frac{x}{\xi}\right) \ln\left(\frac{\mu^2}{\kappa^2}\right), \tag{4.10}$$

with

$$\frac{F_2(x,Q^2)}{x} = e_i^2 \int_x^1 \frac{d\xi}{\xi} q_i(\xi, \mu^2)\left[\delta(1-\frac{x}{\xi}) + \right.$$
$$\left. \frac{\alpha_s}{2\pi} P_{qq}\left(\frac{x}{\xi}\right) \ln\left(\frac{Q^2}{\mu^2}\right) + \frac{\alpha_s}{2\pi} C\left(\frac{x}{\xi}\right)\right], \tag{4.11}$$

where the non-singular piece has been reinstated.[3] These last few steps require a bit of care as they are only correct to order α_s. The exact definition of the parton density and the non-singular piece depends on the renormalization and factorization schemes. For example, in the DIS scheme the $C(z)$ contribution is absorbed into the renormalized parton density and

---

[3] Differences between orders in $\ln Q^2$ and α_s are discussed in Section 4.2.1.

with $\mu^2 = Q^2$ as the choice of factorization scale, the above equation reduces to
$$F_2(x, Q^2) = e_i^2 x q_i(x, Q^2), \tag{4.12}$$
which has the form of the QPM result but with $q_i(x) \to q_i(x, Q^2)$. This relationship for F_2 is maintained to all orders and defines the DIS factorization scheme.

The renormalized parton density $q_i(x, \mu^2)$ cannot be calculated perturbatively, but its variation with $\ln \mu^2$ is given by
$$\frac{\partial q_i(x, \mu^2)}{\partial \ln \mu^2} = \frac{\alpha_s}{2\pi} \int_x^1 \frac{d\xi}{\xi} q_i(\xi, \mu^2) P_{qq}(x/\xi). \tag{4.13}$$

In addition to gluon bremsstrahlung modifying the QPM results, the process of $g \to q\bar{q}$ will also change the quark momentum densities. The diagrams are those of the BGF process and are shown again together with the parton CM frame variables in Fig. 4.3. In terms of the parton-level

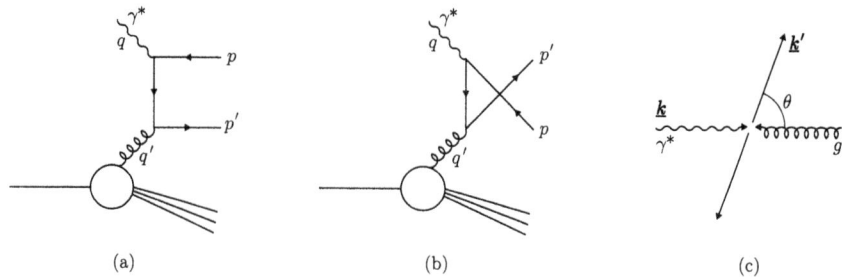

Fig. 4.3 The parton-level diagrams for BGF production of $q\bar{q}$, together with the parton CM frame. Note that p_i of Fig. 4.1 or 4.2 is denoted by q' here, for easier comparison with the tree-level calculation of Section 3.2.

variables, the cross-section from Eq. (3.15) reads:
$$\left. \frac{d\hat{\sigma}}{d\Omega} \right|_{BGF} = \frac{1}{4} \frac{e_i^2 \alpha \alpha_s}{\hat{s}} \left[\frac{\hat{u}}{\hat{t}} + \frac{\hat{t}}{\hat{u}} - \frac{2\hat{s}Q^2}{\hat{t}\hat{u}} \right], \tag{4.14}$$

where $\hat{s} = (q+q')^2$, $\hat{t} = (q-p)^2$, $\hat{u} = (q-p')^2$ and $z = Q^2/2q \cdot q' = Q^2/(\hat{s}+Q^2)$. Introducing the CM scattering angle θ as shown in the figure, the terms in [] become
$$[] \to \left[2\frac{1+c^2}{1-c^2} - 8\frac{z(1-z)}{1-c^2} \right],$$

where $c = \cos\theta$. Both terms are singular as $\theta \to 0, \pi$, which correspond to either the outgoing q or $\bar{q}$ being collinear with the incoming gluon. Writing the angular integral in terms of the parton p_t and approximating $1+c^2 \approx 2$ in the numerator of the first term

$$\hat{\sigma} \simeq \hat{\sigma}_0 e_i^2 \frac{\alpha_s}{2\pi} \frac{1}{2}[z^2 + (1-z)^2] \int_0^{p_t^2(\max)} \frac{dp_t^2}{p_t^2}.$$

Proceeding as above, the p_t integral gives a leading $\ln(Q^2/\kappa^2)$ for large Q^2, where κ^2 regulates the divergence. To relate this result to γ^*p scattering, $\hat{\sigma}$ is convoluted with a gluon density function $g^0(x)$ to give

$$\frac{F_2^g(x, Q^2)}{x} = \int_x^1 \frac{d\xi}{\xi} e_i^2 g^0(\xi) \frac{\alpha_s}{2\pi} P_{qg}\left(\frac{x}{\xi}\right) \ln\left(\frac{Q^2}{\kappa^2}\right), \quad (4.15)$$

where

$$P_{qg}(z) = \frac{1}{2}[z^2 + (1-z)^2] \quad (4.16)$$

is the g $\to$ q(z)q̄(1−z) splitting function and only the leading $\alpha_s \ln(Q^2/\kappa^2)$ term has been retained. This additional contribution to F_2 has a number of consequences. The $\kappa^2 \to 0$ singularity is handled as above by renormalization at the factorization scale μ^2 leading to an extra contribution to $q_i(x, \mu^2)$ of a similar form to Eq. (4.10). The variation of $q_i(x, \mu^2)$ with $\ln \mu^2$ picks up a term depending on the gluon density (which is also renormalized by $g^0(x) \to g(x, \mu^2)$), so Eq. (4.13) becomes

$$\frac{\partial q_i(x, \mu^2)}{\partial \ln \mu^2} = \frac{\alpha_s}{2\pi} \int_x^1 \frac{d\xi}{\xi} [q_i(\xi, \mu^2) P_{qq}(x/\xi) + g(\xi, \mu^2) P_{qg}(x/\xi)]. \quad (4.17)$$

In the DIS scheme, F_2 continues to be given by Eq. (4.12), for other schemes the situation is more complicated and this will be discussed in Section 4.2. Having seen how gluon processes in QCD give rise to a $\ln Q^2$ breaking of exact QPM scaling in DIS, it is now time to pull the results together into a form that can be confronted with data.

An alternative and instructive approach to the physics of parton splitting can also be built on the QED cross-sections for soft or collinear radiation as outlined in Problems 2–5 at the end of the chapter.

4.1.1 Splitting Functions

The essence of the DGLAP approach is contained in Eq. (4.17) which shows how the parton density 'evolves' through the two splitting functions P_{qq} and P_{qg} as the scale μ^2 changes. To form a closed set of equations, two more splitting functions are needed which describe the probabilities for q $\to$ g(z)q(1 − z) and g $\to$ g(z)g(1 − z) splitting. Just as the quark density evolves, so will the gluon density with an equation of similar form to Eq. (4.17)

$$\frac{\partial g(x, \mu^2)}{\partial \ln \mu^2} = \frac{\alpha_s}{2\pi} \int_x^1 \frac{d\xi}{\xi} \left[\sum_i q_i(\xi, \mu^2) P_{gq}(x/\xi) + g(\xi, \mu^2) P_{gg}(x/\xi) \right]. \quad (4.18)$$

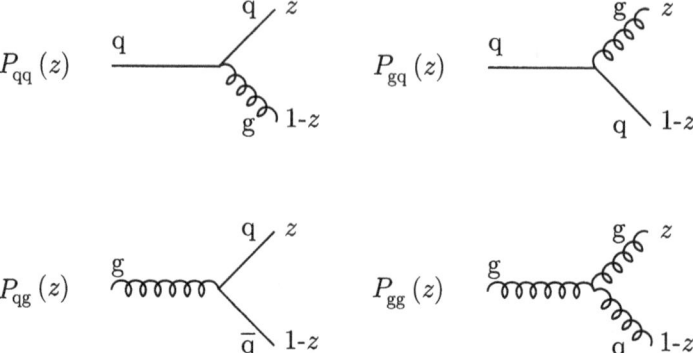

Fig. 4.4 The DGLAP splitting functions.

The splitting functions are summarised graphically in Fig. 4.4. The leading order expressions for P_{qq} and P_{qg} are given in Eqs (4.7) and (4.16) respectively. From the figure it is clear that $P_{gq}(z) = P_{qq}(1-z)$, which gives

$$P_{gq}(z) = \frac{4}{3}\left[\frac{1+(1-z)^2}{z}\right]. \qquad (4.19)$$

P_{gg} requires a bit more calculation but follows ultimately from the form of the 3-gluon vertex and is

$$P_{gg}(z) = 6\left[\frac{1-z}{z} + \frac{z}{1-z} + z(1-z)\right]. \qquad (4.20)$$

Since the QCD Lagrangian conserves fermion number and flavour, the following sum rules must be obeyed by the renormalized parton densities at least to $O(\alpha_s)$:

$$\int_0^1 dx[q_i(x,Q^2) - \bar{q}_i(x,Q^2)] = v_i, \qquad (4.21)$$

where $v_i = 2, 1, 0, \ldots$ for the u, d, s,... flavours in the proton. Overall momentum conservation gives

$$\int_0^1 dx\, x[\Sigma(x,Q^2) + g(x,Q^2)] = 1, \qquad (4.22)$$

where $\Sigma = \sum_i(q_i + \bar{q}_i)$ and the sum runs over all active flavours. Because these equations are independent of Q^2, the following constraints apply to the splitting functions

$$\int_0^1 dz\, P_{qq}(z) = 0, \qquad (4.23)$$

$$\int_0^1 dz\, z[P_{qq}(z) + P_{gq}(z)] = 0, \qquad (4.24)$$

$$\int_0^1 dz\, z[2n_f P_{qg}(z) + P_{gg}(z)] = 0, \qquad (4.25)$$

where the last two integrals correspond to the conservation of momentum and flavour in quark and gluon splittings, respectively.

There is a technical problem with P_{qq} that must be addressed and that is how to cope with the apparent singularity as $z \to 1$. This limit is associated with the emission of soft gluons. As gluon bremsstrahlung is essentially the same as photon bremsstrahlung in QED, the QCDC cross-section is an infra-red safe quantity and thus the singularity will be cancelled by $O(\alpha\alpha_s)$ interference terms between the γ^*parton diagram and the virtual loop corrections to it (see Section 3.3). The required correction terms in P_{qq} are concentrated at $z = 1$ and give a term proportional to $\delta(1 - z)$. As P_{qq} is in a mathematical sense a distribution (i.e. it always appears integrated with a function in any physical quantity) the $1/(1-z)$ singularity is regularized by the so-called '+ prescription'

$$\frac{1}{1-z} \to \frac{1}{(1-z)_+} \quad \text{where} \quad \int_0^1 dz\, \frac{f(z)}{(1-z)_+} = \int_0^1 dz\, \frac{f(z) - f(1)}{1-z}$$

and $1/(1-z)_+ = 1/(1-z)$ for $z < 1$. Rather than calculate the additional delta function term directly, it can be found by using Eq. (4.23) and the '+ prescription' to fix the coefficient of the $\delta(1-z)$ term. The resulting regularized splitting function is

$$P_{qq}(z) = \frac{4}{3}\left[\frac{1+z^2}{(1-z)_+} + \frac{3}{2}\delta(1-z)\right]. \qquad (4.26)$$

Similarly the $1/(1-z)$ singularity in P_{gg} is regularized by applying Eq. (4.25) to give

$$P_{gg}(z) = 6\left[\frac{1-z}{z} + \frac{z}{(1-z)_+} + z(1-z)\right] + \frac{33 - 2n_f}{6}\delta(1-z). \qquad (4.27)$$

4.1.2 Running coupling and the OPE

There is one further important point to be considered before the DGLAP formalism is complete and that is the validity of the $O(\alpha_s)$ results for large Q^2. The leading order QCD corrections to the parton densities are all weighted by $\alpha_s \ln(Q^2/\mu^2)$ (e.g. refer back to Eq. (4.11)) and higher order terms will involve powers of this combination. As it stands it looks as though the QCD corrections will diverge for $Q^2 \gg \mu^2$. The $\alpha_s \ln(Q^2/\mu^2)$ need to be summed and this is achieved by the simple expedient of the substitution $\alpha_s \to \alpha_s(Q^2)$ (i.e. replacing the fixed α_s by the running coupling). To see that this is plausible, refer to the discussion at the end of Section 3.6. More formally this step may be checked against the corresponding result from the renormalization group and operator production expansion. To keep things simple, consider the evolution of a quark density corresponding to the QCD Compton process only (Eq. (4.13), with splitting function P_{qq}). Repeating here the OPE result (Eq. (3.44) from Section 3.8.1 of the previous chapter)

$$\int_0^1 dx'\, x'^{n-1} W(q^2, x') = C_n(Q^2, \mu^2) \bar{N}_n(\mu^2)/4 + O(1/Q^2),$$

identify the matrix element $\bar{N}_n(\mu^2)/4$ with the nth moment of the parton density at scale μ^2, $\tilde{q}_f(n, 0)$, where

$$\tilde{q}_f(n, t) = \int_0^1 dx\, x^{n-1} q_f(x, Q^2),$$

and $t = \ln(Q^2/\mu^2)$. Defining $M_n(Q^2) = \int_0^1 dx'\, x'^{n-1} W(q^2, x')$, the above OPE equation becomes

$$M_n(Q^2) = C_n(Q^2, \mu^2) \tilde{q}_f(n, 0). \tag{4.28}$$

Now bring in the renormalization group equation (3.45) for C_n

$$C_n(Q^2, \mu^2) = C_n(\mu^2) \exp\left(\int_0^t \gamma(n, t') dt'\right),$$

here written in terms of t rather than α. Equation (4.28) may now be rewritten as

$$M_n(Q^2) = C_n(\mu^2) \tilde{q}_f(n, t).$$

where

$$\tilde{q}_f(n, t) = \tilde{q}_f(n, 0) \exp\left(\int_0^t \gamma(n, t') dt'\right). \tag{4.29}$$

The Q^2 dependence from $C_n(Q^2, \mu^2)$, given by the anomalous dimension $\gamma(n, t)$, has been transferred to the the parton density moment. If $\gamma(n, t)$ is expanded as a power series in $\alpha_s(t)$,

$$\gamma(n, t) = \gamma^{(0)}(n) \frac{\alpha_s(t)}{2\pi} + \cdots$$

the differential form of Eq. (4.29) becomes, at LO,

$$\frac{\partial \tilde{q}_f(n, t)}{\partial t} = \frac{\alpha_s(t)}{2\pi} \gamma^{(0)}(n) \tilde{q}_f(n, t). \tag{4.30}$$

This equation should have the same physics content as the evolution given by Eq. (4.13). Taking moments of that equation one finds that the anomalous dimensions[4] are the moments of the splitting function

---

[4] Definitions of the anomalous dimensions differ by factors of π and overall sign. Here we follow the conventions of Ellis et al. (1996). The LO anomalous dimensions are collected in the appendix to this chapter.

$$\gamma^{(0)}(n) = \int_0^1 dz\, z^{n-1} P_{qq}^{(0)}(z). \tag{4.31}$$

Finally, $C_n(\mu^2)$ is identified with the nth moment of the coefficient function linking the parton density with the structure function.

The 'leading logs' that are summed in the DGLAP equations correspond to diagrams with multiple gluon emission as shown in Fig. 4.5 with an important restriction. To pick out the leading $\alpha_s \ln(Q^2/\mu^2)$ terms requires

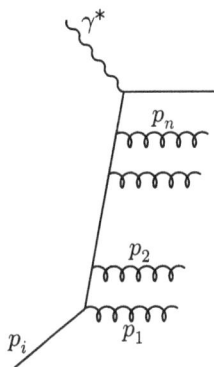

Fig. 4.5 Multiple gluon emissions from a quark line.

that the transverse momenta of the successive steps shown in the diagram must be ordered in the transverse momentum of the gluons

$$\mu^2 < p_{1t}^2 < p_{2t}^2 < \cdots < p_{nt}^2 < Q^2, \tag{4.32}$$

(to understand why, follow Problem 7 at the end of the chapter). This approximation is often known as the leading log approximation (LLA), it is well founded over a large region of phase space but is not the only possibility and other summations with be considered in Chapter 9.

4.2 The DGLAP equations

Taking the results for the quark and gluon densities together (Eqs (4.17) and (4.18)), the DGLAP equations are coupled equations for the change of the quark, antiquark and gluon densities as $\ln Q^2$ changes

$$\frac{\partial}{\partial \ln Q^2} \begin{pmatrix} q_i(x, Q^2) \\ g(x, Q^2) \end{pmatrix} = \frac{\alpha_s(Q^2)}{2\pi} \sum_j \int_x^1 \frac{d\xi}{\xi}$$

$$\begin{pmatrix} P_{q_i q_j}(\frac{x}{\xi}, \alpha_s(Q^2)) & P_{q_i g}(\frac{x}{\xi}, \alpha_s(Q^2)) \\ P_{g q_j}(\frac{x}{\xi}, \alpha_s(Q^2)) & P_{gg}(\frac{x}{\xi}, \alpha_s(Q^2)) \end{pmatrix} \begin{pmatrix} q_j(\xi, Q^2) \\ g(\xi, Q^2) \end{pmatrix}, \tag{4.33}$$

80 QCD improved parton model

where the q_i, q_j are taken to include both quarks and antiquark distributions. The splitting functions are expanded as power series in $\alpha_s(Q^2)$

$$P_{q_i q_j}(z,\alpha_s) = \delta_{ij} P_{qq}^{(0)}(z) + \frac{\alpha_s}{2\pi} P_{q_i q_j}^{(1)}(z) + \cdots$$

$$P_{qg}(z,\alpha_s) = P_{qg}^{(0)}(z) + \frac{\alpha_s}{2\pi} P_{qg}^{(1)}(z) + \cdots$$

$$P_{gq}(z,\alpha_s) = P_{gq}^{(0)}(z) + \frac{\alpha_s}{2\pi} P_{gq}^{(1)}(z) + \cdots$$

$$P_{gg}(z,\alpha_s) = P_{gg}^{(0)}(z) + \frac{\alpha_s}{2\pi} P_{gg}^{(1)}(z) + \cdots$$

The δ_{ij} in front of the leading order term in expansion of P_{qq} follows immediately from the fact that flavour is conserved at a single quark gluon vertex, one must go to higher orders for a change in flavour. Because of charge conjugation and the flavour independence of the QCD Lagrangian, P_{qg} and P_{gq} are independent of quark flavour and the same for q and q̄. The $P_{q_i q_j}$ satisfy $P_{q_i q_j} = P_{\bar{q}_i \bar{q}_j}$ and $P_{q_i \bar{q}_j} = P_{\bar{q}_i q_j}$ and the leading order term vanishes unless $q_i = q_j$. The leading order splitting functions calculated in the previous sections correspond to $P^{(0)}$ in the above expansion. The integral constraints of Eqs (4.23–4.25) apply to the $P^{(0)}$. Beyond leading order it is necessary to take account of flavour dependence in $P_{q_i q_j}$ and $P_{q_i \bar{q}_j}$. This is best done by writing the quark densities and splitting functions in terms of flavour 'singlet' and 'non-singlet' combinations. This is also convenient because the gluon terms in the DGLAP equations do not carry flavour indices. The singlet quark density has already been introduced in the context of the momentum sum rule and is

$$\Sigma(x,Q^2) = \sum_i (q_i(x,Q^2) + \bar{q}_i(x,Q^2)) \qquad (4.34)$$

where the sum runs over all active flavours.

It is useful to have a shorthand notation for the convolutions that occur in the DGLAP equations, so

$$[q \otimes P](x,Q^2) \equiv \int_x^1 \frac{d\xi}{\xi} q(\xi,Q^2) P\left(\frac{x}{\xi}\right) = \int_x^1 \frac{d\xi'}{\xi'} q\left(\frac{x}{\xi'},Q^2\right) P(\xi'). \qquad (4.35)$$

Often the arguments are dropped as well unless they are not obvious from the context. The evolution of Σ is coupled to that of the gluon density

$$\frac{\partial}{\partial \ln Q^2} \Sigma(x,Q^2) = \frac{\alpha_s(Q^2)}{2\pi} \left([\Sigma \otimes P'_{qq}] + [g \otimes 2n_f P_{qg}]\right) \qquad (4.36)$$

$$\frac{\partial}{\partial \ln Q^2} g(x,Q^2) = \frac{\alpha_s(Q^2)}{2\pi} \left([\Sigma \otimes P_{gq}] + [g \otimes P_{gg}]\right), \qquad (4.37)$$

where P'_{qq} is the same as P_{qq} at leading order and at higher orders is constructed from P_{qq} and $P_{q\bar{q}}$. A non-singlet combination is the difference of q_i and q_j (where q_j is any quark or antiquark density other than q_i). There are two combinations that are particularly useful

$$\begin{aligned} q_i^-(x,Q^2) &= q_i(x,Q^2) - \bar{q}_i(x,Q^2), \\ q_i^+(x,Q^2) &= q_i(x,Q^2) + \bar{q}_i(x,Q^2) - \frac{1}{n_f}\Sigma(x,Q^2), \end{aligned} \qquad (4.38)$$

where q_i^- is the valence quark distribution for flavour i. The evolution of non-singlet distributions does not involve the gluon density, so

$$\frac{\partial q_i^\pm(x,Q^2)}{\partial \ln Q^2} = \frac{\alpha_s(Q^2)}{2\pi}[q_i^\pm \otimes P_\pm], \qquad (4.39)$$

where the splitting functions $P_\pm$ are again constructed from P_{qq} and $P_{q\bar{q}}$, reducing to P_{qq} at LO.

The DGLAP equations give a formalism for calculating the changes to the parton densities as Q^2 changes, however they do not allow a calculation of the distributions at the starting scale Q_0^2. This information has to come either from non-perturbative methods or by parameterizing the x dependence of the parton density at Q_0^2 and determining the parameters by fitting to data. A function of the form

$$xf(x,Q_0^2) = Ax^\alpha(1-x)^\beta$$

where A and β are postive constants might seem a reasonable first guess. The valence quark density is expected to peak at x values a bit below 0.3 and to vanish as $x \to 0, 1$. The above expression has the virtue of simplicity and being easy to manipulate analytically. When modified by a low order polynomial in x or $\sqrt{x}$ it is the form used in QCD fits.

Can one say more about the exponents α and β? For α this is the domain of low x physics and is discussed at length in Chapter 9. It will turn out that for valence functions α is expected to be around 0.5 but for the sea and gluon distributions α can be either negative or positive. A more detailed discussion of functional forms for $f_i(x,Q_0^2)$ is given in Section 6.3.

4.2.1 F_2 and F_L

To connect the parton densities to structure functions, one needs the 'coefficient functions' appropriate to the chosen factorization scheme (the $C(z)$ of Eq. (4.11)), in general for F_2

$$\frac{F_2(x,Q^2)}{x} = \int_x^1 \frac{d\xi}{\xi}\left[\sum_i e_i^2 q_i(\xi,Q^2) C_q\left(\frac{x}{\xi},\alpha_s\right) + \bar{e}^2 g(\xi,Q^2) C_g\left(\frac{x}{\xi},\alpha_s\right)\right], \qquad (4.40)$$

where $\bar{e}^2 = \sum_i e_i^2$ and the sums run over all active quark and antiquark flavours. C_q and C_g are the coefficient functions, which are expanded as power series in $\alpha_s(Q^2)$,

$$C_q(z, \alpha_s) = \delta(1-z) + \frac{\alpha_s}{2\pi} C_q^1(z) + \cdots$$
$$C_g(z, \alpha_s) = \frac{\alpha_s}{2\pi} C_g^1(z) + \cdots.$$

At LO, and to all orders in the DIS scheme, $C_q(z, \alpha_s) = \delta(1-z)$ and $C_g(z, \alpha_s) = 0$, thus giving $F_2 = \sum_i e_i^2 \left(q_i(x, Q^2) + \bar{q}(x, Q^2)\right)$. Note that this 'QPM style' relation is only true for F_2, xF_3 and F_L will require convolutions with coefficient functions. In the $\overline{\text{MS}}$ scheme

$$C_q^1(z) = \frac{4}{3}\left[\frac{4\ln(1-z)-3}{2(1-z)_+} - (1+z)\ln(1-z) - \frac{1+z^2}{1-z}\ln z\right.$$
$$\left. +3+2z - \left(\frac{\pi^2}{3}+\frac{9}{2}\right)\delta(1-z)\right], \qquad (4.41)$$

$$C_g^1(z) = \frac{1}{2}\left[(z^2+(1-z)^2)\ln\left(\frac{1-z}{z}\right) - 8z^2 + 8z - 1\right]. \qquad (4.42)$$

In the expressions for the structure functions, the terminology LO, NLO, etc, refers to the behaviour with respect to $\ln Q^2$. One must be careful to account for the $\ln Q^2$ behaviour implicit in α_s. At LO, this gives only the $\ln Q^2$ from the $O(\alpha_s)$ 'one loop' contribute to the splitting functions.[5] At NLO, the one and two loop contributions to the splitting functions are included and the coefficient function from the one loop level. At NNLO, one then needs splitting functions at three loops and coefficient functions at two loops, etc.

It must also be emphasized that the parton densities, the splitting functions and the coefficient functions are all renormalization and factorization scheme dependent. Since physical quantities should be scheme independent conversion formulae exist to move between schemes, but care must be exercised to ensure consistency in any given calculation.

In the QPM, transverse momentum of the partons is assumed to be zero and one of the consequences of this for spin-$\frac{1}{2}$ quarks is that the longitudinal structure function ($F_L = F_2 - 2xF_1$) is zero. The QCDC and BGF processes (and higher order diagrams) that give rise to the $\ln Q^2$ scaling violations also give rise to partons with non-negligible p_t. In turn, this means that the Callan-Gross relation is no longer satisfied and $F_L \neq 0$. The QCD improved parton model gives an expression for F_L, with $O(\alpha_s)$ leading terms. A similar physical approach to that followed in the previous section but projecting out the longitudinal component of the parton level cross-sections $\hat{\sigma}_L$ gives the coefficient functions for F_L (Altarelli and Martinelli 1978). They are factorization scheme independent at $O(\alpha_s)$ giving

---

[5] The 'loops' referred to here come from the forms that the corresponding Feynman diagrams take.

$$\frac{F_L(x,Q^2)}{x} = \frac{\alpha_s}{2\pi} \int_x^1 \frac{d\xi}{\xi} \left[\sum_i e_i^2 \frac{8}{3} \left(\frac{x}{\xi}\right) q_i(\xi, Q^2) + \bar{e}^2 4 \left(\frac{x}{\xi}\right) \left(1 - \frac{x}{\xi}\right) g(\xi, Q^2) \right] \quad (4.43)$$

or the first term involving the quark densities may be replaced by $F_2(x, Q^2)$ to give

$$F_L(x,Q^2) = \frac{\alpha_s}{2\pi} \int_x^1 \frac{d\xi}{\xi} \left[\frac{8}{3} \left(\frac{x}{\xi}\right)^2 F_2(\xi, Q^2) + \bar{e}^2 4 \left(\frac{x}{\xi}\right)^2 \left(1 - \frac{x}{\xi}\right) \xi g(\xi, Q^2) \right]. \quad (4.44)$$

4.2.2 A useful approximation

Already by $Q^2 = 100 \, \text{GeV}^2$, $\alpha_s(Q^2)/(2\pi) \approx 0.03$ which is small enough to indicate that some useful approximations may be possible in the relations between the parton densities and structure functions. Looking at Eq. (4.40) for F_2 and the expresssions for the coefficient functions one sees that

$$F_2(x, Q^2) \approx \sum_i^{n_f} e_i^2 \left(q_i(x, Q^2) + \bar{q}_i(x, Q^2) \right). \quad (4.45)$$

In the DIS factorization scheme this relation is exact.

A modified from of Eq. (4.36) at leading order may be derived for the coupled evolution of F_2 and xg

$$\frac{\partial F_2(x, Q^2)}{\partial \ln Q^2} = \frac{\alpha_s(Q^2)}{2\pi} \{ P_{qq} \otimes F_2 + 2\bar{e}^2 P_{qg} \otimes xg \},$$

where $\bar{e}^2 = \sum_i^{n_f} e_i^2$. Prytz (1993) realised that for $x < 0.01$ the contribution of the P_{qq} term from quark splitting was negligible in comparison to the gluon term. Dropping the first term in the above equation and with an obvious change of integration variable,

$$\frac{\partial F_2(x, Q^2)}{\partial \ln Q^2} \approx \bar{e}^2 \frac{\alpha_s(Q^2)}{\pi} \int_0^{1-x} G\left(\frac{x}{1-z}\right) P_{qg}(z) dz,$$

where $G(x) = xg(x)$. The splitting function $P_{qg}(z) = \frac{1}{2}(z^2 + (1-z)^2)$ is symmetrical and slowly varying over the range $(0, 1)$ so the above convolution may be approximated by expanding $G(x/(1-z))$ about $z = 1/2$,

$$G(x/(1-z)) = G(2x) + (z - 1/2)G'(2x) + \cdots$$

Because $P_{qg}(z)$ is symmetric about $z = 1/2$ the second term in the expansion will integrate to zero and a good approximation is obtained by keeping only the first term (with $1 - x \to 1$ as the upper limit of integration)

$$\frac{\partial F_2(x, Q^2)}{\partial \ln Q^2} \approx \bar{e}^2 \frac{\alpha_s}{\pi} G(2x) \int_0^1 P_{qg}(z) dz$$

84 *QCD improved parton model*

$$\approx \frac{\bar{e}^2 \alpha_s}{3\pi} G(2x). \tag{4.46}$$

The last equation may be inverted to give

$$xg(x) \approx \frac{3\pi}{\bar{e}^2 \alpha_s} \frac{\partial F_2(x/2, Q^2)}{\partial \ln Q^2}. \tag{4.47}$$

The Prytz method gives a simple way of estimating $xg(x)$ at low x, but note that the derivative of F_2 is required at $x/2$ for the extraction of xg at x. The method has been extended to NLO but it loses some of its simplicity.

Given the existence of efficient numerical codes for full NLO DGLAP evolution and fitting, the approximate inversion of the DGLAP equation is no longer a preferred method for accurate extraction of the gluon density. However, the Prytz approximation and Eq. (4.45) are useful for a qualitative understanding of the behaviour of the structure functions, particularly at low x.

4.3 DGLAP evolution at LO

In this section, some simple features of the Q^2 evolution predicted by the DGLAP equations are examined. The form of the input distribution is restricted to a non-singular function of the form $x(1-x)^3$.

4.3.1 Valence quark evolution

The DGLAP evolution equation for a valence quark density (or more generally a non-singlet density) is

$$\frac{\partial q^{NS}(x, Q^2)}{\partial \ln Q^2} = \frac{\alpha_s}{2\pi} [q^{NS} \otimes P_{qq}].$$

Substituting the LO expression for $P_{qq}^{(0)}$ gives

$$\frac{\partial q^{NS}(x, Q^2)}{\partial \ln Q^2} = \frac{2\alpha_s}{3\pi} \int_x^1 \frac{dz}{z} q^{NS}(x/z, Q^2) \left[\frac{(1+z^2)}{(1-z)_+} + \frac{3}{2}\delta(1-z) \right] \tag{4.48}$$

where α_s means $\alpha_s(Q^2)$. The first term is integrated using

$$\int_x^1 \frac{f(z)}{(1-z)_+} dz = \int_x^1 \frac{dz}{1-z} [f(z) - f(1)] + f(1) \ln(1-x)$$

with $f(z) = \frac{1+z^2}{z} q^{NS}(x/z, Q^2)$, to give the expression

$$\frac{\partial q^{NS}(x, Q^2)}{\partial \ln Q^2} = \frac{2\alpha_s(Q^2)}{3\pi} \left\{ \left[\frac{3}{2} + 2\ln(1-x) \right] q^{NS}(x, Q^2) + \int_x^1 \frac{dz}{1-z} \left[\frac{1+z^2}{z} q^{NS}(x/z, Q^2) - 2q^{NS}(x, Q^2) \right] \right\}. \tag{4.49}$$

This equation takes an almost identical form for the NS contribution to

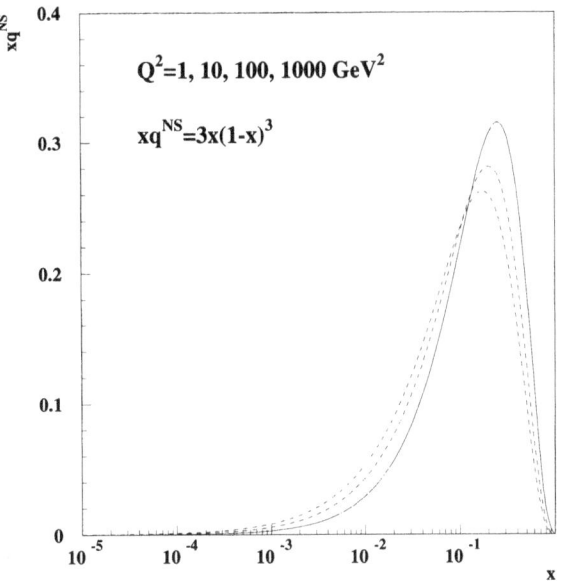

Fig. 4.6 Evolution of non-singlet valence like quark distribution with the LO DGLAP equation. Full line, input distribution at $Q^2 = 1\,\text{GeV}^2$, dashed $Q^2 = 10\,\text{GeV}^2$, dot-dashed $Q^2 = 100\,\text{GeV}^2$, dotted $Q^2 = 1000\,\text{GeV}^2$.

$F_2^{NS} = xq^{NS}$

$$\frac{\partial F_2^{NS}(x,Q^2)}{\partial \ln Q^2} = \frac{2\alpha_s(Q^2)}{3\pi}\left\{\left[\frac{3}{2} + 2\ln(1-x)\right]F_2^{NS}(x,Q^2) + \int_x^1 \frac{dz}{1-z}\left[(1+z^2)F_2^{NS}(x/z,Q^2) - 2F_2^{NS}(x,Q^2)\right]\right\}.$$
(4.50)

The term outside the integral multiplying F_2^{NS} is positive for $x < 0.53$, whereas the integral always gives a negative contribution. For typical q^{NS} the two contributions cancel at values $x \approx 0.2\text{--}0.3$ and are positive at smaller x and negative at larger x. The partial cancellation of the two terms explains why the scaling violations for a typical valence quark momentum distribution are small. Figure 4.6 shows the LO evolution of $xq^{NS}(x) = 3x(1-x)^3$ from a starting scale of $Q_0^2 = 1\,\text{GeV}^2$ to $Q^2 = 10, 100, 1000\,\text{GeV}^2$.[6] The momentum distribution becomes smeared slightly more towards lower x values as Q^2 increases, but the effect is very small.

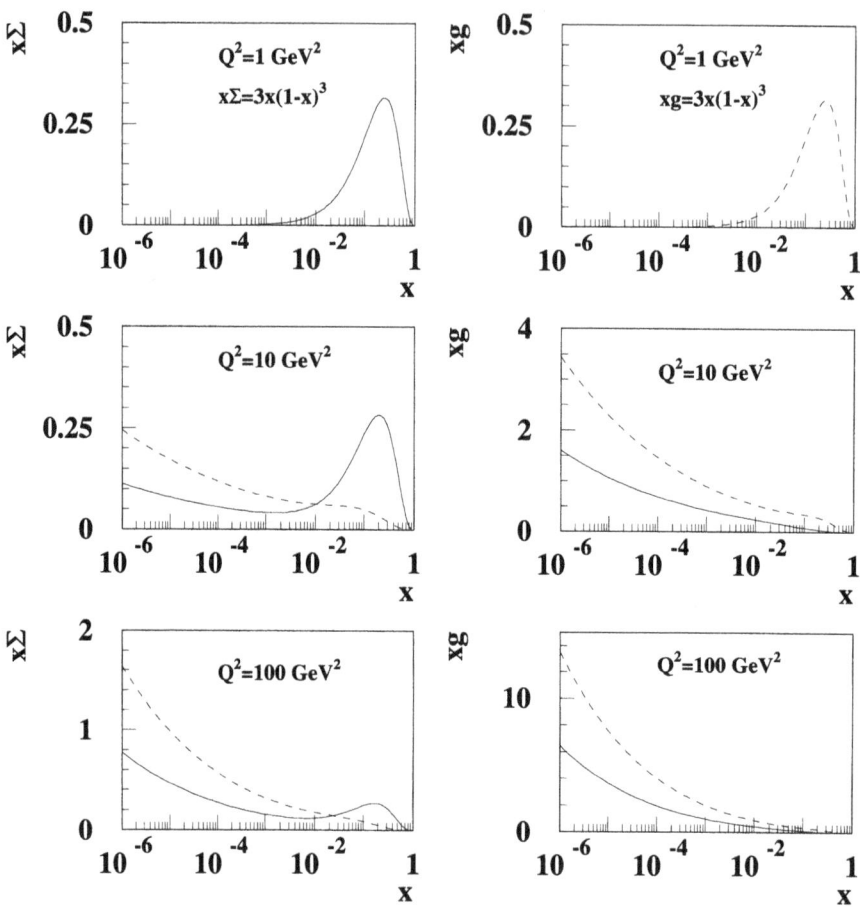

Fig. 4.7 Evolution of the gluon and singlet quark momentum distributions with the LO DGLAP equation. The full curves are for a 'valence-like' singlet input distribution at $Q^2 = 1\,\text{GeV}^2$ and the dashed curves for a 'valence-like' gluon input distribution at $Q^2 = 1\,\text{GeV}^2$. The input distributions are shown in the top two plots.

4.3.2 Singlet and gluon evolution

The evolution of the singlet and gluon densities are both more complicated and more interesting as they are coupled. The singlet quark density is

---

[6]The numerical DGLAP evolutions in this chapter have been performed using the QCDNUM code of Botje (1997).

$$\Sigma(x, Q^2) = \sum_i (q_i(x, Q^2) + \bar{q}_i(x, Q^2))$$

and coupled evolution equations are

$$\frac{\partial}{\partial \ln Q^2} \Sigma(x, Q^2) = \frac{\alpha_s(Q^2)}{2\pi} ([\Sigma \otimes P_{qq}] + [g \otimes 2n_f P_{qg}]) \quad (4.51)$$

$$\frac{\partial}{\partial \ln Q^2} g(x, Q^2) = \frac{\alpha_s(Q^2)}{2\pi} ([\Sigma \otimes P_{gq}] + [g \otimes P_{gg}]) \quad (4.52)$$

Although the LO splitting functions take rather simple forms, the coupled nature of the singlet equations makes it difficult to solve them analytically. Instead, the nature of the solutions are explored with some simple numerical examples. Figure 4.7 shows the evolution of the singlet and gluon distributions for two cases, both with 'valence-like' inputs at $Q_0^2 = 1\,\text{GeV}^2$. The full curves are for singlet input only, so $x\Sigma(x) = 3x(1-x)^3$ and $xg(x) = 0$. The dashed curves correspond to gluon input only, so $x\Sigma(x) = 0$ and $xg(x) = 3x(1-x)^3$. Both input distributions have a maximum value of 0.32 at $x = 0.25$. For the singlet input, already by $Q^2 = 10\,\text{GeV}^2$, the $1/z$ factors in the gluon splitting functions P_{gq} and P_{gg} are having an effect not only on the gluon density but through the coupled evolution on the singlet term as well. By $Q^2 = 100\,\text{GeV}^2$ the values of $x\Sigma$ and xg at $x = 10^{-5}$ are roughly 0.5 and 4 respectively. With gluon only input, the effect is even larger at small x. At $Q^2 = 100\,\text{GeV}^2$ and $x = 10^{-5}$, the values of $x\Sigma$ and xg are 1 and 8 respectively. Note that for both the singlet and gluon inputs at the starting scale, the gluon density rises rapidly at low x almost independently of the shape of input distributions.

4.4 Higher twist

Higher twist is the technical term for $1/Q^2$ corrections to the leading DGLAP or OPE expressions for DIS structure functions. Formally the even moments of a structure function may be written as power series $\sum (1/Q^2)^{\tau/2-1} C_\tau(Q^2) \cdots$ where $\tau \geq 2$ is the twist (Eq. (3.43) of Section 3.8.1). Twist is related to the spin and dimensionality of the operators appearing in the OPE. The leading twist ($\tau = 2$) terms have no other Q^2 than the $\ln Q^2$ dependence implicit in the coefficients $C(Q^2)$ — or equivalently those arising in DGLAP approach. Higher twist terms arise from more complex operators with sub-leading behaviour in Q^2. Only a few have been studied in any detail. At large x 4-quark operator terms will probably be the most important. Recently, some estimates have been made using renormalon techniques.[7]

---

[7] The renormalon method allows one to sum certain classes of multiloop insertions in Feynman diagrams and it has been used to estimate $1/Q$ corrections to event shape variables in both e^+e^- annihilation and ep scattering.

Higher twist contributions are likely to be non-negligible at large x and lowish Q^2, which corresponds to a low $\gamma^* p$ CM energy W. In the absence of calculations, groups fitting the data have taken a pragmatic approach with a modification to the structure function of the form

$$F_i^{\text{HT}}(x, Q^2) = F_i^{\text{QCD}}(x, Q^2)[1 + D_2(x, Q^2)/Q^2], \quad (4.53)$$

where F_i^{QCD} is the twist-2 QCD prediction and $D_2(x, Q^2)$ represents the contributions of twist-4 terms (i.e. first power correction in $1/Q^2$) relative to the leading twist-2 structure function. The function D_2 is often simplified by ignoring Q^2 dependence and modelling the x dependence either by constant for each x bin in the data sample or by parameterizing it as a low order polynomial in x with the coefficients determined from the fit. At the rather low ep energies, with low Q^2 and large x, of the early fixed target experiments some higher twist contributions have a known origin. They come from the neglect of $O(m^2/Q^2)$ terms in the expressions for the kinematic variables — so called 'target mass' corrections. Sometimes these are taken into account by modifying the scaling variable x

$$x \to \xi = \frac{2x}{1+r} \quad \text{where} \quad r = \sqrt{1 + \frac{4m^2 x^2}{Q^2}},$$

with the remaining 'dynamic higher twist' contribution parameterized as above. Figure 4.8 shows some results on twist-4 terms at fixed target en-

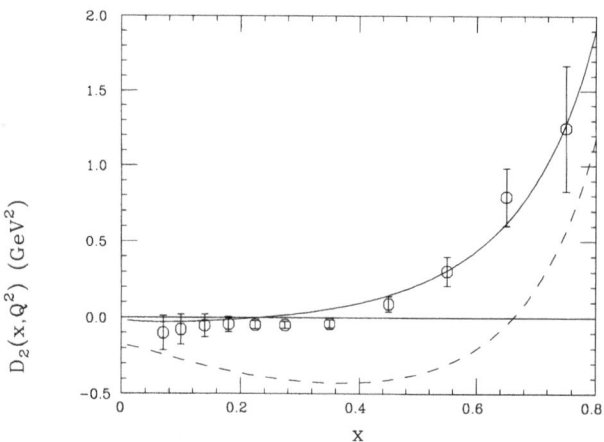

Fig. 4.8 The twist-4 correction term D_2 as defined in Eq. (4.53). The data points the contribution to F_2 from an analysis of SLAC and BCDMS data by Virchaux and Milsztajn and the curves are renormalon predictions for F_2 (full line) and xF_3 (dashed line). (From Dasgupta and Webber 1996).

ergies. The data points are the fitted constants in x bins from the analysis of SLAC and BCDMS data by Virchaux and Milsztajn (1992). The curves are from the renormalon calculation of Dasgupta and Webber (1996) at $Q^2 = 10\,\text{GeV}^2$, but the Q^2 dependence is found to be negligible. The calculated twist-4 terms for the EM structure function F_2 (full line) are positive at all x whereas those for the parity violating structure function in neutrino scattering xF_3 (dashed line) are positive only at the largest x values. In all cases the contributions are small except at the very largest x values.

Apart from parameterizing the higher twist effects at large x, one may reduce their significance by applying cuts on Q^2 and W^2 (see Section 6.6.3). In addition to the possible effects of higher twist at large x, there may also be non-negligible effects at very low x (see also Section 9.9).

4.5 Heavy quarks

At the x and Q^2 values probed at HERA heavy quarks may contribute up to 30% of the structure function F_2. The most important mechanism for heavy quark production in DIS is through boson–gluon fusion (see Fig. 4.3) which is directly sensitive to the gluon momentum density. The introduction of massive quarks requires an extension of the DGLAP formalism discussed so far. This is a somewhat technical issue. The main points are summarised in the following two sub-sections covering the impact on the running of α_s and a brief summary of the approaches for heavy flavours. A closer look at the problem in the context of so-called 'variable flavour number schemes' follows, but is not essential reading for later chapters.

4.5.1 Matching prescriptions for α_s

There are no mass terms in the expansion of the β function given in Eq. (3.30) but n_f will change by one unit when a quark threshold is crossed. In the renormalization schemes that are commonly used in DIS, it is conventional to use the massless expressions for the b_i and put all the terms depending on m_Q^2/μ^2 into the coefficient functions. In addition one needs a prescription for taking α_s across a threshold — or equivalently changing Λ_{QCD}. There is no unique solution and it is important to check which method is being used in a particular calculation. Some of the more popular choices are:

- Define $\alpha_s^{(5)}(\mu^2)$ to be the numerical solution of the renormalization group equation to a given order (e.g. Eq. (3.32)) with $n_f = 5$ in terms of the reference value $\alpha_s(M_Z^2)$. Then $\alpha_s^{(5)}(m_b^2)$ is also known. For $m_c^2 < \mu^2 < m_b^2$, $\alpha_s^{(5)}(m_b^2)$ is used as the new reference value for the numerical solution of the RGE with $n_f = 4$ to give $\alpha_s^{(4)}(\mu^2)$ and in particular $\alpha_s^{(4)}(m_c^2)$. For $\mu^2 < m_c^2$, $\alpha_s^{(4)}(m_c^2)$ is used, in turn, as the new reference value for the RGE with $n_f = 3$ to give $\alpha_s^{(3)}(\mu^2)$. Thus the value of α_s for any $\mu^2 < M_Z^2$ is known. This procedure is not difficult to implement numerically.

- The parameterizations of α_s in terms of Λ_{QCD} are convenient but less well defined. At a threshold Λ will change to reflect the change in n_f, but this is clumsy. A better method is to exploit the freedom in the definition of Λ to introduce an extra constant that is used to fix the matching of α_s at a threshold. For simplicity, the procedure will be outlined using the LO expressions. Using the freedom in the choice of Λ, rewrite Eq. (3.22) as

$$[\alpha_s(\mu^2, n_f)]^{-1} = b_0(n_f)\ln(\mu^2/\Lambda^2) + C(n_f)$$

where $C(n_f)$ is a constant to be determined and Λ is independent of n_f. At a threshold the values of α_s must match, $\alpha_s(m_Q^2, n_f) = \alpha_s(m_Q^2, n_f + 1)$. Define Λ to be that for $n_f = 4$ so that $C(4) = 0$, then $\alpha_s(m_c^2, 3) = \alpha_s(m_c^2, 4)$ fixes

$$C(3) = b_0(4)\ln(m_c^2/\Lambda^2) - b_0(3)\ln(m_c^2/\Lambda^2) = -\frac{1}{6\pi}\ln(m_c^2/\Lambda^2), \quad (4.54)$$

similarly, $\alpha_s(m_b^2, 4) = \alpha_s(m_b^2, 5)$ fixes $C(5)$. With more complicated expressions, the same procedure may be applied to the NLO expressions. In both LO and NLO, this choice of matching gives convenient expressions for numerical evaluation.

- Although in both the previous examples the matching condition is applied at $\mu^2 = m_Q^2$, this is also arbitrary and sometimes $4m_Q^2$ is preferred. Beyond NLO the matching becomes yet more complicated and α_s itself may not necessarily be continuous at a threshold. The subject has been studied in detail by Bernreuther (1993) and a prescription based on his work is recommended by the particle data group. At the centre of his analysis is the process independent relation between $\alpha_s^{(-)}$, the coupling for $n_f - 1$ light quark flavours, and α_s, the coupling for the full theory with an additional heavy quark,

$$\alpha_s^{(-)}(\mu^2) = \alpha_s(\mu^2)\left(1 + \sum_{k=1}^{\infty} C_k(\xi)\left(\frac{\alpha_s(\mu^2)}{\pi}\right)^k\right)$$

where $\xi = \ln(m_Q^2(\mu^2)/\mu^2)$ and $m_Q^2(\mu^2)$ is the running heavy quark mass.[8] The above relation is considerably simplified by choosing the scale $\mu^2 = m_Q^2(\mu^2)$, so that $\xi = 0$. For $\overline{\text{MS}}$ it is

$$\alpha_s^-(m_Q^2) = \alpha_s(m_Q^2) + 7\alpha_s^3(m_Q^2)/72\pi^2.$$

Full details on how to match the Λ parameters are given by Bernreuter.

The numerical differences between α_s calculated by the different prescriptions at NLO and beyond are small, although the differences between Λ parameters can be quite large.

---

[8]The running heavy quark mass parameter is another consequence of the renormalization group, quark masses were ignored for simplicity in the RGE discussion of Section 3.5, but they run in an analogous way to the coupling constant.

4.5.2 Heavy quark production in DIS

Charm production is described as it is the one for which most data exists, but a similar formalism will also describe beauty production. For simplicity the possibility of intrinsic charm in the proton is ignored and $\bar{c}(x, Q^2) = c(x, Q^2)$ is assumed throughout this section. The most frequent approaches to the inclusion of heavy quarks (e.g. charm of mass m_c) within the DGLAP framework are:

- ZM-VFNS (zero-mass variable flavour number scheme) in which the charm parton density $c(x, Q^2)$ satisfies $c(x, Q^2) = 0$ for $Q^2 \leq \mu_c^2$ and $n_f = 3 + \theta(Q^2 - \mu_c^2)$ in the splitting functions and β function. The threshold μ_c^2, which is in the range $m_c^2 < \mu_c^2 < 4m_c^2$, is chosen so that $F_2^c(x, Q^2) = 2e_c^2 x c(x, Q^2)$ gives a satisfactory description of the data and α_s is matched by one of the prescriptions described in the previous section. The advantage of this approach is that the simplicity of the massless DGLAP equations is retained. The disadvantage is that the physical threshold $\hat{W}^2 = Q^2(\frac{1}{z} - 1) \geq 4m_c^2$ is not treated correctly ($\hat{W}$ is the $\gamma^* g$ CM energy).

- FFNS (fixed flavour number scheme) in which there is no charm parton density and all charmed quarks are generated by the BGF process. The advantage of the FFNS scheme is that the threshold region is correctly handled, but the disadvantge is that large $\ln(Q^2/m_c^2)$ terms appear and charm has to be treated ab initio in each hard process.

- VFNS (variable flavour number scheme) in which F_2^c is calculated by interpolating between the FFNS approach at low Q^2 and the ZM-VFNS scheme for large Q^2. The goal here is to combine the correct features of the FFNS scheme near threshold with the appearance of a charm parton density at large Q^2.

4.5.3 Variable flavour number schemes

The essence of the problem may be understood by looking at the full expression for F_2^c in the FFNS at order α_s

$$F_2^c(x, Q^2) = 2 \int_x^1 \frac{dz}{z} x g_3(x/z, \mu^2) C_g^{FF}(z, m_c^2/Q^2, \mu^2) \tag{4.55}$$

where g_3 is the gluon density for three light quark flavours (u,d,s). Two choices of scale are used, $\mu^2 = Q^2$ and $\mu^2 = 4m_c^2$ (the threshold for $c\bar{c}$ pair production). In either case, expanding the massive fixed flavour coefficient function gives

$$C_g^{FF}(z, m_c^2/Q^2, \mu^2) = \frac{\alpha_s(\mu^2)}{2\pi} e_c^2 C_g^{1,FF}(z, m_c^2/Q^2) + \cdots,$$

with

$$C_g^{1,FF}(z, m_c^2/Q^2) =$$

$$\frac{1}{2}\left\{\left[z^2+(1-z)^2+\frac{4m_c^2}{Q^2}z(1-3z)-8z^2\left(\frac{m_c^2}{Q^2}\right)^2\right]\ln\left(\frac{1+\beta}{1-\beta}\right)\right.$$
$$\left.+\left[8z(1-z)-1-\frac{4m_c^2}{Q^2}z(1-z)\right]\beta\right\}\theta(\hat{W}^2-4m_c^2), \qquad (4.56)$$

where β is the velocity of the charm quark and $\beta^2 = 1 - 4m_c^2/\hat{W}^2$. Note that the θ function guarantees correct threshold behaviour. For $Q^2 \gg m_c^2$, $\beta \to 1$ and

$$\frac{1+\beta}{1-\beta} \to \frac{Q^2}{m_c^2}\frac{1-z}{z}$$

so

$$C_g^{1,FF}(z, m_c^2/Q^2) \to \frac{1}{2}\left\{[z^2+(1-z)^2]\ln\left(\frac{1-z}{z}\right)+8z(1-z)-1\right.$$
$$\left.+[z^2+(1-z)^2]\ln\left(\frac{Q^2}{m_c^2}\right)\right\}. \qquad (4.57)$$

The first two terms in the above expression are the same as the gluon coefficient function C_g^1 (Eq. (4.42)) of the massless DGLAP formalism. It is the last term with the $\ln(Q^2/m_c^2)$ that is the problem. Comparing the above expressions with Eq. (4.15) and the discussion preceeding it for the massless BGF case, one sees that m_c^2 is replacing the infrared cutoff κ^2 in the integration over p_t^2. Since m_c^2 is non-zero the heavy quark contribution is finite. For massless partons in the DGLAP framework the large infrared logs are absorbed by the renormalization of the parton density at the factorization scale and the large Q^2 logs are resummed by the renormalization group procedure. Although the massive BGF coefficient functions have been calculated to NLO, they are not available for an all orders resummation. It is of course precisely this problem that the ZM-VFNS scheme solves, but at the price of incorrect threshold behaviour.

The conditions that must be met in any VFNS were derived by Buza et al. (1996). Write the contribution of the heavy quark (taken here to be charm) to F_2 as

$$F_2^c(x, Q^2) =$$
$$2e_c^2 \int_x^1 dz\frac{x}{z}\left[c(x/z, \mu^2)C_q^{VF}(z, Q^2/\mu^2) + g_4(x/z, \mu^2)C_g^{VF}(z, Q^2/\mu^2)\right],$$

where c and g_4 are charm parton and gluon densities for four flavours (u,d,s,c) and the coefficient functions have the expansions

$$C_q^{VF} = C_q^{0,VF} + \frac{\alpha_s}{2\pi}C_q^{1,VF} + \ldots$$

$$C_g^{VF} = \frac{\alpha_s}{2\pi} C_g^{1,VF} + \dots$$

For $Q^2 < m_c^2$ the coefficient functions and parton densities are those of the FF scheme given by Eq. (4.55) and for $Q^2 \gg m_c^2$ they should tend to the massless DGLAP forms given in Eqs (4.40–4.42). At a scale in the vicinity of m_c^2 the FF and VFNS expressions are matched with expressions of the form

$$\begin{aligned}
c(x,\mu^2) &= A^{cg}(\mu^2/m_c^2) \otimes g_3(\mu^2) \\
g_4(x,\mu^2) &= A^{gg}(\mu^2/m_c^2) \otimes g_3(\mu^2) \\
C_b^{VF}(z,Q^2/\mu^2) &= C_a^{FF}(Q^2/\mu^2) \otimes \left[A^{ba}(\mu^2/m_c^2)\right]^{-1},
\end{aligned} \quad (4.58)$$

where A^{cg}, etc are elements of a 5×4 coefficient matrix matching the three and four massless flavour (plus gluon) parton densities. At order α_s the first two of these equations take the form (with $\mu^2 = Q^2$)

$$\begin{aligned}
c(x,Q^2) &= \frac{\alpha_s}{2\pi} \ln\left(\frac{Q^2}{m_c^2}\right) P_{qg}^{(0)} \otimes g_3(x,Q^2) \\
g_4(x,Q^2) &= \left(1 - \frac{\alpha_s}{6\pi} \ln\left(\frac{Q^2}{m_c^2}\right)\right) g_3(x,Q^2)
\end{aligned} \quad (4.59)$$

and matching the coefficient functions gives

$$C_g^{1,FF}(z,Q^2/m_c^2) = $$
$$C_g^{1,VF}(z,Q^2/m_c^2) + C_q^{0,VF}(z,Q^2/m_c^2) \otimes P_{qg}^{(0)} \ln\left(\frac{Q^2}{m_c^2}\right). \quad (4.60)$$

In addition α_s for three and four massless flavours has to be matched. If $C_q^{0,VF}$ is known then Eq. (4.60) serves as a definition of $C_g^{1,VF}$.

A number of ways of satisfying these conditions have been proposed. Methods widely used are those of the CTEQ group (the ACOT scheme — see Krämer et al. (2000) for a summary and references) and Roberts and Thorne (1998,2001). The latter is outlined briefly here. Roberts and Thorne remove some of the arbitrariness in the choice of $C_q^{0,VF}$ by imposing the extra condition of requiring that $dF_2^c/d\ln Q^2$ also matches smoothly. Matching at m_c^2 simplifies the resulting condition which is

$$C_q^{0,VF}(Q^2/m_c^2) \otimes P_{qg}^{(0)} = \frac{\partial C_g^{1,FF}(z,Q^2/m_c^2)}{\partial \ln Q^2}. \quad (4.61)$$

The righthand side of this equation may be found by direct differentiation of Eq. (4.56) and in the limit of large Q^2 it tends to $P_{qg}^{(0)}(z)$. Using the extra condition Eq. (4.60) becomes

$$C_g^{1,VF}(z,Q^2/m_c^2) = C_g^{1,FF}(z,Q^2/m_c^2) - \frac{\partial C_g^{1,FF}(z,Q^2/m_c^2)}{\partial \ln Q^2} \ln\left(\frac{Q^2}{m_c^2}\right). \quad (4.62)$$

This equation has a number of virtues, it gives $C_g^{1,VF}$ the same threshold

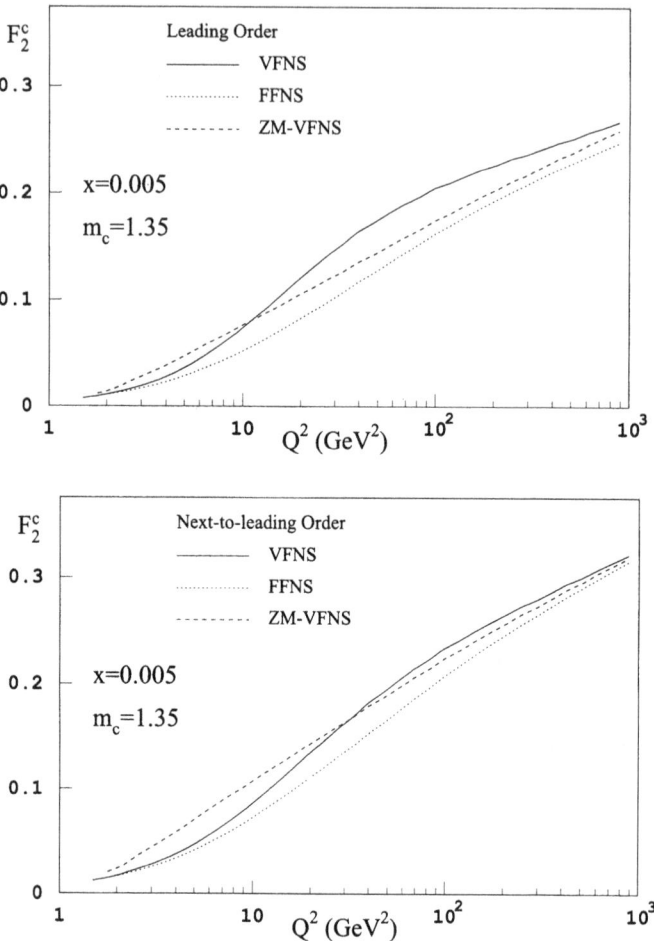

Fig. 4.9 F_2^c at LO and NLO from the VFNS scheme of Roberts and Thorne (1998).

behaviour as $C_g^{1,FF}$ (by construction) and the second term removes the $P_{qg}^{(0)} \ln(Q^2/m_c^2)$ term from $C_g^{1,FF}$ at large Q^2 to leave the correct massless coefficient function. The general solution of Eq. (4.61) is not easily found, however, what is actually required is the convolution $C^{0,VF} \otimes c$ and this has been calculated. The procedure has been extended to NLO and also for F_L. In all cases it produces satisfactory and smooth interpolation between the FFNS near threshold and the zero mass evolution at large Q^2. This is shown in Fig. 4.9 for F_2^c at $x = 0.005$ as a function of Q^2. The differences between the different approaches are quite small.

4.6 Beyond NLO

As data on hard scattering processes get more precise the theoretical uncertainties from having to work at NLO may become comparable to, or even larger than, the experimental systematic uncertainties. The usefulness of parton densities extracted from such processes, for example for calculating standard model cross-sections at the LHC, may be compromised by the resulting level of uncertainty. It is thus important to extend the hard-scattering formalism beyond NLO. This is most advanced for DIS, but still incomplete. As summarised by van Neerven and Vogt (2000), the QCD β function is known to NNNLO, the two-loop coefficient functions for DIS are known at NNLO, but the three-loop splitting functions ($P_{ij}^{(2)}$) are incomplete. They have taken the available information and derived approximations for the complete NNLO splitting functions and NNNLO for non-singlet combinations. The resulting uncertainty on the NNLO contribution to the derivatives $d \ln q / d \ln \mu^2$ of the parton densities is estimated to be less than 2% for $x > 0.001$. The authors have used the sensitivity to α_s of the F_2^{NS} to estimated the effect of additional terms in the perturbation expansion. They estimate that the uncertainty on $\alpha_s(M_Z^2)$ decreases from 0.005 at NLO, to 0.002 at NNLO and reaches 0.001 at NNNLO. The main effect on parton evolution is to reduce scale uncertainty, but the effect on the determination of parton densities is not straightforward as it involves an interplay between the flexibility of the parton density parameterizations, the regions in which the data is most accurate and the extra terms in the evolution equations. van Neerven and Vogt also note that the quark coefficient functions contain large soft gluon emission contributions giving a larger change in the structure function than the NNLO term in the splitting function for $x > 0.01$. For very large $x > 0.8$, it will probably be necessary to use the results of soft-gluon resummation. These matters will be considered briefly in Chapter 6. A corresponding effort to complete the NNLO framework for hard hadron-hadron processes is also well underway (Glover 2002).

4.7 Summary

This chapter has given an account of the QCD improved parton model, which is the paradigm for how perturbative QCD is used to calculate hard scattering processes. The crucial insight, which is underpinned by the factorization theorems based on the operator product expansion, is that a hard scattering cross-section can be separated into a calculable process dependent part and universal non-perturbative parton densities. The exposition has concentrated on the physical processes of gluon radiation and g $\to$ q$\bar{\text{q}}$ leading to the DGLAP equations at LO and NLO for the Q^2 evolution of parton densities. The relation of the DGLAP equations to the more formal approach of the operator product expansion has been outlined and a brief account of higher twist terms given. The problem and some practical

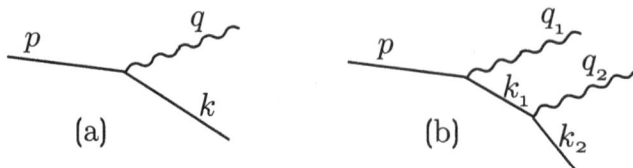

Fig. 4.10 (a) Single soft photon emission. (b) Emission of two soft photons.

approaches to including heavy flavours within the DGLAP formalism have been summarised. Prospects for going beyond NLO have been outlined.

Problems 2–5 below are longer than most and offer an interesting alternative approach to the basic ideas underlying the QCD-enhanced parton model.

4.8 Problems

1. Using the Feynman rules for QED, calculate the $\gamma^* q$ cross-sections given in Eq. (4.3). Start from the matrix element $-iee_q \varepsilon_\mu \bar{u}(p')\gamma^\mu u(p)$ for $\gamma^*(q)\mathrm{q}(p) \to \mathrm{q}(p')$, where $p' = q + p$.

2. Soft or collinear photon emission kinematics. Consider the process shown in Fig. 4.10(a): an electron of energy p emits a photon with energy zp and transverse momentum q_t ($q_t \ll p$), leaving an electron with energy $(1-z)p$. At high energies all particles may be taken as massless. One may write

$$p = (p, 0, 0, p), \quad q = (zp, q_t, 0, zp), \quad k = ((1-z)p, -q_t, 0, (1-zp)$$

with $p^2 = q^2 = k^2 = 0$ to order q_t^2. For cross-section and other calculations that will involve either k or q as a propagator, one needs to refine the calculation slightly.
   - Real photon emission, electron propagator. Show that, to $O(q_t^4)$, writing $q = (zp, q_t, 0, zp - q_t^2/(2zp))$ gives $q^2 = 0$ and $k^2 = -q_t^2/z$.
   - Real electron emission, photon propagator. Similarly, to $O(q_t^4)$, if $q = (zp, q_t, 0, zp + q_t^2/(2(1-z)p))$, $k^2 = 0$ and $q^2 = -q_t^2/(1-z)$.

   Consider next successive emissions of two photons from an electron with the momenta as shown in Fig. 4.10(b). Use arguments similar to the above to show that if $q_1^2 = q_2^2 = 0$, $k_2^2 = -(q_{1t}+q_{2t})^2 - (1-z_2)q_{2t}^2/z_2$, where q_{1t}, q_{2t} are the photon transverse momenta (w.r.t. p) and z_2 is the fraction of the energy of electron k_1 carried off by q_2. Hence k_2^2 will only be $O(q_{2t}^2)$ if $q_{1t}^2 \ll q_{2t}^2$.

3. Matrix element for nearly collinear photon emission. Use the momenta and kinematics of Fig. 4.10(a) and assume that at high energies the EM coupling preserves helicity. Consider first the matrix element for $e_L^- \to e_L^- \gamma$ which may be written

$$\mathcal{M} = \bar{u}_L(k)(-ie\gamma_\mu)u_L(p)\varepsilon_T^{*\mu}(q),$$

where $u_L(p) = (1-\gamma_5)u(p)/2$ and $\varepsilon^\mu_T(q)$ is the (transverse) polarization 4-vector of the photon. Define the energy fraction z and transverse momentum q_T of the photon as in the previous question. First show that (ignoring masses)

$$u_L(p) \to \sqrt{p}\begin{pmatrix}\xi_-\\-\xi_-\end{pmatrix}, \quad u_L(k) \to \sqrt{(1-z)p}\begin{pmatrix}\eta_-\\-\eta_-\end{pmatrix}$$

$$\text{where } \xi_- = \begin{pmatrix}0\\1\end{pmatrix}, \quad \eta_- = \begin{pmatrix}\frac{q_t}{2(1-z)p}\\1\end{pmatrix},$$

$$\text{and } \varepsilon^\mu_R = \frac{1}{\sqrt{2}}(0, 1, i, -\frac{q_t}{zp}), \quad \varepsilon^\mu_L = \frac{1}{\sqrt{2}}(0, 1, -i, -\frac{q_t}{zp}).$$

Show next that the contributions to the matrix element are:

$$\mathcal{M}(e^-_L \to e^-_L \gamma_R) = (-ie)\frac{\sqrt{2(1-z)p}}{z}q_t$$

$$\mathcal{M}(e^-_L \to e^-_L \gamma_L) = (-ie)\frac{\sqrt{2(1-z)p}}{z(1-z)}q_t$$

Parity invariance says that the matrix elements must be the same if all helicities are flipped so one does not need to calculate $e^-_R \to e^-_R \gamma$ separately and the the spin summed and averaged matrix element is

$$\frac{1}{2}\sum_{\text{spins}}|\mathcal{M}(e^- \to e^-\gamma)|^2 = \frac{2e^2q_t^2}{z(1-z)}\left[\frac{1+(1-z)^2}{z}\right].$$

4. Nearly collinear pair production $\gamma(q) \to e^+(k_+)e^-(k_-)$. The spin averaged cross-section for this process may be calculated in a similar way to the previous question. For an electron with LH helicity the positron must have RH helicity, because of helicity conservation at the vertex. Then using parity conservation one only needs to calculate the matrix element

$$\mathcal{M}(\gamma \to e^+_R e^-_L) = \bar{u}_L(k_-)(-ie\gamma_\mu)v_R(k_+)\varepsilon^\mu_T(q).$$

Taking the 4-momenta to be

$$q = (q, 0, 0, q), \quad k_+ = (zq, k_t, 0, zq), \quad k_- = ((1-z)q, -k_t, 0, (1-z)q).$$

show that

$$\mathcal{M}(\gamma_L \to e^-_L e^+_R) = (-ie)\frac{\sqrt{2z(1-z)p}}{z}k_t$$

$$\mathcal{M}(\gamma_R \to e^-_L e^+_R) = (+ie)\frac{\sqrt{2z(1-z)p}}{z(1-z)}k_t$$

and thus

$$\frac{1}{2}\sum_{\text{spins}}|\mathcal{M}(\gamma \to e^-e^+)|^2 = \frac{2e^2k_t^2}{z(1-z)}[z^2+(1-z)^2].$$

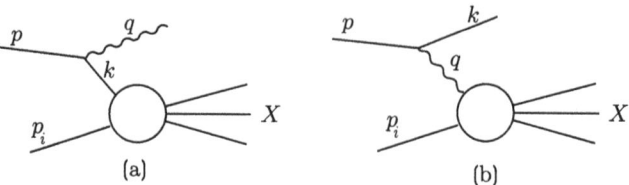

Fig. 4.11 (a) Soft or collinear radiation before scattering. (b) Equivalent photon approximation.

5. Processes with soft or collinear radiation. The results of Problems 2–4 may be used to calculate the factorized forms of the cross-sections for the processes shown in Fig. 4.11. First for (a) in a which a real (soft or collinear) photon is emitted by the electron before it iteracts. The kinematics are as in Problem 2 for $q^2 = 0$, $k^2 = -q_t^2/z$ and using the formulae given in Appendix B one may write

$$\sigma(ep \to \gamma X) = \frac{1}{4pE_i(1+v_i)} \int \frac{d^3q}{(2\pi)^3 2q^0}$$
$$\times \int d\Pi_X \frac{1}{2} \sum |\mathcal{M}(e \to e\gamma)|^2 \left(\frac{1}{k^2}\right)^2 \sum_X |\mathcal{M}(ep \to X)|$$

where $\int \Pi_X$ is the final state phase-space for X. Next rewrite the photon phase-space integrand as $d^3q = pdz\pi dq_t^2$, substitute for k^2 and insert the result of Q3 for the $e \to e\gamma$ matrix element squared to get

$$\sigma(ep \to \gamma X) = \int_0^1 dz \int \frac{dq_t^2}{q_t^2} \frac{\alpha}{2\pi} \left[\frac{1+(1-z)^2}{z}\right] \sigma(ep \to X),$$

where

$$\sigma(ep \to X) = \frac{1}{4(1-z)pE_i(1+v_i)} \int d\Pi_X \sum_X |\mathcal{M}(ep \to X)|^2.$$

Note that one factor of q_t^2 in the propagator has been cancelled by the q_t^2 from the $e \to e\gamma$ matrix element squared. Finally, the integral over the transverse momentum may be performed to give

$$\sigma(ep \to \gamma X) = \int_0^1 dz \frac{\alpha}{2\pi} \ln \frac{\tilde{s}}{\mu^2} \left[\frac{1+(1-z)^2}{z}\right] \sigma(ep \to X),$$

where $\tilde{s}$ is of the order of the ep CM energy squared and μ here would be the electron mass. For a comparison with the parton model, note that the intermediate electron is carrying a fraction $(1-z)$ of the incident electron's energy. Thus, a substitution $x = (1-z)$ gives something very similar to the calculation leading to Eq. (4.8) for the probabilty of finding an electron with momentum fraction x in the original electron of

$$p_{ee}(x) = \frac{\alpha}{2\pi} \ln \frac{\tilde{s}}{\mu^2} \left[\frac{1+x^2}{1-x}\right].$$

One must also take into account the case in which no radiation occurs giving $\delta(1-x)$ and the singularity at $x=1$ must be regularized. Similarly, the cross-section for the process shown in Fig. 4.11(b) may be written

$$\sigma(ep \to eX) = \int_0^1 dz \frac{\alpha}{2\pi} \ln \frac{\tilde{s}}{\mu^2} \left[\frac{1+(1-z)^2}{z}\right] \sigma(\gamma p \to X),$$

where

$$\sigma(\gamma p \to X) = \frac{1}{4zpE_i(1+v_i)} \int d\Pi_X \sum_X |\mathcal{M}(\gamma p \to X)|^2$$

and the intermediate photon carries a fraction z of the energy of the incident electron. This is the *equivalent photon approximation* of Weizsacker and Williams. One may identify a probability function $p_{\gamma e}$ similar to P_{qq} of the parton model. Using a similar factorization of the cross-sections and the matrix elements from Problem 4 one may also calculate $p_{e\gamma}$ the analog of P_{qg}. The one parton splitting function that cannot be derived using QED results and colour factors is P_{gg}. The reader interested in this approach to the DGLAP equations should consult, for example, Peskin and Schroeder (1995), Chapter 17 for more details.

6. Verify the regularisation of P_{gg}, Eq. (4.27).
7. Strong ordering of p_t^2 in DGLAP, Eq. (4.32) and Fig. 4.5. Consider just the first two gluon emissions as shown in Fig. 4.10. For the DGLAP (leading log approximation) one needs the double collinear limit to give

$$\left(\frac{\alpha}{2\pi}\right)^2 \int_{\mu^2}^{Q^2} \frac{dp_{2t}^2}{p_{2t}^2} \int_{\mu^2}^{p_{1t}^2} \frac{dp_{1t}^2}{p_{1t}^2} = \frac{1}{2}\left(\frac{\alpha}{2\pi}\right)^2 \ln^2\left(\frac{Q^2}{\mu^2}\right).$$

Using the results of Problems 2 and 5, show that the approximations used in calculation of the matrix elements and the propagator will not cancel for the second photon emission unless $q_{1t}^2 \ll q_{2t}^2$.

8. Using the definition given in Eq. (3.33) for $\Lambda_{\rm QCD}$ at NLO, find the equivalant condition to that given in Eq. (4.54) for matching α_s at the charm threshold.

4.9 Appendix: LO splitting functions and anomalous dimensions

The LO DGLAP splitting functions are:

$$P_{qq}(z) = \frac{4}{3}\left[\frac{1+z^2}{(1-z)_+} + \frac{3}{2}\delta(1-z)\right],$$

$$P_{qg}(z) = \frac{1}{2}[z^2 + (1-z)^2],$$

$$P_{gq}(z) = \frac{4}{3}\left[\frac{1+(1-z)^2}{z}\right],$$

$$P_{gg}(z) = 6\left[\frac{1-z}{z} + \frac{z}{(1-z)_+} + z(1-z)\right] + \frac{33-2n_f}{6}\delta(1-z).$$

The '+ prescription' is

$$\frac{1}{(1-z)_+} = \frac{1}{(1-z)} \quad \text{for } z < 1, \quad \text{and}$$

$$\int_0^1 dz \frac{f(z)}{(1-z)_+} = \int_0^1 dz \frac{f(z) - f(1)}{1-z}.$$

The LO expressions for the anomalous dimensions are:

$$\gamma_{qq}^{(0)}(n) = \frac{4}{3}\left[\frac{1}{n(n+1)} - \frac{1}{2} - 2\sum_2^n \frac{1}{k}\right],$$

$$\gamma_{qg}^{(0)}(n) = \frac{1}{2}\left[\frac{2+n+n^2}{n(n+1)(n+2)}\right],$$

$$\gamma_{gq}^{(0)}(n) = \frac{4}{3}\left[\frac{2+n+n^2}{n(n^2-1)}\right],$$

$$\gamma_{gg}^{(0)}(n) = 6\left[\frac{1}{n(n-1)} + \frac{1}{(n+1)(n+2)} - \frac{1}{12} - \sum_2^n \frac{1}{k}\right] - \frac{n_f}{3}.$$

Note that the term $2\sum_2^n \frac{1}{k}$ comes from the $1/(1-z)_+$ term in the splitting functions and behaves like $\sim \ln n$ for large n. The signs and conventions agree with those of Ellis et al. (1996), Chapter 4, where expressions for the NLO splitting functions may also be found.

5
DIS experiments and data

Since the first experiments on deep inelastic scattering, the use of leptonic probes ($e^\pm$, $\mu^\pm$, ν, $\bar{\nu}$) has been the dominant experimental technique. Roughly the first half of the chapter covers in turn: kinematics for fixed target and HERA experiments; essential features of the detectors; an outline of how DIS events are selected and the raw data corrected to give cross-section measurements. The second half starts with a summary of nucleon structure function data (mainly F_2) from the classic DIS experiments, including HERA data for $Q^2 < M_Z^2$ (data for high Q^2 NC and CC processes at HERA is considered in Chapter 8) and neutrino beam fixed target data. Then follows an account of how the longitudinal structure, F_L, is measured and the available data. Using their capability to analyse the hadronic final state, both the HERA general purpose detectors have measured the charm contribution to F_2. The importance of such data and some questions for the future are outlined before the chapter ends with a brief account of structure function data in the 'transition region' between photoproduction at $Q^2 = 0$ and deep inelastic scattering with $Q^2 > 1\,\text{GeV}^2$.

5.1 Some numbers

Most high energy physics experiments involve counting events and the rate at which events are expected for an ideal measurement of a total cross-section, σ, is given by the product $\mathcal{L} \times \sigma$ where $\mathcal{L}$ is the beam flux (the number of beam particles per second per unit target area) or luminosity. In more detail, the double differential cross-section for deep-inelastic scattering is related to $N(x, Q^2)$, the number of events measured in the bin of size $\Delta x \Delta Q^2$ at (x, Q^2), by

$$\Delta x \Delta Q^2 \frac{d^2\sigma}{dx\,dQ^2} = \frac{N(x, Q^2)}{\langle \mathcal{L} \rangle A(x, Q^2)}, \quad (5.1)$$

where $\langle \mathcal{L} \rangle$ is the integrated luminosity or beam flux and $A(x, Q^2)$ is the correction function that accounts for finite resolution, efficiencies and acceptance. Strictly one should integrate over the bin size, and this may be necessary for large bins if the cross-section is varying rapidly. At the CM energy of HERA (300–314 GeV), the dynamic range of cross-sections

102 *DIS experiments and data*

of interest is large, over at least 10 orders of magnitude, from 10^6 pb for $ep \to e'X$ at $Q^2 \approx 0$ to 10^{-4} pb for neutral and charged current cross-sections at $Q^2 \approx 10^4$ GeV^2. This follows directly from the dominant $1/Q^4$ behaviour of the NC cross-section. For the design luminosity of HERA-II of 7×10^{35} m^{-2}s^{-1}, it takes around 17 days to accumulate 100 events from a cross-section of 1 pb (assuming perfect acceptance). To cope with the large dynamic cross-section range and to remove background, DIS experiments have to have flexible and sophisticated trigger systems to select events online.

The appendix at the end of this chapter provides some more comments on the magnitude of DIS cross-sections, fixed target beam properties and collider luminosity.

5.2 Kinematics

The various Lorentz invariant variables such as x, y, Q^2 have been introduced and used extensively in earlier chapters. Their determination from quantities measured in DIS experiments is discussed in this section. Understanding the connections is crucial both for the design of DIS detectors and accurate measurement of cross-sections. Expressions for fixed target and HERA experiments are given and limitations on the accessible regions of the (x, Q^2) kinematic plane are outlined. The physical region in x and Q^2 is a triangle bounded by $0 < Q^2 < s$, $0 < x < 1$ and the line $y = 1$, $Q^2 = sx$. This follows from the relation $Q^2 = sxy$ and the ranges $0 < x, y < 1$, which also give the maximum value of Q^2 to be s (the ℓp CM energy squared). The physical region for the HERA collider is that below the diagonal $y = 1$ on the x, Q^2 plot shown in Fig. 5.1.

5.2.1 Fixed target

The classic fixed target experiments were performed at SLAC using electron beams and at CERN and Fermilab using muon and neutrino beams. For a fixed target experiment with an incident charged lepton beam of energy E, the detector for an inclusive measurement needs only to measure the energy and angle of the scattered beam electron or muon. The incident beam direction defines the axis from which the scattered lepton angle, θ, is measured, E' is the energy of the scattered lepton and m_N is the nucleon mass. Then, ignoring mass terms,

$$s = 2m_N E \quad \text{and} \quad y = (E - E')/E, \tag{5.2}$$

$$Q^2 = 4EE' \sin^2 \frac{\theta}{2}, \tag{5.3}$$

$$x = \frac{2EE' \sin^2 \frac{\theta}{2}}{m_N(E - E')}. \tag{5.4}$$

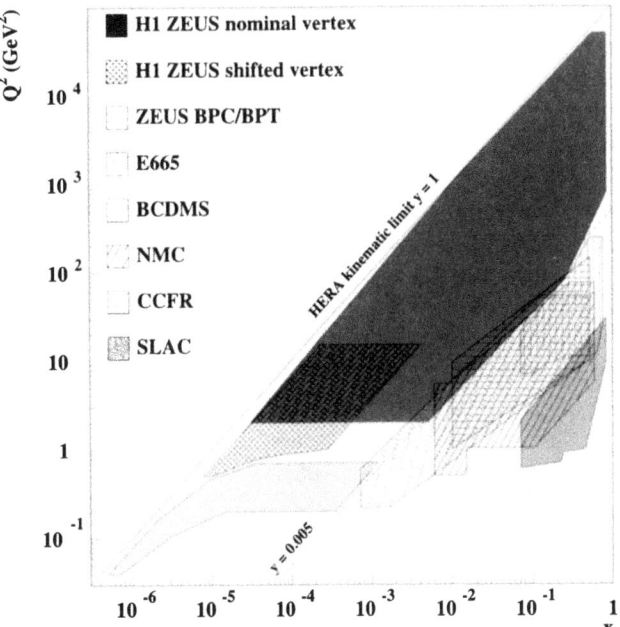

Fig. 5.1 Regions in the (x, Q^2) plane covered by DIS experiments. The lines of constant y are for the HERA centre-of-mass energy.

For fixed target experiments the relation $x = Q^2/(2\nu)$ is often used, where $\nu \equiv p \cdot q = m_N(E - E')$.[1] In the (x, Q^2) plane lines of constant scattering angle θ are given by $Q^2 = sxD/(x+D)$ where $D = 2E \sin^2 \frac{\theta}{2}/m_N$ and lines of constant E' are given by straight lines $Q^2 = 2m_N(E - E')x$. A real detector cannot measure θ to zero degrees or E' below some minimum value. Both these constraints will limit the accessible (x, Q^2) region. Further restrictions may follow from the finite resolution with which a given detector can measure θ and E'. This can be seen from the expressions for the relative errors

$$\frac{\delta Q^2}{Q^2} = \frac{\delta E'}{E'} \oplus \cot\left(\frac{\theta}{2}\right)\delta\theta$$
$$\frac{\delta x}{x} = \frac{1}{y}\frac{\delta E'}{E'} \oplus \cot\left(\frac{\theta}{2}\right)\delta\theta \qquad (5.5)$$

where $\oplus$ indicates addition in quadrature. Apart from the obvious dependence of the resolutions on $\delta E'$ and $\delta\theta$, both δQ^2 and δx get magnified

---

[1] Note that ν is often defined as $(p \cdot q)/m_N$ in earlier works on DIS.

at small scattering angles by the $\cot(\theta/2)$ factor. This is not such a severe constraint as very small angles cannot usually be measured. However, for x the error in the energy gets magnified at small values of y and this does give an additional limitation on the region in which accurate measurements can be made. Apart from control of $\delta\theta$ and $\delta E'$, a good understanding of the relative calibration of E and E' is important. In a fixed target experiment E' cannot be larger than the incident beam energy E.

In charged current DIS with an incident neutrino, the neutrino energy is not known but has to be inferred from the final state measurements. The detector has to be more complicated and able to measure the final state hadronic energy, E_{had}, in addition to the high energy muon from the leptonic vertex. The variables x and Q^2 are first estimated from the measured ('visible') quantities by:

$$Q_{\text{vis}}^2 = 4 E_{\text{vis}} E_\mu \sin^2 \frac{\theta_\mu}{2}, \qquad x_{\text{vis}} = \frac{Q_{\text{vis}}^2}{2 m_N E_{\text{had}}}, \qquad (5.6)$$

where $E_{\text{vis}} = E_\mu + E_{\text{had}}$. The visible quantities are then corrected for detector effects using Monte Carlo simulation.

5.2.2 HERA collider

HERA is the first ep collider and consists of two separate rings of circumference 6.3 km, one a warm magnet electron (or positron) ring with energy 27.5 GeV and the other a superconducting magnet proton ring of energy up to 920 GeV. The two general purpose detectors, H1 and ZEUS, are multi-purpose devices designed to investigate all aspects of high energy ep collisions and provide measurements of both the scattered electron and the hadronic system, the latter allowing one to estimate the energy and angle of the struck quark. Taken together, this gives an overdetermined system and there are many ways to reconstruct x and Q^2. Three methods will be covered here: the electron (E), double angle (DA) and sigma (Σ) methods. More details are covered in the exercises and a comprehensive summary with references to original papers is given by Bassler and Bernardi (1999).

First the electron (E) method. The proton beam axis is conventionally taken to define the positive z direction from which all angles are measured. At low Q^2 this means that the electron scattering angle approaches 180°. If E_e, E_p are the electron and proton beam energies, E' and θ_e the energy and angle of the scattered electron, then (ignoring mass terms):

$$s = 4 E_e E_p \quad \text{and} \quad y_e = \left(1 - \frac{E'}{E_e} \sin^2 \frac{\theta_e}{2}\right), \qquad (5.7)$$

$$Q_e^2 = 4 E_e E' \cos^2 \frac{\theta_e}{2} = \frac{E'^2 \sin^2 \theta_e}{1 - y_e}, \qquad (5.8)$$

$$x_e = \frac{E_e E' (1 - \sin^2 \frac{\theta_e}{2})}{E_p (E_e - E' \sin^2 \frac{\theta_e}{2})}. \qquad (5.9)$$

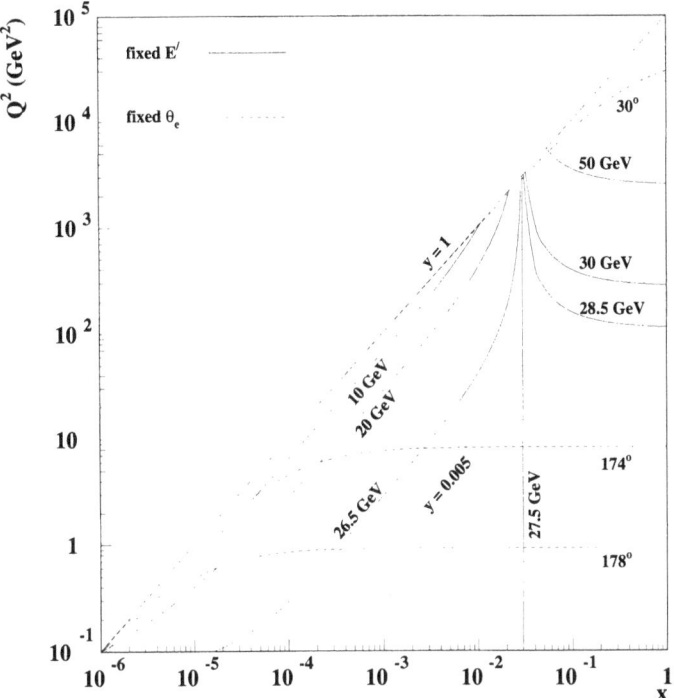

Fig. 5.2 Contours of constant E' and θ_e for the HERA with $E_e = 27.5\,\text{GeV}$ and $E_p = 920\,\text{GeV}$. The vertical line at $x_0 = E_e/E_p$ divides the kinematic peak region.

Note that if $E' = E_e$ then $x = x_0 \equiv E_e/E_p$, the position of the so-called kinematic peak. In the (x, Q^2) plane, lines of constant scattering angle θ_e are given by $Q^2 = sxC/(sx + C)$ where $C = 4E_2^2 \cot^2 \dfrac{\theta_e}{2}$ and lines of constant E' are given by $Q^2 = (4xE_e(E_e - E'))/(x_0 - x)$. Figure 5.2 shows contours of constant E' and θ_e. The contours of large θ_e show clearly how the minimum electron scattering angle of a detector restricts the accessible phase space. The large region around the kinematic peak, in which E' is roughly constant is evident. Kinematic peak events, which can be selected to be almost independent of the structure function samples, are used for calibration purposes by both H1 and ZEUS. The figure also shows that on the high x high Q^2 side of the kinematic peak the scattered $e^\pm$ has energies $E' > E_e$.

As in a fixed target experiment, E' is required to be greater than a minimum value and θ_e less than a maximum angle for the scattered electron to be properly measured. Although there are differences of detail, the formulae for the resolutions of x and Q^2 in terms of $\delta E'$ and $\delta\theta_e$ are similar to those of Eq. (5.5).[2] The conclusion is the same, that at small y (large x) the resolution in x deteriorates.

In the double angle (DA) method[3], only angles are used to reconstruct x and Q^2. In addition to θ_e one constructs the angle γ from the hadronic energy flow using:

$$\cos\gamma = \frac{(\sum_h p_x)^2 + (\sum_h p_y)^2 - (\sum_h (E - p_z))^2}{(\sum_h p_x)^2 + (\sum_h p_y)^2 + (\sum_h (E - p_z))^2} \quad (5.10)$$

where $\sum_h$ runs over all calorimeter energy deposits (with momentum vectors (p_x, p_y, p_z)) not assigned to the scattered electron. In the QPM, γ is the polar angle of the struck quark. The variables x and Q^2 are then determined by:

$$Q^2_{DA} = 4E_e^2 \frac{\sin\gamma(1 + \cos\theta_e)}{\sin\gamma + \sin\theta_e - \sin(\gamma + \theta_e)} \quad (5.11)$$

$$x_{DA} = x_0 \frac{\sin\gamma + \sin\theta_e + \sin(\gamma + \theta_e)}{\sin\gamma + \sin\theta_e - \sin(\gamma + \theta_e)} \quad (5.12)$$

The method is insensitive to hadronization and, to first order, is independent of the detector energy scales. At small values of θ_e or γ, the resolution in x_{DA} and Q^2_{DA} worsens. For the hadronic system to be well measured, it is necessary to require a minimum of hadronic activity away from the beampipe. A suitable quantity for this purpose is the Jacquet–Blondel hadronic estimator of y:

$$y_{JB} = \frac{\sum_h (E - p_z)}{2E_e} \quad (5.13)$$

At low y (< 0.04) and low Q^2, there is low hadronic activity in the detector and the DA method becomes sensitive to noise in the calorimeter.

A variation on the DA method, introduced by ZEUS, is the PT method. In this the expression for γ given in Eq. (5.10), is modified in two ways:

- first, the hadronic transverse momentum $(\sum_h p_x)^2 + (\sum_h p_y)^2$ is replaced by that of the electron $(\mathbf{p}_T^e)^2$ as it more accurately measured;
- second, the hadronic longitudinal (momentum – energy) difference $\sum_h (E - p_z)$ is corrected for energy loss through the forward beamhole using the Monte Carlo simulation of the detector.

---

[2] See Bassler and Bernardi (1999) for details.

[3] The derivation of the DA formulae is followed in Problems 2–4 at the end of the chapter.

To improve the resolution of x at small y and to reduce radiative corrections H1, use the Σ method of reconstruction. Returning to Eq. (5.13), $\sum_h (E - p_z)$ is referred to as Σ, and the essence of the idea is to remove explicit dependence on E_e from the kinematic reconstruction of the electron method. This is done by noting that energy and momentum conservation gives $2E_e = \Sigma + E'(1 - \cos\theta_e)$. Using this to replace $2E_e$ in the denominator of y_{JB} gives

$$y_\Sigma = \frac{\Sigma}{\Sigma + E'(1 - \cos\theta_e)}, \qquad Q_\Sigma^2 = \frac{E'^2 \sin^2\theta_e}{1 - y_\Sigma}. \qquad (5.14)$$

The variable x is then calculated from $Q^2 = sxy$. Apart from allowing an extension of measurements to lower y values than the E method, the Σ method has a reduced sensitivity to initial state radiative corrections because E_e no longer appears explicitly.

In practice to get the best possible resolution over the complete (x, Q^2) space available at HERA, both H1 and ZEUS use combinations of the above methods.

For the charged current process at HERA in which the outgoing scattered lepton is a neutrino, only final state hadronic information is available for the reconstruction of the kinematic variables. It is for this type of measurement that hemiticity of the HERA detectors is essential. For a CC interaction in a completely hermetic detector, the neutrino is the only undetected particle and its 4-momentum can be deduced by using 4-momentum conservation since the initial 4-momenta are known. However, in reality, although the HERA detectors cover more than 99% of 4π, energy escapes through the hole for the beam pipe — particularly in the forward (proton) direction. The Jacquet–Blondel method of estimating y (Eq. (5.13)) is insensitive to loss of energetic particles at very forward angles for which $E \approx p_z$. Once y is known, Q^2 is given from Eq. (5.8) using conservation of transverse momentum by

$$Q^2 = \frac{(\sum_h p_x)^2 + (\sum_h p_y)^2}{1 - y_{\text{JB}}}. \qquad (5.15)$$

Finally x is determined using $x = Q^2/sy$. The regions in the x, Q^2 plane in which structure functions have been measured at HERA are shown in Fig. 5.1.

5.3 Detectors

How well can the requirements for measuring DIS kinematic variables in fixed target and collider frames be realised in experiments? In this section the essential experimental techniques will be outlined by reference to the E665 and CCFR spectrometers and the ZEUS and H1 experiments at HERA. A good general reference for the techniques and physics of detectors in high energy physics is Grupen (1998).

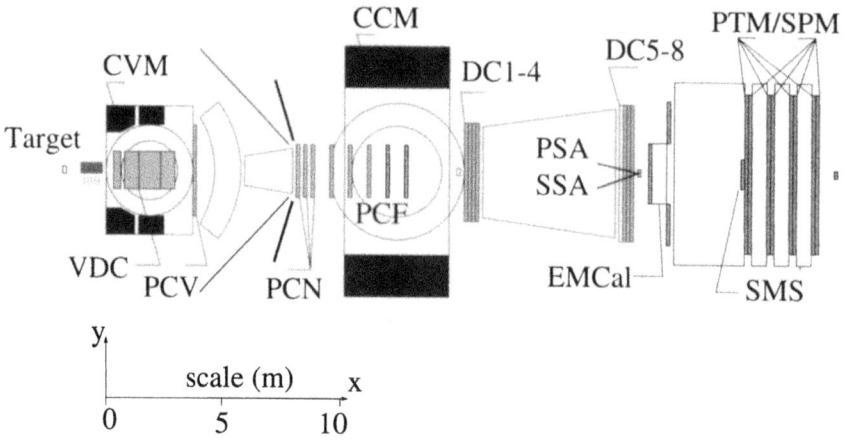

Fig. 5.3 The E665 detector.

5.3.1 The E665 muon scattering experiment

The Experiment E665 (Adams et al. 1990) was located at the end of the NM beamline at Fermilab. The NM beam provided muons of average energy 470 GeV with a spread of 50 GeV. The design of the muon beam is itself an interesting problem and a few more details are given in the appendix at the end of the chapter. The experiment took data in 1987–88 and 1990–92. The detector, shown in Fig. 5.3, was designed to measure beam and scattered muons with high precision and to provide measurement of charged and neutral particles in the final state. The technique used to analyse the charged particles, particularly the scattered muon, is bending in a powerful magnetic field. Note the extended linear scale of the apparatus, roughly 30 m between the target and the final muon detection system (PTM/SPM). This is required because the large Lorentz boost — for the NM beam on a stationary proton target the CM system has a γ factor of 15.84 — gives a strong forward collimation of the reaction products. Aspects of the detector important for measuring DIS are:

Beam spectrometer. This was positioned between the end of the NM beamline and the main detector. It consisted of four measuring stations each equipped with beam hodoscopes and multiwire proportional chambers, two before a dipole magnet and two after, each lever arm of approximately 27 m resulting in a resolution on the beam momentum of $\delta(p^{-1}) \sim 8 \times 10^{-6}\,\mathrm{GeV}^{-1}$.

Target. The target assembly was placed in the field free region in front of the first spectrometer magnet (CVM). It consisted of three identical target cells on a precision table that moved the targets laterally into the beam following a regular cycle. The three target cells were identical and of active length 1 m, two were filled with liquid hydrogen and liquid deuterium, respectively. The third was empty and was used to provide data for subtraction of off-target scatters.

Main spectrometer. The charged particle spectrometer was constructed around two large magnets with reversed polarities (CVM 4.315 Tm and CCM −6.734 Tm) arranged so that the position of the scattered muon at a focusing plane in the muon detector was independent of momentum and depended only on the scattering angle. Tracking was performed using planar drift chambers placed inside both magnets and before and after the CCM. For tracks that traversed the full length of the spectrometer a momentum resolution of $\delta(p^{-1}) \sim 2 \times 10^{-5}\,\text{GeV}^{-1}$ was achieved, which gave resolutions of about 5% on x at low x and about 4% on Q^2. Muons were identified by four sets of wire planes (PTM) placed behind a 3 m iron absorber, with the sets separated by 1 m thick concrete absorbers.

EMcal. The electromagnetic calorimeter was placed just in front of the muon absorber and was a 20 radiation length lead-plate gas sampling device. In addition to measuring photons from neutral hadron decays, it also provided information on elastic muon–electron scatters.

Trigger. The large angle trigger required a beam muon and a scattered muon with hits in at least three of the four scintillator planes (SPM) of the muon detector after the absorber. The muon beam passes through a 20 cm × 20 cm hole in the SPM array, giving an angular cut of 3.3–4.7 mrad, which corresponds to $Q^2 = 2.7\text{--}5.5\,\text{GeV}^2$ for 500 GeV muons. In order to study DIS at smaller values of x than in earlier fixed target experiments, the trigger had to be extended to include events in which the scattered muon was within the envelope of the beam. This was achieved by a veto trigger based on a finely segmented scintillator array (SMS) placed in the beam immediately after the absorber. The small angle trigger was a combination of a beam muon and the absence of a signal in the SMS at the position predicted for that muon (without interaction). It was efficient down to angles as small as 1 mrad ($Q^2 = 0.5\,\text{GeV}^2$ for 500 GeV muons).

Flux. The integrated incident muon flux was measured by assuming that the beam spectrometer response was the same for random and physics triggers. The number of usable beam muons was determined by counting the number of random beam triggers with a good muon and multiplying by the pre-scale factor. The latter was determined by comparing the number of random beam triggers with the actual number counted by scalers.

Calibration. The calibration and resolution of the E665 spectrometers was checked using the following techniques. Primary protons at $800.6 \pm 2\,\text{GeV}$ (determined from the Tevatron magnet currents) were directed through the apparatus. The relative calibration of the beam and main spectrometers was checked using non-interacting muons and comparing the difference of the momenta measured in the two spectrometers in data and simulation.

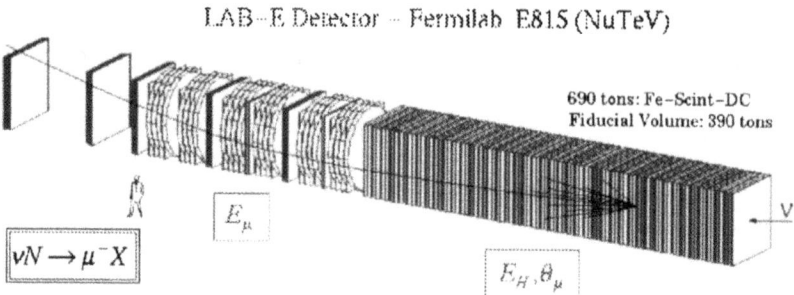

Fig. 5.4 The CCFR-NuTeV detector. Note that the incident neutrino beam is from the right.

This leads to an estimate of 0.13% relative error at the nominal muon beam energy of 470 GeV. Elastic muon–electron scattering events also provided information on the calibration and resolution of the spectrometer. Using signals from the EMcal, the events have a distinctive signature and are constrained by kinematics to have $x = m_e/m_N$.

5.3.2 The CCFR neutrino scattering experiment

The CCFR neutrino scattering experiment is shown schematically in Fig. 5.4. It was exposed to the Tevatron Quad-Triplet wide-band neutrino beam, which was composed of neutrinos with average energy 185 GeV and antineutrinos of average energy 143 GeV. The maximum beam energy was about 600 GeV and the ratio of $\nu : \bar{\nu}$ was about 2.5 : 1. The CCFR detector (Sakumoto et al. 1990, King et al. 1991) consists of a 17.7 m long 690 ton unmagnetized steel-scintillator target calorimeter, which is instrumented with drift chambers for muon tracking. The hadronic energy resolution of the calorimeter was $\sigma/E = 0.85/\sqrt{E(\text{GeV})}$. The calorimeter energy scale was calibrated to 1% using momentum analysed hadron beams with energies between 15 and 450 GeV. The target is followed by a 17.8 m long solid iron toroidal magnetic spectrometer for muon identification and momentum measurement. The spectrometer was calibrated to about 0.5% using a momentum analysed muon beam with energies of 50, 75, 120 and 200 GeV. The muon momentum resolution is $\Delta p/p = 0.11$ and it is limited by multiple Coulomb scattering in the iron. The detector provides measurements of the visible hadronic energy E_{had}, the momentum, p_μ, and angle with respect to the neutrino beam line, θ_μ, of the scattered muon. The relative neutrino flux at different energies and the relative $\bar{\nu}/\nu$ flux are obtained from events with low hadron energy, $E_{\text{had}} < 20$ GeV. The absolute normalization is determined so that the total νN cross-section equals the average value for Fe from the CHDSW and CCFR experiments of $\sigma^{\nu N}/E = (0.677 \pm 0.014) \times 10^{-42} \text{ m}^2\text{GeV}^{-1}$ per nucleon and

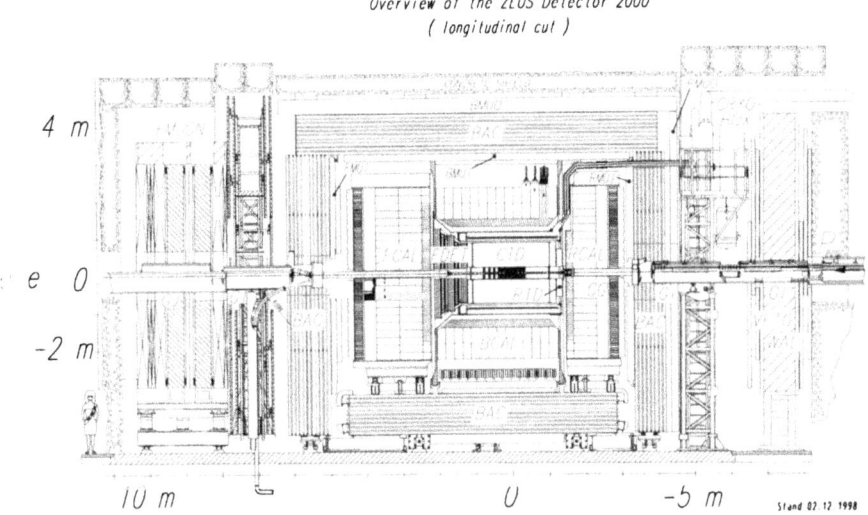

Fig. 5.5 The ZEUS detector.

$\sigma^{\bar{\nu}N}/\sigma^{\nu N} = 0.499 \pm 0.005$.

5.3.3 Detectors for the ep collider HERA

A cross-section of the ZEUS detector is shown in Fig. 5.5. The first thing to note is the very different layout compared to a fixed target detector. The ZEUS and H1 detectors are designed to measure as much as possible of all final states in ep collisions at HERA, which means NC and CC DIS processes and almost-real photon interactions. Although there is a large disparity in the proton (920 GeV) and $e^{\pm}$ (27.5 GeV) beam energies, the γ factor of the ep CM system in the HERA frame is only 2.98. The proton remnant is thrown forward, but the scattered beam $e^{\pm}$ and other products of the hard interaction (the quark or 'current' jet system) will emerge at larger angles. If the detectors are to have good efficiency for reconstructing the final state and to be able to measure CC DIS with a missing neutrino then the sensitive components must be nearly hermetic, particularly the calorimeters. The HERA detectors differ from e^+e^- or $p\bar{p}$ detectors in being asymmetric, both H1 and ZEUS have deeper calorimeters in the proton direction and both have sophisticated forward muon detectors. The main differences between the ZEUS and H1 detectors are in the size of the solenoidal coils for the central tracking systems and the calorimeter technology. ZEUS has a 2 m diameter coil providing a 1.43T field with the calorimeter outside the coil, H1 has a 6 m diameter coil giving a 1.15T field with the lead liquid-argon calorimeter inside the coil. The main components of the H1 detector are shown in the event display of Fig. 5.6.

The ZEUS calorimeter is a compensating uranium-scintillator sandwich design (UCAL) which gives an equal response to electromagnetic and hadronic particles of the same energy and is thus optimized for measuring hadronic jets at large Q^2. The polar angle coverage is 2.6°–176.2°. Under test beam conditions the energy resolution is $\sigma_E = 0.35\sqrt{E(\text{GeV})}$ for hadrons and $\sigma_E = 0.18\sqrt{E(\text{GeV})}$ for electrons. The UCAL also provides a time resolution of better than 1 ns for energy deposits greater than 4.5 GeV, which is used for background rejection. The H1 liquid argon calorimeter (LAR) is used to measure the hadronic final state and scattered electrons for $Q^2 > 120\,\text{GeV}^2$. The LAR covers the angular range 3°–155° and has an EM section with lead absorber plates of 20–30 radiation lengths and a hadronic section with steel absorber plates giving a total depth of 4.5–8 interaction lengths. For EM particles the energy resolution is $12\%/\sqrt{E(\text{GeV})}$, and for hadrons $50\%/\sqrt{E(\text{GeV})}$. In the rear direction, which is important for low Q^2 ($< 120\,\text{GeV}^2$), H1 has a lead-scintillating fibre calorimeter (SPACAL) covering the angular range 153°–178.5° with high granularity giving an angular resolution of 1–2 mrad. Energy resolution, measured during experimental operations, is $7.5\%/\sqrt{E(\text{GeV})} \oplus 2.5\%$.

Outside the main calorimeters, both experiments have instrumented iron to catch energy leaking from the calorimeters and to provide absorbers before the barrel muon chambers.

H1 and ZEUS have extensive tracking systems inside the main calorimeters and within the magnetic volume. Around a thin beam pipe are silicon microstrip vertex detectors, then cylindrical drift chambers (CTD). For a full length track in the ZEUS CTD, the resolution in transverse momentum is $\sigma_{p_T}/p_T = 0.005 p_T \oplus 0.016$ (for p_T in GeV). In the forward direction there are further drift chambers and straw tube chambers for charged particle tracking. The vertex detectors consist of two or three layers of microstrip modules around the beampipe and three or four 'wheels' in the forward direction. Typical individual hit resolutions are better than 20 μm. Apart from providing a large improvement in momentum resolution and primary vertex position, the vertex detectors improve the efficiency for the identification of charm and beauty states. Such states are recognized on an individual event basis by looking for secondary vertices displaced from the primary 'event vertex' or statistically by looking for an asymmetric distribution of the distance of closest approach of charged particle tracks to the measured interaction point in the event.

Because of the interest in measuring the behaviour of F_2 or the total $\gamma^* p$ cross-section as $Q^2 \to 0$, both ZEUS and H1 have extended their detectors to be able to measure θ_e very close to 180°. For example the ZEUS beam pipe calorimeter (BPC) is located at 2.9 m from the interaction point adjacent to the beam pipe on the (rear) electron side and covers $e^\pm$ scattering angles in the range 15–43 mrad. It is a small tungsten-scintillator sampling calorimeter with energy resolution $\Delta E/E = 17\%/\sqrt{E(\text{GeV})}$ and was installed in ZEUS in 1995. Later, the BPC was augmented by two planes of silicon microstrip tracking detectors — the beam pipe tracker (BPT). This improved the acceptance for very low angle scattered positrons and the BPC/BPT combination gave an angular resolution of $\Delta\theta_e = 0.2$ mrad. In addition, data have been taken with special runs of the HERA collider in which the primary interaction vertex is shifted by 70 cm in the proton beam direction. The interaction is thus further from the rear calorimeters which are used to identify the scattered electron and hence smaller angles can be reconstructed. The regions in the x, Q^2 plane at the very small values of Q^2 reached by these techniques are shown in Fig. 5.1. These detectors were removed in 2000 as they are incompatible with the extra final focus magnets inserted for the HERA upgrade.

The interval between bunch crossings in HERA is 96 ns which necessitates sophisticated multi-level trigger systems. For ZEUS and H1, at the first level data is stored temporarily ('pipelined') while hardware specific trigger processors, operating synchronously with HERA beam crossings, arrive at a decision in about 2–5 μs. The higher trigger levels operate asynchronously and involve more sophisticated calculations, culminating in an 'event filter' which uses a fast version of the offline reconstruction code running on a 'farm' of dedicated processors. The overall reduction achieved is from a raw interaction rate of many 100 kHz (which is very sensitive to beam conditions in HERA) to about 5–10 Hz written to tape.

At HERA, ep luminosity is measured using the Bethe–Heitler process, $ep \to ep\gamma$, which has a large and accurately known cross-section. Photons emerging from the electron–proton interaction point (IP) at angles $\theta_\gamma \leq$ 0.5 mrad with respect to the electron beam axis hit the photon calorimeter at 107 m (103 m) from the IP for ZEUS (H1). Electrons emitted from the IP at scattering angles less than or equal to 6 mrad and with energies $0.2E_e < E'_e < 0.9E_e$ are deflected by beam magnets and hit the electron calorimeter placed 35 m (33 m) from the IP. The systematic error on the luminosity measurement is typically less than 2%.

5.4 Measurement of the cross-section

The first step in the measurement of the double differential cross-secton is to identify the DIS events. The problems faced by fixed target and HERA experiments are somewhat different. In the case of a fixed target muon scattering experiment, the event selection criteria fall into two categories. First, those based on information from the beam spectrometer to select events with a single well measured muon on the target. Second, those based on information from the main detector which define a well measured scattered muon originating from the fiducial volume of the target. Typical cuts require a minimum momentum (p_μ) and a minimum energy energy loss ($\tilde\nu$). Further cuts are then applied to ensure that the events lie in the 'well-measured' region of kinematic plane as discussed in Section 5.2 above.

At HERA the signature of a DIS NC event is a scattered beam electron recorded in the EM section of the calorimeter balanced in transverse

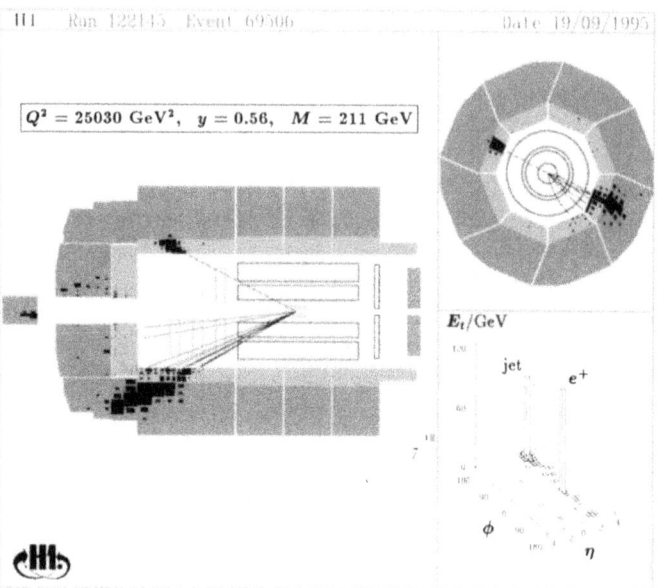

Fig. 5.6 A DIS NC event in the H1 detector at HERA. The figure also shows the main active components, central and forward tracking detectors as open rectangles surrounded by the liquid argon calorimeter. At the rear is the SPACAL and in the forward beam hole is the plug calorimeter.

momentum by the hadronic jet or jets. An example of a high Q^2 NC event in the H1 detector is shown in Fig. 5.6. For this type of event the electron and hadron energy deposits are well separated and there is little chance of confusion in electron identification. At low x and Q^2, both the scattered electron and the hadronic system emerge at small angles to the rear beam pipe and may well overlap in the calorimeter. In such events track reconstruction may also be difficult. At HERA, there are two large backgrounds to be overcome. The first is from beam–gas interactions, mainly beam protons striking rest gas molecules in the long straight section leading up to the interaction region. This background is reduced very effectively by precise timing — a genuine ep scattering event will give signals at roughly the same time in all parts of the detector whereas an upstream p-gas interaction will produce debris which hits the rear detectors before the forward sections. Good reconstruction of the the primary vertex position along the beam line by the tracking system is also important for reducing the level of beam–gas background. The second large background is from fake electron signals in photoproduced events. This happens because the photoproduction cross-section is large and in some events an overlap of a π^0 and a charged particle in the final state may produce an EM cluster with a track pointing at it. In principle, the size of this effect can be reduced by good granularity in the EM calorimeter. A variable that is effective in reducing the photoproduction background is

$$\delta = \sum_i (E_i - E_i \cos\theta_i) = \sum_i (E - p_z)_i, \qquad (5.16)$$

where the sum runs over all calorimeter deposits E_i with polar angle θ_i. From conservation of longitudinal momentum, in an ideal detector, δ sums to twice the electron beam energy $2E_e$. For photoproduction background events, δ peaks well below $2E_e$ because the genuine scattered electron remains in the beam pipe. In a real detector, the deep inelastic scattering peak at $2E_e$ is smeared out by finite resolution and becomes asymmetric because of initial state radiation. For $E_e = 27.5\,\text{GeV}$, typical cuts on δ are $38 < \delta < 65\,\text{GeV}$.

Once the DIS sample has been selected and the corresponding luminosity measured, a number of corrections have to be applied to give the Born level cross-section.

5.4.1 Radiative corrections

A full discussion of all electroweak radiative corrections is beyond the scope of this book, only the important features of electromagnetic corrections for neutral current reactions will be covered here. The corrections are of two sorts, virtual corrections and the infrared part of real photon emission which give an overall modification to the Born cross-section and non-infrared real photon emission in which the photon is not identified. The radiative process $e(k)p(p) \to e(k')\gamma(\ell)X$ (symbols in brackets indicate the corresponding 4-momenta) is dominated by photon emission from the initial and final lepton. In principle, there is also interference between the two. Consider first the situation when the kinematics are reconstructed using the scattered lepton only and the radiation is from the initial lepton (ISR). The standard definitions of the kinematic variables are (where ℓ indicates that they have been calculated from the lepton side),

$$Q_\ell^2 = -(k-k')^2 \qquad x_\ell = \frac{Q_\ell^2}{2p \cdot (k-k')} \qquad y_\ell = \frac{p \cdot (k-k')}{p \cdot k}. \qquad (5.17)$$

The effect of the radiation of a photon of 4-momentum l, is to lower the actual centre of mass energy in the ep interaction from $s = (k+p)^2$ to $s' = (k-l+p)^2$. The true kinematic variables are correctly reconstructed from the hadron side

$$Q_h^2 = -(k-k'-l)^2 \qquad x_h = \frac{Q_h^2}{2p \cdot (k-k'-l)} \qquad y_h = \frac{p \cdot (k-k'-l)}{p \cdot (k-l)}. \qquad (5.18)$$

For collinear ISR, the lepton side and true variables are related by

$$Q_h^2 = z_i Q_\ell^2 \qquad x_h = \frac{x_\ell y_\ell z_i}{y_\ell + z_i - 1} \qquad (5.19)$$

where z_i is the fractional energy loss from the initial state lepton $z_i = (E - E_\gamma)/E$. Similar expressions may be deduced for the case of collinear radiation from the final state lepton (FSR). The size of the radiative contributions is characterized by $\ln(m_\ell^2/Q^2)$ where m_ℓ is the electron or muon mass, and are thus generally more severe for ep than μp scattering. For a measurement at (x_ℓ, y_ℓ), the radiative correction is calculated by integrating over the region $x_\ell \leq x \leq 1$, $0 \leq y \leq y_\ell$ which implies prior knowledge of the cross-section being measured.

The procedure adopted for fixed target experiments is to use a phenomenological parameterization of $F_2(x, Q^2)$ chosen to give a good representation of the data over a wide range of Q^2 and particularly at very small values. The radiative corrections are then estimated using the parameterized F_2. As the data from the experiment will be used in the parameterization, the whole procedure is iterated. One of the biggest uncertainties comes from lack of knowledge of R, the ratio of cross-sections for longitudinally and transversely polarized virtual photons. The radiative corrections are sizeable for large y and small x, for example in the lowest x bins of the E665 experiment the corrections reach 40% at the largest Q^2 values.

At HERA, even though the lepton beam is composed of electrons or positrons, the radiative corrections turn out to be less severe. There are two reasons for this. The first is that since the scattered electron is measured by calorimetric methods the energy lost through FSR will usually be deposited in the same calorimeter cell as the electron itself. The second is that kinematic information is measured from both the electron and hadron sides. The variable δ introduced in the previous section plays an important role. In the case of ISR, the energy of the electron in the ep interaction is less than E_e by the amount E_γ and thus δ will sum to $2E_e - 2E_\gamma$. The requirement of a minimum value for δ, used to reduce photoproduction background, will also remove ISR events and limit the size of the radiative correction. Over the region of the (x, Q^2) plane used for F_2 measurements at HERA, radiative corrections do not exceed 10% and are typically a factor two smaller.

The experiments H1 and ZEUS at HERA have also used initial state radiative events as a way of accessing ep scattering at lower centre of mass energies than that given by the nominal values of the HERA beam energies. This is done by using the luminosity detectors to tag ISR photons giving, in effect, events from a 'broad band' electron beam with energies $E_e^{ISR} = E_e - E_\gamma^{lumi}$. The interest in these events is that it allows the kinematic reach of HERA to be extended at moderate Q^2 to overlap with the fixed target region. Referring to Fig. 5.1 the region is roughly between $Q^2 = 0.3$ and $10\,\text{GeV}^2$, x between 10^{-5} and 0.01, filling the 'hole' between the shifted vertex data and the fixed target region. In principle, the data also offer the possibility of measuring F_L (see Section 5.6.3 below). The main problem with this data is the large background from overlaid $ep \rightarrow ep\gamma$ bremsstrahlung events.

5.4.2 Acceptance and other corrections

The final step to get from binned numbers of events, $N(x, Q^2)$, to the cross-section is the calculation of the correction function $A(x, Q^2)$ defined in Eq. (5.1). This is calculated by Monte Carlo simulation in two stages: first DIS events are generated and second the simulated events are passed through a simulation of the detector. Here the procedure for HERA experiments will be outlined.

The event generator is based on a full simulation of the NC or CC matrix elements including electromagnetic radiative effects. For the hadronization of the final state quarks and gluons, procedures using either the Lund string approach or angular ordering as implemented in HERWIG are used.[4] The output of the event generator is a set of 4-vectors describing the 'stable' final state particles (that is particles with lifetimes of order 10^{-8} s and longer), together with a complete history of the event. The event is then input to the detector simulation. Many simulations of high energy physics detectors are based on the GEANT package developed at CERN. It is a suite of programs that allow detector elements with different geometrical shapes (planes, cylinders etc) to be put together in a consistent way. The program also takes care of transporting the particles through the geometry and material of the detectors, allowing for the different interactions of electromagnetic particles, hadrons and charged leptons. The response of each type of detector (calorimeter, drift chamber, etc) is simulated to a sufficient level of detail to model the actual signals. At each stage of the detector simulation, the history of the interactions is recorded so that it is possible to relate the 'signals' to the input particles from the event generator. The final output of the complete DIS event simulation consists of three parts: the 4-vectors of the physics event, the simulated detector signals in the same format as a real event and a history record linking the two.

An important check of the whole process of simulation is to demonstrate that the Monte Carlo data does describe the key primary measured variables. Some examples of 'control plots' are shown in Fig. 5.7 (the data are from a ZEUS medium Q^2 e^+p NC cross-section measurement). The number of events generated is matched to the luminosity of the data sample. The overall level of agreement between simulation and data is good. Plots (a) and (b) show the energy and angle of the scattered positron and (c) Z_{vertex} the position of the primary event vertex along the beamline. An accurate description of this last distribution is particularly important for a correct calculation of the geometrical acceptance, as the NC cross-section is a very rapidly varying function of angle. The shape of the vertex distribution is typically measured from an unbiased sample of general ep interactions and it is dependent on the beam optics of the HERA collider. Plot (d) shows the ratio of hadronic transverse momentum to that of the electron for a low range of the hadronic angle γ (defined by Eq. (5.10)). It is a control plots

---
[4]More details on a range Monte Carlo packages are given in Appendix E.

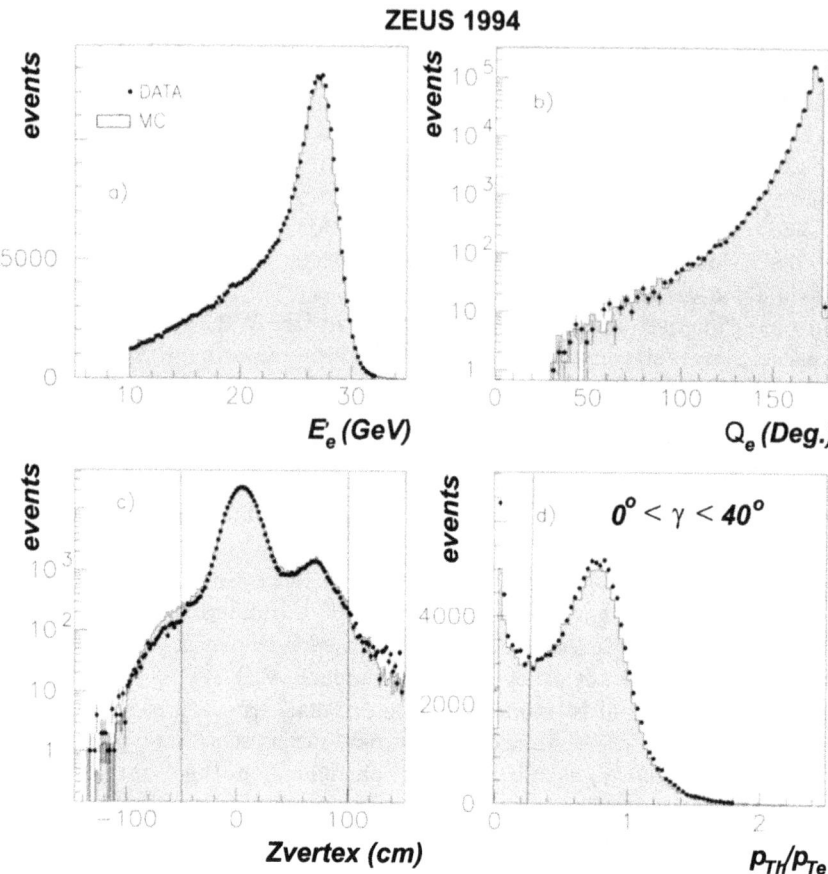

Fig. 5.7 Examples of control plots: (a) E'; (b) θ_e; (c) Z_{vertex}; (d) p_T^h/p_T^e for a range of the hadronic angle γ. The vertical lines in plots (c) and (d) show the positions of selection cuts. The Monte Carlo distributions are normalized to the integrated luminosity of the data. (From ZEUS 1996.)

for the PT kinematic reconstruction procedure (see the discussion following Eq. (5.13) above).

The Monte Carlo data is used to provide information on a range of indicators that are used to define the bins in x and Q^2 into which the data is sorted. The basic idea is to choose bins of a size commensurate with the local resolutions of x and Q^2 to avoid large corrections for data migrating from one bin to another. At very large Q^2 bins may need to be larger than the resolution would indicate simply to reduce statistical uncertainties from finite luminosity and very small cross-sections. The indicators that are used are:[5]

Acceptance — this is the ratio of the number of events generated in a bin and passing the selection criteria (N_{MC}) to the number of events generated in the bin (N_{true}). Typically one should expect a high figure (80% or better) except near the edges of phase space or near the limit of detector acceptance (for example the minimum measurable scattering angle). The minimum acceptance was 20%.

Purity — this is the ratio of the number of events generated and measured in a bin to the number of events measured in the bin. It should be better than 50% for most of the kinematic region, better at high Q^2 ($> 1000\,\mathrm{GeV}^2$) and worse at low values ($< 10\,\mathrm{GeV}^2$). The minimum purity was 30%.

Efficiency — this is the number of events measured in a bin to the number of events generated in the bin. It should be high (95% or better) for most of the kinematic region, being significantly lower only for very large ($> 10^4\,\mathrm{GeV}^2$) or very small ($< 10\,\mathrm{GeV}^2$) Q^2. The minimum efficiency was 30%.

Bins that have indicators below the minimum values are excluded from the final measurement. The advantage of defining and selecting bins in this way is that a very straight-forward overall correction or 'unfolding' procedure may be used — the so-called bin-by-bin method. Returning to Eq. (5.1) at the start of this chapter and noting that the acceptance $A = N_{\mathrm{MC}}/N_{\mathrm{true}}$, the equation may be rewritten as

$$\frac{d^2\sigma}{dx\,dQ^2} = \frac{N(x_m, Q_m^2)}{N_{\mathrm{MC}}} \left[\frac{N_{\mathrm{true}}}{\langle \mathcal{L} \rangle \Delta x \Delta Q^2} \right],$$

where x_m, Q_m^2 are the measured values of the variables. Now the ratio of factors within the square brackets on the right-hand side is none other than the theoretical cross-section with which the MC events were generated, thus

$$\left.\frac{d^2\sigma}{dx\,dQ^2}\right|^{\mathrm{meas.}} = \frac{N_{\mathrm{data}} - N_{\mathrm{bkgnd}}}{N_{\mathrm{MC}}} \left.\frac{d^2\sigma}{dx dQ^2}\right|^{\mathrm{theory}},$$

where $N(x_m, Q_m^2) = N_{\mathrm{data}} - N_{\mathrm{bkgnd}}$ relates the number of genuine events from those measured in the bin minus the background (for NC cross-sections this is small and comprises residual mis-identified beam–gas and photoproduction events). A modified form of the above equation is frequently used, in which the double differential cross-section is replaced by the dominant structure function, F_2^{em}, together with a number of numerical small calculated correction factors

$$F_2^{em} = \frac{Q^4 x Y_+}{2\pi\alpha^2} \frac{d^2\sigma}{dx dQ^2} \left[1 + \delta_{RC} + \delta_{F_L} + \delta_Z\right], \qquad (5.20)$$

---

[5]The numbers quoted are taken from the high statistics ZEUS measurement of NC cross-sections and F_2, ZEUS (2001a).

where δ_{RC}, δ_{FL}, δ_Z are the corrections for radiative effects, the longitudinal structure function and the contribution of Z^0 exchange at large Q^2.

The alert reader will have noticed a serious flaw in this procedure — at the centre of the event simulation is the cross-section that is being measured. This is overcome by iteration using

$$F^{i+1}(x,Q^2) = \frac{N_{\text{data}} - N_{\text{bkgnd}}}{N^i_{\text{MC}}(x_m,Q^2_m)} F^i(x,Q^2)$$

to improve the correction, where the distribution of simulated events $N^i_{\text{MC}}(x_m,Q^2_m)$ has been calculated from $F^i_2(x,Q^2)$. The F^{i+1} values are then used to re-weight the Monte Carlo events for the next iteration. The procedure is repeated, usually only two to three times, until a convergence criterion is satisfied (typically that the change in F is less than 0.2–0.5%). The input for the first iteration is F^0_2 calculated using a recent set of parton density functions, for example, from the CTEQ or MRST teams (see Appendix G).

5.5 Nuclear effects

The study of DIS from nuclear targets is an interesting subject in its own right and it has provided some surprises, for example, the 'EMC effect'. In recent years, there has been a revival of interest because of high energy heavy-ion collision experiments and related new semi-classical approaches to QCD for regimes with very high gluon densities. Also, it is becoming possible to study the propagation of quarks and gluons in extended hadronic matter with sophisticated modern detectors. The aim of this section is limited to a brief summary of the relationship between nuclear and nucleon structure functions and how this information may be used to correct structure function data from nuclear targets to that for a single nucleon.

The nuclear structure function $F^A(x,Q^2)$ is defined *per nucleon*, so that if there were no nuclear effects at all the ratio of F^A to that for a single isocalar nucleon would be unity at a given x and Q^2. Nuclear effects in DIS are encapsulated by studying the ratio of structure functions from nucleus A to that of deuterium, $R^A(x) = F^A_2(x)/F^d_2(x)$, as a function of x. To a good approximation, the effects show little dependence on Q^2. The data for $R^A(x)$ is summarised in Fig. 5.8. The lower plot shows the data, averaged over Q^2, from experiments on a variety of nuclear targets as a function of x. The curve in the upper plot shows the trend of the data with the various regions delineated and described.

Before the EMC experiment of 1982, it was assumed that nuclear effects would only be seen at low x where nuclear shadowing occurs and as $x \to 1$ where additional smearing from nuclear Fermi motion is expected. The EMC experiment showed that for x roughly in the range 0.3–0.8, R^A was significantly less than 1. Both the NMC and E665 experiments were designed in part to study R^A in detail. Special attention was paid to target design to reduce systematic errors in the measurement of cross-section

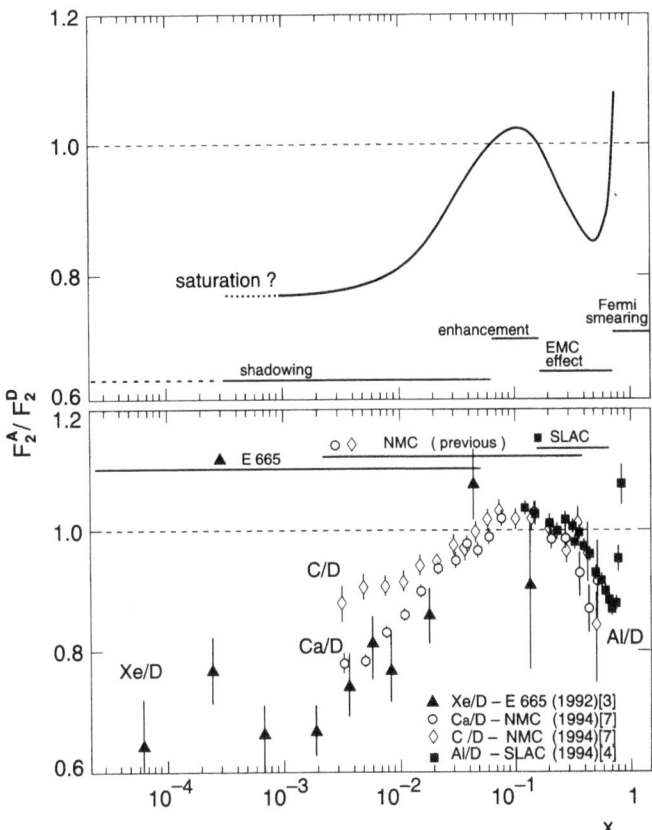

Fig. 5.8 The structure function ratio $R^A(x) = F_2^A(x)/F_2^d(x)$ as a function of x. Upper plot: the trend of the data; lower plot ratio data for a variety of nuclei. (From Arneodo et al. 1996.)

ratios. There are many models for the EMC effect but most fall into one of two categories: nuclear models in which the partons are unchanged but nuclear binding effects give rise to a rescaling in x; models in which the quark confinement size in nuclear matter is larger than for a free nucleon giving a decrease in the Fermi momentum and leading to a rescaling in Q^2. The subject of the EMC effect is well covered in Chapter 8 of Roberts (1990) and in the review article by Arneodo (1994). A number of parameterizations of $R^A(x)$, giving very good representations of the data are available for use in global fits to determine parton distribution functions (this point will be discussed further in Section 6.2).

5.6 Structure function data

The key parameters of DIS experiments are summarised in Table 5.1. For most of those listed, the measurement of the NC cross-section has been essentially that of F_2^{em} (i.e. without Z^0 contributions) up to small corrections and unless otherwise stated, F_2 will be taken to mean F_2^{em} for the proton. Although a number of other DIS experiments are of great historical importance, the experiments considered here are the ones that provide the bulk of the data for the determination of parton densities. References to the original measurements and other information on how to get access to structure function data are given in Appendix F.

The double differential differential neutral current cross-section at high energy is expressed in terms of the structure functions by[6]

$$\frac{d^2\sigma(\ell^\pm N)}{dx\, dQ^2} = \frac{2\pi\alpha^2}{Q^4 x} \left[Y_+ \, F_2^{lN}(x, Q^2) - y^2 \, F_L^{lN}(x, Q^2) \mp Y_- \, xF_3^{lN}(x, Q^2) \right], \quad (5.21)$$

where $Y_\pm = 1 \pm (1-y)^2$ and mass terms have been ignored. For Q^2 values well below that of the Z^0 mass squared, the parity violating structure function xF_3 is negligible and the structure functions F_2, F_L are given purely by γ^* exchange. This is true of all fixed target data and for HERA data with $Q^2 < 1000\,\text{GeV}^2$. For $Q^2 > 1000\,\text{GeV}^2$, the contributions from Z^0 exchange cannot be ignored and in the extraction of F_2^{em} from the cross-section data the Z^0 exchange terms are removed using a calculated correction based on a recent set of parton densities. For data at large x and small Q^2 (e.g. the SLAC data) some terms involving the nucleon mass become non-negligible and Eq. (5.21) becomes

$$\frac{d^2\sigma(\ell^\pm N)}{dx\, dQ^2} = \frac{4\pi\alpha^2}{Q^4 x} \left[1 - y - \frac{m_N^2 x^2 y^2}{Q^2} + \frac{y^2}{2} \frac{1 + 4m_N^2 x^2/Q^2}{1+R} \right] F_2^{em}(x, Q^2), \quad (5.22)$$

written now in terms of F_2 and R. R, the ratio of the longitudinally to transversely polarized virtual-photon absorption cross-sections (σ_L/σ_T), is given in terms of the structure functions by $R = F_L/(F_2 - F_L)$ (ignoring mass terms). To disentangle F_2 and F_L or R at fixed (x, Q^2) requires data collected at different centre of mass energies. This has been done by most of the fixed target experiments and more details are given in Section 5.6.3. For HERA data, which has been collected at essentially a fixed CM energy, F_L is usually a calculated correction. In fact for most of the experiments considered here, the effect of F_L on the cross-section is small because it appears multiplied by y^2, which is a small quantity. However, given the intrinsic accuracy of the structure function data, a consistent treatment of R is important and the lack of such has caused apparent inconsistencies between data sets in the past.

---

[6]More details on the derivation of this expression are given in Appendix C and Chapter 8.

Structure function data

Table 5.1 Key parameters of DIS experiments.

Beam(s)	Targets	Experiment	Q^2 (GeV^2)	x	R	Process
-	p,d,A	SLAC	0.6 – 30	0.06 – 0.9	✓	NC
-	p,d,A	BCDMS	7 – 260	0.06 – 0.8	✓	NC
-	p,d,A	NMC	0.5 – 75	0.0045 – 0.6	✓	NC
-	p,d,A	E665	0.2 – 75	$8 \cdot 10^{-4}$ – 0.6	-	NC
$\nu, \bar{\nu}$	Fe	CCFR	1. – 500.	0.015 – 0.65	✓	CC
$e^{\pm}$, p	-	H1	$0.35 - 3 \cdot 10^5$	$6 \cdot 10^{-6}$ – 0.65	✓	NC,CC
$e^{\pm}$, p	-	ZEUS	$0.045 - 3 \cdot 10^5$	$6 \cdot 10^{-7}$ – 0.65	-	NC,CC

The regions in (x, Q^2) covered by the experiments are shown in Fig. 5.1, the Q^2 scale is that of the HERA collider. Because of the very large CM energy available at HERA, new regions of phase have been opened up at both large Q^2 and small x. The fixed target experiments occupy the lower right hand corner. For all experiments, there is broad correlation between x and Q^2 — large x is reached only at large Q^2. The constraints outlined in Section 5.2 are also evident in the figure. For a given experiment, the region at low Q^2 is limited by the angular acceptance for the scattered beam lepton. Worsening resolution at large x (small y) accounts for the limited overlap of the regions covered by the fixed target and HERA experiments. This is particularly true if the electron method is used for kinematic reconstruction. At HERA, other methods, such as the double angle and sigma methods (see Section 5.2.2), have allowed a significant extension to smaller y, giving a larger region of overlap. At large y, the region important for F_L measurements, the limit is set by the minimum value of the scattered $e^{\pm}$ that can be identified.

5.6.1 F_2

An overview of the proton data for F_2 is shown in Fig. 5.9, data are plotted at fixed x as functions of Q^2. The values of x range over five orders of magnitude and those for Q^2 over four. For x bins larger than 0.01, the data show only a weak dependence on Q^2. This is Bjorken scaling, as expected from the simple quark–parton model (Chapter 2). For smaller values of x, F_2 rises quite strongly with Q^2. Another important feature is the consistency of the data from the different experiments, both fixed target and HERA collider. The good agreement in shape and normalization between the F_2 data sets is shown in more detail as a function of Q^2 in Fig. 5.10 for six bins of x. Although there is apparent Bjorken scaling over quite a range of x values, the precision of the F_2 data sets allows this to be studied in detail. Exact scaling in Q^2 is rather the exception than the rule. Of the x bins shown in Fig. 5.10, scaling is only evident for $x = 0.18$. In general, there is a systematic breaking of exact scaling, at large values of x, F_2 decreases as Q^2 increases and at small values of x, the structure function increases with Q^2. In both cases, the variation with Q^2 is approximately logarithmic, as expected from the QCD enhanced parton model at leading

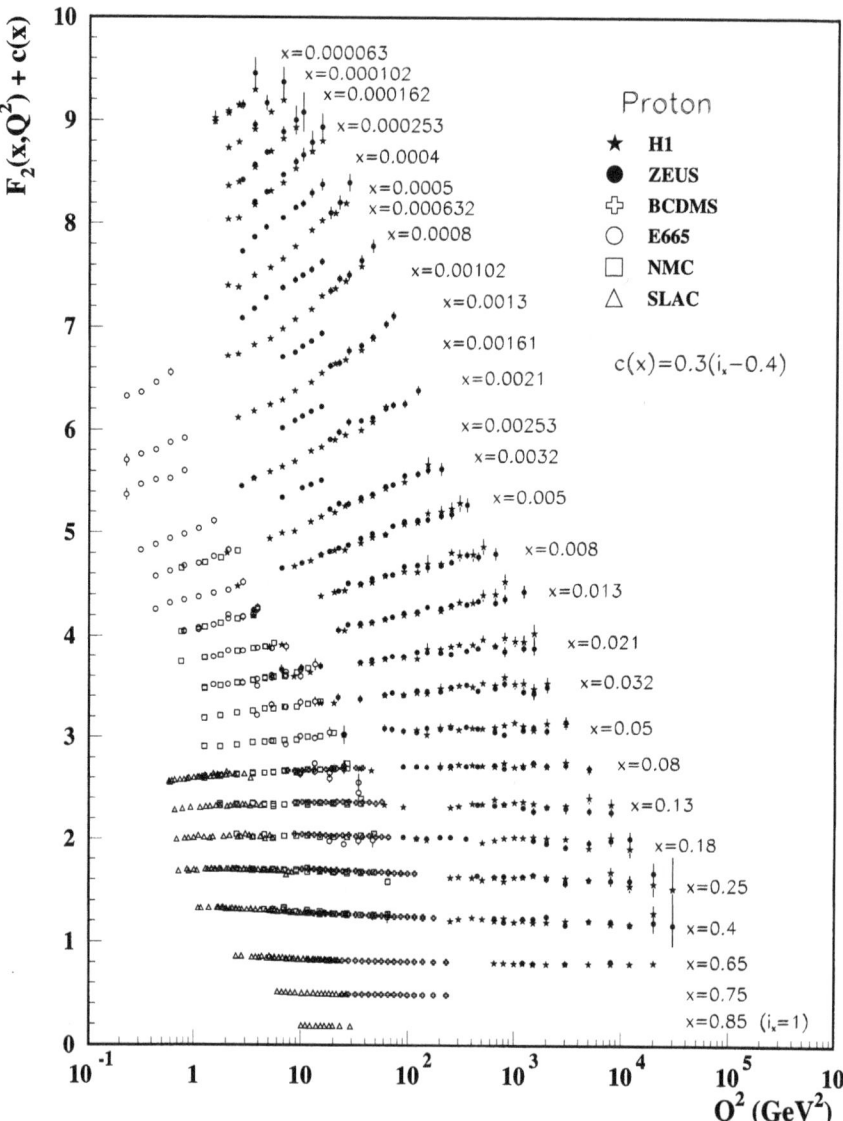

Fig. 5.9 F_2^p data from HERA and fixed target experiments at fixed x as a function of Q^2. Note that for clarity a constant $c(x) = 0.3(i_x - 0.4)$ has been added to F_2 where i_x labels the 28 bins x bins in decreasing order. The plot is from the PDG tables (PDG 2002) and full references to the data are given in Appendix F.

Structure function data   125

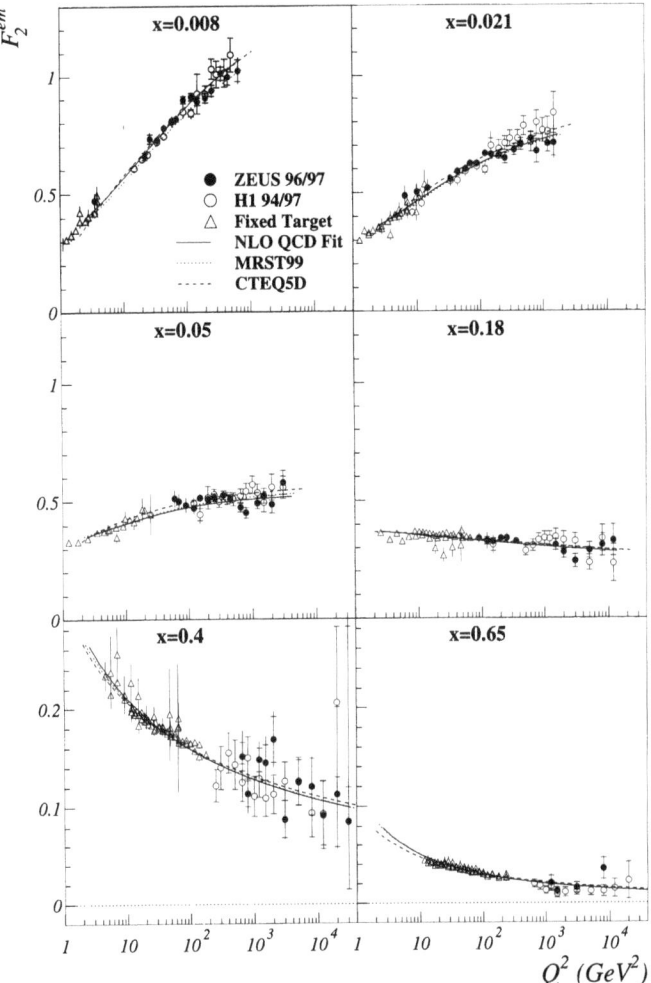

Fig. 5.10 F_2^p in six x bins showing the good agreement in shape and normalization between fixed target and HERA data. (From ZEUS 2001a).

order (Chapter 4). The data is now precise enough in some x bins to show that, even on the compressed scale of the figure, higher order terms are needed.

For most of the fixed target data and the HERA data for Q^2 up to almost $1000 \, \text{GeV}^2$, systematic errors are larger than statistical. Systematic errors are typically at the level of 2% or less and are dominated by uncertainties in absolute energy calibrations and beam flux or luminosity measurements. For the HERA data, F_L was calculated using the QCD ex-

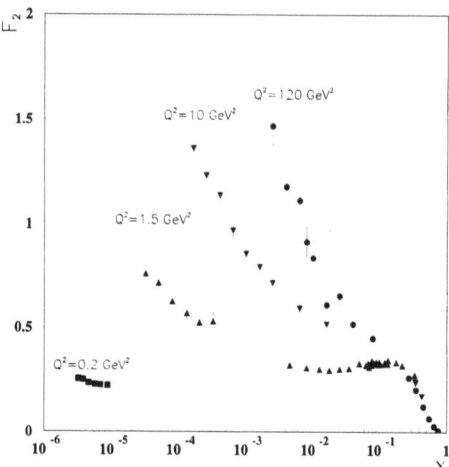

Fig. 5.11 F_2^p data showing the strong rise at small x. The HERA data are from the 1994 runs and the data for $Q^2 = 1.5\,\text{GeV}^2$ and $x > 10^{-3}$ (triangles) are from fixed target experiments. (From Abramowicz and Caldwell 1999).

pression (Eq. (4.43)) with a consistent set of parton densities. Most of the fixed target experiments made use of a parameterization of R produced in the course of reanalysis of the original DIS data from SLAC. However as R was usually measured in the same experiment, the direct measurement provided an alternative approach and certainly a check on the parameterization.

So far the data has been plotted in a way that illustrates immediately the essential prediction of pQCD, namely logarithmic breaking of scaling with Q^2. Perturbative QCD and particularly the standard DGLAP approach has much less to say about the x dependence of the structure functions. To illustrate one of the most striking features of the F_2 results from HERA, it is necessary to plot the data versus x at fixed Q^2. This is done for three representative values of Q^2 in Fig. 5.11 and it shows that F_2 rises steeply as x decreases and that the strength of the rise increases as Q^2 increases. The rise in x is strong, at $Q^2 = 120\,\text{GeV}^2$ F_2 increases by almost a factor of 3 as x decreases from 0.08 to 0.003. Possible explanations for this behaviour are discussed in Chapter 9. It is worth remarking that while a range of possible behaviours had been predicted for F_2 at small x, including a strong rise, when the first preliminary HERA data showing a rising F_2 was presented at the Durham Workshop in 1993, it caused great excitement. The data for $Q^2 = 1.5\,\text{GeV}^2$ and $x > 10^{-3}$ (triangles) in Fig. 5.11 shows why — before HERA the data showed a fairly flat behaviour as x

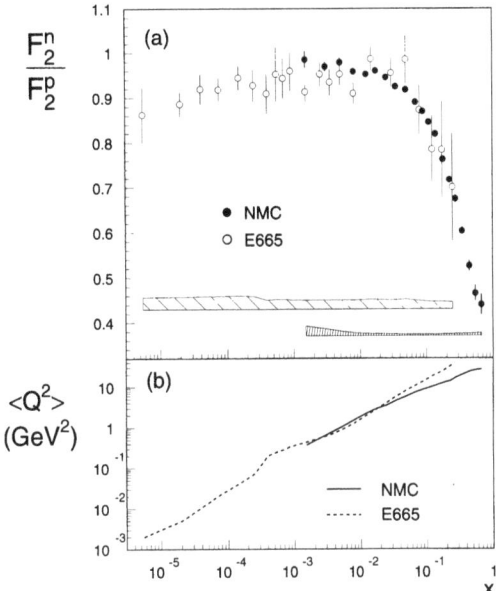

Fig. 5.12 (a) F_2^n/F_2^p averaged over Q^2 as a function of x from the NMC and E665 experiments. (b) $\langle Q^2 \rangle$ for the above data as a function of x. (From NMC 1997).

decreased with only the merest hint of a possible rise.

At very low Q^2, F_2 no longer rises steeply. Although it perhaps cannot be described as deep inelastic scattering, the behaviour of F_2 for very small Q^2 is very interesting and has been investigated in some detail at HERA. The measurements are summarised in Section 5.7.

A measurement that is of importance for the extraction of nucleon parton densities is the ratio F_2^n/F_2^p. Both the E665 and NMC experiments have deduced the ratio from primary measurements on the ratio of deuterium to proton cross-sections. A summary of the data, averaged over Q^2, is shown in Fig. 5.12(a), the lower part of the figure (b) shows how the average value of Q^2 varies with x. The size of the systematic errors are also indicated. The F_2^n/F_2^p ratio approaches 1, within errors, as x decreases and appears to be approaching 0.4 as x tends to 1.

5.6.2 $F_2^{\nu N}$ and $xF_3^{\nu N}$

Most DIS experiments with incident neutrino beams have been performed on heavy targets to increase the interaction rate. The CCFR/NuTeV ex-

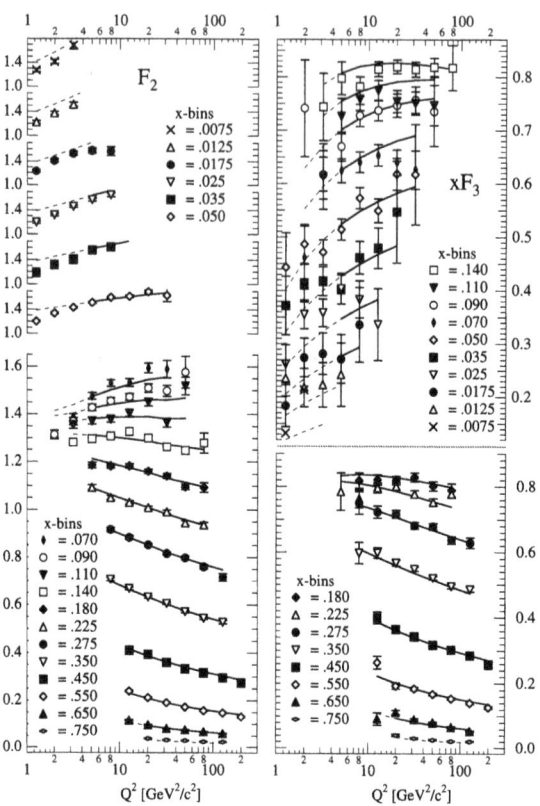

Fig. 5.13 CCFR(97) F_2 and xF_3 data as functions of Q^2 at fixed x. The results of a NLO QCD are also shown (full line), the dashed line is the extrapolation of the fit to lower Q^2. (From CCFR 1997.)

periment is the most recent and probably definitive such experiment to date. Because of the handedness of the $\nu, \bar\nu$ couplings, the parity violating structure function xF_3 can be determined as well as F_2. This is important because the scaling violations of xF_3 in QCD do not involve the gluon density and hence offer, in principle, a direct way of measuring α_S (see Section 7.1.1). CCFR data for F_2 and xF_3 have been measured for x between 0.0075 and 0.75 and Q^2 between 1.3 and 126 GeV^2 and are shown in Fig. 5.13. The major difference between this data and the other data considered in this chapter, is that significant nuclear corrections must be applied to the CCFR before it can be compared with DIS data from proton or deuterium targets or vice-versa. The $F_2^{\nu N}$ data (per nucleon) show a similar pattern of scaling violations to that already discussed. The curves shown in Fig. 5.13 are from an NLO QCD fit to determine α_S.

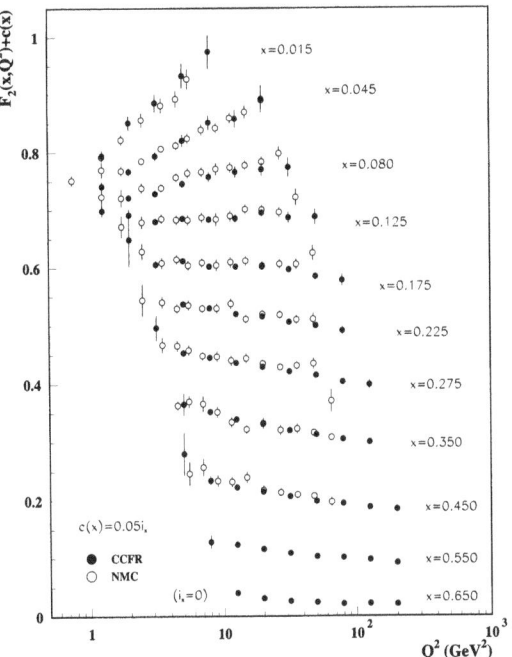

Fig. 5.14 Comparison of $F_2^{\mu d}$ from NMC and $F_2^{\nu d}$ derived from CCFR data using Eq. (5.23). For clarity a constant $c(x) = 0.05(i_x - 1)$ has been added to F_2 where i_x labels the 11 bins x bins in decreasing order. From PDG data tables (PDG 2002).

From the QPM, one expects $F_2^{\ell N} \approx \frac{5}{18} F_2^{\nu N}$ and this was an important early test of the model (Section 2.4). CCFR have used the relation (Eq. (2.51))

$$F_2^{\mu d} = \frac{5}{18} F_2^{\nu d} - \frac{1}{6}(s + \bar{s} - c - \bar{c}) \qquad (5.23)$$

to compare their data with that from NMC and other fixed target e and μ induced DIS experiments. Nuclear corrections are applied to the Fe target neutrino data to produce $F_2^{\nu d}$ equivalent data. Figure 5.14 shows the relationship between the neutrino and muon data are in good agreement over a wide range of x and Q^2. To achieve this level of agreement, matching the precision of the available data, requires careful treatment of the charm quark. Heavy quark effects are different in μ and ν induced DIS, a single charm quark may be produced from the strange sea $\nu s \to \mu^- c$ for example, whereas in μN $c\bar{c}$ pairs have to be produced through the boson–gluon fusion mechanism. These differences are particularly important in the charm threshold region spanned by the data. The analysis avoids model dependent assumptions on the behaviour of $\Delta x F_3$[7] by using the CCFR data and NLO massive quark expression for the charm threshold, the Thorne–Roberts formulation giving the best fit (Section 4.5.2). This brief outline shows both that the QCD enhanced parton model is in good shape and that the level of precision of structure function data demands careful attention to many details.

5.6.3 F_L

Although F_L or R is usually treated as a small correction in the extraction of F_2 from the cross-section, it is an important quantity because of its rather direct relation to xg — the momentum density of the gluon in the proton. Starting from Eq. (5.21) and ignoring xF_3 and mass terms, one may express the cross-section as

$$\frac{d^2\sigma}{dx\,dQ^2} = \frac{2\pi\alpha^2}{Q^4 x} Y_+ \left(1 - \frac{y^2}{Y_+} \frac{F_L}{F_2}\right) F_2(x, Q^2) \qquad (5.24)$$

$$= \frac{2\pi\alpha^2}{Q^4 x} Y_+ \left(\frac{1 + \varepsilon R}{1 + R}\right) F_2(x, Q^2), \qquad (5.25)$$

where in the second expression the polarization parameter of the virtual photon has been introduced, $\varepsilon = \dfrac{2(1-y)}{1+(1-y)^2}$ and $R = F_L/(F_2 - F_L)$. To measure R or F_L directly at a fixed value of (x, Q^2) requires data at two different values of y and hence at two different centre of mass energies (because of the constraint $Q^2 = sxy$). Since R is determined by the ratio of subtracted quantities, one needs very small statistical errors and very good control of the systematic errors at the two different energies. Another way of looking at the problem is to examine the 'leverage' from varying ε as a function of y, the largest variation in ε, for a given change in $\sqrt{s}$, is at large y. This, in turn, means being able to measure small scattered lepton energies (small E'), which is usually difficult. The fixed target experiments SLAC, BCDMS and NMC have used data taken at different values of s to extract R.

The HERA beam energies have been essentially constant throughout the period of HERA-I data taking (apart from the increase of the proton energy from 820 to 920 GeV), thus it has not been possible to make a direct measurement of F_L from data with 'standard' DIS triggers. As already mentioned, data with tagged ISR photons does allow the centre of mass energy to be varied for fixed (x, Q^2), but the measurement requires a very large integrated luminosity and very careful treatment of the large bremsstrahlung background.

The H1 collaboration (H1 1997a) have taken a different approach to estimate F_L from their large low Q^2 DIS neutral current data sample (H1

---

[7]The difference between the values for ν and $\bar{\nu}$ beams.

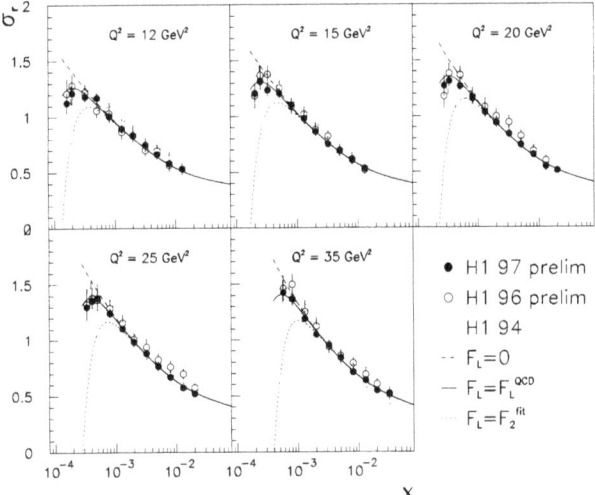

Fig. 5.15 The QCD subtraction method used by H1 to extract F_L. (From H1 1997a.)

2001c). The basis of the method follows from Eq. (5.24), which shows the suppression of the F_L at small y, and the fact that NLO QCD evolution describes the HERA F_2 data very well. F_2 is first determined by applying a NLO QCD fit to H1 data satisfying $y < 0.35$, that is in a region where the effect of F_L in the cross-section is small. The fit is then extrapolated to the high y (small x) region and used to subtract the contribution of F_2 from the cross-section at high y (≈ 0.7), as shown in Fig. 5.15. For this analysis, the minimum cut on E' was lowered from 6.9 to 3 GeV, thus extending the y range to $y \leq 0.89$. The identification of very low energy scattered positrons in the H1 SPACAL was made possible by associating charged particle tracks from the central tracker with SPACAL energy clusters. A summary of the resulting data on F_L from H1 and from fixed target experiments is shown in Fig. 5.16. Although the errors are large, the data are well consistent with the behaviour expected from QCD as shown by the curve in the figure which is from the H1 NLO QCD fit.

5.6.4 F_2^c

The structure functions studied so far are inclusive objects and thus contain contributions from both valence and sea quarks. The ability of a detector to provide identification of a particular quark flavour opens up the possibility of studying the contribution of that flavour to F_2. This is particularly interesting in the case of heavy flavours, as they are likely to be produced in the hard scattering and not in the subsequent hadronisation of the struck parton. It is estimated that charm could account for as much as 30% of F_2 for $x < 0.001$ and $Q^2 > 10\,\text{GeV}^2$. At order α_S heavy quark production in

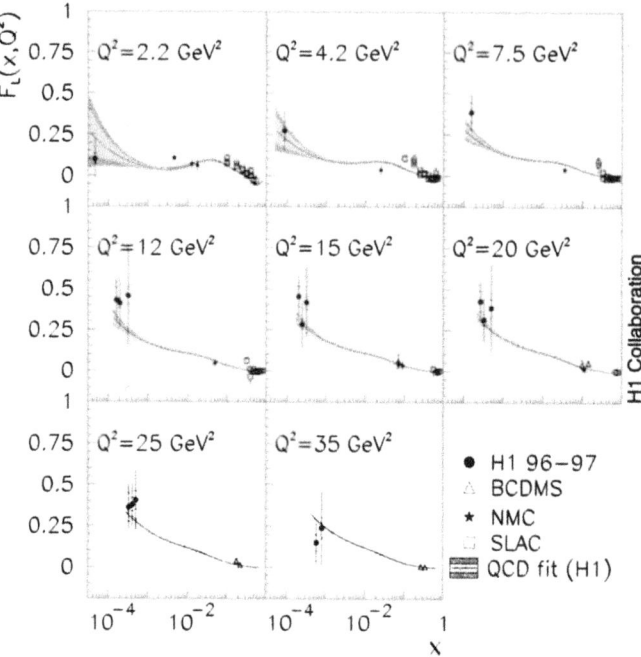

Fig. 5.16 F_L as a function of x in bins of Q^2 as obtained by H1 using the QCD fit and extrapolation method explained in the text, together with measurements from fixed target experiments at larger x. The curve and error band shows the result of a QCD by H1. (From H1 2001c).

DIS occurs through the boson–gluon fusion process. This process involves the gluon density xg directly so it gives another experimental handle on this quantity. As heavy flavour production by BGF occurs at relatively small x, a signal at large x might indicate other production mechanisms or even be evidence for intrinsic heavy flavours in the proton.

The first measurements of F_2^c were made by fixed target experiments at FermiLab and CERN. For example the EMC collaboration measured F_2^c, in the range $0.0042 < x < 0.422$, $1 < Q^2 < 70\,\text{GeV}^2$, from di- and trimuon events produced from 250 GeV muon scattering on an iron target. The advent of HERA, with a reach to much smaller x values, together with detectors capable of measuring details of the DIS final state gives the prospect of accurate data on F_2^c over a much wider range of x and Q^2. The second major step forward since the early measurements is the improved theoretical treatment of heavy quark production to order α_s^2, as described in Section 4.5. Expressions for the rapidity and transverse momentum distributions of the charm quark have been worked out to the same order and incorporated into a Monte Carlo code describing the production and fragmentation of heavy quarks in ep scattering (HVQDIS, Harris and Smith 1998). Heavy quark production has also been incorporated into the CAS-

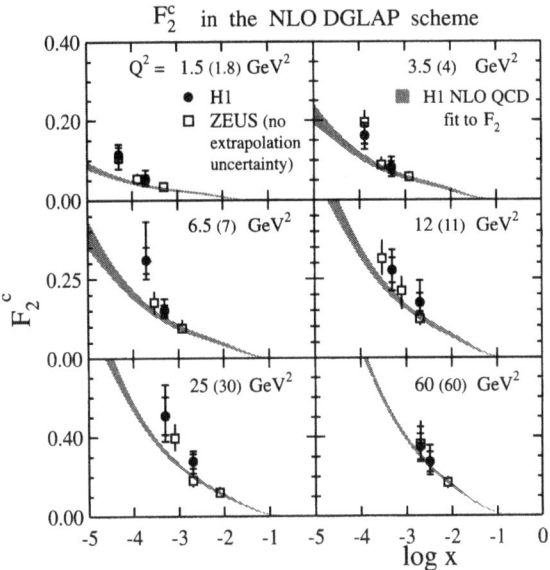
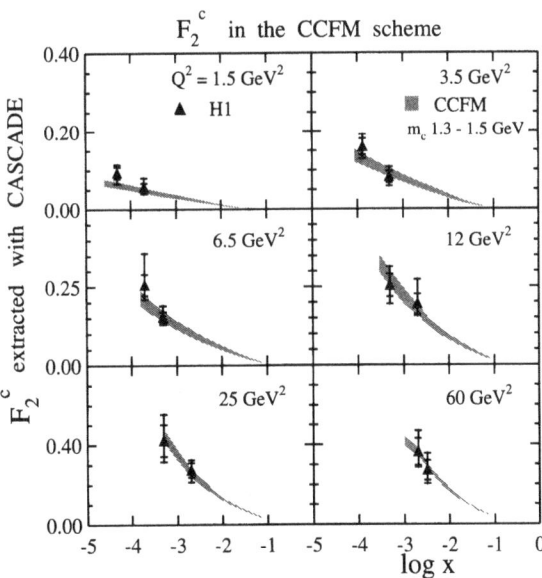

Fig. 5.17 F_2^c. The upper plots show data from the ZEUS and H1 corrected with HVQDIS and compared to NLO QCD calculation using the H1 gluon density. The lower plots show the H1 data corrected using the CASCADE Monte Carlo. For both analyses the spread in the calculated curves is dominated by the uncertainty in the charm quark mass ($1.3 < m_c < 1.5$ GeV). (From H1 2002).

CADE Monte Carlo code (Jung and Salam 2001), which is based on the CCFM equation (which effectively resums certain higher order terms at low x — see Section 9.5). Heavy quark events may be identified by a variety of techniques: D^* and D decays for charm; high p_T electrons and muons; secondary vertices and broadening of charged track impact parameter distributions. Most of the data on charm production in DIS from HERA-I has come from the first technique. Heavy quark physics is one of the major goals of the HERA-II programme.

Both the H1 and ZEUS experiments have searched for D^* production in DIS using the well-established technique which exploits the accurately known mass difference $\Delta M = M(K\pi\pi_s) - M(K\pi)$ (from the decay chain $D^{*+} \to D^0 \pi_s^+ \to K^-\pi^+\pi_s^+$) as the primary signal. The D^* is only just above threshold and its decay produces a 'slow' pion, π_s. H1 have also identified $D^0 \to K^+\pi^-$ decays directly and ZEUS have added the $D^0 \to K^+ 3\pi$ decay channel. Although the signal is a very clean one, the combined branching ratios are small (for the $D^* \to K\pi\pi$ it is $2.62 \pm 0.10\%$) which means that high luminosity is required. For example in the analysis of the $37\,\text{pb}^{-1}$ of data collected by ZEUS in 1996–7, the data sample of D^* events amounted to 2064 ± 72 ($K2\pi$) and 1277 ± 124 ($K4\pi$). From these events, the $ep \to eD^*X$ cross-section is measured in a restricted region of D^* phase space, defined by cuts on $p_T(D^*)$ and $\eta(D^*)$.

To get from the measured $ep \to eD^*X$ cross-section to the inclusive charm production cross-section one first extrapolates to the full D^* phase space. Then using the probability that the charm quark will fragment into a D^* meson ($P(c \to D^*)$) and correcting for the small fraction (ξ) of indirect D mesons produced through B meson decay or fragmentation gives

$$\sigma(ep \to e c\bar{c} X) = \frac{1}{2} \frac{\sigma(ep \to eDX)}{P(c \to D)(1+\xi)}. \tag{5.26}$$

From measurements at e^+e^- colliders, $P(c \to D^{*\pm}) \sim 25\%$ and ξ is about 2%. Finally F_2^c is related to the $ep \to e c\bar{c} X$ cross-section by

$$\frac{d^2\sigma(c\bar{c})}{dx\,dQ^2} = \frac{2\pi\alpha^2}{Q^4 x} \left(Y_+ F_2^c - y^2 F_L^c \right), \tag{5.27}$$

where the small contribution from F_L^c is calculated from QCD.

The upper plots in Fig. 5.17 show the ZEUS and H1 F_2^c data derived from D^* decays using the HVQDIS approach for extrapolation to the full phase space. The size of the error bars reflects the limited statistics and the model uncertainties. The data from the two experiments are in good agreement with each other. The curves shown are NLO massive quark DGLAP predictions using a gluon density derived from fitting H1 F_2 scaling violations. The uncertainty comes largely from the uncertainty in m_c. Although the overall agreement with data is quite good, there is a tendency for the theory to undershoot the data at low x. This has been investigated further by using the CCFM CASCADE code for the extrapolation — the results are shown in the lower plots of Fig. 5.17. A comparison of the upper and lower plots shows a steeper rise in the predicted F_2^c at small x using CCFM evolution than that obtained from NLO DGLAP. Using the acceptances and efficiencies calculated from the CASCADE program, the measured values of the H1 F_2^c are found to be systematically smaller than those determined with the HVQDIS program. The largest differences (up to $\approx 20\%$) are observed at small x values. Clearly high statistics measurements of F_2^c at HERA-II are going to be an important tool for furthering the understanding of low x parton dynamics.

5.7 F_2^p at very low Q^2

A remarkable feature of the proton structure function F_2 is its rapid rise at low x — shown in Fig. 5.11. First observed for Q^2 values above $10\,\text{GeV}^2$, the persistence of this rise down to Q^2 values as small as $0.65\,\text{GeV}^2$ challenges our understanding of QCD. For smaller values of Q^2, the rise of the structure function at small x diminishes. It is not surprising that the behaviour of F_2 changes as $Q^2 \to 0$. At small x, F_2 is related to the $\gamma^* p$ total cross-section

$$\sigma_{tot}^{\gamma^* p}(W^2, Q^2) \equiv \sigma_T + \sigma_L = \frac{4\pi^2 \alpha}{Q^2} F_2(x, Q^2),$$

where $W^2 \approx Q^2/x$ (the derivation of this relation is given in Appendix C). The above equation shows that $F_2(x, Q^2) \to \text{const.} Q^2$ as $Q^2 \to 0$, as it must to satisfy conservation of the electromagnetic current. The surprising features of the HERA data are that NLO QCD appears to describe the data to really quite low values of Q^2 and that the strong rise of F_2 at small persists to even smaller Q^2 values. HERA has opened a new arena in which to investigate the limits of pQCD.

Both the H1 and ZEUS collaborations have investigated the region of very small Q^2 using either special detectors very close to the beam line to tag the scattered $e^\pm$ at very small angles or the shifted vertex technique. For example the ZEUS BPC/BPT measurements cover $0.045 < Q^2 < 0.65\,\text{GeV}^2$ and $6 \times 10^{-7} < x < 1 \times 10^{-3}$. An instructive way to plot the data is shown in Fig. 5.18 which shows F_2 data as a function of Q^2 at fixed values of y. The data below $Q^2 \sim 1\,\text{GeV}^2$ show that F_2 is becoming proportional to Q^2 as it must as $Q^2 \to 0$. For $Q^2 > 1\,\text{GeV}^2$ the dependence on Q^2 is much less pronounced as approximate Bjorken scaling takes over. The full curves (ZEUS Regge fit) are based on a general approach to the energy dependence of hadronic cross-sections and incorporate the constraint from current conservation (details are given in Section 9.3). At larger Q^2, the dashed curves are from a standard NLO QCD fit to be described in detail in the next chapter.

136 DIS experiments and data

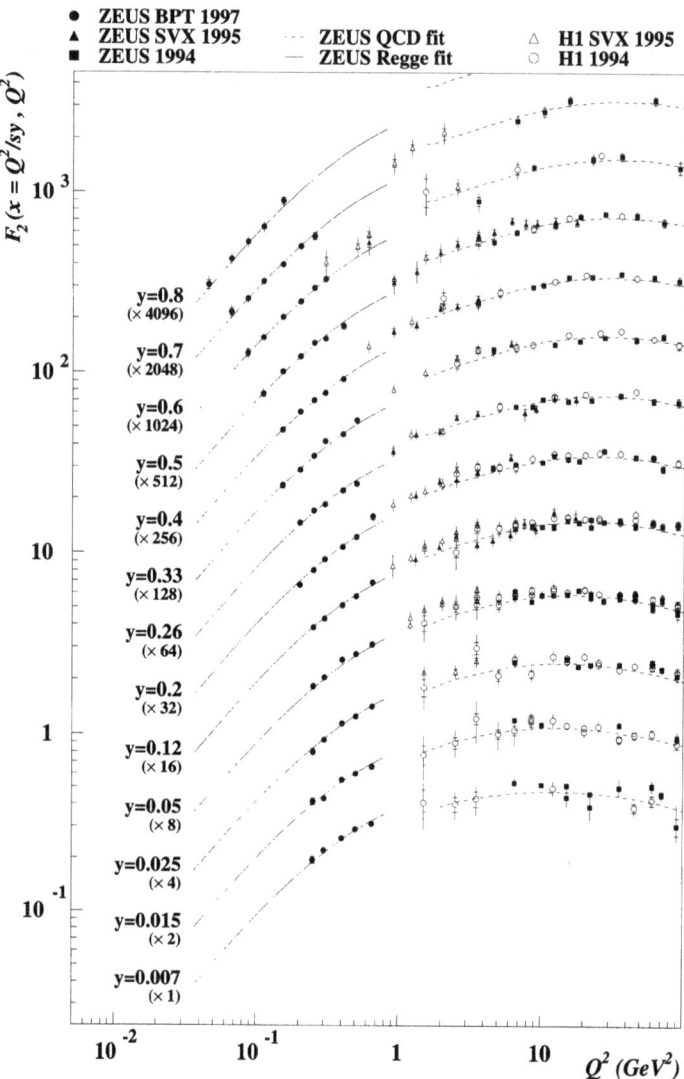

Fig. 5.18 F_2 as a function of Q^2 in bins of y at low Q^2. The solid circles show the data from the ZEUS BPC/BPT and data from other ZEUS and H1 measurements as indicated. The solid line is a Regge parameterization and the dashed line a ZEUS NLO QCD fit. (From ZEUS 2000a).

5.8 Summary

This chapter has given an account of the detectors used for deep inelastic scattering with fixed targets and the HERA collider. The principal methods for reconstruction of the kinematic variables, x, y and Q^2, and the extraction and correction of the cross-sections and structure functions from the measured event rates has been explained. The main features of structure function data for $Q^2 < M_Z^2$ from the 'classic' fixed target and HERA experiments have been outlined, particularly the rise of F_2^p at low x and the precision with which deviations from Bjorken scaling have now been measured. The importance of neutrino induced DIS has been touched upon and the promise of learning more about parton dynamics at low x from accurate measurements of F_2^c elaborated. The chapter ends with an outline of the precise measurements of F_2 in the 'transition region' as $Q^2 \to 0$. The measurement of consistent data on F_2 in both fixed target and collider regions with systematic errors at the level of 2% or better is a major achievement and a necessary precursor to the accurate determination of the proton's parton momentum density functions to be described in the next chapter.

5.9 Problems

1. Use the equations of Section 5.2.1 to work out the region in the x, Q^2 plane in which a fixed target detector with the given characteristics could measure x to within 10% and Q^2 to within 5%. [$E_\mu = 280\,\text{GeV}$; $E'_{min} = 40\,\text{GeV}$; $\theta_{min} = 15\,\text{mrad}$; $\delta E'/E' \sim 10^{-4}$; $\delta \theta \sim 1\,\text{mrad}$.]

2. The hadronic angle γ for DIS at HERA. Assume that in the hard ep interaction, the electron strikes a quark with momentum xp to produce the scattered electron (E', θ_e) and a passless object of energy F at an angle γ. Assume further that the proton remnant has zero transverse momentum. Show that: $F = \sum_h E - (1-x)p$; $F\cos\gamma = \sum_h p_z - (1-x)p$; $F\sin\gamma = \sqrt{\sum_h p_t^2}$; where $\sum_h$ is the sum over all final state hadrons. From these results find an expression for F and verify that $\cos\gamma$ is given by Eq. (5.10).

3. Derive the Jaquet-Blondel formula for y at HERA (Eq. 5.13).

4. Using the results of the previous two questions, complete the derivation of the double angle formulae for HERA kinematics (Eqs (5.11) and (5.12)). [You may find it helpful to derive an expression for y first, starting from y_Σ.]

5. Design a beam line to provide high energy ν and $\bar\nu$ for DIS experiments. Assume that high intensity 400 GeV charge selected $\pi^\pm$ and $K^\pm$ beams are available. What will the range of neutrino energies be?

6. Taking the dimensions of the ZEUS detector (shown in Fig. 5.5) as typical, estimate the timing resolution needed to reject cosmic ray muons and beam–gas interactions. [Assume that all particles travel at the speed of light.]

7. Show that
$$F_2^{\ell N} = \frac{5}{18}\left(1 - \frac{3}{5}\frac{s+\bar s-c-\bar c}{\sum(q+\bar q)}\right)F_2^{\nu N}$$
is valid for isoscalar nucleon structure functions ($F_2^N = 0.5(F_2^p + F_2^n)$). What assumptions have to be made and why is the expression only

valid beyond the leading order in QCD in the DIS renormalization scheme?

8. Verify Eq. (5.25) for R and the expression $\varepsilon = \dfrac{2(1-y)}{1+(1-y)^2}$. Sketch ε as a function of y.

5.10 Appendix: cross-sections and luminosity

5.10.1 The size of DIS cross-sections

A typical hadron–hadron (strong interaction) cross-section may be estimated using πR^2 where R is the hadron radius, for example taking $R \approx 0.8$ fm gives $\sigma \approx 20$ mb to be compared with the measured $\sigma(\pi^\pm p)$ of about 25 mb at high energies. The total cross-section for γp scattering is of order $\alpha_{qed}\sigma(\pi^\pm p) \approx 180\,\mu$b. To estimate $\sigma(ep \to e'X)$ at $Q^2 \approx 0$ one may use the 'equivalent photon approximation' (see Problem 5 at the end of the previous chapter) which gives $\sigma(ep) = f_\gamma \sigma(\gamma p)$, where f_γ is the photon flux integrated over the acceptance of the detector used to measure the electrons scattered at almost zero degrees. The scale of f_γ is set by another factor of α_{qed}, for a ZEUS measurement at HERA (at 300 GeV ep CM energy) f_γ was 0.005, giving $\sigma(ep) \approx 850$ nb. This gives the order of magnitude upper limit on the cross-sections expected in electron or muon deep inelastic scattering. For neutral-current scattering, dominated by virtual photon exchange, the cross-section decreases as $1/Q^4$. At a given Q^2 the cross-section is largest at low x — for a typical bin at $Q^2 = 10\,\text{GeV}^2$, $x = 1.6 \times 10^{-4}$ $\sigma \sim 200$ pb.

At large Q^2, it is interesting to compare the NC and CC single differential cross-sections $d\sigma/dQ^2$ — see Fig 8.1 of Chapter 8. At $Q^2 \approx 10^3\,\text{GeV}^2$, $d\sigma^{NC}/dQ^2 \sim 0.3\,\text{pb}/\text{GeV}^2$ and $d\sigma^{CC}/dQ^2 \sim 0.02\,\text{pb}/\text{GeV}^2$. By $Q^2 \approx 10^4\,\text{GeV}^2$, both NC and CC cross-sections have similar values of $\sim 7.5 \times 10^{-4}\,\text{pb}/\text{GeV}^2$, this is a direct consequence of electroweak unification as will be explained in Chapter 8.

5.10.2 Beam flux and luminosity

The other crucial parameter needed to estimate a counting rate is the beam flux or luminosity. As an example of a fixed target beam consider the NM beamline of the E665 experiment at Fermilab. The primary beam is an 800 GeV proton beam extracted from the Tevatron, a typical 'spill' contains about 4×10^{12} protons from the last 23 s of the 60 s cycle of filling, acceleration and extraction. The protons then interact with a Be target producing secondary pions and kaons which are transported down a 1.1 km decay channel. About 5% of the particles decay to muons, the remainder are absorbed in a beam stop. The total length of the beamline is 1.5 km which includes space for the final focussing optics in addition to the decay channel. Positive muons are selected and the beam profile and phase space adjusted to give a beam with $\langle E_\mu \rangle \sim 465$ GeV, $\Delta E_\mu \sim 50$ GeV, a cross-sectional area of about $6 \times 4\,\text{cm}^2$ and containing of order 2×10^{11} μ^+ per spill. In a fixed target experiment, the target is frequently a liquid or solid which gives a large increase in rate over a collider from the much larger target density.

For a collider, taking HERA as an example, the luminosity is given by

$$\mathcal{L} = \frac{N_e N_p N_B f_{rev}}{4\pi \sigma_x \sigma_y},$$

where N_e, N_p, N_B are the numbers of particles in an $e^{\pm}$ bunch, a proton bunch and the number of bunches respectively, f_{rev} is the revolution frequency and σ_x, σ_y are the transverse dimensions of the beams at the interaction point (the values have to be same for the two beams to maximise the luminosity). The design figures for HERA-II are $N_e = 4.18 \times 10^{10}$, $N_p = 10 \times 10^{10}$, $N_B = 174$ $f_{rev} = 4.7 \times 10^4$ Hz, $\sigma_x = 118\,\mu$m, $\sigma_y = 32\,\mu$m, giving $\mathcal{L} = 7 \times 10^{35}$ m^{-2}s^{-1}. This is nearly five times larger than the maximum design luminosity of HERA-I and is obtained largely be reducing the beam size at the interaction point.

6
Extraction of parton densities

The chapter discusses how parton momentum densities are extracted from data. In Sections 6.1–6.3, the general framework based on NLO pQCD is outlined. The concept of the *global fit* is introduced and the choice of reactions and data sets discussed. In Section 6.4, a detailed account of the results of some global fits is given. Particular attention is paid to the extraction of the gluon distribution since most cross-sections at hadron colliders, present and future, are dominated by gluon induced processes. In Section 6.5, sources of information on PDFs from non-DIS processes are considered. Theoretical and experimental sources of uncertainty on PDFs are considered in detail in Section 6.6–6.7. The problems of defining a reasonable measure of the error on a parton density and how to estimate it are addressed. The approach of the GRV group to the determination of 'dynamically generated' PDFs is explained in Section 6.8. Prospects for improving our knowledge of the gluon distribution are presented in Section 6.9. Finally, the appendix to this chapter compares different approaches to the technical challenges involved in solving the QCD evolution equations.

6.1 Determining parton distribution functions

Perturbative QCD predicts the Q^2 evolution of the parton distributions, but not the x dependence. Ideally one would like to find analytic parameterizations of parton distributions, which are consistent with the Q^2 dependence predicted by pQCD. The problem is to perform a Mellin inversion of the exact predictions of pQCD for the moments of structure functions in order to find suitable analytic expressions for the structure functions themselves. However, if one requires consistency with pQCD beyond leading order, one cannot find exact analytic expressions which are valid for more than a limited x, Q^2 range. For this reason, the most commonly used method of extracting parton distributions is to perform a direct numerical integration of the DGLAP equations at NLO. The technique is broadly as follows. An analytic shape for the parton distributions (valence, sea and gluon) is assumed to be valid at some starting value of $Q^2 = Q_0^2$. This starting value is arbitrary, but should be large enough to ensure that

$\alpha_s(Q_0^2)$ is small enough for perturbative calculations to be applicable. Then the DGLAP equations (see Sections 4.2–4.3) are used to evolve the parton distributions up to a different Q^2 value, where they are convoluted with coefficient functions, appropriate to the chosen renormalization scheme, in order to make predictions for the structure functions. These predictions are then fitted to the data sets described in Chapter 5. Typically a χ^2 fit is made using between 10 and 20 fit parameters to describe around 1500–2000 data points distributed over the x, Q^2 plane illustrated in Fig. 5.1. The fit parameters are those necessary to specify the input analytic shape, and Λ_{QCD}, or equivalently $\alpha_s(M_Z^2)$. Thus obtaining an acceptable χ^2 is not only a stringent test of the validity of pQCD in the DGLAP formalism, but also provides a way of determining the shapes of the PDF distributions and the value of $\alpha_s(M_Z^2)$. Determinations of $\alpha_s(M_Z^2)$ are discussed in Chapter 7. In the present Chapter the focus will be on the PDF distributions.

Note that the input analytic form assumed for the parton distributions is only valid at the starting scale Q_0^2. For other Q^2 values the distributions must be obtained from the DGLAP evolution equations. Thus the parton distributions resulting from the fits are usually supplied on x, Q^2 grids which can be used to obtain the value of the PDFs at any x, Q^2 point by interpolation (see Appendix G).

Many groups now produce PDF sets using the method outlined above. The theoretical groups MRST and CTEQ have traditionally concentrated on estimating uncertainties from the input model assumptions and the theoretical framework (see Section 6.6). Accordingly they have each issued several different sets of PDFs each extracted under slightly different conditions. Detailed descriptions of the procedures of these groups may be found in Martin et al. (1998) for MRST and Tung et al. (2002) for CTEQ. The experimental groups have concentrated on estimating the uncertainties from experimental sources. Recently, techniques which address the problem of how the correlated experimental uncertainties on the measurements are to be propagated into experimental uncertainties on PDFs have been developed. These are described in Section 6.7.

Sources for the PDF sets which successfully describe modern data are gathered together in Appendix G. Eigenvector PDF sets which can be used to calculate the uncertainties on the PDFS are also compiled in Appendix G.

6.2 Treatment of data sets in global analyses

The most reliable parton distributions are obtained by making global analyses of a wide range of data from DIS and other hard processes. There are various pitfalls to be avoided when combining data from different sources.

6.2.1 Nuclear binding corrections

Firstly, data must be taken on comparable targets. As discussed in Section 5.5, a heavy nuclear target cannot simply be treated as an additive combination of nucleons, and data taken on such a target have to undergo corrections for nuclear shadowing, binding and Fermi motion effects, before they can be used to extract information on parton distributions within the free nucleon. Modern data are taken on hydrogen or deuterium targets to avoid the need for such corrections. However, in current global fits ν, $\bar{\nu}$ data is still very important in determining the valence distributions and for such beams the event rate on proton/deuterium targets is very low. Accordingly, the CCFR $\nu,\bar{\nu}$ scattering experiment uses a heavy target and the data are corrected for nuclear effects before they are compared to the electroproduction data.

Thus high luminosity HERA data will be very valuable, since the information on valence distributions which is currently obtained from CCFR data can also be obtained from HERA high Q^2 data, as discussed in Chapter 8

Furthermore, for precision measurements, the need for correction even extends to deuterium targets. Uncertainty in the size of these corrections leads to uncertainty in the d_v/u_v ratio at high x. The high luminosity HERA charged current data will constrain this ratio without need of such corrections.

6.2.2 Data consistency

In order to extract meaningful results from a χ^2 fit, the input data sets must be consistent with each other. Historically, some of the DIS data sets failed reasonable consistency tests, and for this reason only modern DIS data sets are used in current global analyses. These are: NMC F_2 data on muon scattering from p and D targets, including the special extraction of the ratio $F_2^{\mu d}/F_2^{\mu p}$ with minimal systematic error; E665 and BCDMS F_2 data on muon scattering from p and D targets; CCFR F_2 and xF_3 data on ν, $\bar{\nu}$ scattering on an Fe target; the reanalysed SLAC data on electron scattering on a proton target; and HERA data from ZEUS and H1 on electron/positron scattering on a proton target. The data sets are presented in Chapter 5 and their kinematic ranges are given in Table 5.1 and shown in Fig. 5.1.

These data sets have been used since they represent the highest precision data sets available, for which detailed information is available on systematic errors and their correlations. The importance of this and further questions concerning data consistency are explored in Section 6.7.

6.3 Global fits: the general formalism

This section gives a general description of the parton parameterizations and the considerations which motivate the choice of these forms.

There is some flexibility as to which parton distributions one choses to parameterize. A typical choice is u_v, d_v, S, g, $\bar{d} - \bar{u}$ but u, $\bar{u}$, d, $\bar{d}$, g is also used. The quark distributions must be combined to form singlet and non-singlet distributions. The singlet quark distribution is

$$x\Sigma = xu_v + xd_v + xS \qquad (6.1)$$

and the non-singlets are the xq^- distributions, namely xu_v and xd_v, and the xq^+ distributions which are constructed as

$$xq_i^+ = x(q_i + \bar{q}_i) - \frac{x\Sigma}{n_f} \qquad (6.2)$$

where n_f is the number of active flavours. The evolution of non-singlet distributions is independent of the gluon distribution, whereas that of the singlet distributions is coupled to the evolution of the gluon distribution as detailed in Eqs (4.34–4.39). The singlet and non-singlet quark distributions are then combined in the combinations appropriate to form the structure functions at LO. For example, the combination for the proton distribution in charged lepton scattering is

$$q^{lp} = \sum_i e_i^2 q_i^+ + \langle e^2 \rangle \Sigma \qquad (6.3)$$

where $\langle e^2 \rangle$ is the average of the square of the quark charges appropriate for the number of flavours, and the discontinuity in $\langle e^2 \rangle$ at flavour thresholds is compensated by the discontinuity in the q^+ distributions. These combinations are then convoluted with the coefficient functions to form the NLO predictions for the corresponding structure functions, as shown in Eq. (4.40).

The parton distributions must be parameterized by a suitably flexible analytic form at the starting scale $Q^2 = Q_0^2$. Before HERA, the choice was $Q_0^2 \sim 4\,\text{GeV}^2$, since a perturbative approach was not expected to work for lower Q^2, but the inclusion of the HERA data in fits stimulated the MRST group to drop their starting scale to $Q_0^2 = 1\,\text{GeV}^2$, in response to the fact that the rise of F_2 data at small x is still apparent at such low Q^2 values, see Section 5.7. The quality of the fit and the resulting parton parameterizations should not be sensitive to the choice of Q_0^2. QCD evolution can be performed both upwards and downwards in Q^2. The only proviso is that the miniumum Q^2 considered be large enough to ensure that $\alpha_s(Q^2)$ is small enough for pQCD to be applied . The choice of the minimum Q^2 appropriate for pQCD fits is discussed further in Section 6.4.

The form of the input parameterization is usually

$$xu_v = A_u x^{\lambda_u}(1-x)^{\eta_u} P(x,u) \qquad (6.4)$$

$$xd_v = A_d x^{\lambda_d}(1-x)^{\eta_d} P(x,d) \qquad (6.5)$$

$$xS = A_S x^{-\lambda_S}(1-x)^{\eta_S} P(x,S) \qquad (6.6)$$

$$xg = A_g x^{-\lambda_g}(1-x)^{\eta_g} P(x,g) \qquad (6.7)$$

where $P(x,i)$ is a polynomial in x or $\sqrt{x}$, for example $(1 + \epsilon_i \sqrt{x} + \gamma_i x)$, or $P(x,i) = (1 + \gamma_i x^{\epsilon_i})$, or sums of Chebyshev polynomials. Not all of the normalizations A_i are free parameters: A_u, A_d are determined by the need to satisfy flavour sum rules and A_g is determined in terms of the other three by the momentum sum rule.

The distributions are usually defined within the $\overline{\text{MS}}$ renormalization and factorization scheme. Conventionally, analyses are done assuming that the factorization scale, μ_f^2, and the renormalization scale, μ_r^2, are both equal to Q^2. This is the obvious choice of scale for the DIS process.

6.3.1 The form of the parameterization

Clearly, the choice of the form of parameterization represents a model uncertainty. In global fits this is usually not too severe since the PDFs extracted for $Q^2 > Q_0^2$ soon lose sensitivity to the exact form of the parameterization at Q_0^2. However, it is more significant if restricted data sets are used since some of the parameters can be insufficiently constrained. This is explored further in Section 6.6.1.

Some comments on the standard choice of parameterization are in order. The simple form $x^a(1-x)^b$ respects the limits at $x \to 0, 1$ as suggested by Regge theory (see Section 9.3) and the constituent counting rules (Brodsky and Farrar 1973), respectively. The values of a and b for each distribution can also be predicted from these considerations. However, it is unclear at what value of Q^2 such predictions should apply and, since the parameterizations describe the data only at the value Q_0^2, such predictions can only be taken as a rough guide to the sorts of value to be expected. Furthermore, whereas the valence parameters are not very sensitive to the chosen value of Q_0^2, the sea and especially the gluon parameters are sensitive because the shapes of these distributions evolve rapidly with Q^2, as discussed in Section 4.3.

The counting rules provided an estimate of how rapidly the distribution for each type of parton tends to zero as $x \to 1$. In this limit there can be no momentum left for any of the partons other than the struck quark, and these partons thus become 'spectators'. The prediction is that, $xq(x) \sim (1-x)^{2n_s-1}$, where n_s is the minimum number of spectator partons. Thus for valence quarks one has $n_s = 2$ and so the valence parameters η_u, η_d are expected to have a value around 3. Measurements indicate values of roughly this magnitude with $\eta_d > \eta_u$, such that $d_v/u_v \to 0$, as $x \to 1$. The counting rules also lead to the expectations $\eta_g \sim 5$, $\eta_S \sim 7$ for the high-x behaviour of the gluon and sea distributions. These values are only very broadly in agreement with the fitted values. Note that there is a correlation between the η_i parameters which control the high-x behaviour of a parton distribution (via $(1-x)^{\eta_i}$) and the corresponding values of γ_i. Thus to understand results in detail, one should consider the whole shape of the high x distributions, not just a single parameter.

The parameters λ_u, λ_d for the valence (non-singlet) distributions are expected to have values near 0.5, close to the value $1 - \alpha_R(0)$ deduced from the non-singlet Regge trajectory intercept (see Section 9.3), and this is roughly true for Q_0^2 in the range 1–4 GeV^2.

The value of λ_g which describes the gluon shape at small x is also suggested by Regge behaviour. Since the gluon and singlet quark distributions mix, a value related to the singlet Pomeron trajectory $1 - \alpha_P(0)$ is expected. The conventional Regge exchange is that of the soft Pomeron, giving $\lambda_g \sim 0.08$. However, for $Q_0^2 \gtrsim 4\,\text{GeV}^2$ data indicate the need for a significantly larger value. This may reflect the need for a contribution from a hard Pomeron, with $\lambda_g \sim 0.5$, or it may simply be a feature of the fast evolution of the gluon distribution such that the smaller value, $\lambda_g \sim 0.08$, would be appropriate at a lower starting scale. The variation of the value of λ_g with the value of Q_0^2 for a typical fit is illustrated in Fig. 6.8. The theoretical interpretation of the low x behaviour of the gluon distribution is an active research topic which is discussed in detail in Chapter 9.

The value of λ_S which describes the sea shape at small x should be similar to λ_g since, at small x, the singlet distribution is dominated by the sea and the process $g \to q\bar{q}$ dominates the evolution of the sea quarks. Hence one expects $\lambda_S = \lambda_g - \epsilon$, but only if Q_0^2 is large enough for the effect of DGLAP evolution to be seen. This expectation is not fulfilled until Q_0^2 is in the range 5–10 GeV^2, as will be discussed further in Section 6.4 (see also Fig. 6.8).

Finally the extracted values of $\lambda_{S,g}$ should not be interpreted without also considering the values of the parameters $\epsilon_{S,g}$ which are correlated with them. The steepness of the low x shape of the data should not be judged from the values of $\lambda_{S,g}$ alone.

6.3.2 The flavour composition of the sea

The flavour composition of the sea also needs some discussion. The sea distribution refers to all flavours $S = 2(\bar{u} + \bar{d} + \bar{s} + \bar{c} + \bar{b})$ where one has assumed $q_{\text{sea}} = \bar{q}$ as usual. The contribution of the heavy quarks needs a detailed theoretical treatment which will be discussed in Section 6.6.2. The strange sea is not treated with such sophistication. It is suppressed relative to the u and d seas because of its larger mass and this is accounted for by introducing a simple suppression factor: for example, $\bar{s} = (\bar{u} + \bar{d})/4$, such that the strange sea is suppressed by about 50% compared to the u and d sea distributions at Q_0^2. The justification for this comes from CCFR opposite sign dimuon data (CCFR 1995). Briefly, opposite sign dimuon events dominantly arise in $(\bar{\nu})\,\nu$ scattering when the struck quark is an $(\bar{s})s$ quark which becomes a $(\bar{c})c$ quark through the flavour changing weak current. This $(\bar{c})c$ quark then decays through the muon channel to $(\bar{s}\mu^-\bar\nu_\mu)\,s\mu^+\nu_\mu$ yielding a muon of opposite sign to that at the original lepton scattering vertex. The rate for the process thus gives a measure of the size of the strange sea, which is consistent with $\sim 50\%$ suppression at low $Q^2 \sim Q_0^2$. The description of the dimuon data is not very sensitive to the exact value of Q_0^2.

Early parameterizations assumed that the u, d content of the sea is flavour symmetric, but there is no necessity for this and in 1992 the NMC collaboration gave the first evidence that this is not the case when they observed a violation of the Gottfried sum rule. This sum rule is given by

$$\int_0^1 \frac{dx}{x}(F_2^p - F_2^n) = \frac{1}{3}\int_0^1 dx(u_v - d_v) + \frac{2}{3}\int_0^1 dx(\bar{u} - \bar{d}) \qquad (6.8)$$

such that if $\bar{u} = \bar{d}$, the value of the sum is 0.33. Thus the NMC observation of a value ~ 0.23 for this sum-rule indicated that $\bar{d} > \bar{u}$. Hence, the structure of the sea at Q_0^2 may be expressed as

$$2\bar{u} = 0.4(1-\delta)S - \Delta, \quad 2\bar{d} = 0.4(1-\delta)S + \Delta, \quad 2\bar{s} = 0.2(1-\delta)S, \quad 2\bar{c} = \delta S \quad (6.9)$$

where δ gives the small contribution of the charmed sea at Q_0^2 and $x\Delta$ is given by

$$x\Delta = x(\bar{d} - \bar{u}) = A_\Delta x^{\lambda_\Delta}(1-x)^{\eta_S}P(x,\Delta) \qquad (6.10)$$

with the normalization A_Δ fixed to agree with data on the Gottfried sum-rule.

6.3.3 Global fits: the relationship of the measurements to the parton distributions

The relationship of the measured structure functions to the parton distributions is not straightforward at NLO since the evolved parton distributions must be convoluted with coefficient functions and all types of parton may contribute to a particular structure function through the evolution. However, the simple QPM formulae of Section 2.4 still give a good guide to the major contributions.

- (Anti-)neutrino CCFR data on xF_3 gives information on the valence shapes for all x, with u_v and d_v contributing equally (Eq. 2.52). This information is most reliable at medium x, since the nuclear corrections are less well determined for $x \lesssim 0.1$ and $x \gtrsim 0.6$.
- NMC data on the ratio $F_2^{\mu d}/F_2^{\mu p}$ gives information on the ratio d_v/u_v at large x (Eq. 2.55) and is thus the only data set currently used which strongly constrains the difference in the shapes of d_v and u_v.
- The F_2^{lD}, F_2^{lp} data from NMC, BCDMS, E665, SLAC and the F_2 data from CCFR, give information on singlet combinations of quarks (see Eqs 2.42 and 2.44), which thus gives accurate information on the sea distributions throughout the x, Q^2 ranges covered by each of these experiments.
- The data on F_2 also give information on combinations of the u and the d valence distributions at high x. Since these combinations are weighted by the quark charges squared, the u_v distribution is the dominant contribution for proton targets, whereas u_v and d_v contribute equally for deuterium targets. Overall this means that the u_v distribution is much better determined than the d_v distribution.

- The data on F_2 have also been used to constrain the gluon distribution. Since the gluon does not couple to the photon it does not enter the expressions for the structure functions at all in the QPM. It is constrained by the momentum sum rule, and by the way the DGLAP equations feed its evolution into the sea distribution (see Eqs 4.36 and 4.37). The shape of the gluon distribution extracted from a DGLAP QCD fit will be correlated with the value of α_s, since an increase in α_s increases the negative contribution from the P_{qq} term but this may be compensated by a positive contribution from the P_{qg} term if the gluon is made harder. Hence, a fixed value of α_s, as determined from independent data, is often assumed.
- Information on the difference $\bar{d} - \bar{u}$ comes from the different combinations of $\bar{u}$ and $\bar{d}$ entering into the isoscalar and proton target structure function data (see Eqs 2.42 and 2.44).
- CCFR dimuon data gives information on $\bar{s}$ as explained in Sec 6.3.2.
- Current HERA reduced cross-section data give information on the shape of the sea distribution way down into the low x region. Post upgrade HERA-II data will also give precision information on valence distributions, particularly the less well known d_v distribution, see Chapter 8.
- HERA data are invaluable in constraining the gluon distribution, since at small x QCD evolution becomes gluon dominated and the uncertainties referred to above are considerably reduced. This is because F_2 is essentially given by the singlet sea quark distribution for $x \lesssim 0.01$, and this in turn is driven by the gluon through the P_{qg} term in Eq. (4.36). Hence, in this kinematic region, the gluon distribution may be obtained almost directly from the F_2 scaling violation data.[1]

6.4 Results on PDF extraction

In the present section the parton distribution functions extracted from NLO QCD fits are presented. The fits made by experimental collaborations to data from DIS experiments are considered first.

A major problem in the combination of data from different experiments concerns the treatment of correlated systematic errors. Traditionally point to point systematic errors have been accounted for by adding them in quadrature to the statistical errors in the definition of a χ^2 for the fit. Most global fits have not taken full account of correlations between experimental systematic errors (partly because this information was not available for all data sets). However, modern deep inelastic scattering experiments have very small statistical uncertainties, so that the contribution of systematic uncertainties becomes dominant and consideration of point to point correlations between systematic uncertainties is essential.

---

[1] See also Section 4.2.2 for a relevant approximate relationship.

148 *Extraction of parton densities*

An obvious example of a correlated uncertainty is the overall normalization of the data, but there other more complex correlated effects which affect the evaluation of kinematic variables. For example, a small change in the calorimeter energy scale can move events between x, Q^2 bins and thus induce correlations which modify the shape of the extracted structure function. The correct procedure when combining different data sets with diverse sources of correlated systematic uncertainties is still a subject of active discussion. Section 6.7 is devoted to discussing the procedures which are currently used.

6.4.1 Results from the experimental collaborations

The ZEUS Collaboration have made an NLO QCD fit in the DGLAP formalism using the conservative Offset Method to evaluate the contribution of correlated experimental uncertainties to the PDF error bands (ZEUS 2003b). A global fit to ZEUS reduced cross-section data together with all the modern precision fixed target DIS structure function data from BCDMS, NMC, E665 and CCFR was made. Data from deuteron targets and the heavy target (anti-)neutrino data was used in order to have sufficient information to constrain all parameters. The aim was to extract the valence, sea and gluon PDFs, the value of $\alpha_s(M_Z^2)$ and their experimental uncertainties.

The ZEUS gluon distribution for a fit with fixed $\alpha_s(M_Z^2) = 0.118$ is shown in Fig. 6.1, for several Q^2 values. For $Q^2 \gtrsim 7\,\text{GeV}^2$, the gluon density rises dramatically towards low x and this rise increases with increasing Q^2. This rise is one of the most striking discoveries of HERA. Prior to 1993, the NMC data had suggested that the gluon was flattening towards low x, but the x range only extended down to $x \sim 0.01$, at Q^2 values around $Q^2 \sim 20\,\text{GeV}^2$.

The error bands on the ZEUS gluon analysis are illustrated beneath the distributions in terms of the fractional differences of the upper and lower error from the central value. The total error includes that from correlated systematic uncertainties. Further uncertainties due to model choices (see Section 6.6 are negligible by comparison, however there is additional uncertainty on the gluon PDF due to the variation of $\alpha_s(M_Z^2)$. This correlation can be taken into account by allowing $\alpha_s(M_Z^2)$ to be a parameter of the fit and the resulting uncertainty on the gluon is also illustrated in Fig. 6.1. The gluon is determined to within $\sim 15\%$ for $Q^2 > 20\,\text{GeV}^2$, $10^{-4} < x < 10^{-1}$, and its uncertainty decreases as Q^2 evolves upwards. The narrowing of the error band around $x \sim 0.1$ reflects the feed through of uncertainty at high x to uncertainty at low x, which is introduced by the momentum sum-rule, and the fact that $x \sim 0.1$ is the region of x where Bjorken scaling of the F_2 measurements holds, so that there is least uncertainty in the derived gluon distribution at this value of x.

At low Q^2 the gluon shape flattens at low x. Figure 6.1 shows the ZEUS fit extrapolated back to $Q^2 = 1\,\text{GeV}^2$, where the gluon shape be-

Results on PDF extraction 149

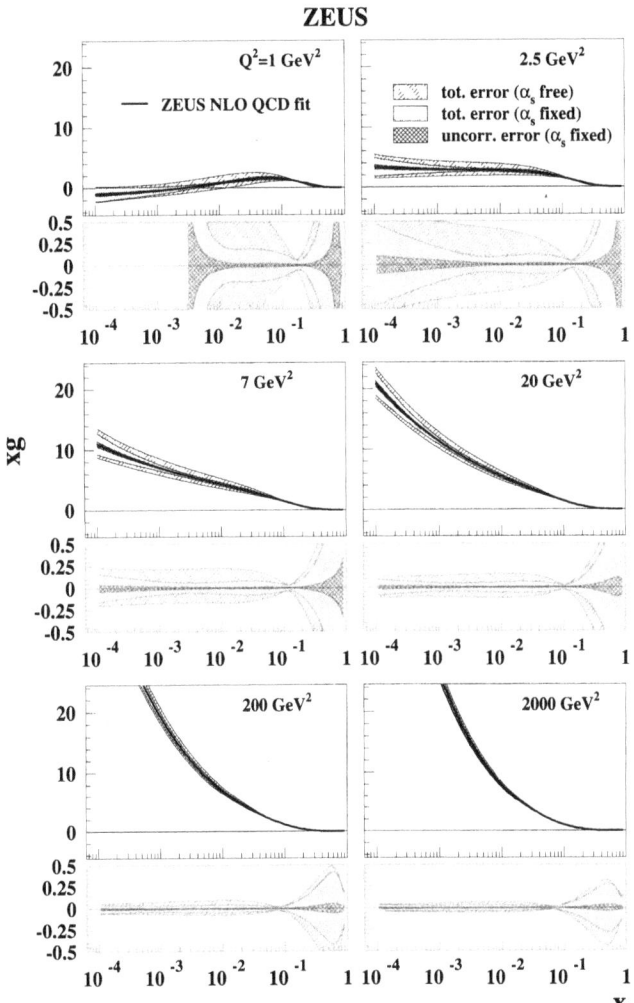

Fig. 6.1 Gluon distribution from an NLO QCD fit to ZEUS plus fixed target data, for various Q^2. The inner cross-hatched error bands shown represent the uncorrelated errors (from statistical and uncorrelated systematic experimental sources). The middle grey error bands represent total experimental errors (from correlated and uncorrelated sources) and the outer hatched error band gives the additional error due to the variation of α_s. The size of these uncertainties is more clearly illustrated beneath each distribution, where the error bands are displayed as fractional differences from the central value. (From ZEUS 2003b).

comes valence like, with low x values that are negative within errors. This unexpected prediction of the ZEUS fit is also observed in the fits of the theoretical team MRST, see Section 6.4.2.

The shapes of the gluon and the sea distributions are compared in Fig 6.2. The uncertainty in the sea distribution is considerably less that that of the gluon distribution. It is less than $\sim 5\%$ for $Q^2 \gtrsim 2.5\,\text{GeV}^2$ and $10^{-4} < x < 10^{-1}$. For $Q^2 \gtrsim 5\,\text{GeV}^2$, the low-$x$ slope of the gluon density is larger than the that of the sea density, but for lower Q^2, the sea density continues to rise at low x, whereas the gluon density flattens and turns over, giving a 'valence-like' shape for $Q^2 \lesssim 2\,\text{GeV}^2$. This may be a signal that the application of the DGLAP NLO formalism for $Q^2 \lesssim 5\,\text{GeV}^2$ is questionable.

ZEUS have also compared their NLO QCD fit to very low Q^2 data (see Section 5.7). Figure 6.3 illustrates this comparison. The fit is unable to describe the very precise BPT data for $Q^2 \lesssim 1\,\text{GeV}^2$, even when the full error bands due to the correlated systematic errors are included. It is the need to fit the extreme steepness of the ZEUS data at intermediate Q^2 ($2.7 < Q^2 \lesssim 200\,\text{GeV}^2$) which prevents a fit to the very low Q^2 BPT data. Thus the DGLAP formalism is clearly inapplicable for $Q^2 \lesssim 1\,\text{GeV}^2$.

Further evidence comes from the predictions for F_L at these low Q^2 values, as illustrated in Fig 6.4. F_L is the structure function most closely related to the gluon distribution at small x (see Eq. 4.43), so that the unexpected behaviour of the gluon distribution at low x, Q^2 may be reflected in the behaviour of F_L. Figure 6.4 shows that F_L is predicted to be significantly negative for $Q^2 \lesssim 1\,\text{GeV}^2$.

Furthermore, although the formalism of NLO QCD fits using the DGLAP equations works technically down to $Q^2 \sim 1\,\text{GeV}^2$, two features, already noted, make its application questionable in the Q^2 region, 1–5 GeV^2. First, the fact that the gluon density falls below that of the sea density for the lowest Q^2, taking a valence-like shape and becoming negative at low x. Second, the fact that even when the magnitude of the gluon density has become larger than that of the sea density, at $Q^2 \gtrsim 2\,\text{GeV}^2$, its slope remains less than that of the sea density until $Q^2 \gtrsim 5\,\text{GeV}^2$. The breakdown of the applicability of the DGLAP equations at low x and low Q^2 is discussed further in Section 6.4.2, Section 6.6 and Chapter 9.

The valence distributions extracted from the ZEUS fit are shown in Fig. 6.5. Note that the abscissa is linear and the ordinate logarithmic in order to illustrate the behaviour of these distributions at higher x ($x \gtrsim 0.1$), where they are directly constrained by the fixed target data. These distributions are discussed further in Chapter 8 where the role of high-Q^2 HERA data is discussed.

The H1 collaboration have also made an NLO QCD fit in the DGLAP formalism with the aim of extracting the gluon distribution and the value of α_s, using just the H1 reduced cross-section data, plus the BCDMS proton target F_2 data to give information at high x (H1 2001c). Thus uncertain-

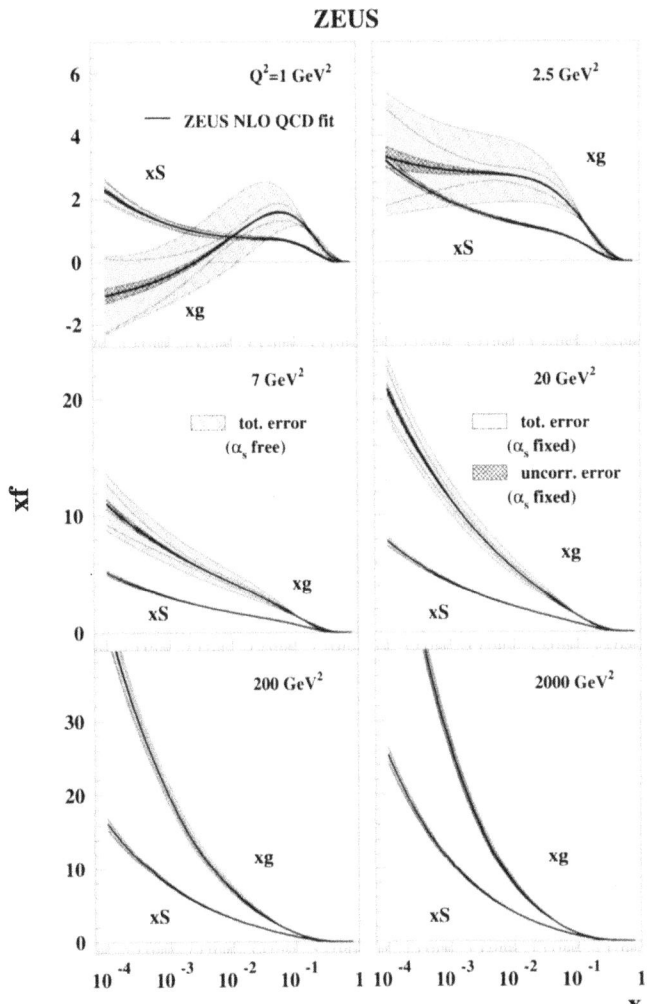

Fig. 6.2 The gluon and sea distributions from an NLO QCD fit to ZEUS plus fixed target data, for various Q^2. The error bands have been described in the caption to Fig 6.1. (From ZEUS 2003b).

ties due to nuclear effects and the combination of different data sets are minimized. The extracted gluon distribution, for $\alpha_s(M_Z^2) = 0.115$ fixed, is shown in Fig. 6.6 for three Q^2 values. The H1 analysis also sees the strong rise of the gluon distribution at small x and the total uncertainty on the gluon distribution is constrained to an accuracy similar to that of the ZEUS analysis.

The H1 collaboration have used the Hessian method to evaluate the

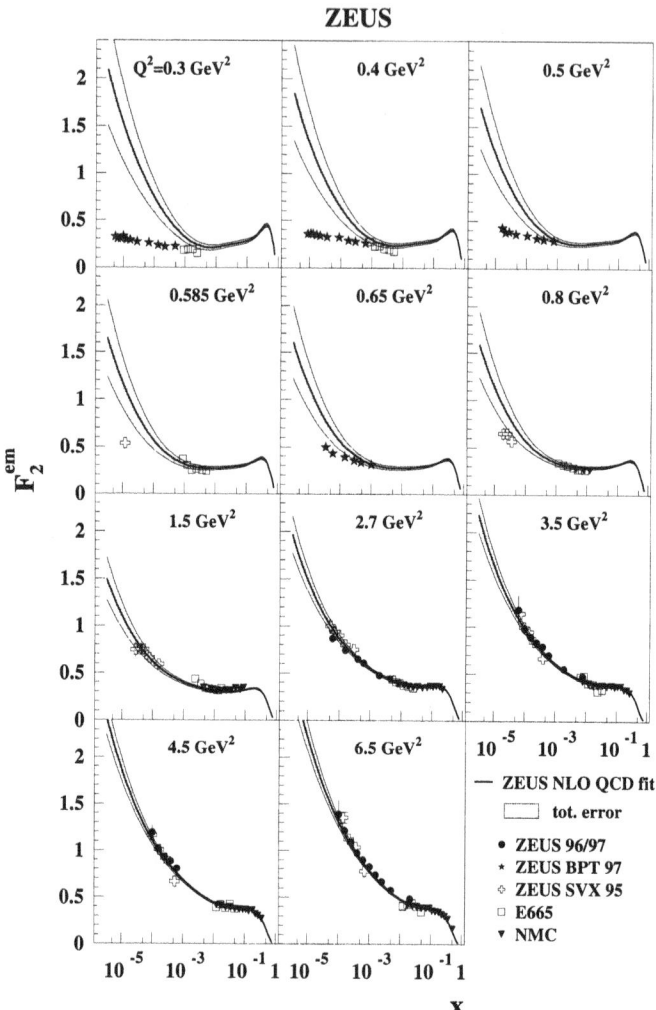

Fig. 6.3 F_2 data down to the very low Q^2 (including BPT data) compared to the standard NLO QCD fit backward extrapolated. The error band illustrated accounts for statistical plus correlated and uncorrelated systematic experimental errors. (From ZEUS 2003b).

contribution of the correlated experimental uncertainties to the gluon error band. A detailed comparison of the contributions to the gluon error bands in the ZEUS and H1 fits is given in Section 6.7 since it illustrates many of the salient points concerning the different methods of treating correlated systematic errors.

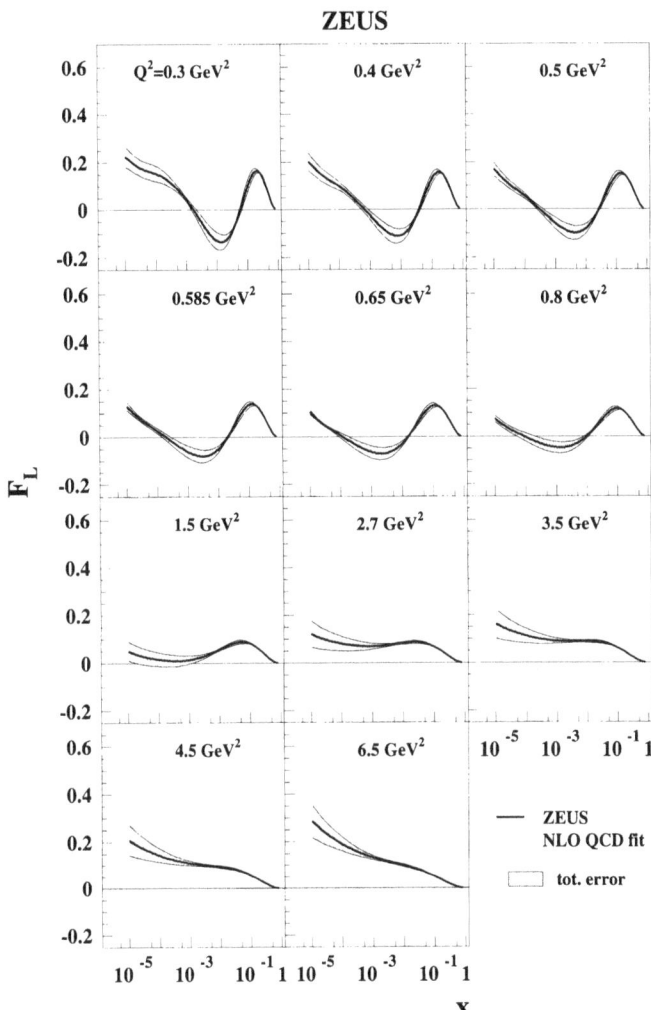

Fig. 6.4 The prediction of the standard NLO QCD fit for F_L backward extrapolated to very low Q^2. The error band illustrated accounts for statistical plus correlated and uncorrelated systematic experimental errors. (From ZEUS 2003b).

6.4.2 Results from the theoretical groups

A comparison of the PDFs from the ZEUS fit (with $\alpha_s(M_Z^2) = 0.118$) to those of the MRST2001 (Martin et al. 2001) and CTEQ6M (Tung et al. 2002) PDFs is shown in Fig 6.7. The resulting PDFs are broadly similar to the ZEUS PDFs and to each other.

The MRST group have also observed that when the starting scale of the

154 *Extraction of parton densities*

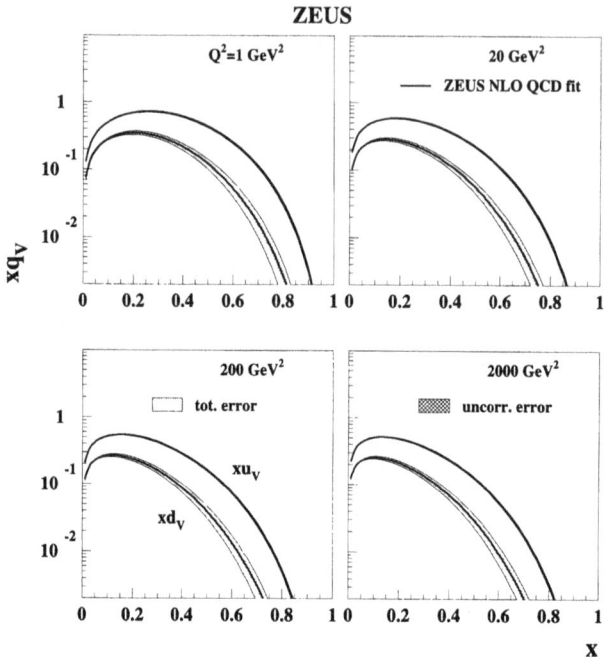

Fig. 6.5 The xu_v and xd_v valence distributions from an NLO QCD fit to ZEUS plus fixed target data for various Q^2. The error bands illustrated account for experimental uncorrelated errors from statistical and systematic sources and for the total experimental errors including those from correlated systematic sources. (From ZEUS 2003b).

fits is dropped to $Q_0^2 \sim 1\,\text{GeV}^2$, or lower, the gluon becomes 'valence-like' in shape, or even negative. In fact, the GRV group predicted such behaviour. The GRV approach is rather different than that of MRS, CTEQ or the experimental teams. A full discussion of it is given in Section 6.8.

The shape of the gluon distribution changes quickly as Q^2 increases. The MRST group studied this by fitting the analytic forms of Eqs (6.6) and (6.7) to their evolved distributions for Q^2 values from 1 to $10^3\,\text{GeV}^2$ and plotting the extracted values of λ_S, λ_g vs. Q^2, (see Fig. 6.8). At low Q^2, $\lambda_g < \lambda_S$; for Q^2 around $10\,\text{GeV}^2$ these values become equal and for larger Q^2, $\lambda_g > \lambda_S$, as expected from $g \to q\bar{q}$ splitting. This behaviour is typical of NLO QCD fits performed within the DGLAP formalism. It is consistent with the behaviour of the gluon and the sea distributions already noted in the ZEUS analysis (Section 6.4.1). It is disturbing that these fits predict that $\lambda_g < \lambda_S$ for $1 < Q^2 \lesssim 10\,\text{GeV}^2$, since it indicates that the naive idea that the rise in the sea distribution is driven by the gluon $g \to q\bar{q}$ splitting cannot be applicable in this region.

It is interesting to compare the slope of the sea distribution, λ_S, shown in Fig. 6.8 to the effective slope of F_2, $\lambda_{\text{eff}} = \partial \ln F_2/\partial \ln(1/x)$, obtained

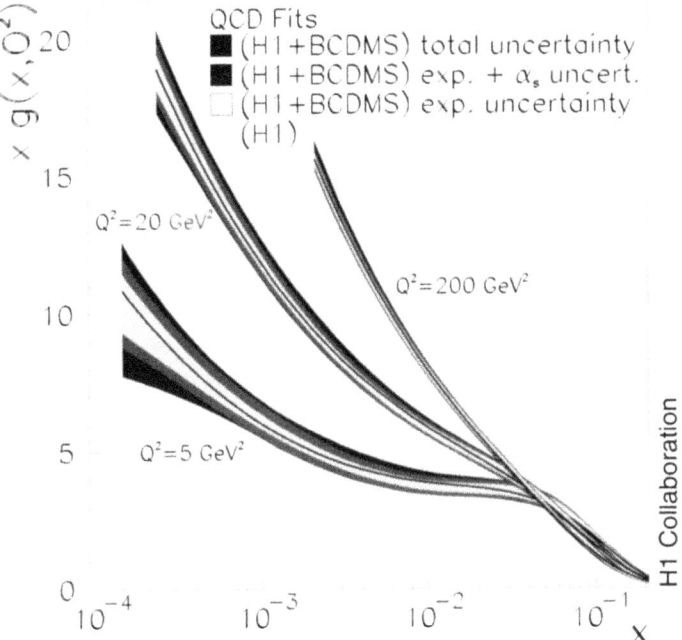

Fig. 6.6 Gluon distribution from an NLO QCD fit to H1 plus BCDMS data, for various Q^2. The inner error band results from the statistical and correlated systematic errors, for $\alpha_s(M_Z^2) = 0.115$ fixed. The central error band shows the additional error induced by letting α_s be a parameter of the fit. The outer error band represents various model errors, such as varying the cuts on the data entering the fit, the value of Q_0^2, and the form of the parameterizations at Q_0^2. (From H1 2001c).

directly from the data and shown in Fig. 9.8. The variation of the measured slope λ_{eff} with Q^2 is compatible with the predictions of the DGLAP formalism for $Q^2 \gtrsim 1\,\text{GeV}^2$.

There is no comparable measured slope which relates to the gluon distribution. The structure function F_L is closely related to the gluon (see Eq. 4.43), but measurements of F_L are not sufficiently accurate to make a direct check of the DGLAP predictions for λ_g. The fact that at low Q^2 the low x gluon is tending to become negative is much more disturbing than the behaviour of the gluon and sea slopes but, since parton distributions have to be convoluted with coefficient functions to produce observable quantities such as structure functions, a negative gluon distribution is not necessarily an indication that the perturbative approach has broken down. A definitive breakdown would be indicated if the structure function F_L were predicted to be negative. An example from the ZEUS fit, indicating such a breakdown for $Q^2 < 1\,\text{GeV}^2$, has already been shown in Fig. 6.4.

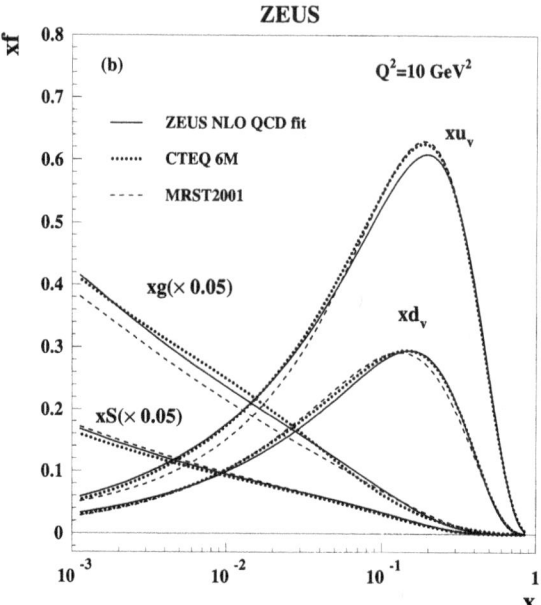

Fig. 6.7 The PDF distributions at $Q^2 = 10 \text{GeV}^2$, from an NLO QCD fit to ZEUS plus fixed target data, compared to the central values of the MRST2001 and CTEQ6M distributions. Note that the gluon and sea distributions are scaled down by a factor of 20. The error band illustrated is that of the ZEUS fit accounting for statistical plus correlated and uncorrelated systematic experimental errors. (From ZEUS 2003b).

A major concern of the the theoretical groups has been to investigate the sensitivity of the extracted PDFs to variations in the model and theoretical assumptions going into the PDF extraction. Accordingly the question of a breakdown in the DGLAP formalism is revisited in Section 6.6.

Before turning to the discussion of model and theoretical assumptions an outline is given of how further information on PDFs can be gained from non-DIS processes, since both the MRST and CTEQ theoretical groups include such data in their analyses in addition to the DIS data.

6.5 Information on PDFs from non-DIS processes

Valuable information on PDFs may also be gained from a variety of non-DIS processes. The data which constrain quark distributions will be considered first.

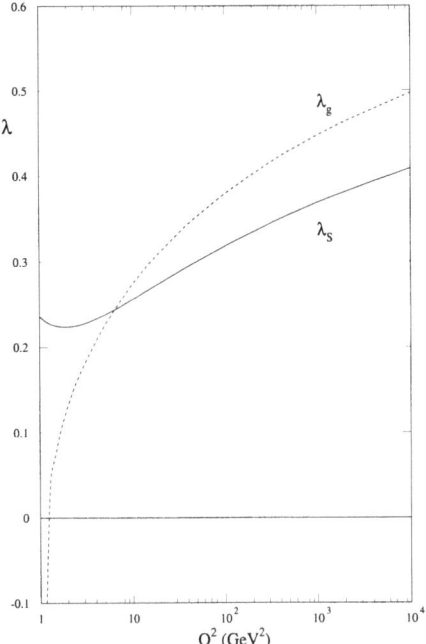

Fig. 6.8 The evolution of the parameters λ_S, λ_g with Q^2 for the MRST98 central fit. (From Martin et al. 1998).

6.5.1 Quark distributions

As discussed in Section 10.3 Drell–Yan dilepton production in the process $pN \to \mu^+\mu^- X$ can be a sensitive probe of the sea quark distribution since the dominant subprocess is $q\bar{q} \to \gamma^*$, and thus the cross-sections are directly proportional to the antiquark distributions. The E605 data have been used to constrain the sea distribution at medium to large x values, $0.15 < x < 0.4$, and the more recent data from E772 have extended the range down to $x \simeq 0.025$. Data on the asymmetry, A_{DY}, between the differential cross-sections $\dfrac{d^2\sigma}{dM\,dy}$ at $y = 0$ for the processes $pp \to \mu^+\mu^- X$ and $pn \to \mu^+\mu^- X$ (where M and y are the invariant mass and rapidity of the lepton pair) can give information on the ratio $\bar{d}/\bar{u}$. The dominant subprocesses are $u\bar{u}, d\bar{d} \to \gamma^*$, and the partons are to be evaluated at $x = M/\sqrt{s}$. Eqs (10.12) and (10.13) give the relationships of the measurements to the parton distributions. NA51 data constrained $\bar{d}/\bar{u}$ at a single x value, $x \sim 0.18$, and more recent data from the E866 experiment has measured the shape of the $\bar{d} - \bar{u}$ in the x range $0.03 < x < 0.35$ (see Fig. 10.5).

As discussed in Section 10.4, data on W production may also be used to investigate quark distributions. The processes $p\bar{p} \to W^+(W^-)X$ proceed via the subprocesses $u\bar{d} \to W^+ (d\bar{u} \to W^-)$ and the cross-sections are thus sensitive to the u and the d distributions at $x \sim M_W/\sqrt{s}$, i.e. $x \sim 0.13$ at CERN, where UA2 data are taken, and $x \sim 0.05$ at FNAL, where CDF and D0 data are taken. The $W^\pm$ charge asymmetry, $A_W(y)$, (discussed in Section 10.4.3) probes the slope of the d/u ratio, since the u quarks carry more momentum on average than the d quarks and so the W^+ tend to follow the direction of the incoming proton and the W^- that of the antiproton. These measurements give information on the d/u distribution which is less sensitive to the contribution of the antiquarks than the F_2^d/F_2^p ratio, at intermediate x values.

6.5.2 The gluon distribution

A common problem in extracting the gluon using only DIS structure function or reduced cross-section data is the limited sensitivity to the gluon at large x, as illustrated clearly by error bands on the ZEUS gluon (see Fig 6.1). The global analyses of MRS, CTEQ and GRV have always included additional data from non-DIS processes in their fits in order to constrain the gluon. Data on prompt photon production and single inclusive jet data from hadronic collisions have been used to date.

The prompt photon process is discussed in Section 10.6. Prompt photon data have been used for the x range: $0.02 < x < 0.5$. The process $pN \to \gamma X$ should provide information on the gluon distribution since the dominant subprocess is $gq \to \gamma q$, at leading order. However, one must also account for non-direct γ production in fragmentation processes from other partons. There is also some uncertainty from factorisation and renormalization scale dependence. Low energy data from WA70, UA6 and E706 on $pp \to \gamma X$ should constrain the gluon in the x range, $0.3 \leq x \leq 0.5$. ISR data extend to lower x, $0.15 \leq x \leq 0.3$, and data from UA2 and CDF extend into the medium-small x region, $0.02 \leq x \leq 0.15$. These higher energy data also indicate that the shape of the E_T distribution of the photons with respect to the beam axis is steeper than that predicted by pQCD calculations even after correction for fragmentation effects. More recent precision data from E706 establishes that this is also the case for lower energy data ($\sqrt{s} = 31.5, 38.8$ GeV).

These data can be described by including the effects of 'intrinsic' k_T, i.e. k_T broadening of initial state partons due to soft gluon radiation with $\langle k_T \rangle \approx 1$ GeV. Such broadening would have to vary with $\sqrt{s}$ (as $\sim \ln(\sqrt{s})$ in order to describe all the data and there is some evidence for this (see Fig. 10.21). E706 data are particularly sensitive to such effects because at higher energy the photon p_T is large and k_T enhancement affects only the lowest end of the spectrum, whereas at lower energies (UA6, WA70) the k_T enhancement can be masked by scale or normalization uncertainties. For E706 data both the shape and the normalization of the p_T spectra are affected. The MRST group addressed this problem by providing several versions of the MRST99 PDFs with more or less k_T broadening, as far as can be consistent with the WA70 data. This results in gluon distributions which are softer or harder at high x, respectively. Unfortunately, not all implementations of the effects of k_T broadening agree with each other so that currently both the CTEQ and the MRST groups consider that, until resummation of multi-gluon radiation is rigorously included in QCD calculations, it is unreliable to use prompt photon data to pin down the gluon.

Further information on the gluon may come from high E_T jet production. There are high statistics jet measurements from the CDF and D0 collaborations at $\sqrt{s} = 1.8$ TeV as discussed in Section 10.5. The single jet inclusive cross-sections in hadro-production, for jets with transverse energies $E_T \sim 100$ GeV, are dependent on the gluon via gg, gq and $g\bar{q}$ initiated subprocesses. For jets with transverse energy $50 \lesssim E_T \lesssim 200$ GeV pQCD calculations are considered reliable. Soft gluon effects are not important because of the high energy scale. The slope of the E_T distribution constrains the combination $\alpha_s(\mu^2) g(x, \mu^2)$ at the scale $\mu^2 = E_T^2$ and $x = 2E_T/\sqrt{s} \sim 0.1$. The CTEQ group, and more recently the MRST group, have included this information in global fits to constrain the high x gluon. A correct treatment of correlated systematic errors is essential because CDF and D0 data are inconsistent if statistical errors alone are considered. The CTEQ group have provided special versions of their fits with the high x gluon enhanced. The initial motivation for this was to describe early measurements of an excess of very high $E_T > 200$ GeV jets at CDF. Even though the Tevatron jet measurements are now consistently described by the standard fits, it remains true that they are better fit by these special fits with a harder high-x gluon.

6.6 Theoretical and 'model' uncertainties

We make a distinction between sources of uncertainty within the theoretical framework of leading-twist NLO QCD (called 'model' uncertainties) and the theoretical uncertainty due to the use of this framework. An obvious example of an uncertainty within the framework is the value of $\alpha_s(M_Z^2)$, since fixed values have been used for many of the fits. There are variants of both CTEQ and MRST fits which allow for different fixed values of $\alpha_s(M_Z^2)$. This uncertainty has also been accounted for by allowing $\alpha_s(M_Z^2)$ to be a parameter of the fit.

Further model uncertainties within the theoretical framework are discussed in Sections 6.6.1–6.6.2, whereas Sections 6.6.3–6.6.5, concern the uncertainty due to the framework. Finally, a small amount of model dependence can be introduced by the method of solving the evolution equations. A discussion of this 'experts' topic is left for the appendix to this chapter.

6.6.1 Model assumptions

There are various common assumptions made in PDF extractions in order to limit the number of parameters. All of these are very reasonable within the framework of the parton model, but all may be challenged since the starting shapes of the parton distributions at Q_0^2 involve unknown non-perturbative input.

- The mathematical form of the parameterization, particularly the polynomial $P(x,i)$, is arbitrary. In principle, it is possible to avoid using a parameterization of the PDFs in x, if the DGLAP equations are solved in n-space (see the Appendix). However, in practice, parameterizations are used to impose smoothness constraints. The CTEQ group have tried various different forms of parameterization for flexibility. However, one should avoid introducing a larger number of parameters than the data can constrain. The technique of searching for the eigenvector basis of the PDF parameters (see Section 6.7.3) can be used to optimize the form of the parameterization, and the number of free parameters, so that they are well constrained by the data.
- It is usually assumed that strong isospin swapping relates the parton distributions in the proton and the neutron such that $d_{\text{proton}} = u_{\text{neutron}}$ and vice versa. However, some authors have considered charge symmetry violation (CSV). There is as yet no evidence for this.
- The breaking of the flavour symmetry of the sea has already been discussed in Section 6.3.2. It is no longer assumed that $\bar{d} = \bar{u} = \bar{s}$. These ratios are determined by the data, and the heavy quarks are treated as explained in Section 6.6.2. However, there is some model uncertainty left in the choice of the heavy quark masses, the parameterization of $\bar{d} - \bar{u}$, and the exact value of the ratio $\bar{s}/(\bar{u} + \bar{d})$, at Q_0^2.
- The violation of $\bar{d} = \bar{u}$ should be expected from Pauli suppression. However the size of the effect is larger than anticipated and meson cloud models involving proton fluctuations into $N\pi$ have been invoked to explain it. Such models would also suggest that the proton may fluctuate into a meson and a baryon such as $K\Lambda$. Thus the s and the $\bar{s}$ quarks may have different momentum distributions, although their numbers must cancel when integrated over all x. This challenges the assumption that $q_{sea} = \bar{q}$, for all flavours. However, the evidence for an $s, \bar{s}$ asymmetry from current $\nu, \bar{\nu}$ scattering data is not compelling (Barone et al. 2000).
- Historically, it was assumed that $d_v/u_v = 1/2$ for all x, but this relationship need only be true for the overall number integrals. The u_v valence shape is not the same as the d_v valence shape as established by the measurements of the ratios $F_2^{\nu p}/F_2^{\bar{\nu} p}$ and $F_2^{\mu D}/F_2^{\mu p}$, which indicate that $d_v/u_v \ll 1/2$ as $x \to 1$ (see Fig. 5.12). This is now built into the PDF extractions such that most parameterizations have $d_v/u_v = (1-x)^p$ $(p > 0)$ as $x \to 1$. Recently, it has been suggested that the nuclear corrections applied to the deuterium target data are inadequate, and that a re-evaluation of these corrections would yield $d_v(x)/u_v(x) \to 0.2$ as $x \to 1$ (Bodek and Yang 2000). Since the cur-

rent $\nu, \bar{\nu}$ scattering data on proton targets are insufficently accurate to confirm this, high Q^2 HERA-II data on $e^+p \to \bar{\nu}X$, $e^-p \to \bar{\nu}X$, will help to clarify this situation.
- All PDF extractions have to make choices such as: the value of Q_0^2; the minimum values of Q^2, W^2, x. All of these can have small systematic effects on the PDF shapes extracted

All groups vary some or all of these input assumptions in evaluating the model uncertainties on their fits. In particular, the MRST group issue different versions of their fits which account for some of these variations. There are variants of the MRST fits to account for larger or smaller amounts of u and d quarks at low x. These derive from varying the normalizations of the HERA data sets up and down within their quoted normalization errors. There are variants with more or less s quarks which derive from changing the strangeness suppression applied at Q_0^2 by $\pm 10\%$ and there are variants with more or less c quarks which derive from varying the mass of the charmed quark. A special variant with a finite d/u ratio at large x is also available, because the usual parameterizations of d_v and u_v are not optimal for estimating the error on a prediction such as $d_v/u_v = 0.2$.

6.6.2 Heavy quark production schemes

The treatment of heavy quark thresholds, needs careful consideration. The theoretical background to this has been given in Section 4.5. Historically, the global fits used a zero-mass variable flavour number (ZM-VFNS) scheme, assuming $c(x, Q^2) = 0$ for $Q^2 \leq m_c^2$. The charm content of the nucleon at higher Q^2 is then generated by the boson–gluon fusion process, as embodied in the DGLAP equations for massless partons. Clearly such a treatment of heavy quarks can only be valid far above the relevant thresholds, $W^2 \gg 4m_c^2$. Near threshold, a proper treatment of quark mass effects is required. The fixed flavour number schemes (FFNS) treat the threshold correctly, generating charm by BGF for $W^2 > 4m_c^2$, but $\ln(Q^2/m_c^2)$ terms are not resummed for higher Q^2. One requires a consistent treatment of charm from threshold to high Q^2. The formalism of such general mass variable flavour number schemes (GM-VFNS), has been described in Section 4.5.2 and is used in many of the modern global fits.

The contributions of bottom and top quarks are treated similarly to charm, and but they turn out to be negligible for present data. A further small uncertainty comes from the exact values used for the heavy quark masses and thresholds.

One should also be careful when comparing data from neutrino and muon/electron scattering, since the effect of mass thresholds will be different. In neutrino scattering, charmed quarks can be produced directly from scattering off the strange sea, as well as in the boson–gluon fusion process (flavour excitation, $W^*s \to c$, as well as flavour creation, $W^*g \to s\bar{c}$) and the former process is dominant at the energies at which present day neutrino data have been taken. Historically, neutrino data were corrected with a 'slow rescaling' prescription to allow for the charm mass threshold in the flavour excitation process. However, this is not really adequate when performing a NLO analysis. This was the source of a long standing disagreement between F_2 measurements in muon (NMC) and neutrino (CCFR) scattering. A full GM-VFN treatment of the flavour creation process has resolved much of this discrepancy, see Section 5.6.2.

The CTEQ group provide versions of their fits using different heavy quark production schemes: zero mass variable flavour number, general mass variable flavour number and fixed flavour number with either three or four flavours. Such variants are needed since the GM-VFN schemes have not been fully worked out for all non-DIS processes of interest and a consistent set of PDFs worked out in the same scheme may be needed when comparing predictions to data.

6.6.3 Higher twist contributions

The DGLAP equations embody the predictions of pQCD in the NLLA at leading twist only. Thus one must be sure that the data which are fitted are not likely to be subject to strong higher twist corrections.

As explained in Section 4.4, kinematic higher twist terms, or target mass corrections, arise from the neglect of $O(m^2/Q^2)$ terms in the formalism. These may be accounted for by modifying the scaling variable x. Dynamical higher twist terms have been parameterized by multiplying QCD prediction for the leading twist structure function by a factor $(1 + h_i/Q^2)$, where h_i are parameters fitted separately in each x bin i for which one has data. Fits to the early data (see also Fig. 4.8) indicated that such contributions are large only for high x and low Q^2. Thus the traditional consensus is that one may cut out the need for higher twist contributions by cutting out the kinematic region in which they are important. A standard choice is to include only data for which $Q^2 \gtrsim 4\,\text{GeV}^2$, $W^2 \gtrsim 10\,\text{GeV}^2$.

However note that: (i) F_L may be subject to higher twist corrections for all x, at low Q^2; (ii) the renormalon predictions indicate that xF_3 may have small higher twist corrections for all x; (iii) if the Q^2 cut is lowered to $1\,\text{GeV}^2$, as in many of the recent global fits, one should be more cautious.

In HERA, very low x data one might expect to see the effects of unconventional higher twist effects which contribute to parton shadowing. The MRST group have investigated whether there is any evidence for such low x higher twist contributions in the HERA data and in particular whether such contributions could account for the more disturbing features of the DGLAP fits at low Q^2. Low W^2 data ($W^2 < 10\,\text{GeV}^2$) were included in these special fits and data were fitted down to $Q^2 = 1\,\text{GeV}^2$. Multiplicative higher twist terms of the form suggested above were fitted in different regions of x. It was found that the higher twist contribution is negligible for $x < 10^{-2}$. This is in agreement with some of the low x higher twist calculations which suggest cancellations in the contributions to F_2. The MRST group also tried forcing the gluon at $Q_0^2 = 1\,\text{GeV}^2$ to be flat or singular to see whether a contribution from higher twist at low x could mimic the effect of the 'valence-like' gluon found in the standard fits. This possibility is strongly excluded.

At larger x, MRST found that small negative higher twist terms do improve the χ^2 for $10^{-2} < x < 0.5$ (where it is the χ^2 for NMC data which improves the most) and large positive higher twist terms are needed at large x because target mass effects are not included in the formalism. The 'feedback' of these higher twist terms into the standard MRST fits for higher W^2 represents a only small shift in the standard PDF parameters.

6.6.4 The need to go beyond NLO and scale uncertainty

Conventionally, analyses are done assuming that the factorization scale, μ_f^2, and the renormalization scale, μ_r^2, are both equal to Q^2, the natural choice of scale for the DIS process. The shapes of the extracted PDFs are not very sensitive to the choice of scale because the PDF parameters adjust themselves to reproduce very similar shapes, well within the experimental errors on the PDFs. The fit parameter which is most sensitive to scale choice is $\alpha_s(M_Z^2)$ and this is discussed further in Section 7.2.1.

However, although NLO DGLAP QCD fits give good descriptions of F_2 data down to Q^2 values in the range 1–2 GeV^2. Such fits assume the validity of the NLO DGLAP QCD formalism even for low Q^2, where α_s is becoming larger such that NNLO corrections are increasingly important. Varying the renormalization amd factorization scales has been used to obtain a crude estimate of the importance of NNLO corrections which are not included in the analyses.

It is no longer necessary to rely on such estimates since calculations to NNLO are now available and although the calculations of the three-loop splitting functions are as yet incomplete, the results are considered reliable to within 2% for $x > 10^{-4}$. The impact of such contributions has been considered by van Neerven and Vogt (2000) who have examined the effects of LO, NLO and NNLO evolution on a fixed set of partons (see Section 4.6).

There have always been variants of the standard MRST and CTEQ fits done at LO rather than NLO. More recently NNLO variants have become available. Rather than considering the differences between LO, NLO and NNLO on a fixed set of partons, as considered by van Neerven and Vogt, MRST have considered a complete re-fit of the data at each order. Figure 6.9 displays the gluon distributions for fits done to LO, NLO and two versions of NNLO which represent extremes of fast and slow evolution compatible with the current level of our knowledge. It is evident that the gluon PDF is very sensitive to the difference between LO/NLO and NNLO for low Q^2, and this sensitivity persists at low x even for relatively high Q^2. The NNLO predictions for the gluon PDF are negative at low x even for Q^2 as high as $Q^2 = 2\,\text{GeV}^2$. The residual scale dependence of PDF shapes and $\alpha_s(M_Z^2)$ at NNLO is considerably reduced.

164 *Extraction of parton densities*

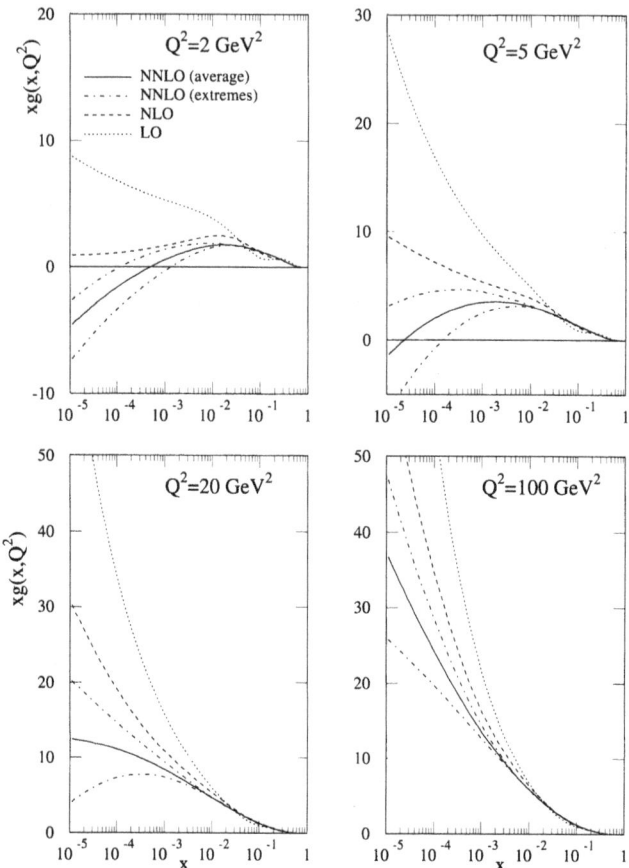

Fig. 6.9 Comparison of gluon PDFs at various Q^2 from MRST DGLAP fits performed to LO, NLO and two extreme versions of NNLO. (From Martin et al. 2000).

MRST have also investigated higher twist terms within an NNLO analysis. Strong positive higher twist contributions are still necessary at large x and low W^2 to account for target mass effects. However, the small negative higher twist contributions at intermediate x are much reduced, confirming the idea from renormalon theory that the higher twist terms are related to the truncation of the perturbative series.

In addition to NNLO corrections, soft gluon resummations of $\alpha_s \ln(1-x)$ terms may be necessary at high $x > 0.8$. Current data do not reach such large x values, with the exception of the very low Q^2 SLAC data, which are subject to large higher twist effects and are usually excluded from QCD fits for this reason.

6.6.5 Alternatives to the DGLAP evolution equations?

It is implicitly assumed that the DGLAP evolution equations can be applied in the kinematic range studied. However, at low x, it may be necessary to consider the BFKL or CCFM equations, or at least to include $\alpha_s \ln(1/x)$ resummation terms, as discussed in Chapter 9. The use of the DGLAP equations down to low $Q^2 \sim 1\,\text{GeV}^2$ is also questionable, not just because non-perturbative physics is important at low Q^2, but also because at low Q^2 and low x high parton densities may require the use of non-linear evolution equations (see Section 9.6).

The gluon extraction is particularly vulnerable to the assumption that the DGLAP evolution equations can be applied in the kinematic range studied. The difference between the standard DGLAP formalism and an alternative analysis including $\ln(1/x)$ resummation terms (discussed in Sections 9.1.1 and 9.4.4) is illustrated in Fig. 6.10 (Thorne 2002) for the prediction for F_L. At NLO, F_L is predicted to be negative at small x and Q^2. At NNLO, the coefficient functions have compensated for this, resulting in the peculiar shape of F_L. This shape becomes much more reasonable for the analysis including $\ln(1/x)$ resummation terms.

6.7 Treatment of correlated systematic uncertainties

The validity of a χ^2 fit and the errors on PDF parameters and quantities deduced from them depends on the assumption that input experimental errors are Gaussian distributed and that the shape of the χ^2 function around its minimum is well represented by the quadratic approximation, such that parameter errors are symmetric. Correlated systematic errors are frequently not Gaussian so that their treatment needs some care.

Traditionally point-to-point correlated systematic errors have not been correctly treated. They have been added in quadrature to the uncorrelated errors such that the χ^2 was formulated as

$$\chi^2 = \sum_i \frac{\left[F_i^{\text{NLO QCD}}(p) - F_i(\text{meas})\right]^2}{\sigma_i^2 + \Delta_i^2} \tag{6.11}$$

where $F_i^{\text{NLO QCD}}(p)$ represents the prediction from NLO QCD in terms of the theoretical parameters p for data point i, $F_i(\text{meas})$ the measured data point, σ_i the one standard deviation uncorrelated error on this data point from both statistical and systematic sources and Δ_i the one standard deviation correlated systematic error on the data point — treated as uncorrelated.

The correct formulation of the χ^2 including correlated systematic uncertainties is

$$\sum_i \sum_j \left[F_i^{\text{NLO QCD}}(p) - F_i(\text{meas})\right] V_{ij}^{-1} \left[F_j^{\text{NLO QCD}}(p) - F_j(meas)\right] \tag{6.12}$$

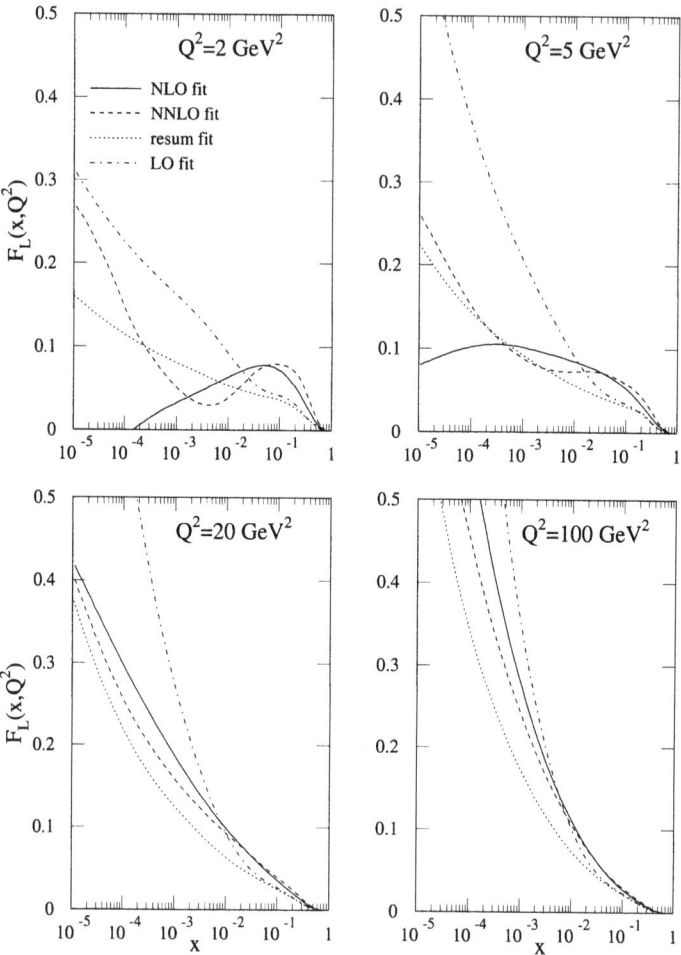

Fig. 6.10 Comparison of F_L at various Q^2 from MRST01 DGLAP fits performed to LO, NLO and NNLO, to the result of a fit including $\ln(1/x)$ resummation terms (Thorne 2002).

where

$$V_{ij} = \delta_{ij}\sigma_i^2 + \sum_\lambda \Delta_{i\lambda}^{\text{sys}} \Delta_{j\lambda}^{\text{sys}}$$

is the covariance matrix and the symbol $\Delta_{i\lambda}^{\text{sys}}$ represents the one standard deviation correlated systematic error on data point i due to correlated error source λ. Treating correlated errors as uncorrelated is equivalent to constructing $\Delta_i^2 = \sum_\lambda (\Delta_{i\lambda}^{\text{sys}})^2$

An alternative definition of χ^2, which gives equivalent results to Eq. (6.12), is constructed as follows. The correlated uncertainties are included in the theoretical prediction such that

$$F_i(p, s) = F_i^{\text{NLOQCD}}(p) + \sum_\lambda s_\lambda \Delta_{i\lambda}^{\text{sys}}$$

where the parameters s_λ represent independent variables for each source of systematic uncertainty. They have zero mean and unit variance by construction. The χ^2 is then formulated as

$$\chi^2 = \sum_i \frac{[F_i(p, s) - F_i(\text{meas})]^2}{\sigma_i^2} + \sum_\lambda s_\lambda^2 \qquad (6.13)$$

There are then two different ways to proceed. The systematic uncertainty parameters s_λ can be fixed to zero for minimisation, but used in error calculations ('Offset methods'), or these parameters can be fitted together with the theoretical parameters ('Hessian methods').

6.7.1 Offset methods

Traditionally, experimentalists have used 'offset' methods to account for correlated systematic errors. The χ^2 is formulated without any terms due to correlated systematic errors ($s_\lambda = 0$ in Eq. (6.13)) for evaluation of the central values of the fit parameters. However, the data points are then offset to account for each source of correlated systematic error in turn (i.e. set $s_\lambda = +1$ and then $s_\lambda = -1$ for each source λ) and a new fit is performed for each of these variations. The resulting deviations of the theoretical parameters from their central values are added in quadrature. (Positive and negative deviations are added in quadrature separately.) This method does not assume that the correlated systematic uncertainties are Gaussian distributed.

An equivalent (and much more efficient) procedure for performing the offset method has been given by Pascaud and Zomer (1995). The systematic uncertainty parameters s_λ are fixed to zero for minimisation. However, the s_λ are varied for the error analysis, such that in addition to the usual Hessian matrix, M_{jk}, given by

$$M_{jk} = \frac{1}{2} \frac{\partial^2 \chi^2}{\partial p_j \, \partial p_k},$$

which is evaluated with respect to the theoretical parameters, a second Hessian matrix, $C_{j\lambda}$, given by

$$C_{j\lambda} = \frac{1}{2} \frac{\partial^2 \chi^2}{\partial p_j \, \partial s_\lambda}$$

is evaluated. The M matrix expresses the variation of χ^2 with the theoretical parameters, accounting for uncorrelated errors from both statistical and systematic sources. It is the inverse of the covariance matrix $V^p = M^{-1}$ which accounts for uncorrelated sources of error. The C matrix expresses the variation of the χ^2 with respect to theoretical and systematic uncertainty parameters. The covariance matrix which accounts for correlated systematic errors is then given by $V^{ps} = M^{-1}CC^TM^{-1}$ and the total covariance matrix by $V^{tot} = V^p + V^{ps}$. Then the standard propagation of errors to any distribution F which is a function of the theoretical parameters is given by

$$<\sigma_F^2> = T \sum_j \sum_k \frac{\partial F}{\partial p_j} V_{jk} \frac{\partial F}{\partial p_k} \qquad (6.14)$$

by substituting V^p, V^{ps} or V^{tot} for V, to obtain the uncorrelated (from both statistical and systematic sources), correlated systematic or total experimental error band, respectively. The quantity T represents the χ^2 tolerance, which will be explained in Section 6.7.6. For the offset method, $T = 1$.

This is a conservative method of error estimation compared to the Hessian methods which will be described below. It gives fitted theoretical predictions which are as close as possible to the central values of the published data. It does not use the full statistical power of the fit to improve the estimates of s_λ, since it choses to mistrust the systematic error estimates, but it is correspondingly more robust.

6.7.2 Hessian methods

An alternative procedure is to allow the systematic uncertainty parameters s_λ to vary in the main fit when determining the values of the theoretical parameters. The errors on the theoretical parameters, and any distribution which depends on them, are then calculated from Eq. (6.14) using the covariance matrix V, which is the inverse of a single Hessian matrix expressing the variation of χ^2 with respect to both theoretical and systematic offset parameters. Effectively, the theoretical prediction is not fitted to the central values of the published experimental data, but allows these data points to move collectively, according to their correlated systematic uncertainties. The theoretical prediction determines the optimal settings for correlated systematic shifts of experimental data points such that the most consistent fit to all data sets is obtained. Thus systematic shifts in one experiment are correlated to those in another experiment by the fit. It is necessary to check that the fitted values for the systematic error parameters s_λ do not imply that data points move far outside their one standard deviation errors. Such shifts can indicate inconsistency of a data set, or part of a data set, with respect to the rest of the data. Provided that this is not the case, a further check should be made that superficial changes to the model choices of the fit (such as the choice of Q_0^2, the form of the parameterization at Q_0^2, etc.) do not result in large changes to the values of the systematic error parameters, since this can also indicate data inconsistency.

The formulation of the Hessian method, using Eq. (6.13), becomes a cumbersome procedure when the number of sources of systematic uncertainty is large. Recently CTEQ (Tung et al. 2002) have given an elegant analytic method for performing the minimization with respect to systematic–uncertainty parameters. They show that the χ^2 can be written as

$$\chi^2 = \sum_i \frac{\left[F_i^{\text{NLO QCD}}(p) - F_i(\text{meas})\right]^2}{\sigma_i^2} - BA^{-1}B \qquad (6.15)$$

where

$$B_\lambda = \sum_i \Delta_{i\lambda}^{\text{sys}} \frac{\left[F_i^{\text{NLO QCD}}(p) - F_i(\text{meas})\right]}{\sigma_i^2}$$

and

$$A_{\lambda\nu} = \delta_{\lambda\nu} + \sum_i \Delta_{i\lambda}^{\text{sys}} \Delta_{i\nu}^{\text{sys}} / \sigma_i^2,$$

such that the contributions to the χ^2 from correlated and uncorrelated sources can be evaluated separately. The problem of large systematic shifts to the data points, which can be encountered in the first formulation of the Hessian method, now becomes manifest as a large value of the correlated contribution to the χ^2 in Eq. (6.15), such that a small overall value is achieved by the cancellation of two large numbers. Different authors have chosen to approach this sort of problem in different ways. Some restrict the data sets entering the fits to use only sets which are sufficiently consistent with each other that such problems do not arise. Others use all data sets but use a larger tolerance in Eq. (6.14) as discussed in Section 6.7.6.

6.7.3 Diagonalization and eigenvector PDF sets

In either the Hessian or the offset method, the Hessian matrices and covariance matrices are not, in general, diagonal. The variation of χ^2 with respect to some parameters is much more rapid than that of others, but because the parameters are correlated with each other the effect of each parameter is not clear. When evaluating uncertainties on physical observables, it can be an advantage to use an eigenvector basis of PDFs, which provide an optimized representation of parameter space in the neighbourhood of the minimum. The eigenvalues of the covariance matrix represent the squares of the errors on the combination of parameters which gives the corresponding eigenvector. Thus for a stable fit these values must always be positive. Given that this is true, for a typical fit their values span an enormous range. An eigenvector with a small eigenvalue corresponds to a steep direction in parameter space, i.e. a direction in which χ^2 rises rapidly, so that the combination of parameters represented by this eigenvector are tightly constrained by the data. Conversely, large eigenvalues correspond to combinations of parameters which are weakly constrained. In such cases, the quadratic approximation for the shape of χ^2 around its minimum may break down. It is not useful to introduce more parameters into the fit than the data can reliably constrain. Very often an eigenvector combination of parameters is dominated by only one or two of the parameters. Thus identifying eigenvectors tells us which parameters are best constrained and adding or removing data sets from the fit tells us which data are responsible for constraining which parameters. Hence useful constraints and efficient ways of parameterizing the PDFs can be identified.

An eigenvector basis of PDFs may also be used to summarize the results of a PDF analysis including error estimates. In order to use Eq. (6.14) to calculate the predicted uncertainty on any physical quantity of interest one has to know the covariance matrix. This can be made simpler if an eigenvector basis of several PDF sets is supplied. Two sets of PDF parameters must be supplied for each eigenvalue, representing displacements up and down the eigenvector direction by the χ^2 tolerance T. The error on a quantity F which is a function of the PDF parameters is then simply given by

$$<\sigma_F^2> = \sum_j \left[\frac{F(p_j^+) - F(p_j^-)}{2}\right]^2$$

where $F(p_j^+)$, $F(p_j^-)$ are the values of F evaluated up and down the direction of eigenvector j.

6.7.4 Normalizations

Normalization errors are usually treated separately from other systematic errors since, for many experiments, quoted normalization uncertainties represent the limits of a box-shaped distribution rather than the standard deviation of a Gaussian distribution. To account for this, the normalizations of the data sets are introduced as parameters of the fit, which are applied multiplicatively to the measurement and its statistical error. If the normalization uncertainties are to be treated as Gaussian, then a penalty term equal to the number of standard deviations by which the fitted normalization of each experiment differs from its nominal value, is added to the χ^2. If the normalization uncertainties are considered to represent the limits of a box–shaped distribution then the normalization parameters can be restricted within limits given by their quoted uncertainties, and/or the penalty χ^2 can be augmented to account for this.

6.7.5 Comparison of offset and Hessian methods

The ZEUS Collaboration has compared the offset method to the Hessian method by performing the error analysis in their global fit by both methods. Figure 6.11 compares the gluon distributions extracted by these methods, for several Q^2 values. The Hessian method gives much reduced 'optimal' experimental error estimates. The shape of the gluon PDF is also shifted to be somewhat steeper than that obtained by the offset method for low

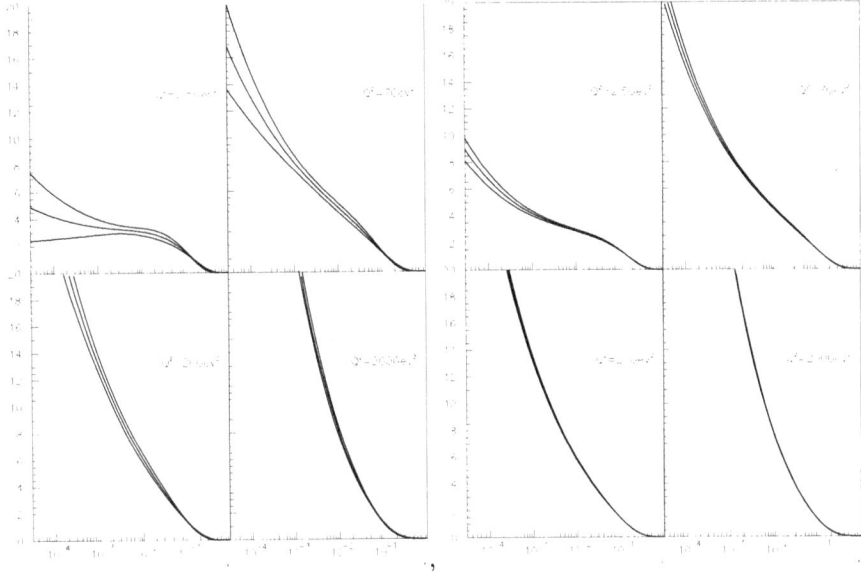

Fig. 6.11 Left plot: the gluon PDF from the ZEUS analysis, using the offset method to account for experimental correlated systematic uncertainties. Right plot: the gluon PDF from the ZEUS analysis, using the Hessian method to account for experimental correlated systematic uncertainties, with χ^2 tolerance $T = 1$.

Q^2, but this difference is within the error estimates quoted for the offset method. The other PDFs are not significantly shifted. It is significant that the values of some of the systematic error parameters in the Hessian method do imply that data points will move outside their one standard deviation errors. Hence questions of data set inconsistency arise. A further pragmatic modification to the Hessian method has been suggested to account for this. This involves consideration of the χ^2 tolerance when assigning errors.

6.7.6 χ^2 tolerance

Assuming that the experimental uncertainties which contribute have Gaussian distributions, the errors on theoretical parameters which are fitted within a fixed theoretical framework are derived from the criterion for 'parameter estimation' $\chi^2 \to \chi^2_{\min} + 1$. On the other hand the goodness of fit of a theoretical hypothesis is judged on 'hypothesis testing' criterion, such that its χ^2 should be approximately in the range $N \pm \sqrt{(2N)}$, where N is the number of degrees of freedom.

Fitting DIS data for PDF parameters and $\alpha_s(M_Z^2)$ is not a clean situation of either parameter estimation or hypothesis testing. Within the theoretical framework of leading-twist-NLO QCD, many model inputs such as the form of the PDF parameterizations; the minimum Q^2, x and W^2 of data entering the fit; the value of Q_0^2; the choice of data sets used in the fit; the choice of the heavy-quark production scheme etc., can be varied. These represent different hypotheses and they are accepted, provided the fit χ^2 fall within the hypothesis-testing criterion. The theoretical parameters obtained for these different hypotheses can differ from those obtained in the standard fit by more than their errors as determined from the experimental uncertainties. In this case, the model uncertainty on the parameters exceeds the uncertainty due to experimental sources. This does not generally happen for the offset method in which the uncorrelated experimental uncertainties on the parameters are augmented by the contribution of the correlated experimental systematic uncertainties as explained in Section 6.7.1. However, when fits are performed using the Hessian method, the shifts in theoretical parameter values for different model hypotheses are often outside the estimates of the errors on the parameters due to experimental sources.

A comparison of the ZEUS and H1 analyses of the uncertainties on their gluon extractions illustrates this point. Compare Figs 6.1 and 6.6 for ZEUS and H1, respectively. The contribution to the error band from uncorrelated and correlated experimental sources are labelled 'exp. uncertainty' in the H1 plot of Fig. 6.6 and labelled 'tot. error(α_s fixed)' in the ZEUS plot of Fig. 6.1. These experimental errors are somewhat more restrictive for the H1 analysis than for the ZEUS analysis, reflecting the use of the Hessian and the offset methods, respectively. However, the use of the more conservative offset method in the ZEUS analysis ensures that model uncertainties are negligible by comparison, whereas the H1 analysis has extra model uncertainties which contribute significantly to error band labelled 'total uncertainty'. However, since this band also includes the additional uncertainty on the gluon PDF due to the variation of $\alpha_s(M_Z^2)$, one can only compare like with like once this uncertainty is taken into account in the ZEUS analysis. Thus one may compare the total error bands of both analyses which are labelled 'total uncertainty' in Fig. 6.6 and 'tot. error (α_s free)' in Fig. 6.1). The size of this total error on the gluon for each analysis is similar: the gluon is determined to within $\sim 15\%$ for $Q^2 > 20\,\text{GeV}^2$, $10^{-4} < x < 10^{-1}$. Thus the error estimates are similar once model uncertainty has been accounted for.

One of the motivations behind making NLO QCD fits to DIS data is to estimate errors on the PDF parameters and $\alpha_s(M_Z^2)$ within a general theoretical framework, rather than as specific to particular model choices within this framework. Thus, one needs to consider the additional error due to model choices very carefully.

First one should consider whether the strict application of the χ^2 tolerance for parameter estimation is appropriate. The CTEQ collaboration have investigated this problem in detail. They consider that $\chi^2 \to \chi^2 + 1$ is not a reasonable tolerance on a global fit to ~ 1500 data points from diverse sources, with theoretical and model uncertainties which are hard to quantify and experimental uncertainties which may not be Gaussian distributed. They have tried to formulate criteria for a more reasonable setting of the tolerance T, such that $\chi^2 \to \chi^2 + T^2$ becomes the variation on the basis of which errors on parameters are calculated. In setting this tolerance, they have considered that all of the current world data sets must be acceptable and compatible at some level, even if strict statistical criteria are not met, since the conditions for the application of strict criteria, namely Gaussian error distributions, are also not met. The level of tolerance they suggest is $T \sim 10$. The situation is illustrated in Fig 6.12, where the dis-

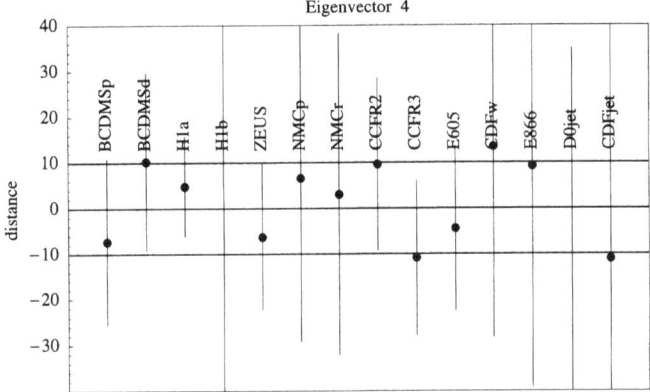

Fig. 6.12 Distance along a parameter combination (eigenvalue 4, CTEQ6) from the χ^2-minimum of an individual data set to the global minimum. The error bar for each individual data set represents the acceptable range of the eigenvalue for this experiment within a 90% confidence level. In the neighborhood of the global minimum a distance of 1 in the ordinate corresponds to $\Delta\chi^2_{\text{global}} \approx 1$. Similar plots are found for all the eigenvalues of the CTEQ6 fit. (From Tung et al. 2002.)

tances from the χ^2-minima of individual data sets to the global minimum for all the data sets is illustrated. These distances by far exceed the range allowed by the $\Delta\chi^2_{\text{global}} = 1$ criterion. Strictly speaking such variations can indicate that data sets are inconsistent and some authors have chosen to work with restricted data sets in order to reduce this variation. The CTEQ collaboration consider that it is not possible to simply drop "inconsistent" data sets, as then the partons in some regions would lose important constraints. On the other hand, the level of "inconsistency" should be reflected in the uncertainties of the PDFs. This is achieved by modification of the χ^2 tolerance to $T = 10$, as shown by the horizontal lines in Fig. 6.12.

A further motivation for using a tolerance of this size comes from considering the comparison of the offset and Hessian methods of Section 6.7.5. Interestingly, if the errors for the Hessian method are re-evaluated for various values of the tolerance, then it is found that for $T = 7$ the errors on the

PDF parameters are comparable to those of the offset method. If increased tolerances are used in the Hessian method, then the model uncertainties on the PDFs are generally much less than the experimental uncertainties, just as for the offset method. In both cases this then gives PDF sets and their errors robust against model changes. Nevertheless the use of any other tolerance than $T = 1$ must be regarded as controversial.

6.7.7 Uncertainties in predicting high energy cross-sections

A realistic estimate of PDF uncertainties is vital when using PDFs as input for calculations of high energy scattering processes in order to constrain the parameters of the Standard Model, or even to investigate physics beyond the Standard Model. It is necessary to consider variants on the standard PDFs in which the parameters most sensitive to the derived result are varied as widely as possible, while maintaining a reasonable global fit quality. An example of such a study is discussed in Section 10.7.2.

CTEQ have developed a general technique for such studies using Lagrange multipliers. The method uses the variation of the effective χ^2 along a specific direction in PDF parameter space — that of maximum variation of the physical variable of interest. The result is an optimized sample of PDFs — optimized to give low or high values of the cross-section of interest-while remaining consistent with current experiments. The method starts with a standard set of PDFs from a global fit, which represent the best estimate consistent with current experiment and theory. The global χ^2_{global} for this set is denoted χ^2_0 and the parameters for this set are called $\{p_0\}$. If X is a physical quantity of interest then the estimate of this quantity from the standard PDF set is denoted $X(p_0)$. The Lagrange multiplier method assesses the error on this quantity by considering the minimisation of the function

$$\Psi(\lambda, p) = \chi^2_{global}(p) + \lambda X(p)$$

with respect to the PDF parameters, p, for fixed values of λ. For each value of λ there is a set of parameters $\{p_\lambda\}$, a value of $\chi^2_{global,\lambda}$ and a value of X_λ, for the minimum of Ψ. This gives a parametric relationship between X and χ^2_{global}, such that one may plot the curve $\chi^2_{global}(X)$ vs. X. This will have its minimum at $X = X_{(p_0)}$ and one can define the errors on X from the chosen tolerance on $\Delta\chi^2_{global}$ around this minimum. This method is more robust than traditional error propagation because it does not approximate X or χ^2_{global} by a linear or quadratic dependence on $\{p\}$ around the minimum. It can be generalised to many observables by introducing a Lagrange multiplier for each observable.

6.7.8 Alternative statistical techniques

These techniques go further than the method suggested in the previous section in trying to remove the assumptions of Gaussian approximations and the linearization of uncertainties. The method of χ^2 fit analysis is abandoned altogether.

The essential problem in PDF analysis is how to determine a function from a finite number of data points? One needs a probability density measure over the functional space of possible PDFs. The mean and deviation of an observable O are then given by

$$<O> = \sum_i O(PDF_i) P(PDF_i)$$

and

$$<\sigma_O^2> = \sum_i (O(PDF_i) - <O>)^2 P(PDF_i)$$

where $O(PDF_i)$ is the value of O based on PDF set i and $P(PDF_i)$ is the weight of PDF set i according to the probability density measure.

Monte-Carlo methods are used to construct the measure. A functional form for the PDF distribution at Q_0^2 (such as those given in Section 6.3) is chosen and many such PDF sets can be generated with different parameter values such that the prior probability distribution of the parameter values is flat. This prior is then improved by using Bayesian statistical inference to update it iteratively based on the available data. The expectation values of of the mean and deviation for each measurement must eventually reproduce the central experimental values, with covariance matrix and error matrix equal to the corresponding experimental quantities. Arbitrarily many such Monte Carlo sets can be generated such that relevant properties of the data set such as the correlations between systematic errors are correctly reproduced by the whole sample. The aim is to optimally combine statistical errors while reproducing the correlated systematic errors.

A variation on this approach is to train a neural network to learn the shape of each of the data sets thus reducing dependence on assumed functional forms, but applying requirements such as smoothness in the training. The final set of networks provides a representation of the probability measure. The success of this technique requires strict application of criteria for compatibility of data sets and thus it has so far only been applied to restricted data sets.

6.8 Dynamically generated partons, the GRV approach

Whereas the PDFs provided by the MRST and CTEQ groups have some dependence on the non-perturbative input parameterization at Q_0^2, the PDFs of the GRV group are designed to be almost independent of their inputs (Glück et al. 1992, 1994, 1998).

The original idea behind these parameterizations is that at some VERY LOW scale $Q^2 = \mu^2$ the nucleon consists only of constituent valence quarks. As Q^2 increases, one generates the gluons and sea quarks in the nucleon dynamically from these valence quarks, through the conventional DGLAP equations for the processes $q \to qg$, $g \to q\bar{q}$. It did not prove possible to describe all relevant data using such a picture, and it was accordingly modified to include gluons and sea quarks at the starting scale μ^2, but these distributions have a valence-like shape (i.e. non-singular at small x). One may even interpret this picture by saying that the gluons and sea quarks are frozen upon the valence current quarks for $Q^2 < \mu^2$, and these composite objects form the constituent quarks.

The scale μ^2 is set as the scale at which the gluon distribution is as hard as the valence quark distribution u_v. The splitting $q \to qg$ naturally generates a softer distribution in the split products than that of the original quark, hence a gluon distribution which is harder than the valence quark distribution which generates it is considered physically unreasonable. Thus the scale μ at which $g \sim u_v$ is assumed to be the lowest scale at which the DGLAP equations may be used.

It turns out that $\mu^2 \sim 0.4\,\text{GeV}^2$ at NLO, which seems a very low scale at which to use perturbative QCD. However, GRV consider that it is the value of $\alpha_s(\mu^2)/\pi \sim 0.2$, which determines the relative size of higher order corrections, not the value of $\alpha_s(\mu^2)$, and that their predictions are perturbatively stable between LO and NLO, if one considers measurable quantities like structure functions, rather than parton distributions. They emphasize that their calculation is only applicable to the leading twist operators of QCD, and clearly at very low Q^2 higher twist operators can be important. Hence GRV state that the calculation can be made consistently at low Q^2, but they only expect the resulting parameterization to describe reality for somewhat higher Q^2 ($\gtrsim 1\,\text{GeV}^2$).

Whereas at larger x, $x > 0.01$, the approach is similar to that of MRS and CTEQ, at small x, $x < 0.01$, it loses sensitivity to the form of the initial distributions at $Q^2 = \mu^2$. This is because of the long evolution length between μ^2 and the Q^2 values at which the comparison with data is made, taken together with the fact that the initial distributions are valence-like and thus vanishing at small x. Thus the resulting steepness of the gluon and sea quark distributions at small x and larger Q^2 is dynamically generated by pQCD. HERA data were not input to the original GRV parameterization so that it represented a successful prediction for the steep behaviour of F_2 in the HERA region. The flattening of F_2 as Q^2 decreases is also predicted, but the recent DGLAP fits described in this Chapter indicate that whereas the gluon PDF becomes valence-like at low Q^2 the sea quark PDF does not. This presents a potential difficulty for the GRV approach, however currently (GRV98 PDF set) a small amount of fine tuning of the original choice of the scale μ and the input valence-like shapes can produce a reasonable fit to HERA data, provided $\alpha_s(M_Z^2) \lesssim 0.114$.

6.9 Future prospects for information on the gluon

Despite tremendous improvements in our knowledge of the gluon PDF from the early 1990's, the gluon remains the least well known parton. Fig. 6.13 illustrates which data determine the gluon for various ranges of x. Our

Future prospects for information on the gluon 177

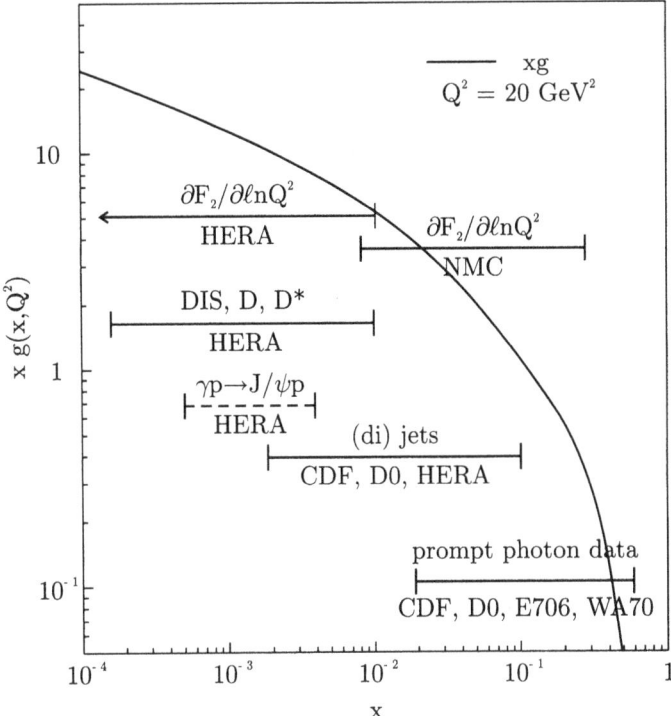

Fig. 6.13 Ranges of x which are determined by different data sets, with a representative gluon PDF at $Q^2 = 20 \text{GeV}^2$

knowledge of the gluon can be improved by various measurements which can be performed at HERA-II.

- The longitudinal structure function depends directly on the gluon distribution at low x. At $x \lesssim 10^{-3}$ the dominant contribution to F_L comes from the gluon regardless of the exact shape of the gluon distribution (see Eq. 4.43). A direct measurement (without the model dependence of the H1 extrapolation method) requires data at two, or more, different beam energies as discussed in Section 5.6.3. With a very large integrated luminosity, it may also be possible to use ISR events.

- Dijet production initiated by the DIS processes $\gamma^* g \to q\bar{q}$ and $\gamma^* q \to qg$ can also been used to extract the gluon distribution. The NLO corrections are known and the effects of scheme dependence and jet algorithms have been quantified. At present the results are not competitive with those from scaling violation data and there are still some unresolved theoretical questions on the absolute normalization of the jet rates, but the outlook is promising. Such measurements would constrain the gluon in the x range, $0.001 < x < 0.1$. (see Section 7.4).

- Open charm production, which tags the boson–gluon fusion process and hence the gluon density directly. The charm data in DIS have been discussed in Section 5.6.4, and Fig. 5.17 shows the comparison of the HERA-I measurements of $F_2^{c\bar{c}}$ with the predictions using the gluon distribution from the H1 NLO QCD fit. The agreement is good but the data errors are still too large to make a competitive extraction. High luminosity at HERA-II and improved detectors (e.g. use of microvertex detectors) will facilitate accurate measurements of the gluon density in the x range, $0.0001 < x < 0.01$.

- Elastic (diffractive) J/ψ production in DIS and in photoproduction may yield information on the gluon distribution, since the cross-section depends on $[xg(x, Q_V^2)]^2$. Similarly, diffractive production of ρ and ϕ in DIS at higher Q^2, depends on the square of the gluon density. These data could give information on the gluon distribution in the region $0.0001 < x < 0.01$.

6.10 Summary

This chapter has covered the practical aspects of how parton momentum densities are extracted using the technique of NLO QCD based global fits. Experimental uncertainties and model and theoretical uncertainties have been discussed. A summary of the resulting parton densities from the 'theory teams' and experimental groups has been given. Overall the results are in reasonable agreement with each other, but there is still a range of opinion on how to define and measure the error on a parton density, particularly that arising from experimental systematic errors. The two most frequently used approaches have been described.

It has been established that the leading twist NLO DGLAP formalism works technically down to $Q^2 \sim 1\,\mathrm{GeV}^2$, much lower than had been anticipated. However, there are indications that all is not well within the NLO DGLAP analyses: the gluon becomes valence-like and even negative for $x \lesssim 0.005$ and $Q^2 \lesssim 2\,\mathrm{GeV}^2$; and the low x slope of the gluon is flatter than that of the sea for $Q^2 \lesssim 7\,\mathrm{GeV}^2$. It is not clear that the formalism should really be applied for $Q^2 \lesssim 10\,\mathrm{GeV}^2$, not only because Q^2 is low and thus the use of a perturbative formalism is questionable, but also because, within the kinematic reach of present experiments, low Q^2 means low x ($x < 0.01$) and at such low x values the perturbative formalism may be in need of modification to resum large $\ln(1/x)$ terms. Furthermore, the data indicate that at low x parton densities are increasing dramatically, such that non-linear dynamics may be needed to describe parton evolution. These matters are explored further in Chapter 9.

6.11 Appendix: Comparability of evolution programs

Since various groups make PDF extractions the compatibility of the codes must be ensured. Two approaches have been widely used for the solution of the DGLAP integro-differential equations. Solution in x-space or in moment space (n-space).

In a typical x-space code, the Q^2 evolution is calculated on a grid in x and Q^2. Starting from the parameterization of the x dependence of the PDFs at Q_0^2, the slopes of each PDF in Q^2 are calculated and the PDFs at the next grid point ($Q^2 > Q_0^2$, or $Q^2 < Q_0^2$) are deduced. These new distributions are then used to continue evolution over the whole grid. It is assumed that the PDFs can be interpolated from one x grid point to the next, so that the integrals can be evaluated as weighted sums. The accuracy of the procedure depends on the density of the grid, particularly in x.

The alternative approach is to transform the evolution equations into equations involving Mellin moments. The convolutions are then reduced to simple products so that the evolution equations are ordinary differential equations which can be solved analytically. The x-space results are then recovered by contour integration in the complex n-plane (inverse Mellin transform). This step is usually performed numerically.

Detailed comparisons between codes using both of these approaches have been made (Blümlein et al. 1996) and very small differences ($< 1\%$) in the NLO results can be attributed to the procedures for truncation of the contributions from NNLO.

Various techniques have been suggested for the improving the speed and accuracy of these procedures.

The method of numerical integration described for the x-space method maybe replaced by analytic methods. An early method for performing the convolutions in the x-space formalism involved expanding both the PDF functions and the splitting functions and coefficient functions in terms of a functional basis, for which the convolution of two functions is a simple operation. Laguerre polynomials in $\ln(1/x)$ have this property, since the convolution of two such polynomials is simply given by the difference of two other Laguerre polynomials. However, any such expansion has to be truncated and this can lead to inaccuracy in the sensitive regions, $x \to 0$ and $x \to 1$. More recently, various semi-analytical methods have been suggested. These are analytic in Q^2, although discrete in x. The discretisation of x can also be exploited to avoid the use of a functional form in x at Q_0^2. A parameter for each x bin of the data may be used instead.

In the n-space methods the Mellin inversion can be performed using Jacobi polynomials. The x-space behaviour of a structure function, or parton distribution, can be reconstructed from a sum over a finite number of Mellin moments weighted by shifted Jacobi polynomials.

Alternatively, one may compare results directly with the moments of structure function measurements, rather than with the measurements themselves, in order to avoid the final Mellin inversion. Such comparisons have always suffered from the fact that measurements exist over only a limited range in x. Strategies to minimize this include the use of Bernstein moments, and of truncated moments.

Bernstein moments involve weighting the integrals over the structure functions by Bernstein polynomials in x rather than by simple powers.

These polynomials are chosen to weight the integral such that only the region of x in which the structure function is measured contributes strongly to the integral. The Bernstein moment can be evaluated theoretically as a weighted sum over the usual moment predictions.

Truncated moments are Mellin moments with the integration range restricted to $x_0 < x < 1$, avoiding the need for measurements at very low x (which are not available for all Q^2, or which maybe regarded as in need of an alternative theoretical treatment, see Chapter 9). Truncation at low x is more important than truncation at high x since the high-x part of the moment has little weight. The evolution equations obeyed by truncated moments are a set of coupled ordinary linear differential equations, which can be truncated to finite accuracy.

In principle it is possible to avoid using a parameterization of the PDFs in x, when using the n-space methods. In practice is is often necessary to use one in order to impose smoothness constraints, when performing the integrals. A promising approach to this problem is the use of neural networks to train fits to learn the shape of the data without the constraint of an analytic form.

7
α_s from scaling violations and jets at high Q^2

Measurement of the magnitude of the strong coupling constant and the demonstration that it 'runs', that is, follows the predictions of scale dependence given by the renormalization group equations for the exact SU(3) of colour have been touchstones of QCD since it was accepted as the most likely theory of the strong interaction in the late 1970s. Scaling violations in deep inelastic scattering provided some of the earliest estimates of the QCD scale parameter Λ_{QCD} and more recently, with the advent of the high energies and momentum transfers available at HERA, DIS is contributing to both types of measurement.

Before the advent of HERA, the values of $\alpha_s(M_Z^2)$ extracted from DIS data were in the region of $\alpha_s \sim 0.113$. This contrasted with the values of $\alpha_s \sim 0.120$ extracted from LEP data: n-jet production rates and the hadronic width of the Z^0. Data from the Tevatron on the single jet inclusive E_T distribution are also best described by $\alpha_s \sim 0.120$. Much thought went into understanding the origins of this discrepancy, including the suggestion that new physics maybe responsible for a process dependent discrepancy. However, when the uncertainties involved in the extractions are considered it is not clear that there is any significant discrepancy. More recent global fits to DIS processes, now including HERA data, obtain $\alpha_s \sim 0.117$.

This chapter covers the determination of $\alpha_s(M_Z^2)$ from structure function data and measurement of the running of α_s, as well as its magnitude, from jet production at large Q^2. Sections 7.1–7.3 cover the use of structure function data. Section 7.1 reviews methods of analysis, in particular the limitations of approaches which attempt to exploit scaling violations without explicit dependence on the shapes of the parton distributions and Section 7.2 reviews the conventional method in which α_s is determined in an NLO QCD fit in the DGLAP formalism in which the parton momentum densities are determined simultaneously. The estimation of experimental and theoretical uncertainties is discussed in detail. Various possible extensions to the theoretical framework are summarised in Section 7.3.

The remaining sections in the chapter consider jet production in DIS, which is interesting in its own right as two hard scales are involved, Q^2 and the transverse energy of the jets. The subject is introduced in Section 7.4, which also covers the importance of the Breit frame and gives a brief outline of the pQCD calculation of the dijet cross-section. In Section 7.5 jet measures, which provide the link between the final state hadrons and the underlying partonic structure, are outlined. Section 7.6 establishes that NLO QCD gives a good overall description of dijet data in DIS before the jet measurements of α_s are discussed in Section 7.7 and Section 7.8. An appendix at the end of the chapter gives more details of how to construct the Lorentz transformation between the HERA laboratory frame and the Breit frame.

7.1 Methods of determining $\alpha_s(M_Z^2)$ from structure function data

The simplest predictions of pQCD concern the Q^2 evolution of the non-singlet structure functions in moment space. In principle, one can extract a value of $\alpha_s(M_Z^2)$ independent of assumptions as to the shape of the non-singlet quark PDFs. However, predictions involving moments of structure functions require integrals of the structure function over the whole interval $0 < x < 1$, for a large enough range of Q^2 to give measurable scaling violations. As illustrated in Fig. 5.1, this is not practically possible so measurements must be extrapolated outside the measured region. This usually involves the use of an assumed functional form which introduces model uncertainty into the procedure. Techniques such as the use of truncated moments, which use a restricted integration range $x_0 < x < 1$, or Bernstein moments, which weight the integrals to emphasize the measured region, have been developed to evade this problem.

The most widely used moment space prediction is that for the Gross Llewellyn-Smith (GLS) sum-rule (the $N = 1$ moment of the xF_3 structure function). This is discussed below in Section 7.1.1.

A further difficulty with the use of pQCD predictions concerning non-singlet structure functions is that data on non-singlet functions (such as xF_3 in neutrino scattering and $F_2^p - F_2^n$ in muon scattering) are much less precise than those on singlet functions. Unfortunately predictions for the scaling violations of the more accurately measured singlet structure functions are more complicated because the evolution of the singlet structure functions involves mixing of quark-singlet PDFs with the gluon PDF. There is then a coupling between the value of $\alpha_s(M_Z^2)$ extracted and the shape of the gluon distribution, such that increasing $\alpha_s(M_Z^2)$ can be compensated by making the gluon distribution harder. Hence $\alpha_s(M_Z^2)$ determinations from singlet measurements have often been performed within the framework of NLO QCD fits to structure function data using the DGLAP formalism, in which the PDFs are fitted at the same time as $\alpha_s(M_Z^2)$. Such determinations are considered in Section 7.2.

7.1.1 Determinations of $\alpha_s(M_Z^2)$ from GLS sum-rule

The GLS sum-rule has been calculated to NNNLO

$$\int_0^1 xF_3(x, Q^2)\frac{dx}{x} = $$
$$3\left[1 - \frac{\alpha_s(Q^2)}{\pi} - a(n_f)\left(\frac{\alpha_s(Q^2)}{\pi}\right)^2 - b(n_f)\left(\frac{\alpha_s(Q^2)}{\pi}\right)^3\right] - \Delta HT \qquad (7.1)$$

where a, b are known calculable functions of the appropriate number of flavours n_f, and ΔHT represents the higher twist contribution. Higher twist contributions to the low n moments can be more reliably estimated than contributions to the structure functions. Thus the GLS sum-rule presents a method of measuring α_s very accurately, in principle. However, currently the theoretical uncertainty on the measurement is still large. The higher twist term has been estimated by several techniques to be in the range $(0.05 - 0.27)/Q^2$. The central values of $\alpha_s(M_Z^2)$ extracted from from the CCFR xF_3 data vary from 0.118 to 0.112, accordingly. The experimental error on the measurement due to the need to extrapolate outside the measured region in x, is also substantial giving $\Delta\alpha_s(M_Z^2) \sim 0.010$. Hence measurements from the GLS sum-rule are not yet competitive with other techniques.

7.2 Determinations of $\alpha_s(M_Z^2)$ from structure function data: DGLAP NLO QCD fits

Analyses of structure function data, to extract α_s from the scaling violations, are most usually done by numerical solution of the DGLAP equations, as discussed in Chapter 6 in the context of PDF extractions. The correlation between the gluon parameters and the value of $\alpha_s(M_Z^2)$, is accounted for by allowing $\alpha_s(M_Z^2)$ to be a free parameter of the fit simultaneously with the PDF parameters. It is essential to take into account correlations between experimental systematic uncertainties, as outlined in Section 6.7, since the resulting experimental uncertainties on $\alpha_s(M_Z^2)$ are significant. Theoretical and model uncertainties are also significant and these will be considered before results are presented.

7.2.1 Theoretical and model uncertainties

As in Chapter 6, we make a distinction between sources of model uncertainty within the theoretical framework of leading-twist NLO QCD and the theoretical uncertainty due to the use of this framework.

Sources of model uncertainty have already been discussed in the context of PDF extraction. Here we merely summarize the model choices to which α_s can be sensitive:

- the specific evolution program used and the different approximations to the solution of renormalization group equation (Eq. 3.33) for α_s at second order;

- the choice of heavy quark production scheme, the values of heavy quark masses and the different ways of dealing with the behaviour of α_s at the heavy quark thresholds;
- the value of Q_0^2 for the input form of the PDF parameterization;
- the analytic form of the input PDF parameterization, the choices for the flavour decomposition of the sea and the further assumptions which reduce the number of parameters, see Section 6.6.1;
- the minimum Q^2, x, W^2 of data entering the fit;
- the choice of data sets entering the fit, and the corrections for deuterium or heavy target data.

Some comments are in order. Choices such as that of the evolution program to be used, or the approximations made, should make no difference to the results, but differences in NLO truncation procedures can introduce some differences. Modern programs give very good agreement and any remaining differences should be regarded as part of the uncertainty involved in using NLO rather than NNLO code.

Heavy quark production should be handled within a general mass variable flavour number scheme, but different schemes have been used historically, so care must be taken when comparing results. This is particularly important when using HERA data in fits, since the correct treatment of charm production makes a significant difference at small x where $F_2^{c\bar{c}}$ is a significant fraction of F_2. If a QCD fit is made without accounting for the charm threshold correctly then threshold behaviour can fake a stronger dependence of F_2 on Q^2 than is truly attributable to QCD scaling violations and thus make the value of α_s seem larger.

The arbitrariness of the choice of the PDF parameterizations reflects a more fundamental theoretical uncertainty which can only be finally resolved by a solution to the problem of confinement within non-perturbative QCD. Presently, this uncertainty can be minimised if a large range of data sets are used to tie down the variation of the PDF parameters. However, such global fits can suffer from systematic experimental uncertainties deriving from the marginal compatibility of data sets (see Section 6.7) and/or from the need for uncertain heavy-target corrections.

The choice of the kinematic cuts on the data sets should be an insignificant model uncertainty, but this will only be the case if the cuts are carefully chosen to avoid regions where the theoretical framework of leading-twist NLO QCD in the DGLAP formalism is questionable.

Sources of theoretical uncertainty are:

- renormalization and factorization scale uncertainties;
- higher order effects (NNLO rather than NLO);
- higher-twist effects;
- the need to resum $\ln(1-x)$ terms as $x \to 1$ (soft gluon resummation);
- the need to resum $\ln(1/x)$ terms as $x \to 0$, or to use the BFKL or CCFM equations rather than the DGLAP equations;

- the need to use non-linear evolution equations as the gluon density increases;
- non-perturbative effects as $\alpha_s(Q^2)$ becomes large.

A proper consideration of these uncertainties involves extending the theoretical framework, as considered in Section 7.3. Remaining within the NLO framework kinematic cuts such as: $Q^2 \gtrsim 1-4$ GeV^2, to avoid the non-perturbative region; $W^2 \gtrsim 10-20$ GeV^2, to avoid large higher twist contributions; $x > 0.01$, to avoid the need for $\ln(1/x)$ resummation; have been applied.

The choice of Q^2 for the renormalization and factorization scale is both obvious and conventional for the inclusive DIS processes. The variation of this scale is used as a crude way of estimating the importance of higher order terms, for example, those involved in extending an NLO to an NNLO analysis. Varying the renormalization and factorization scales from $Q/2$ to $2Q$, both independently and simultaneously, gives an uncertainty $\Delta \alpha_s \sim 0.005$, mostly from the renormalization scale change. It is unclear that such arbitrary scale changes give any reasonable estimate of the effect of higher order terms.

Not all of the results described below have treated all of the sources of uncertainty mentioned in the present section, so care is needed when comparing results.

7.2.2 Results

Some typical results for $\alpha_s(M_Z^2)$, from the experimental collaborations and theoretical groups, are presented in Table 7.1. The experimental uncertainties represent those from statistical and correlated and uncorrelated systematic uncertainties, but the theoretical uncertainties are sometimes separated into model uncertainties and the theoretical uncertainties as defined in Section 7.2.1.

One of the earliest analyses is that of the BCDMS and SLAC F_2 data. This analysis accounted for the correlated systematic errors in the BCDMS data, the relative normalizations of the data sets and the need for target mass corrections and higher twist contributions at high x, low Q^2. Theoretical and model uncertainties from the choice of the renormalization and factorization scales, the position of the flavour thresholds, and the value of R were also considered. The resulting value for α_s is $0.113 \pm 0.003 (\text{exp.}) \pm 0.004 (\text{th.})$.

The early global fits of the MRST and CTEQ teams were consistent with the low value of α_s from the BCDMS/SLAC analysis, but once the HERA data were included in the fits, larger values of α_s were favoured. In principle low-x HERA data should be less sensitive to the correlation between the value of α_s and the shape of the gluon PDF. The point is that although α_s and the small x shape of the gluon distribution remain correlated, the shape of the gluon and the singlet quark distribution are also strongly correlated in the low x region and the singlet quark distribution

Table 7.1 A summary of the results for $\alpha_s(M_Z^2)$ extracted from structure function data. The experimental errors include both statistical and systematic errors.

Authors/Data	$\alpha_s(M_Z^2)$
BCDMS/SLAC	$0.113 \pm 0.003(\text{exp.}) \pm 0.004(\text{th.})$
CCFR(97)	$0.119 \pm 0.002(\text{exp.}) \pm 0.004(\text{th.})$
H1/BCDMS	$0.1150 \pm 0.0017(\text{exp.})^{+0.0009}_{-0.0005}(\text{model.}) \pm 0.005(\text{th.})$
CTEQ6(global)	$0.1165 \pm 0.0065(\text{exp.})$
MRST01(global)	$0.1190 \pm 0.002(\text{exp.}) \pm 0.003(\text{th.})$
ZEUS02(global)	$0.1166 \pm 0.0049(\text{exp.}) \pm 0.0018(\text{model.}) \pm 0.004(\text{th.})$

is directly constrained through the measurement of F_2. Thus the errors on α_s are smaller at low x than at moderate x because at small x the gluon is strongly related to a directly measurable quantity.

The α_s determination from the CCFR neutrino and anti-neutrino data differs from the others shown in the table in that it relies only on data from the one experiment. Correlated systematic errors are taken into account using the Hessian method. The theoretical uncertainty quoted in their analysis covers higher twist corrections and renormalization and factorization scale uncertainties.

The last four $\alpha_s(M_Z^2)$ values in Table 7.1 are all from analyses using high precision F_2 data from HERA plus fixed target data. The CTEQ group global fit result for α_s is $0.1165 \pm 0.0065(\text{exp.})$, where the correlated systematic experimental uncertainties have been treated by the Hessian method using the conservative χ^2 tolerance, $T = 10$, to account for the uncertainties due to the combination of many marginally consistent data sets, as explained in Section 6.7.6. Model uncertainties within the theoretical framework are small compared to this experimental error estimate. The CTEQ group do not consider the global fit as a competitive way to measure α_s and hence they do not evaluate further theoretical uncertainties.

The MRST group global fit result is $0.1190 \pm 0.002(\text{exp.}) \pm 0.003(\text{th.})$, where the systematic and statistical experimental uncertainties have been combined in quadrature without consideration of correlations (except for Tevatron jet data) but the error is quoted using the χ^2 tolerance, $T = \sqrt{20}$, to take into account the problems encountered in the combination of data sets. It is interesting to note that the partial χ^2s for individual data sets in a global fit vary with $\alpha_s(M_Z^2)$ differently from the global χ^2, indicating that different data sets prefer different values for α_s (see Fig. 7.1). Thus a global fit value for α_s represents a compromise. The theoretical uncertainty for the MRST result is obtained from comparison with alternative theoretical treatments including NNLO terms and resummation of $\ln(1/x)$ terms at small x. MRST have also studied the stability of their fit to removing data in the low x, low Q^2 region where the NLO DGLAP formalism may not be appropriate. If data with $x < 0.005$ and $Q^2 < 7$ GeV^2 are cut out

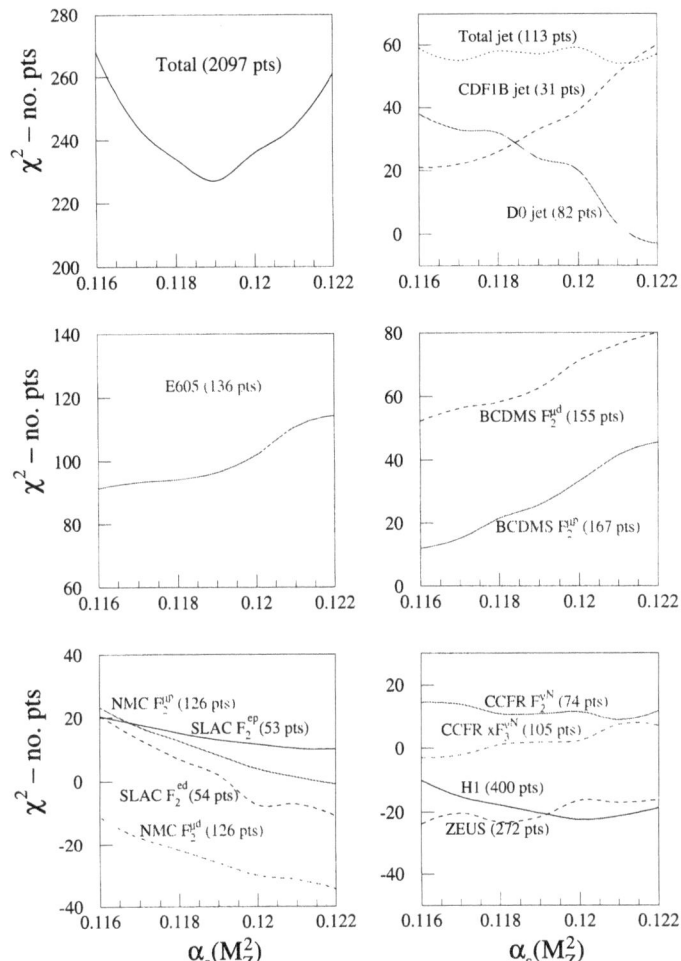

Fig. 7.1 Partial χ^2 for data sets in the MRST01 fit as a function of $\alpha_s(M_Z^2)$. (From Martin et al. 2001.)

of the fit then the value for α_s obtained is, $\alpha_s = 0.116 \pm 0.002(\text{exp.}) \pm 0.002(\text{th.})$, where a smaller χ^2 tolerance $T = \sqrt{5}$ has been applied since the data sets are now much more compatible and the theoretical error is also reduced since the uncertainties appertaining to the formalism are also reduced (Thorne 2003).

The ZEUS global fit result is $0.1166 \pm 0.0049(\text{exp.}) \pm 0.0018(\text{model.}) \pm 0.005(\text{th.})$. The experimental error was evaluated using the conservative offset method, which gives a similar size of error to the Hessian method with tolerance $T = 7$ (see Section 6.7.6). The experimental error maybe broken down into, 0.0008 from statistical uncertainty, 0.0032 from correlated systematic uncertainties excepting normalizations and 0.0036 from normalization uncertainties treated as Gaussian. This indicates that normalization uncertainties between data sets can be very important. The model uncertainty accounts for all the sources summarised in Section 7.2.1 and the theoretical, uncertainty accounts for the effect of residual higher twist terms in the fit and for renormalization and factorization uncertainties.

The H1 analysis uses only H1 and BCDMS proton target data. The resulting value is, $\alpha_s(M_Z^2) = 0.1150 \pm 0.0017(\text{exp.})^{+0.0009}_{-0.0005}(\text{model.}) \pm 0.005(\text{th.})$. The choice of restricted data sets is deliberate in order to minimise the error from the combination of data sets and the need for heavy target corrections. The experimental errors are accordingly evaluated by the Hessian method with χ^2 tolerance $T = 1$. The model errors account for the various sources listed in Section 7.2.1 and the theoretical error is evaluated from the variation of renormalization and factorization scales.

The values of α_s from these various analyses are in agreement, within their quoted errors. However, it is clear that there are many uncertainties involved in extractions of α_s from structure function data.

7.3 Determinations of $\alpha_s(M_Z^2)$ from structure function data: extending the theoretical framework

A number of authors have made determinations of α_s extending the theoretical framework beyond the DGLAP NLO QCD formalism.

A common extension to the theoretical framework is to consider higher twist terms. These are usually introduced as terms of the form $(1 + h_i/Q^2)$ multiplying the leading twist prediction for the structure function. The values of h_i are fitted in different x bins. Historically, neglect of large positive higher twist terms at large x led to overestimation of the value of α_s, since the scaling violations from these terms increases that due to the leading twist $\ln(Q^2)$ terms. However, most of this effect can be accounted for by the correct treatment of target mass effects. In modern analyses, dynamical higher-twist terms are usually excluded by a hard W^2 cut. However, some analyses do not make such a cut, but use the low W^2 data to try to determine the shape of the higher twist terms. These have been found to be broadly in agreement with the renormalon predictions given in Section 4.4, both for F_2 and xF_3. The residual effect of higher twist terms above the hard W^2 cut is not completely negligible, leading to an uncertainty in α_s of $\Delta \alpha_s \sim \pm 0.003$.

Another common extension to the theoretical framework is to perform fits to NNLO. As discussed in Section 4.6, the QCD β function is known to NNNLO, the two-loop coefficient functions for DIS are known at NNLO, but the three-loop splitting functions are incomplete. However, some of their Mellin moments for low N are available and the approximations in the calculations are now just a few percent. Many of the analyses at NNLO make use of alternative techniques such as the use of Jacobi polynomials, Bernstein moments and truncated moments. This is because the NNLO corrections are most easily taken into account by techniques which use the predictions for moments directly. The results of these studies suggest that the change in α_s going from NLO to NNLO is $\Delta\alpha_s \sim \pm 0.003$.

It is important to notice that these two sources of uncertainty in α_s are probably not independent of each other. NNLO fits have often been combined with fits which introduce higher twist terms into the formalism, since calculations from renormalon theory suggest that higher twist terms may be related to the truncation of the perturbative series. This would suggest that higher-twist terms should be less significant at NNLO and the NNLO analyses bear this out. The inclusion of soft gluon resummation (resummation of $\ln(1-x)$ terms at high x) in the formalism also reduces the size of the extracted higher twist terms.

It is also expected that renormalization and factorization scale dependence is considerably decreased at NNLO. Preliminary analyses indicate that a scale dependence of $\Delta\alpha_s \sim \pm 0.005$ is reduced to $\Delta\alpha_s \sim \pm 0.002$. There have been some studies of NNNLO evolution and the need for soft gluon resummation which suggest that the scale uncertainty on α_s can be reduced to less than 1%.

Small-x data are potentially sensitive to the need for making a complete resummation of $\ln(1/x)$ terms, and moreover, low x data may be in need of a more fundamental change to the theoretical formalism than simply resumming $\ln(1/x)$ terms in the splitting functions of the DGLAP formalism. In Chapter 9 the need for a different approach at low x is discussed, and alternative evolution equations are described, including non-linear evolution equations. The consequence for the extraction of α_s is that low x data ($x \lesssim 0.01$) may not give a reliable estimate of α_s when treated conventionally.

7.4 Jet production in DIS

The QCD processes that give rise to scaling violations in DIS at leading order, namely QCD Compton (QCDC) and boson–gluon fusion (BGF), also give rise to distinct jets in the final state provided that the energy and momentum transfer are large enough. For the QCDC and BGF processes, there could be two final-state parton jets in addition to the proton target remnant. This suggests the terminology '$n + 1$' jets, where the n refers to the number of final state parton jets and the '1' to the proton remnant. Beyond leading order, there will be a greater variety of possible final states at the parton level and the distinction between the QCDC and BGF type processes becomes blurred. Here, for clarity, the discussion focuses on the NC LO processes which are illustrated in Fig. 7.2. Jet production is also important in CC scattering but the features of the final state are similar to

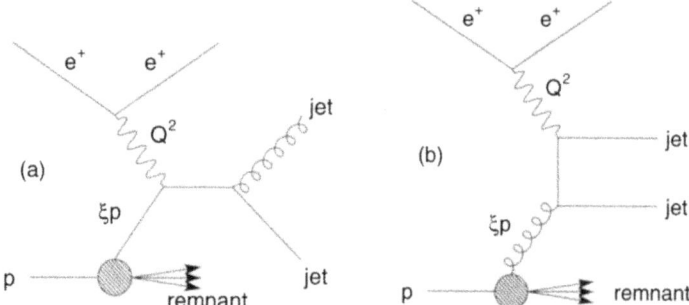

Fig. 7.2 Leading order QCD diagrams from dijet production in DIS. (a) QCD Compton; (b) Boson-gluon fusion.

those for NC and, as the event rates are lower at medium Q^2 values, CC jet rates do not give a competitive determination of α_s.

The leading order processes at the parton level are two-body scattering processes, either $\gamma^* q \to q'g$ or $\gamma^* g \to q\bar{q}$, which allow more detailed exploration of the QCD matrix elements than is possible from scaling violations alone. In addition QCD predicts a non-trivial dependence on the azimuthal angle, Φ, between the planes defined by the incident and scattered beam leptons and that formed from the two partons in the final state. Apart from α_s, the BGF process depends on the gluon density and QCDC depends on quark momentum density functions. The QCDC process dominates at large Q^2, where the most important quark densities are the well-determined u and d valence distributions. Thus measuring jet rates at large Q^2 should allow a determination of α_s with reduced uncertainty from the less well known gluon density. Higher order QCD terms complicate the picture and introduce some dependence on the gluon density, however, it is possible to reduce the importance of this dependence by judicious choice of phase space . In principle, one could also try a more ambitious approach of a simultaneous determination of both α_s and the gluon density by fitting jet rates and scaling violations at the same time. Both will be discussed.

7.4.1 Breit frame kinematics

For both practical and theoretical reasons, the Breit frame is particularly appropriate for the study of DIS jets. Formally the frame is defined as that in which $2x\mathbf{p} + \mathbf{q} = 0$, where $\mathbf{p}$ and $\mathbf{q}$ are the 3-momenta of the proton and virtual photon, respectively. In the Breit frame, the lowest order $\gamma^* q \to q'$ process is such that the virtual photon and initial state quark collide head–on and the quark momentum is reversed (see Fig. 7.11 of the appendix to this chapter). There is no energy transfer and it is convenient to choose the 3-momentum of the virtual photon to lie long the negative z-axis, giving $q = (0, 0, 0, -Q)$ where $Q = \sqrt{Q^2}$. The positive z-axis is then

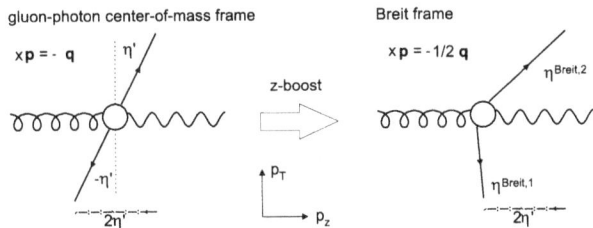

Fig. 7.3 $\gamma^* g \to q\bar{q}$ scattering in the $\gamma^* g$ CM and Breit frames.

that of the incoming proton 3-momentum **p**.[1] With this choice of axes, the incoming quark has 4-momentum $p_a = (Q/2, 0, 0, Q/2)$ and the outgoing scattered quark has 4-momentum $p'_a = (Q/2, 0, 0, -Q/2)$ — thus explaining the alternative name of the frame as the 'brick wall' frame in which the quark is turned around by scattering from the 'brick wall' of the virtual photon. The proton remnant has 4-momentum $((1-x)Q/(2x), 0, 0, (1-x)Q/(2x))$. The $\gamma^* p$ CM frame (defined by $\mathbf{q}+\mathbf{p}=0$) and the Breit frame are related by a boost along the z-axis. To get from the HERA laboratory frame to the Breit frame requires rotations and a boost. The details of how to construct the transformation are given in the appendix to this chapter. At the parton level for $\gamma^* q \to q'$, there is no transverse momentum in either the initial or final state. At the hadron level, there will be limited $\mathbf{k}_T$ from the fragmentation of both the scattered quark ('current' region, backwards in Breit frame) and proton remnant ('target' region, forward in the Breit frame). For the LO QCDC and BGF processes, the two final state partons will balance in transverse momentum, as shown in Fig. 7.3, giving rise to to jets with large $\mathbf{k}_T$ at the hadron level. Thus the Breit frame has practical advantages for the separation of $O(\alpha_s)$ (and above) jet processes, in addition to this, the factorization of the cross-section has been proved in this frame (Webber 1993).

To describe the two-body scattering processes at the parton level, $\gamma^* q \to q' g$ or $\gamma^* g \to q\bar{q}$, two variables are needed (in addition to the usual DIS variables, Q^2, x, y and the angle Φ between the lepton and parton scattering planes). These will be related to the CM energy and scattering angle of the γ^*-parton system, however it is more convenient to choose Lorentz invariant variables. Let the 4-momentum of the initial state parton (q or g) be p_a and $p_{1,2}$ the 4-momenta of the two final state partons (or jets at the hadron level), then $x_p = Q^2/(2 p_a \cdot q)$ is a Bjorken like variable for the parton target and $z_p = (p_a \cdot p_1)/(q \cdot p_a)$ is related to the scattering angle. The components of the 3-momenta of the final state partons parallel and perpendicular to the z-axis in the Breit frame may be expressed in terms

---

[1] The positive direction in the Breit frame is often taken to be along **q**, the direction of the γ^*, but for HERA the proton direction is the 'natural' forward direction and it is more convenient to follow that in the Breit frame as well.

of x_p and z_p as follows

$$(p_1)_\| = \frac{Q}{2x_p}(1 - x_p - z_p), \quad (p_1)_\perp = Q\sqrt{\frac{z_p}{x_p}(1-x_p)(1-z_p)}, \quad (7.2)$$

with the expressions for p_2 obtained by the substitution $z_p \leftrightarrow 1 - z_p$ in the above. Using $(p_i)_\|$ the regions of final state phase space may be characterized as

1. $(p_1)_\| < 0$; $(p_2)_\| < 0$. Both jets in the current region (i.e. backwards in the Breit frame), p_T balanced on this side.
2. $(p_1)_\| < 0$; $(p_2)_\| > 0$ (and vice-versa). One jet in the current region and one in the target region, with p_T unbalanced on both sides.
3. $(p_1)_\| > 0$; $(p_2)_\| > 0$. Both jets in the target region with p_T balanced and the current region 'empty'.

Another useful variable is ξ, the fraction of the proton's momentum carried by the initial state parton, $p_a = \xi p$ and $x_p = x/\xi$. In terms of the invariant mass of the two final state jets, M_{jj}, one finds

$$x_p = \frac{Q^2}{Q^2 + M_{jj}^2}, \quad \text{or} \quad \xi = x\left(1 + \frac{M_{jj}^2}{Q^2}\right). \quad (7.3)$$

7.4.2 Cross-section calculations

Integrating over the angle Φ the four-fold differential cross-section may be written in the form

$$\frac{d^4\sigma}{dx dQ^2 dx_p dz_p} = \frac{\alpha^2 \alpha_s(Q^2)}{x_p Q^4}[I_q + I_g] \quad (7.4)$$

where I_q and I_g correspond to the QCDC and BGF processes respectively. At large Q^2 both γ^* and Z^0 exchange will contribute to the cross-section. Since one is assuming that the masses of the partons may be ignored, there are configurations of the final state that will lead to singularities in the cross-section. To discuss how these are avoided it is enough at this stage to concentrate on the γ^* contribution only. This gives

$$I_q = \sum_{i=1}^{n_f} e_i^2 \left(q_i(x/x_p, Q^2) + \bar{q}_i(x/x_p, Q^2)\right) \tilde{I}_q(x_p, z_p, y) \quad (7.5)$$

$$I_g = \left(\sum_{i=1}^{n_f} e_i^2\right) g(x/x_p, Q^2) \tilde{I}_g(x_p, z_p, y) \quad (7.6)$$

where n_f is the number of active flavours and q_i, $\bar{q}_i$, g are the usual quark, antiquark and gluon parton densities for the proton and the $\tilde{I}$ terms are constructed from the hard scattering cross-sections. The full expressions for the latter are not needed here, but only the terms that lead to singular configurations, these are

$$\tilde{I}_q(x_p, z_p, y) \sim \frac{1 + x_p^2 z_p^2}{(1 - z_p)(1 - x_p)}$$

$$\tilde{I}_g(x_p, z_p, y) \sim \frac{\left(z_p^2 + (1 - z_p)^2\right)\left(x_p^2 + (1 - x_p)^2\right)}{z_p(1 - z_p)}.$$

The QCDC cross-section, $\tilde{I}_q$, is singular when $x_p \to 1$ or $z_p \to 1$. These limits arise as follows: if the final state parton with momentum p_2 is collinear with the initial state parton (p_a) then $z_p \to 1$; if the two final state partons are collinear (p_1, p_2) then $x_p \to 1$; if p_2 is soft $(p_2 \to 0)$ then both $x_p, z_p \to 1$. The BGF cross-section, $\tilde{I}_g$, is singular when $z_p \to 0$ or $z_p \to 1$. These limits arise as follows: if p_2 is either soft or collinear with p_a then $z_p \to 1$; if p_1 is either soft or collinear with p_a then $z_p \to 0$.

These problems have been encountered before, in Section 4.1, the LO corrections to the parton model were constructed by integrating over the contributions from the QCDC and BGF diagrams. The collinear singularities in the initial state were absorbed by the redefinition of the parton densities (which became scale dependent in the process). The soft (infrared) and collinear singularities of the final state cancel through the Kinoshita–Lee–Nauenberg mechanism as described in outline in Section 3.3. The net result is that the '2+1' jet cross-sections are finite.

In practice, the correct treatment of singularities makes calculations for the '2+1' jet processes challenging. One wants to be able to study the 2+1 jet final state as a function of the hadronic final state variables accounting for the finite acceptance of the detectors and the need to apply event and jet selection cuts. This makes an analytic calculation almost impossible. A number of numerical codes are available for calculating the 2+1 jet parton final state at NLO. They are based on the QCD factorization theorems and allow arbitrary infrared safe observables to be calculated by first isolating and cancelling the infrared singularities. Two different methods are available to do this: phase-space slicing and the subtraction technique. For phase-space slicing, the range of integration is split into a small region around the singularity and the remainder by introducing a cut-off parameter. The singular term is isolated and cancelled against a matching singular term from the virtual corrections, the remaining integral may then be performed numerically. While this method is straightforward to apply, it requires quite delicate cancellations leading to potentially large numerical fluctuations. For the subtraction method, a counter term is constructed to remove the singularity locally in phase space, giving a finite integrand over the whole of phase space. This method is more robust numerically, but the construction of the counter term is often non-trivial. Three codes are in common use, MEPJET (Mirkes and Zeppenfeld 1996) which uses phase–

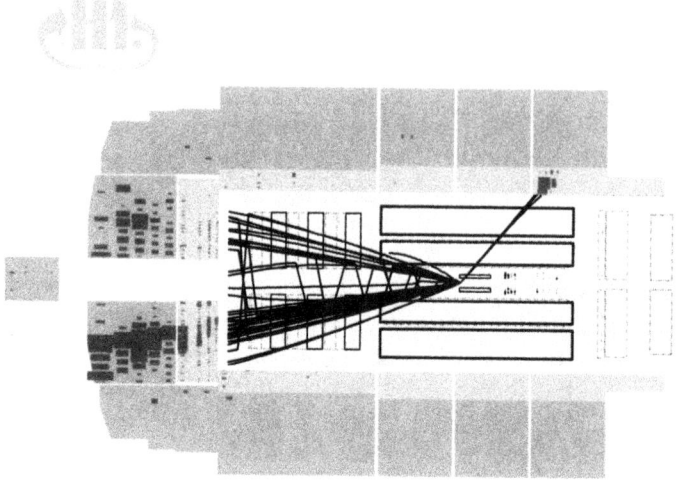

Fig. 7.4 A dijet event at medium Q^2 in the H1 detector.

space slicing and DISENT (Catani and Seymour 1997) and DISASTER++ (Graudenz 1997) which both use the subtraction method.

7.5 Jet measures

In deep inelastic scattering or hadron–hadron collisions, it is often not appropriate to apply global 'event shape' measures such as thrust developed for the 'cleaner' environment of e^+e^- annihilation.[2] The laboratory frame is not the parton–parton CM frame and one has the problem of the proton or hadron remnants largely invisible in the beam pipe. In these circumstances, it is more useful to identify clusterings of final state hadrons, or 'jets', to be identified with the final state partons of the hard process under study. In most events this will not be a unique identification. An example of a DIS NC dijet event recorded in the H1 detector is shown in Fig. 7.4. The event is at a medium Q^2 with the single high momentum track of the scattered positron tending towards the rear direction of the detector.

There are two main classes of jet measures, those that cluster energies into 'cones' in rapidity-azimuthal-angle space and those that cluster using momenta. In one way or another, the jet algorithm exploits the limited transverse momentum expected in the fragmentation or hadronization of a high energy quark or gluon. A jet measure needs the following ingredients: an initiator or seed (often a high momentum 'particle', i.e. either a charged hadron or a large hadronic energy deposit); a 'distance' measure between two particles and a way to combine them; a termination condition. Finally the 4-momentum of the jet is constructed from its constituents.

[2] See, for example, Chapter 3 of Ellis et al. (1996).

Another consideration for the choice of jet measure is that it should be 'infra-red' safe. Higher order processes in QCD are analogous to radiative processes in QED. Both involve massless or almost massless partons which means that singularities occur when the radiated parton is soft or when two partons are collinear. Ideally the jet measure should be compatible with the regularization procedures used so that finite calculations may be compared with the data.

In DIS, a large fraction of the longitudinal momentum is carried by proton remnant particles in, or close to, the beamline and thus invisible to the main detector components. This needs to be allowed for, either by defining a pseudo-particle carrying the missing longitudinal momentum (the beam jet) or by excluding the region around the forward beam pipe. If all particles not associated with the scattered $e^\pm$ are assigned to a jet or the proton remnant, the jet algorithm is said to be *exclusive*. If the algorithm does not require this it is termed *inclusive*.

7.5.1 The cone algorithm

The cone algorithm was until quite recently the jet measure most frequently used in hadron–hadron collider physics. Each particle in an event is described by transverse energy, $E_T = E \sin\theta$, azimuthal angle, ϕ and pseudo-rapidity, $\eta = -\ln\tan(\theta/2)$, where θ is the polar angle with respect to the proton beam direction (in the case of HERA).[3] Particles with $E_T > E_T^{cut}$ (a given threshold) are taken as the initiators. Each initiator defines a direction about which a cone of radius R on the η, ϕ surface is defined. Then all particles with $E_T < E_T^{cut}$ and distance to the cone axis less than R ($\sqrt{(\eta_{cone} - \eta_i)^2 + (\phi_{cone} - \phi_i)^2} < R$) are included in that cone (typical values for R are between 0.7 and 1.0). The cone position is then refined using all its constituent particles

$$\eta_{cone} = \frac{1}{E_{T,cone}} \sum_i E_{T,i}\eta_i; \quad \phi_{cone} = \frac{1}{E_{T,cone}} \sum_i E_{T,i}\phi_i; \quad (7.7)$$

$$\text{with} \quad E_{T,cone} = \sum_i E_{T,i}.$$

Particles are then tested again for inclusion in the cone and the procedure repeated until the cone position converges. The same process is repeated for all initiators. Note that in NC DIS, the 'particles' associated with the scattered beam lepton are excluded as is a region about the forward (proton) beam direction. Further steps are needed to decide if overlapping cones should be merged and for how the common particles should be redistributed to the jets. Finally a cut $E_T^{jet,min}$ may be applied to the cone energies to isolate hard processes. Note this algorithm is inclusive as not all particles will necessarily end up assigned to a jet cone.

---

[3] These coordinates are discussed in more detail in Section 10.1.

7.5.2 The k_T cluster algorithm

For the k_T cluster algorithm, the quantity

$$k_{T,ij}^2 = 2\min[E_i^2, E_j^2](1 - \cos\theta_{ij})$$

where $E_{i,j}$ are the energies of particles i, j and θ_{ij} the angle between their 3-momenta is calculated for all pairs of particles. In addition, the 'distance' of all particles to the beam jet is calculated

$$k_{T,ip}^2 = 2E_i^2(1 - \cos\theta_{ip}),$$

where θ_{ip} is the angle between i and the beam jet. Particles associated with the scattered beam $e^{\pm}$ are excluded. The values of $\{k_{T,ij}, k_{T,ip}\}$ are now tested against a threshold E_T^{min}. Pairs with k_T below the threshold are combined with their nearest neighbour (either another particle or the beam jet). The procedure is then iterated until all values of k_T are above the threshold. The algorithm may be run in two modes either with a fixed value of E_T^{min} producing a variable number of jets or with variable E_T^{min} adjusted to produce a fixed number of jets (e.g. 2+1) in each event. Some authors prefer to scale the k_T^2 values by W'^2 where W' is the invariant mass of the hadronic final state calculated from all particles entering the jet algorithm [4]. The tests and cuts are then defined in terms of $y = k_T^2/W'^2$.

7.5.3 Longitudinally invariant k_T algorithm

This algorithm combines the advantages of both the previous ones and is the one most frequently use at HERA. The particle variables are the set (E_T, η, ϕ) and it proceeds in an analogous way to the k_T algorithm.

1. For every pair of particles, the 'distance' d_{ij} is calculated, where

$$d_{ij} = \min[E_i^2, E_j^2]\left[(\eta_i - \eta_j)^2 + (\phi_i - \phi_j)^2\right]$$

   and for every particle a distance to the beam direction is defined by $d_{ip} = E_{T,i}^2 R^2$, where R is the $\eta - \phi$ radius of the cone algorithm.

2. The minimum of all $\{d_{ij}, d_{ip}\}$ values is found, if it is d_{ij} then the two particles are combined to form a new one, if it is d_{ip} then that particle is regarded as a *protojet* and removed from further consideration.

3. New values of $\{d_{ij}, d_{ip}\}$ are formed and step 2 repeated. This is iterated until all particles are assigned to protojets. [Note that since one particle is removed at each iteration, the procedure is finite.]

Two particles are merged to form a new one using

$$\eta_k = \frac{1}{E_{T,k}}(E_{T,i}\eta_i + E_{T,j}\eta_j); \quad \phi_k = \frac{1}{E_{T,k}}(E_{T,i}\phi_i + E_{T,j}\phi_j);$$

---

[4] Ideally $W' = W$, but this may not always be the case in practice

with $E_{T,k} = E_{T,i} + E_{T,j}$.

This variant of k_T clustering has the advantage that it is less influenced by soft particles than the cone algorithm and is subject to smaller hadronization and detector corrections.

Once the jet algorithm is complete, the jet variables are calculated from their constituent particles using Eq. (7.7) and further cuts on jet E_T may be used to identify events with hard jets. In addition for dijet studies, asymmetric E_T cuts may be applied to the jets to avoid singular regions in the final state phase space.

7.6 Description of jet data at NLO

Before turning to α_s measurements, this section reviews the quality of the description of jet properties given by the NLO matrix element calculations with particular attention on scale uncertainty and dependence on parton densities.

The ZEUS collaboration has undertaken a detailed study of dijet production in neutral current deep inelastic e^+p scattering (ZEUS 2002b). Using an integrated luminosity of 38.4 pb^{-1} with $10 < Q^2 < 10^4$ GeV^2, dijet events were found in the Breit frame using the longitudinally invariant k_T jet algorithm in inclusive mode. The algorithm was applied to all calorimeter energy deposits, assumed to be massless particles and excluding those corresponding to the scattered positron. Events with two or more jets in the Breit frame were further refined using the following selection cuts:

- $E_{T,1}^B > 8$ GeV, $E_{T,2}^B > 5$ GeV, where $E_{T,i}^B$ are the transverse energies of the jets in the Breit frame;
- $E_{T,1} > 5$ GeV, $E_{T,2} > 5$ GeV, where $E_{T,i}$ are the transverse energies of the jets in the HERA laboratory frame;
- $|\eta_{1,2}| < 2$, where η_i are the pseudorapidities of the jets in the HERA laboratory frame;

giving a dijet sample of 39.6k events. Note the asymmetric cuts on the jet energies in the Breit frame which are applied to avoid incomplete cancellation of singular terms if a symmetric energy cut is applied with too low a value. The cuts on E_T and η in the HERA frame are applied to ensure well measured jets within the acceptance of the ZEUS calorimeter. The data are corrected for energy loss and other detector effects to the hadron level. The largest systematic uncertainty is that of the jet-energy scale and this was studied using transverse energy balance in NC DIS events with single jets. The end result is an estimate of a ±10% uncertainty in the measured cross-sections.

Various differential cross-sections are measured and compared with the NLO QCD calculations of the DISENT code, using $\mu_R^2 = \mu_F^2 = Q^2$ as the default choices for the renormalization and factorization scales. The

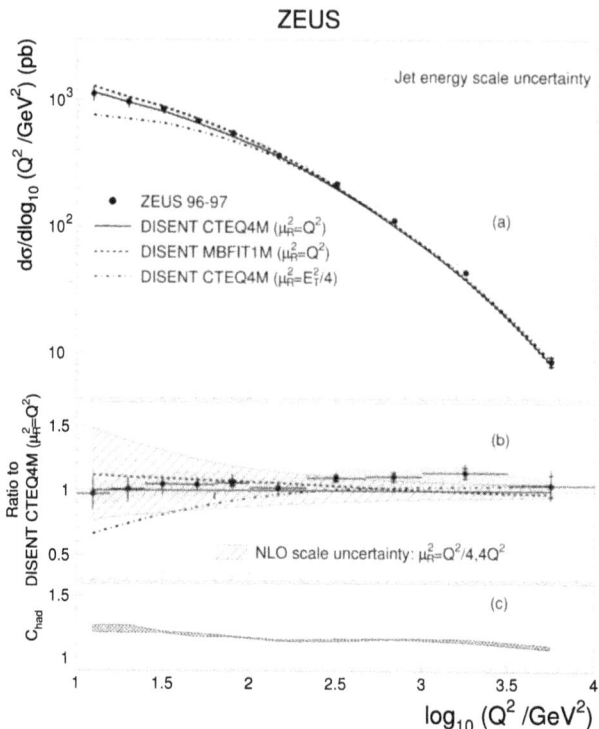

Fig. 7.5 (a) The hadron level dijet cross-section $d\sigma/d\log_{10} Q^2$ measured in the Breit frame. (b) The ratio of the measured cross-section to the DISENT calculation. (c) The parton to hadron correction (with uncertainty) used to adjust the DISENT parton level values to the hadron level of the measurement. (From ZEUS 2002b).

scale uncertainties were estimated by repeating the calculations while varying these scales in the range $(Q^2/4, 4Q^2)$. The renormalization scale uncertainty is the larger of the two, only falling to below 10% for $Q^2 > 2000\,\text{GeV}^2$, whereas the factorization scale uncertainty is only about 5%. A different choice of renormalization scale, $\mu_R^2 = E_T^2/4$ (where E_T is the sum of all final state parton transverse energies), was also investigated. Although this choice gives a somewhat smaller scale uncertainty, the predicted dijet cross-section for $Q^2 < 100\,\text{GeV}^2$ is well below the data. At larger Q^2 calculations with both choices of μ_R^2 give good descriptions of the data. This can be seen in Fig. 7.5(a) and (b) which shows $d\sigma/d\log_{10} Q^2$ and the ratio of the measured cross-section to the DISENT calculation, respectively. The figure also shows the result of varying the input parton densities (CTEQ4M, MBFIT1M), the NLO renormalization scale and jet energy scale uncertainties, and the parton to hadron correction factor. Us-

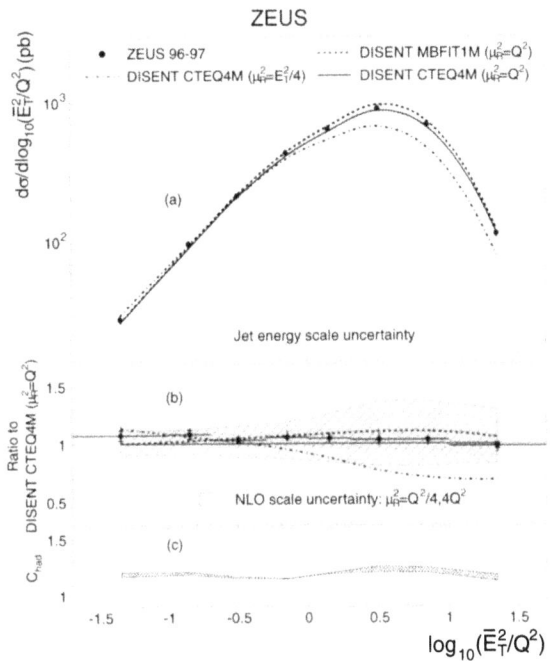

Fig. 7.6 (a) The hadron level dijet cross-section $d\sigma/d\log_{10}(\bar{E}_T^2/Q^2)$ measured in the Breit frame. (b) The ratio of the measured cross-section to the DISENT calculation. (c) The parton to hadron correction (with uncertainty) used to adjust the DISENT parton level values to the hadron level of the measurement. (From ZEUS 2002b).

ing $\mu_R^2 = Q^2$, the general level of agreement over nearly four orders of magnitude in Q^2 is good, without significant dependence on the choice of parton densities.

Dijet production in DIS is a process with two hard scales, Q^2 and the transverse jet energy. The latter is estimated using the mean transverse energy of the two highest energy jets in the event, $\bar{E}_T$. The interplay of the two scales is shown in Fig. 7.6 which shows $d\sigma/d\log_{10}(\bar{E}_T^2/Q^2)$, again with the ratio of measured to calculated values and estimates of the effect of various uncertainties. The DISENT calculation with Q^2 as scale describes the overall trend of the data vary well. The figure shows that the scale uncertainty is much larger when $\bar{E}_T^2 > Q^2$.

Another check on the reliability of the NLO description of the data is shown in Fig. 7.7 which shows the $\log_{10}\xi$ distribution of the dijet cross-section in bins of Q^2 at medium to large Q^2. In a given bin, the peaked

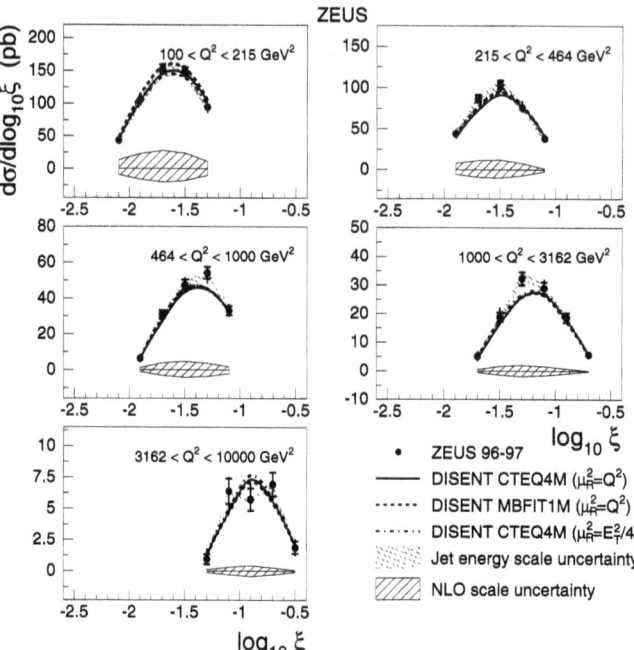

Fig. 7.7 The hadron level dijet cross-section $d\sigma/d\log_{10}\xi$ at medium to large Q^2 measured in the Breit frame in bins of Q^2. Also shown are NLO QCD calculations from DISENT with $\mu_R^2 = Q^2$, $E_T^2/4$ and two different parton densities (CTEQ4M, MBFIT1M). The Jet energy scale is shown as the shaded band and the NLO scale uncertainty as the hatched region at the bottom of each plot. (From ZEUS 2002b).

shape of the distribution is caused by the the decrease in the gluon density at large x (large ξ) and the requirement of quite high minimum transverse jet energies for the sample as a whole together with the lower Q^2 boundary of the bin in question restricting events in the low ξ region. The figure also shows nicely how dijets at large Q^2 are probing regions of large ξ and hence parton densities at large x. The NLO QCD calculations from DISENT describe the data quite well with the NLO renormalization scale uncertainty decreasing as Q^2 increases.

7.7 α_S from DIS jets

Having established that the NLO QCD calculation describes the behaviour of the dijet data over wide ranges of both Q^2 and $\bar{E}_T^2$, one may proceed with some confidence to attempt a measurement of α_s. It is clear, for two reasons, that the data at large Q^2 should give the smallest systematic uncertainty:

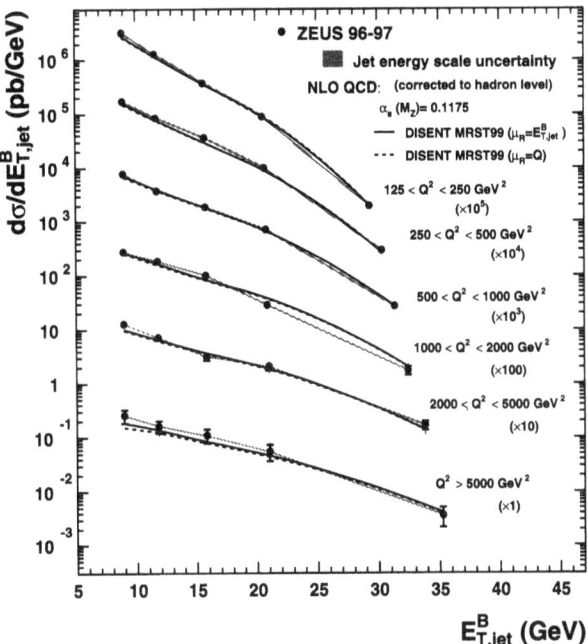

Fig. 7.8 The differential cross-section $d\sigma/dE^B_{T,jet}$ for inclusive jet production with $E^B_{T,jet} > 8\,\text{GeV}$ in the Breit frame in bins of Q^2. The NLO QCD calculations are from DISENT with $\alpha_s(M_Z^2) = 0.1175$. (From ZEUS 2002c).

first the NLO renormalization scale uncertainty is smallest for large Q^2 and second the QCDC process dominates in this region thus reducing the dependence on the gluon PDF. Using essentially the same data sample as for their dijet study, the ZEUS collaboration have measured α_s from the inclusive jet cross-section data using events with $Q^2 > 125\,\text{GeV}^2$ and $E^B_{T,jet} > 8\,\text{GeV}$ (ZEUS 2002c). As before the jets were reconstructed in the Breit frame using the longitudinally invariant k_T cluster algorithm in inclusive mode, and the final sample contained 8523 events with at least one jet satisfying $E^B_{T,jet} > 8\,\text{GeV}$ and $-2 < \eta^B_{jet} < 1.8$. The measured cross-sections were corrected for detector effects and compared to NLO QCD calculations obtained using DISENT. For these calculations the scale choices were, $\mu_F^2 = Q^2$ and $\mu_R^2 = (E^B_{T,jet})^2$, the latter choice being made to minimise the scale uncertainty. Fig. 7.8 shows $d\sigma/dE^B_{T,jet}$ as a function of $E^B_{T,jet}$ in bins of Q^2. The jet energy scale uncertainty is shown as the shaded band and calculations with both choices of renormalization scale, $\mu_R^2 = Q^2$ and $\mu_F^2 = (E^B_{T,jet})^2$, are displayed. Both the shape and normalization of

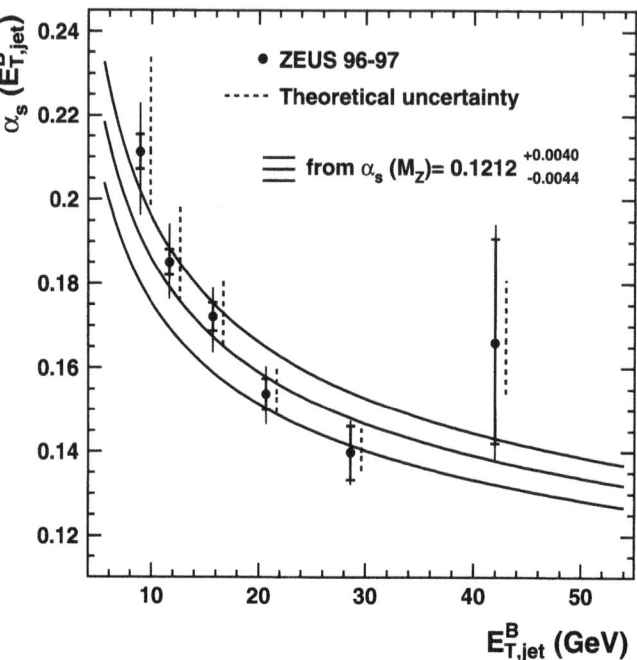

Fig. 7.9 The strong coupling constant α_s as a function of $E^B_{T,jet}$ determined from fits to the inclusive jet cross-section in the Breit frame.

the data are well described.

To determine α_s the measured cross-sections as functions of both $E^B_{T,jet}$ and Q^2 are used together with corresponding NLO QCD calculations performed using the MRST99 PDFs with α_s 'central', 'up' and 'down'. For each bin, i, and each variable A ($A = Q^2$, $E^B_{T,jet}$), the QCD calculations are used to parameterize the dependence on α_s as

$$\left.\frac{d\sigma}{dA}\right|_i = C_1^i \alpha_s(M_Z^2) + C_2^i \alpha_s^2(M_Z^2).$$

The value of $\alpha_s(M_Z^2)$ is then determined by a χ^2 fit to the data using the above expression. The best determination is for $Q^2 > 500 \text{ GeV}^2$, for which the theoretical and overall uncertainties are smallest, giving a χ^2 of 2.1 for four data points and

$$\alpha_s(M_Z^2) = 0.1212 \pm 0.0017(\text{stat.})^{+0.0023}_{-0.0031}(\text{sys.})^{+0.0028}_{-0.0027}(\text{th.}). \tag{7.8}$$

The running of α_s with momentum scale is measured using a wider range of data and parameterizing the α_s dependence of $d\sigma/dE^B_{T,jet}$ in terms of $\alpha_s(\langle E^B_{T,jet}\rangle^2)$ (rather than $\alpha_s(M_Z^2)$), where $\langle E^B_{T,jet}\rangle$ is the mean value of $E^B_{T,jet}$ in a bin. The results are shown in Fig. 7.9. The data points are shown with statistical (inner) and total errors together with the theoretical uncertainty shown as a vertical dashed line. The momentum scale dependence of α_s predicted from the two-loop QCD renormalization group equation with five flavours and $\alpha_s(M_Z^2)$ equal to the ZEUS best fit value is also shown with its uncertainty.

The value of α_s given in Eq. (7.8) above is consistent with the best global determinations, for example, that from the particle data group of 0.1181 ± 0.0020. It also agrees with and has a precision comparable to the best determination in e^+e^- annihilation (Bethke 2000).

7.8 Combined analysis of NC jet and inclusive data

The H1 collaboration have made a similar comprehensive study of jet production in NC DIS using 33 pb^{-1} of data with $5 < Q^2 < 15000\,\text{GeV}^2$ and $7 < E_{T,jet} < 60\,\text{GeV}$ (H1 2001b). To determine α_s they have studied fits to a combination of jet cross-section and NC inclusive cross-section data. In addition to a fit for α_s alone, they have investigated fits to determine the gluon density function and combined fits to determine α_s and the gluon density simultaneously. For the fits, the jet data is restricted to the region $150 < Q^2 < 5000\,\text{GeV}^2$ and the inclusive cross-sections to $150 < Q^2 < 1000\,\text{GeV}^2$ where the theoretical uncertainties are smallest, but Q^2 is not large enough for Z^0 effects to be significant. The cross-section for the hard process may be written as a convolution of coefficient functions (c_a) and parton densities ($f_{a/p}$) summed over the active flavours. The coefficient functions are expanded as a power series in α_s, to give an expression of the form

$$\sigma = \sum_{a,n} \int_0^1 dx'\, \alpha_s^n(\mu_R^2) c_{a,n}(x/x',\mu_R^2,\mu_F^2) f_{a/p}(x',\mu_F^2),$$

where the sum runs over active parton flavours ($a = q, \bar{q}$ and g) and all orders in α_s considered in the expansion. For the purposes of this analysis quark masses are ignored, thus reducing the number of coefficient functions to three, c_g, c_u, c_d with three corresponding independent parton densities. They are written in the following form

$$c_g x g(x) + c_\Sigma x \Sigma(x) + c_\Delta x \Delta(x),$$

where

$$x\Sigma = x \sum_a (q_a(x) + \bar{q}_a(x)); \quad x\Delta = x \sum_a e_a^2 (q_a(x) + \bar{q}_a(x))$$

and

$$c_\Sigma = (4c_d - c_u)/3, \quad c_\Delta = 3(c_u - c_d).$$

At zeroth and first order in α_s, the contributions from the different quark flavours are proportional to their squared electric charges, e_a^2, so that c_Σ

vanishes and Σ contributes first at $O(\alpha_s^2)$. The gluon contributes at $O(\alpha_s)$ and above. At LO the structure is

$$\sigma_{\text{incl.DIS}} \propto \Delta; \quad \sigma_{\text{jet}} \propto \alpha_s(c_g g + c_\Delta \Delta),$$

which shows again the inevitable correlation between α_s and the gluon density.

The results of the three types of fits may be summarised as follows.

α_s alone: Here the data used are the double differential inclusive jet cross-sections with respect to Q^2 and E_T. The PDFs are taken from the CTEQ5M1 set and the renormalization scale is $\mu_R^2 = E_T^2$, with the factorization scale set to the fixed value of $\mu_F^2 = 200\,\text{GeV}^2$. The running of α_s with E_T is demonstrated and the best fit to all 16 data points gives

$$\alpha_s(M_Z^2) = 0.1186 \pm 0.0030(\text{exp.})^{+0.0039}_{-0.0045}(\text{th.}),$$

in agreement with both ZEUS and world average values.

xg alone: For this fit, a combination of inclusive jet data and inclusive DIS cross-sections are fit with α_s fixed at the world average value of $\alpha_s = 0.1184 \pm 0.0031$. The combination of data allows both xg and the quark density $x\Delta$ to be determined. Both xg and $x\Delta$ are parameterized by four-parameter functions of the form $Ax^b(1-x)^c(1+dx)$. The result of the fit for the gluon density together with the uncertainty is shown in the LH plot of Fig. 7.10, where a comparison is also made with other determinations. From the fit result, the integral of the gluon density over the range $(0.01, 0.1)$ gives

$$\int_{0.01}^{0.1} dx\, xg(x, \mu_F^2 = 200\,\text{GeV}^2) = 0.229^{+0.031}_{-0.030}(\text{tot.}),$$

which is in good agreement with results from global fits.

combined: The data are the inclusive jet cross-sections and the inclusive DIS cross-sections, but this time α_s is allowed vary together with the parameters of the gluon and quark PDF parameterizations. The RH plot of Fig. 7.10 shows correlations between xg and α_s for four values of x. The plots show that the product $\alpha_s xg(x)$ is reasonably well determined by the combined data but neither component can be determined very precisely from a combined fit, given the present quality of the data.

7.9 Summary

This chapter has covered two key areas of DIS that provide precision measurements of the strong coupling constant α_s. The first area covers techniques that use scaling violations of nucleon structure functions. The conceptually simplest technique uses low order moments of non-singlet structure functions, for which the pQCD expressions are often known to quite

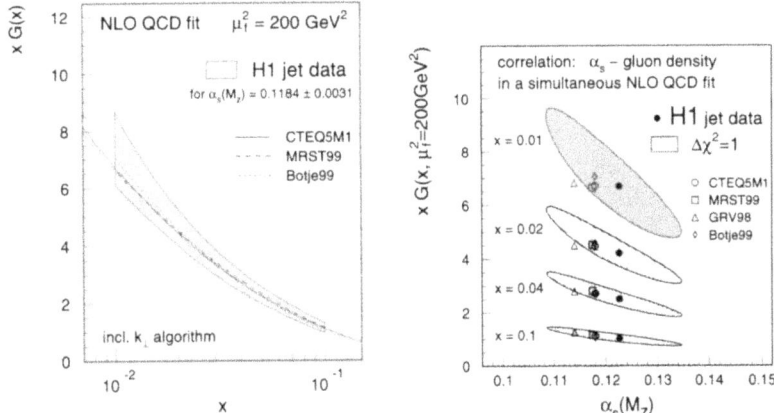

Fig. 7.10 H1 fits to inclusive jet cross-sections and the inclusive NC cross-section. LH plot: the gluon momentum density function $xg(x)$; (b) RH plot: the correlation between xg and α_s. (From H1 2001b).

high orders in α_s. However, it is difficult to get a broad enough coverage of measurements to be able to evaluate these moments accurately, without further assumptions. The alternative technique includes α_s as a parameter in the fits used to extract parton density functions from both singlet and non-singlet structure function data. The increasing sophistication of these fits, particularly in regard to the inclusion and propagation of both experimental and theoretical systematic errors, accounts for the correlation of the value of α_s and the shape of the gluon distribution and allows realistic errors to be assigned to these α_s determinations. Quite a wide range of results for α_s have been published which allow the interplay of data from different regions of the (x, Q^2) plane to be studied. Recent results, including precision data from HERA, give values in the range 0.1150–0.1190 with total experimental error of about 0.003 and a comparable theoretical error. The standard formalism for these global fits is that based on the NLO DGLAP evolution equations. An, as yet, unresolved question is whether increasing precision will require refinements to the calculations from NNLO evolution, or from terms beyond the DGLAP formalism at low x, or both.

The second area is that of large E_T jet production at high Q^2. The cross-sections for such processes depend explicitly on α_s. The importance of the Breit frame, both experimentally and theoretically has been outlined: on the experimental side, it allows a clean separation of high E_T jets from the zeroth order DIS process and on the theoretical side, it is the frame in which factorization of the jet cross-sections has been established. Another crucial ingredient for jet physics is the appropriate choice of jet algorithm to relate the final state hadronic structure to the underlying partonic structure. Although dependence of the calculations on parton densities, in particular the gluon cannot be eliminated completely, it can be minimized. Best fit values for α_s from H1 and ZEUS measurements are in the range 0.1186–0.1212, with errors comparable to the scaling violation determinations. The jet data also allow the running of α_s to be demonstrated in single experiment. For the future, apart from better control of experimental systematic effects, one also needs a better theoretical understanding of scale uncertainties. This would be a likely outcome of extending the current NLO calculations to NNLO, but this will be a formidable undertaking.

In principle, the best way to extract the maximum information on α_s from DIS is to combine the fits to scaling violations and jet rates. H1 have made a pioneering study of this approach, but the present precision of the data limits its competitiveness.

Overall the measurements of α_s from DIS are consistent with various 'world' average values and offer a precision comparable to techniques from other areas such as e^+e^- annihilation.

7.10 Problems

1. Show that in the HERA laboratory frame in which the positive z-axis is along the direction of the proton beam, the components of the virtual photon's 4-momentum may be written as

   $$q_0 = +y(E_e - xE_p), \quad q_1 = -\sqrt{Q^2(1-y)}$$
   $$q_3 = -y(E_e + xE_p), \quad q_2 = 0$$

   where E_e and E_p are the electron and proton beam energies, respectively; Q^2, x, y are the usual DIS Lorentz invariant variables and masses have been ignored.

2. Referring to the discussion at the start of Section 7.4.2, show that $x_p = x/\xi$ and prove the relations given in Eq. (7.3), where parton masses have been ignored.

3. Check that the soft and collinear singularities in the QCDC and BGF cross-sections discussed following Eqs (7.5) and (7.6) do occur for the configurations of partons given. Explain qualitatively the physical origin of the singularities, and how the situation would change if the quark masses were not neglected?
   [It may be helpful to refer to Section 4.1, particular the discussions following Eqs (4.4) and (4.14).]

4. Check the calculations given in the Appendix, particularly Eq. (7.9) and the final form of the matrix $\mathcal{L}(L \to B)$

5. Use $\mathcal{L}(L \to B)$ to show the following

   $$p = (E_p, 0, 0, E_p) \quad \to \quad p^* = \left(\frac{Q}{2x}, 0, 0, \frac{Q}{2x}\right)$$

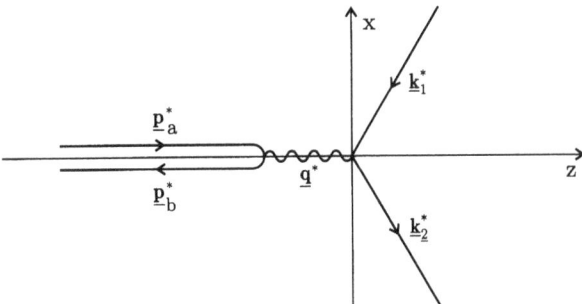

Fig. 7.11 Diagram showing the 3-momenta for elastic eq scattering in the Breit frame, $e(k_1^*) + q(p_a^*) \to e(k_2^*) + q(p_b^*)$.

$$k_1 = (E_e, 0, 0, -E_e) \quad \to \quad k_1^* = \left(\frac{Q(2-y)}{2y}, -\frac{1}{y}\sqrt{Q^2(1-y)}, 0, -\frac{Q}{2}\right)$$

$$k_2 = (E', E'\sin\theta', 0, E'\cos\theta') \quad \to \quad k_2^* = \left(\frac{Q(2-y)}{2y}, -\frac{1}{y}\sqrt{Q^2(1-y)}, 0, \frac{Q}{2}\right)$$

7.11 Appendix: transformation to the Breit frame

The construction of the full transformation from the HERA laboratory frame to the Breit frame is best taken in stages.

First establish the boost parameter, in both magnitude and direction. In the HERA frame, define 4-momenta for $e^{\pm}$-quark scattering by $e(k_1)+q(p_a) \to e(k_2)+q(p_b)$, where $p_i = (\omega_i, \mathbf{p}_i)$ and $p_a + k_1 = p_b + k_2$. Corresponding momenta in the Breit frame are are labelled with *,

$$p_a^* = (\omega^*, \mathbf{p}^*), \; p_b^* = (\omega^*, -\mathbf{p}^*); \; k_1 = (E^*, \mathbf{k}_1^*), \; k_2 = (E^*, \mathbf{k}_2^*),$$

where $|\mathbf{k}_1^*| = |\mathbf{k}_2^*|$ and $q^* = k_1^* - k_2^* = (0, -2\mathbf{p}^*)$. The struck quark is turned around and the $e^{\pm}$ scatters elastically off the 'brick wall'. Figure 7.11 shows the 3-momenta for the eq elastic scattering in the Breit frame after the final transformation to align the virtual photon momentum along the z-axis.

The Lorentz transformation with velocity $\boldsymbol{\beta}$ relates 3-momenta in the two frames

$$\mathbf{p}^* = \mathbf{p} + \gamma\boldsymbol{\beta}\left(\frac{\gamma\boldsymbol{\beta}\cdot\mathbf{p}}{\gamma+1} - \omega\right).$$

Using this

$$\mathbf{p}_a^* + \mathbf{p}_b^* = 0 = \mathbf{p}_a + \mathbf{p}_b + \gamma\boldsymbol{\beta}\left[\frac{\gamma}{\gamma+1}\boldsymbol{\beta}\cdot(\mathbf{p}_a + \mathbf{p}_b) - (\omega_1 + \omega_2)\right]$$

which gives (after some algebra — start by dotting the above equation with $\boldsymbol{\beta}$)

$$\boldsymbol{\beta} = \frac{\mathbf{p}_a + \mathbf{p}_b}{\omega_1 + \omega_2} = \frac{\mathbf{q} + 2x\mathbf{p}}{q_0 + 2xE_p}, \quad (7.9)$$

where $(q_0, \mathbf{q})$ are the components of the virtual photon 4-momentum in the HERA frame, $(E_p, \mathbf{p})$ those of the proton, x is Bjorken x and masses have been ignored. Note that $\boldsymbol{\beta}$ is at an angle with respect to the z-axis in the HERA frame.

Although the problem is formally solved, $\mathbf{q}^*$ does not yet lie along the z-axis in the Breit frame. The complete transformation that takes the z-axis in the HERA frame to the z-axis in the Breit frame may conveniently be written in matrix form as

$$\mathcal{L}(L \to B) = R_y(\alpha')\Lambda(\beta)R_y(\alpha),$$

where R_y are rotations about the y-axis in the respective frames and Λ is a Lorentz transformation along the z-axis. The elements of these matrices will be written in terms of the components of $q = (q_0, q_1, 0, q_3)$ in the HERA frame and here the results of Q1 below will be useful. From Eq. (7.9) above, the Lorentz parameters are

$$\beta \equiv \sqrt{|\boldsymbol{\beta}|^2} = \frac{D_1}{q_0 + Q^2/(q_0 - q_3)}, \qquad \gamma\beta = \frac{D_1}{Q},$$

where

$$D_1^2 = q_1^2 + (q_3 + 2xE_p)^2 = q_1^2 + \left(q_3 + \frac{Q^2}{q_0 - q_3}\right)^2.$$

The rotation about the y-axis in the HERA frame is given by

$$\sin\alpha = -\frac{q_1}{D_1}, \qquad \cos\alpha = \frac{q_3 + Q^2/(q_0 - q_3)}{D_1}.$$

The final rotation about the y-axis in the Breit frame is given by

$$\sin\alpha' = \frac{Qq_1}{D_2}, \qquad \cos\alpha' = -\frac{q_0(q_0 - q_3)}{D_2},$$

where $D_2^2 = Q^2 q_1^2 + q_0^2(q_0 - q_3)^2$. After some algebra, the overall transformation matrix takes a simple form

$$\mathcal{L}(L \to B) = \begin{pmatrix} \frac{q_0}{Q} + \frac{Q}{q_0-q_3} & -\frac{q_1}{Q} & 0 & -\frac{q_3}{Q} - \frac{Q}{q_0-q_3} \\ -\frac{q_1}{q_0-q_3} & 1 & 0 & \frac{q_1}{q_0-q_3} \\ 0 & 0 & 1 & 0 \\ \frac{q_0}{Q} & -\frac{q_1}{Q} & 0 & -\frac{q_3}{Q} \end{pmatrix}$$

where $q = (q_0, q_1, 0, q_3)$ is the 4-momentum of the γ^* in the laboratory frame, $Q^2 = q_1^2 + q_3^2 - q_0^2$, $Q = \sqrt{Q^2}$ and the identity $(q_0 - q_3)D_1 = D_2$ has been used.

8
DIS at high Q^2

This chapter outlines the necessary extensions to the formalism to allow for W and Z exchange at high Q^2, in charged lepton scattering processes. The formalism is also extended to allow for polarization of the $e^\pm$ beams. The relationships of the structure functions to the PDFs via the electroweak couplings are detailed. The presently available high-Q^2 HERA data are presented and their contribution to determining PDFs is considered. The importance of HERA-II data for precision measurements of the PDFs, without many of the uncertainties of the fixed target analyses, is outlined. The contribution of HERA-II data to the precision measurement of electroweak parameters is explored. The situation for unpolarized beams is considered first as it is simpler and there is already some data from HERA-I high Q^2 running. The second half of the chapter then covers the extension of the formalism to include polarized beams.

The $e^\pm$ beams at HERA become transversely polarized through the Sokolov–Ternov effect. At 27.5 GeV polarizations of up to 70% have been observed at HERA-I with risetimes of under two hours. For useful physics, the transverse polarization must be 'rotated' into left- or right-handed longitudinal polarization (and back again). This has been achieved routinely for the fixed target ep experiment HERMES at HERA and will be available for the collider experiments H1 and ZEUS in the HERA-II programme. Because of different beam optics, the maximum polarization expected at HERA-II is about 60%.

8.1 Cross-sections for unpolarized lepton beams
8.1.1 Neutral Current

The form for the differential cross-section for charged lepton–nucleon scattering, mediated by the photon has already been given in Chapter 2, Eq (2.15) and Appendix C. However, at high Q^2, Z^0 exchange can no longer be ignored so that the weak neutral current of Eq (2.30) and the electromagnetic current of Eq. (2.1) both contribute and the cross-section gains terms due to Z^0 exchange and due to γZ interference. Formally, the parity violating term involving W_3 must be included so that the differential cross-section is now given in terms of three structure functions, F_2, F_L, xF_3, as,

$$\frac{d^2\sigma(l^\pm N)}{dx\,dQ^2} = \frac{2\pi\alpha^2}{Q^4 x}\left[Y_+ F_2^{lN}(x,Q^2) - y^2 F_L^{lN}(x,Q^2) \mp Y_- xF_3^{lN}(x,Q^2)\right], \tag{8.1}$$

where $Y_\pm = 1 \pm (1-y)^2$. The LO QCD improved parton model still predicts $F_L = 0$ at high Q^2, but the quark–parton model formula for F_2, given in Eq. (2.41), must be extended as follows:

$$F_2^{lN}(x,Q^2) = \sum_i A_i^0(Q^2) * (xq_i(x,Q^2) + x\bar{q}_i(x,Q^2)), \tag{8.2}$$

where, for unpolarized lepton scattering,

$$A_i^0(Q^2) = e_i^2 - 2e_i v_i v_e P_Z + (v_e^2 + a_e^2)(v_i^2 + a_i^2)P_Z^2 \tag{8.3}$$

The term in P_Z arises from γZ^0 interference and the term in P_Z^2 arises purely from Z^0 exchange, where P_Z accounts for the effect of the Z^0 propagator relative to that of the virtual photon, and is given by

$$P_Z = \frac{Q^2}{Q^2 + M_Z^2}\frac{1}{\sin^2 2\theta_W}. \tag{8.4}$$

The other factors in the expression for A_i^0 are the charge, e_i, NC electroweak vector, v_i, and axial-vector, a_i, couplings of quark i and the corresponding NC electroweak couplings of the electron, v_e, a_e.[1] The parity violating structure function xF_3 is given by

$$xF_3^{lN}(x,Q^2) = \sum_i B_i^0(Q^2) * (xq_i(x,Q^2) - x\bar{q}_i(x,Q^2)), \tag{8.5}$$

where,

$$B_i^0(Q^2) = -2e_i a_i a_e P_Z + 4a_i v_i v_e a_e P_Z^2 \tag{8.6}$$

In NLO QCD, the expressions for F_2 and xF_3 must be modified so that parton distributions are convoluted with appropriate coefficient functions, as explained in Chapter 4. In addition, the contribution from F_L is non-zero at NLO, however for high-Q^2 HERA data the contribution is small, roughly at the percent level (since for $Q^2 > 200\,\text{GeV}^2$ one is restricted to larger x, $x > 0.004$). It is convenient to define a NC reduced cross-section as

$$\tilde{\sigma}_{NC}(\pm) = \left[\frac{2\pi\alpha^2 Y_+}{xQ^4}\right]^{-1}\frac{d^2\sigma^\pm}{dx\,dQ^2} = F_2 \mp \frac{Y_-}{Y_+}xF_3.$$

---

[1] More details on the full NC EW formalism for DIS are given in the appendix to this chapter.

8.1.2 Charged Current

For charged–lepton–nucleon scattering mediated by $W^{\pm}$ exchange (where the final state lepton is a neutrino), the differential cross-sections are given by

$$\frac{d^2\sigma^{CC}(l^{\pm}N)}{dx\,dQ^2} = \frac{G_F^2}{4\pi x}\frac{M_W^4}{(Q^2+M_W^2)^2} \quad (8.7)$$
$$\times \left[Y_+ F_2(x,Q^2) - y^2 F_L(x,Q^2) \mp Y_- xF_3(x,Q^2)\right]$$

and the correspondence to the neutral current case can be seen easily if we express the Fermi coupling constant G_F as

$$G_F = \frac{\pi\alpha}{\sqrt{2}\sin^2\theta_W M_W^2} \quad (8.8)$$

It is convenient to define a CC reduced cross-section as

$$\tilde{\sigma}_{CC}(\pm) = \left[\frac{G_F^2}{2\pi x}\left(\frac{M_W^2}{M_W^2+Q^2}\right)^2\right]^{-1}\frac{d^2\sigma^{\pm}}{dx\,dQ^2}. \quad (8.9)$$

In LO QCD, $F_L = 0$, and

$$F_2(x,Q^2) = \sum_i x(q_i(x,Q^2) + \bar{q}_i(x,Q^2))$$
$$xF_3(x,Q^2) = \sum_i x(q_i(x,Q^2) - \bar{q}_i(x,Q^2)),$$

where the sums contain only the appropriate quarks or antiquarks for the charge of the current. Specifically for $l^- p \to \nu X$

$$F_2 = 2x(u(x,Q^2)+c(x,Q^2)+\bar{d}(x,Q^2)+\bar{s}(x,Q^2)),$$
$$xF_3 = 2x(u(x,Q^2)+c(x,Q^2)-\bar{d}(x,Q^2)-\bar{s}(x,Q^2)), \quad (8.10)$$

and for $l^+ p \to \bar{\nu}X$

$$F_2 = 2x(d(x,Q^2)+s(x,Q^2)+\bar{u}(x,Q^2)+\bar{c}(x,Q^2)),$$
$$xF_3 = 2x(d(x,Q^2)+s(x,Q^2)-\bar{u}(x,Q^2)-\bar{c}(x,Q^2)). \quad (8.11)$$

Using these expressions, one finds

$$\tilde{\sigma}_{CC}(+) = x[\bar{u}+\bar{c}+(1-y)^2(d+s+b)],$$
$$\tilde{\sigma}_{CC}(-) = x[u+c+(1-y)^2(\bar{d}+\bar{s}+\bar{b})], \quad (8.12)$$

where the arguments x, Q^2 have been suppressed. It is assumed that there is no significant top quark content in the nucleon and that energies are above threshold for the production of c and b quarks in the final state.

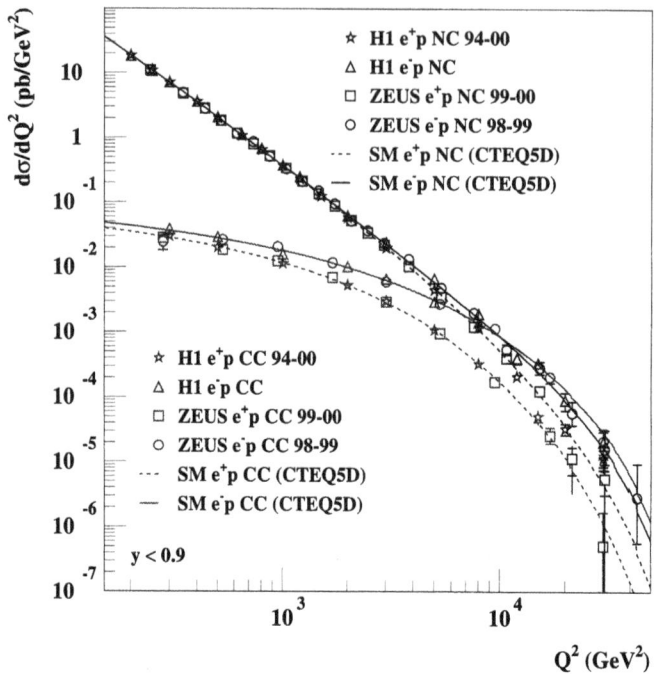

Fig. 8.1 HERA-I data on unpolarized $e^{\pm}p$ CC and NC scattering at high Q^2.

For charged–lepton scattering on neutron targets, these relationships undergo isospin swapping $u \to d, d \to u, \bar{u} \to \bar{d}, \bar{d} \to \bar{u}$, and for an isoscalar target the average $(n+p)/2$ is taken.

At NLO, the parton distributions must be convoluted with their appropriate coefficient functions in order to form the structure functions, just as in the NC case.

8.2 Unpolarized high–Q^2 data

Both the ZEUS and the H1 experiment have published high Q^2 cross-sections from $\sim 16 \, \mathrm{pb}^{-1}$ of $e^- p$ data (H1 2001a, ZEUS 2002a, ZEUS 2003a) taken in the years 1998-1999 and $\sim 60 \, \mathrm{pb}^{-1}$ of $e^+ p$ data (H1 2003b, ZEUS 2003d, ZEUS 2003e) taken in the years 1999-2000, both at $\sqrt{s} = 318 \, \mathrm{GeV}$. These $e^+ p$ data can be combined with the previously published data at $\sqrt{s} = 300 \, \mathrm{GeV}$ taken in the years 1994-1997 to give a total $e^+ p$ sample of $\sim 90 \, \mathrm{pb}^{-1}$ per experiment from HERA-I running.

The differential cross-section data for neutral (NC) and charged current (CC) $e^{\pm}p$ scattering are compared in Fig. 8.1. There is excellent agreement with the Standard Model predictions for electroweak unification at high Q^2 (these predictions have been evaluated using the CTEQ5D PDF set). One can clearly see that the CC cross-sections are suppressed at low Q^2, due

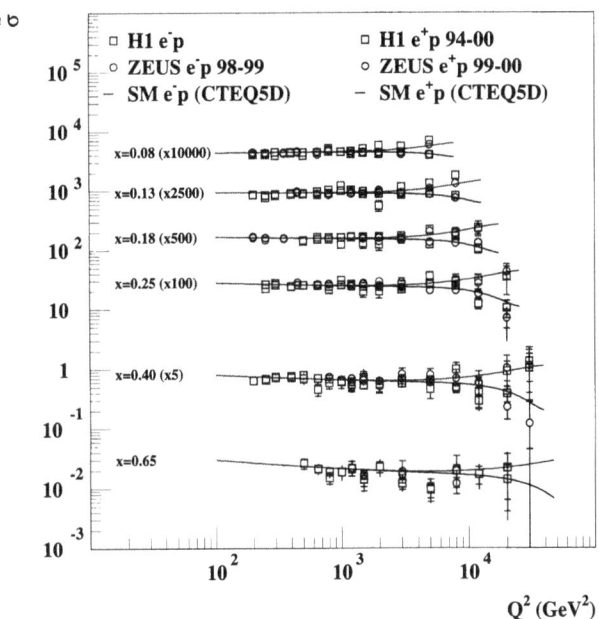

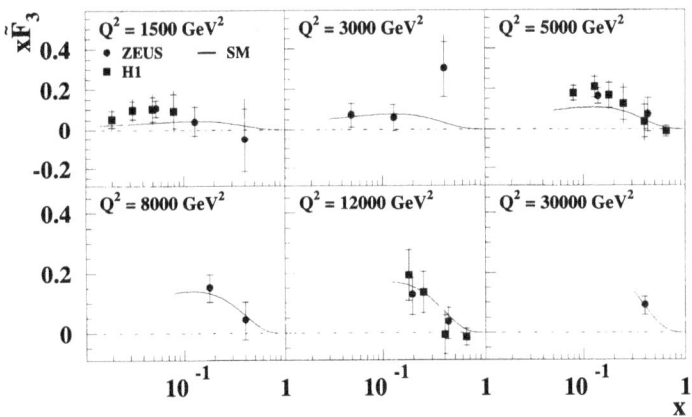

Fig. 8.2 Top plot: $\tilde{\sigma}_{NC}(\pm)$ data from HERA as a function of Q^2. Bottom plots: xF_3 versus x for fixed Q^2 extracted from the ZEUS high Q^2 NC $e^{\pm}p$ data.

to the large mass of the W propagator, but become comparable to the NC cross-sections for $Q^2 \sim M_W^2$. One can also see the difference between the e^+p and e^-p cross-sections for both NC and CC processes.

The top plot of Fig. 8.2 shows the difference between the e^+p and the e^-p NC reduced cross-sections as a function of Q^2. This difference is due to the contribution of the xF_3 parity violating structure function. Since the e^+p and e^-p cross-sections were measured at slightly different CM energies, the kinematic factors removed in the definition of the reduced cross-section must be adjusted to give

$$xF_3 = \left(\frac{Y_-^{300}}{Y_+^{300}} + \frac{Y_-^{318}}{Y_+^{318}}\right)^{-1} (\tilde{\sigma}_{NC}(-) - \tilde{\sigma}_{NC}(+)) - \Delta F_L.$$

Here the superscripts '300', '318' refer to the two CM energies and the term ΔF_L is a small correction to allow for the different values of F_L at the two energies. The resulting data for xF_3 from ZEUS is shown in the bottom plots of Fig. 8.2.

The HERA high Q^2 CC reduced cross-section data are shown in Fig. 8.3. The contributions of the different flavours of PDFs (evaluated at LO from the CTEQ5D PDF set according to Eq. (8.12)) are also indicated so that one can clearly see that the d_v PDF dominates the e^+p cross-section (top set of plots), whereas the u_v PDF dominates the e^-p cross-section at high x (bottom set of plots).

Both the NC and CC high-Q^2 data are very well described by the global PDF fits, described in Chapter 6. However, the high Q^2 HERA data can also be used to to gain information on the high x valence PDFs, independently of the fixed-target data. Figure 8.4 shows the valence distributions extracted from a PDF fit to ZEUS data alone (ZEUS-ONLY fit). The data included are all the data taken in the years 1994–2000 in HERA-I running. This figure should be compared with Fig. 6.5, which shows the valence distributions from the ZEUS global PDF fit, which included the fixed target data and ZEUS NC data from 1996–1997 running, but not the more recent high-Q^2 ZEUS data. The level of precision of the ZEUS-ONLY fit is approaching that of the global PDF fit.

As discussed in Chapter 6, in the global fits the data which most strongly determine the valence distributions are the fixed target data: the xF_3 measurement from CCFR and the F_2^D/F_2^p ratio measurement from NMC. These data suffer from uncertainties associated with nuclear target corrections. This is not just a problem for the CCFR Fe target data, the NMC F_2^D/F_2^p ratio data are also subject to some uncertainty from deuteron binding corrections. Although the PDFs extracted from fits to HERA data alone have not yet reached the precision of the global fits, their importance is that they are performed on a proton target. In particular, the CC e^+p cross-section uniquely gives nuclear target correction independent information on the less well known d_v distribution.

8.2.1 PDF extraction from high–Q^2 data

The precision of the distributions shown in Fig. 8.4 is statistics limited rather than systematics limited, so that improvement can be expected with higher luminosity HERA-II data. The xF_3 measurement shown in Fig. 8.2

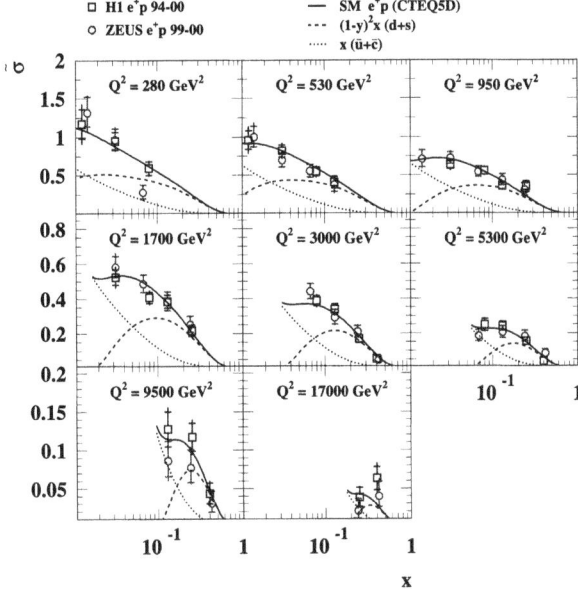

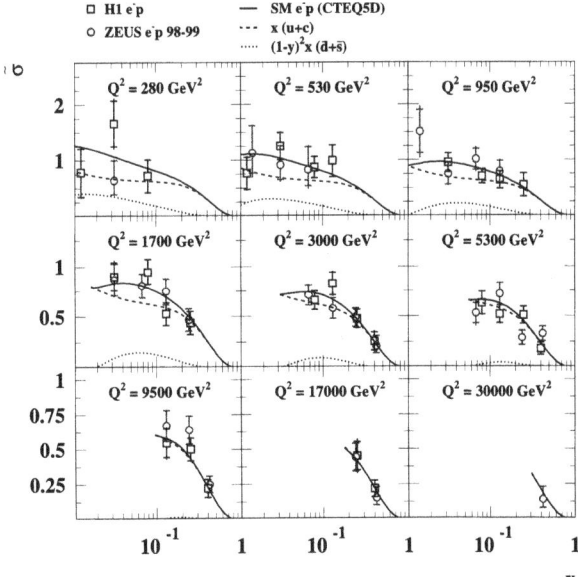

Fig. 8.3 CC $e^{\pm}p$ reduced cross-section data from HERA; top set of plots for e^+p, bottom set for e^-p.

will also improve to become a precision measurement across a broad x range. This will be the only precision measurement of a valence structure function at small x, using a proton target. It should also be possible to use the flavour information in NC and CC e^+p and e^-p scattering to ex-

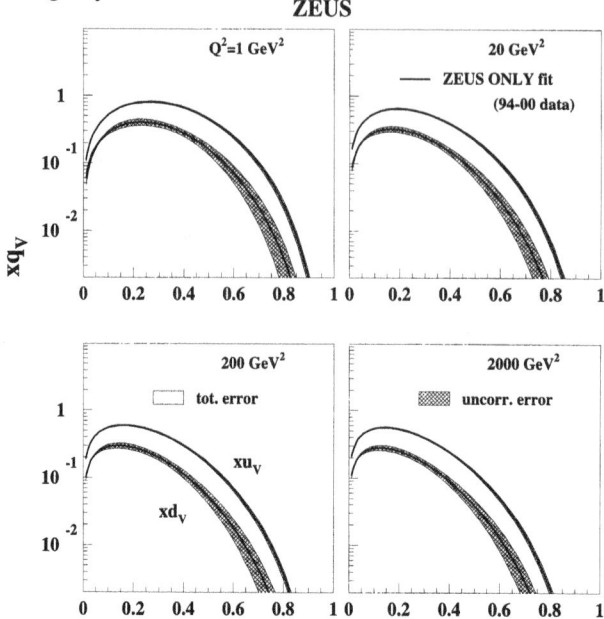

Fig. 8.4 The xu_v and xd_v valence distributions from an NLO QCD fit to ZEUS data alone (including high-Q^2 data from 1998–2000 running) for various Q^2. The error bands illustrated account for experimental uncorrelated errors from statistical and systematic sources and for the total experimental errors including those from correlated systematic sources. (From ZEUS 2003b.)

tract PDFs in a model independent way (i.e. without need of introducing a parametrization dependence in a QCD fit) for the complete x range. The sums and differences of the unpolarized NC and CC reduced cross-sections

$$\Sigma^0_{NC/CC} = \tilde{\sigma}_{NC/CC}(-) + \tilde{\sigma}_{NC/CC}(+),$$
$$\Delta^0_{NC/CC} = \tilde{\sigma}_{NC/CC}(-) - \tilde{\sigma}_{NC/CC}(+). \quad (8.13)$$

can be expressed in LO QCD as

$$\Sigma^0_{CC} = xU + (1-y)^2 xD, \quad (8.14)$$
$$\Delta^0_{CC} = xu_v - (1-y)^2 xd_v, \quad (8.15)$$
$$\Sigma^0_{NC} = 2F_2 = A^0_u xU + A^0_d xD, \quad (8.16)$$
$$\Delta^0_{NC} = 2\frac{Y_-}{Y_+}xF_3 = B^0_u xu_v + B^0_d xd_v, \quad (8.17)$$

where

$$xU = xu + x\bar{u} + xc + x\bar{c}, \qquad xD = xd + x\bar{d} + xs + x\bar{s} + xb + x\bar{b}.$$

Referring to Eqs (8.3) and (8.6), it can be seen that the A and the B coefficients each take only two values: U-type and D-type. Thus combinations of the four cross-sections $\tilde{\sigma}_{NC/CC}(\pm)$ can be used to extract U, D, u_v and d_v distributions. With about 200 pb^{-1} for each e^+ and e^- beam energy it is expected that $\sim 10\%$ accuracy could be obtained (Ingelman and Rückl 1989).

8.3 Cross-sections for polarized lepton beams

8.3.1 Neutral Current

As explained in detail in the appendix at the end of this chapter, the cross-section for NC scattering of polarized lepton beams, with polarization P, off unpolarized protons $(e^\pm(P)p \to e^\pm X)$ may be written as:

$$\left.\frac{d^2\sigma^\pm}{dx\,dQ^2}\right|_{NC} = \frac{2\pi\alpha^2}{xQ^4}\left[H_0^\pm + PH_P^\pm\right], \tag{8.18}$$

with

$$H_{0,P}^\pm = Y_+ F_2^{0,P} \mp Y_- xF_3^{0,P}, \tag{8.19}$$

$$F_2^{0,P} = \sum_i x(q_i + \bar{q}_i) A_i^{0,P}, \tag{8.20}$$

$$xF_3^{0,P} = \sum_i x(q_i - \bar{q}_i) B_i^{0,P}, \tag{8.21}$$

and $P = \dfrac{N_R - N_L}{N_R + N_L}$, where N_R, N_L are the number of right- and left-handed leptons in the beam, respectively. The coefficients A_i^0, B_i^0 are those for unpolarized beams given in Eqs (8.3) and (8.6), respectively. The coefficients for the polarization terms are

$$A_i^P = 2e_i a_e v_i P_Z - 2a_e v_e (v_i^2 + a_i^2) P_Z^2, \tag{8.22}$$

$$B_i^P = 2e_i a_i v_e P_Z - 2a_i v_i (v_e^2 + a_e^2) P_Z^2. \tag{8.23}$$

Before getting into the detail of how the extra polarization dependent information could be used, it is useful to have a simple overview of the effect of the parity violating Z^0 exchange on the NC processes. Figure 8.5 (left hand plot) shows the ratio of NC cross-sections including Z^0 exchange to that from the virtual photon alone for states of definite helicity $e_{L,R}^\pm$. The ratios are shown as functions of Q^2 for $x = 0.2$. By $Q^2 = 10^4\,\text{GeV}^2$ the effect of the weak terms can change the cross-sections by up to a factor of 2. At this value of x, the valence quarks will dominate. It is clear from the sign of the xF_3 contribution why the $e_{L,R}^-$ cross-sections are larger than the $e_{L,R}^+$ cross-sections. The behaviour of these ratios may be understood in detail by using Eq. (8.37) of the appendix. The right-hand plot of the figure shows the absolute values of the NC cross-sections and illustrates very clearly why high luminosity is needed to exploit the discrimination that polarization offers at high Q^2.

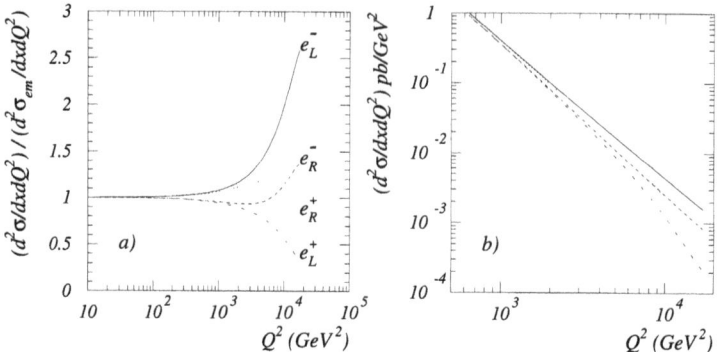

Fig. 8.5 LH plot: ratio of NC cross-sections including Z^0 exchange to that from the virtual photon for states of definite helicity. RH plot: the absolute NC cross-sections. Both plots are shown as functions of Q^2 for $x = 0.2$. (From Cashmore et al. 1996).

8.3.2 Charged Current

Since CC scattering already involves a pure chiral coupling, the changes required to allow for polarized beams are straightforward. In terms of parton densities and with a beam polarization P

$$\left.\frac{d^2\sigma^+}{dx\,dQ^2}\right|_{CC} = (1+P)\frac{G_F^2}{2\pi x}\frac{M_W^4}{(Q^2+M_W^2)^2}x[\bar{u}+\bar{c}+(1-y)^2(d+s+b)],$$

$$\left.\frac{d^2\sigma^-}{dx\,dQ^2}\right|_{CC} = (1-P)\frac{G_F^2}{2\pi x}\frac{M_W^4}{(Q^2+M_W^2)^2}x[u+c+(1-y)^2(\bar{d}+\bar{s}+\bar{b})],$$

for incident positron and electrons, respectively. With the appropriate choice of polarization, the cross-sections may be increased by almost a factor of 2 compared to the unpolarized case. The behaviour of the integrated CC cross-sections as a function of polarization P is shown in Fig. 8.6.

8.4 Extraction of electroweak parameters

8.4.1 M_W measurements

The data are now sufficiently precise to make it worthwhile examining the extent to which HERA can contribute to constraints on the electroweak sector of the Standard Model. For example, the strong dependence of the CC cross-sections on the W propagator can be used to make an extraction of M_W in a space-like process. The results from ZEUS and H1 for e^- and e^+ data are given in Table 8.1. The e^- data give the better determinations, both because of the larger cross-section and because of the reduced uncertainty from the PDFs when the better known u quark distribution is dominant.

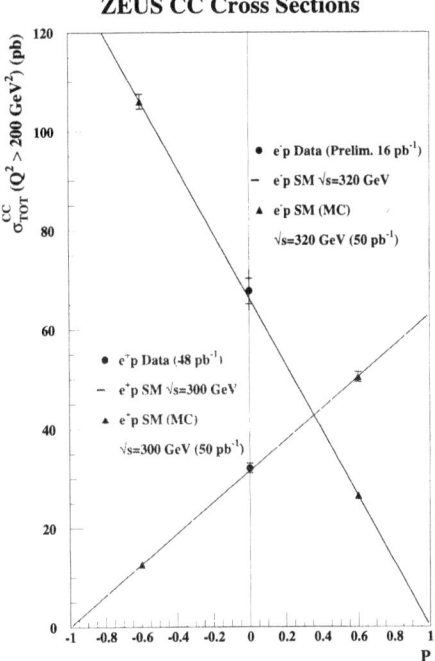

Fig. 8.6 Dependence of the integrated charged current cross-sections for polarized $e^\pm$ beams. The e^+ cross-section increases with P and the e^- cross-section decreases with P. (Figure courtesy of K R Long.)

Table 8.1 Values of M_W extracted from ZEUS and H1 CC data.

Experiment	beam	M_W
ZEUS	e^+	$81.4 \pm 2.7(\text{stat}) \pm 2.0(\text{sys}) \pm 3.0(\text{PDF})$
H1	e^+	$80.9 \pm 3.3(\text{stat}) \pm 1.7(\text{sys}) \pm 3.7(\text{PDF})$
ZEUS	e^-	$80.3 \pm 2.1(\text{stat}) \pm 1.2(\text{sys}) \pm 1.0(\text{PDF})$
H1	e^-	$79.9 \pm 2.2(\text{stat}) \pm 0.9(\text{sys}) \pm 2.1(\text{PDF})$

The precision on such measurements can be improved by using information on normalization as well as on shape. The factor

$$\frac{G_F^2 M_W^4}{(Q^2 + M_W^2)^2}$$

can be replaced by

$$\frac{1}{2}\left(\frac{\pi\alpha}{1-M_W^2/M_Z^2}\frac{1}{Q^2+M_W^2}\right)^2$$

using Eq. (8.8) and the relationship, $\sin^2\theta_W = 1 - M_W^2/M_Z^2$. This expression has a much stronger dependence on M_W and could yield measurements of M_W with greatly improved accuracy, since M_Z is accurately known from LEP data. Such a measurement could also be regarded as a measurement of G_F, made at $Q^2 \sim 400\,\text{GeV}^2$ rather than at $Q^2 = m_\mu^2$, where very accurate measurements of G_F from muon decay already exist.

However, considerations so far have focused on the cross-sections at Born level. It is also necessary to consider electroweak radiative corrections. Through such corrections the values of standard model parameters such as the top mass, m_t, and the Higgs mass, M_H, enter into the high-Q^2 DIS cross-sections. For example, in the CC cross-section, the expression for G_F should be modified to

$$G_F = \frac{\pi\alpha}{\sqrt{(2)}\sin^2\theta_W M_W^2}\frac{1}{1-\Delta r} \qquad (8.24)$$

where Δr is a function of M_W, M_Z, m_t, α and M_H. There are similar modifications for the Z^0 contribution to the NC cross-sections. The full modifications to the formulae, including form factors which multiply the couplings, are specified in Cooper-Sarkar et al. (1999) and Spiesberger et al. (1991). The point is that the standard model parameters (α, M_Z, M_W, M_H, m_t, the light quark masses plus the CKM matrix elements) are not constrained by theory. The HERA experiments can have an impact only on the first five of these parameters. The Higgs mass, M_H, enters only logarithmically via loop corrections to the gauge boson self energies, so it is usually taken as fixed. The values of α and M_Z are already very well measured. This leaves M_W and m_t. In the past the accurate measurements of G_F from muon decay were used together with HERA data to get constraints on M_W and m_t. Now that accurate measurements of m_t from Tevatron Run II data are imminent one can tighten the accuracy on M_W. Comparing values of M_W from HERA with those from LEP provides constraints which have to be simultaneously satisfied. Since radiative corrections enter into the space-like and time-like processes differently, it is a non-trivial test of the standard model that consistent values of M_W and M_t at HERA, G_F from muon decay and M_t from the Tevatron, be obtained.

The accuracy achievable on M_W from HERA measurements is illustrated in Fig. 8.7, as a function of M_t. Quantifying this accuracy in terms of the error on M_W, ΔM_W, one finds that with unpolarized beams a measurement of the CC cross-sections can yield an accuracy of $\Delta M_W \sim 200 MeV$ if normalization as well as shape information is used. Sensitivity is greatest with electrons rather than positrons, simply because the cross-section is greater, and polarization improves accuracy for the same reason. For the NC processes, polarization is much more important. This is because it is

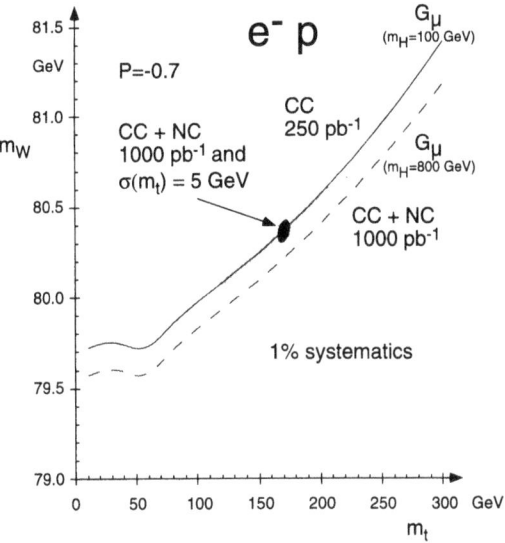

Fig. 8.7 Constraints on the W mass from various e^-p measurement scenarios at HERA. The plot shows 1σ-confidence limits in the M_W, m_t plane for: 250 pb^{-1} CC data (large ellipse); 1 fb^{-1} CC and NC data (shaded ellipse); and 1 fb^{-1} CC and NC data combined with an m_t mass measurement accurate to $\sigma(m_t) = 5$ GeV (full ellipse). The full and dashed diagonal lines show the relationship given by Eq. (8.24) for two values of m_H. (From Beyer et al. 1996).

the shape of the cross-section as a function of Q^2 which is giving indirect dependence on M_W via the γZ interference term. Left handed polarization enhances the size of this term (indeed 70% polarization would be worth a factor four in luminosity) and sensitivity is again greatest with electrons rather than positrons. So with 1 fb^{-1} of NC and CC polarized data one can improve on the accuracy obtainable from CC data alone and achieve $\Delta M_W \sim 55$ MeV, if M_t is known to 5 GeV (Cooper-Sarkar et al. 1999, Beyer et al. 1996). This should be compared to the PDG value $\Delta M_W \sim 49$ MeV, measured in the time-like processes.

8.4.2 Measurements of the quark weak neutral couplings

One may also be able to measure the weak neutral couplings of quarks: v_u, a_u, v_d, a_d. In the DIS processes, it is the light quarks which dominate the cross-sections so such measurements would be complementary to the LEP measurements, where the U-type couplings are got from $c\bar{c}$ production and the D-type couplings from $b\bar{b}$ production. The analysis will be considered at the Born level since the electroweak radiative corrections to the a_i and v_i couplings are not signficant in the HERA kinematic regime (until experimental accuracies of better than 2% are achievable).

The NC couplings of the electron take the values $a_e = -1/2$, $v_e = -1/2 + 2\sin^2\theta_W \sim -0.05$, so that terms involving v_e are small. Using Eq. (8.6) for B_i^0 and Eqs (8.22) and (8.23) for A_i^P and B_i^P, respectively, one also sees that the pure Z^0 terms are suppressed by an extra factor of P_Z (at least until very large values of Q^2 are reached). Thus unpolarized scattering is sensitive to a_i through xF_3^0, whereas polarized scattering is also sensitive to v_i through F_2^P. These sensitivities may be exploited as follows. First, define some additional polarization dependent reduced cross-section sums and differences. For these, it is assumed that both electron and positron beams with right- and left-handed polarizations are available and that the specific choice $P_R = -P_L = P$ is made, where P_R, P_L are the degrees of polarization of the right- and left-handed beams, respectively. For the CC cross-sections, the sum

$$\Sigma_{CC}^P = \tilde{\sigma}_{CC}^{P_L}(-) + \tilde{\sigma}_{CC}^{P_R}(+) = (1+P)\left(xU + (1-y)^2 xD\right), \tag{8.25}$$

using a obvious notation for the reduced CC polarized cross-sections. For the NC case, the differences between cross-sections with the same beam charge but opposite polarizations

$$\tilde{\Delta}_{NC}^{P-} = Y_+\left(\tilde{\sigma}_{NC}^{P_L}(-) - \tilde{\sigma}_{NC}^{P_R}(-)\right) = -2PH_P^-, \tag{8.26}$$

$$\tilde{\Delta}_{NC}^{P+} = Y_+\left(\tilde{\sigma}_{NC}^{P_L}(+) - \tilde{\sigma}_{NC}^{P_R}(+)\right) = -2PH_P^+, \tag{8.27}$$

using an obvious notation for the reduced NC polarized cross-sections. The tilde on the Δ indicates that these differences of polarized NC reduced cross-sections differ from the corresponding differences of unpolarized NC reduced cross-sections by the Y_+ factor.

The axial couplings a_u, a_d are sensitive to the ratio of the unpolarized NC/CC differences

$$R_\Delta^0 = \frac{\Delta_{NC}^0}{\Delta_{CC}^0} = \frac{B_u^0 + B_d^0 \frac{d_v}{u_v}}{1 - (1-y)^2 \frac{d_v}{u_v}} \tag{8.28}$$

The advantage of taking ratios is to reduce sensitivity to assumptions on the shapes of the PDFs, since only the ratio d_v/u_v is involved. For this method sensitivity is greatest to a_u, as the u-quark density is the largest in the valence region.

For the vector couplings combining Eqs (8.26) and (8.27), and using the expressions for $H_P^\pm$ but ignoring terms in v_e, gives

$$2v_u xU - v_d xD = -\frac{3}{8}\frac{\tilde{\Delta}_{NC}^{P-} + \tilde{\Delta}_{NC}^{P+}}{PY_+ a_e P_Z}. \tag{8.29}$$

Then construct the ratio to Σ_{CC}^P to obtain

$$-\frac{3}{8}\frac{\tilde{\Delta}_{NC}^{P-}+\tilde{\Delta}_{NC}^{P+}}{PY_+ a_e P_Z \Sigma_{CC}^P} = \frac{2v_u - v_d \frac{D}{U}}{(1+P)\left(1+(1-y)^2\frac{D}{U}\right)}. \tag{8.30}$$

Again a ratio is taken to minimize sensitivity to assumptions on the shapes of the PDFs, in this case only the ratio D/U is involved. For this method, sensitivity is greatest to v_u.

Finally measurements of the NC cross-sections alone in all four of the lepton charge/polarization combinations may be used in a fit to determine all four couplings. Polarization is essential to achieve reasonable precision. With $\sim 250\,\mathrm{pb}^{-1}$ for each of the four lepton beam charge/polarization combinations one could achieve precisions of 13, 6, 17 and 17% on v_u, a_u v_d and a_d, respectively.

8.5 Summary

In this chapter we have outlined the necessary extensions to the formalism to allow for lepton beam polarization and for the contributions of W and Z exchange at high Q^2. The presently available high-Q^2 HERA data have been presented and their role in determining parton distributions outlined. HERA-II data will allow precision measurements of the PDFs independent of earlier fixed target data. The contribution of HERA-II data to the precision measurement of electroweak parameters has also been discussed. Radiative corrections enter into the space-like and time-like processes differently so that it is a non-trivial test of the Standard Model that consistent values of the electroweak parameters be obtained at HERA, LEP, the Tevatron and in low energy data.

8.6 Problems

1. Check the derivation of Eqs (8.10), (8.11) and (8.12).
2. Check the formulae for the sums and differences of the unpolarized high Q^2 cross-sections, Eqs (8.14)–(8.17), and consider how they might be used to extract the U, D, u_v and d_v PDFs.
3. Using the equations of Section 8.1.2, consider how it may be possible to use HERA data to study the s and c quark content of the sea. How important is particle identification for the final state hadronic system? Does lepton beam polarization make a difference?
4. Consider the possible advantages of lowering the proton beam energy in the study of high-x valence distributions.
5. Using Eqs (8.35)–(8.37) of the appendix, and the numerical values for the quark and lepton axial and vector NC couplings, check that you understand the behaviour of the individual cross-sections for $e_{L,R}^\pm$ at large Q^2 shown in Fig. 8.5.
6. Check the derivations of Eqs (8.28) and (8.29). Estimate the likely accuracy for a_u and v_u, given 5 or 10% measurements of the NC and CC polarized cross-sections.

8.7 Appendix: formalism for NC high–Q^2 DIS

Although the formalism for NC scattering at high Q^2 with polarized lepton beams has been written down in many places, there are a number of potential pitfalls so it is covered here in some detail. The starting point is the article by Klein and Riemann (1984) but the focus will be on the more restricted QCD enhanced parton model formalism developed for the 1996 HERA Workshop by Cashmore et al. (1996).

The general form of the cross-section for scattering of a polarized charged lepton with helicity λ off an unpolarized nucleon $(\ell^{\pm}(\lambda)N \to \ell^{\pm}X)$ is

$$\frac{d^2\sigma^{\pm}(\lambda)}{dx dQ^2} = \frac{2\pi\alpha^2}{xQ^4}\sigma^{\pm}(\lambda), \quad \text{where}$$
$$\sigma^{\pm}(\lambda) = \sigma_0 + \sigma_I^{\pm}(\lambda) + \sigma_Z^{\pm}(\lambda), \qquad (8.31)$$

where $\pm$ indicates the charge of the lepton and radiative corrections have been ignored. The cross-section like terms, $\sigma_0, \sigma_I, \sigma_Z$, contain the contributions of the virtual photon, the γZ interference and the Z exchange, respectively. For the most general theory, there are in principle a total of eight structure functions: F_2, F_L in σ_0; G_2, G_L, xG_3 in σ_I and H_2, H_L, xH_3 in σ_Z. Assuming that the simplest version of the Glashow–Weinberg–Salam electroweak theory is correct,[2] that QCD may be applied in the form of the parton densities following the DGLAP equations and that F_L, G_L, H_L may be ignored at large Q^2, a much simpler picture emerges. One is left with two 'electroweak' structure functions each containing contributions from the three terms defined above.

$$\sigma^{\pm}(\lambda) = Y_+ f_2^{\pm}(\lambda) + Y_- x f_3^{\pm}(\lambda), \qquad (8.32)$$
$$f_2^{\pm}(\lambda) = \sum_i x(q_i + \bar{q}_i) A_i^{\pm}(\lambda), \qquad (8.33)$$
$$x f_3^{\pm}(\lambda) = \sum_i x(q_i - \bar{q}_i) B_i^{\pm}(\lambda), \qquad (8.34)$$

where $Y_{\pm} = 1 \pm (1-y)^2$, $q_i, \bar{q}_i$ are the quark and antiquark densities for flavour i with the arguments x, Q^2 suppressed and all the EW coupling terms are in the coefficients A and B.

$$A_i^{\pm}(\lambda) = e_i^2 + 2e_i v_i(-v_e \mp \lambda a_e)P_Z + (v_e^2 + a_e^2 \pm 2v_e a_e \lambda)(v_i^2 + a_i^2)P_Z^2$$
$$B_i^{\pm}(\lambda) = 2e_i a_i(\pm a_e + \lambda v_e)P_Z + 2a_i v_i(\mp 2v_e a_e - (v_e^2 + a_e^2)\lambda)P_Z^2,$$

where e_i, v_i, a_i are the quark charge (in units of the proton charge) and its NC vector and axial-vector couplings, v_e, a_e are the corresponding NC couplings for the electron and $P_Z = Q^2/((Q^2 + M_Z^2)\sin^2 2\theta_W)$ gives the

---
[2] This is a reasonable assumption as one will, at best, be looking for small deviations from the SM.

effect of the Z^0 propagator relative to that of the virtual photon. The NC fermion couplings are given by $v_f = T_3^f - 2e_f \sin^2\theta_W, a_f = T_3^f$ where T_3^f is the third component of the weak isospin and $\sin^2\theta_W = 1 - M_W^2/M_Z^2 = 0.223$ in the 'on-mass-shell' renormalization scheme. The values of these NC couplings for all quarks and leptons are summarised in Table 8.2.

Table 8.2 EW NC axial and vector couplings for left-handed fermion states.

Particles	T_3^f	e_f	v_f	a_f
ν_e, ν_μ, ν_τ	$+\frac{1}{2}$	0	$+\frac{1}{2}$	$+\frac{1}{2}$
e^-, μ^-, τ^-	$-\frac{1}{2}$	-1	$-\frac{1}{2} + 2\sin^2\theta_W$	$-\frac{1}{2}$
u, c, t	$+\frac{1}{2}$	$+\frac{2}{3}$	$\frac{1}{2} - \frac{4}{3}\sin^2\theta_W$	$+\frac{1}{2}$
d, s, b	$-\frac{1}{2}$	$-\frac{1}{3}$	$-\frac{1}{2} + \frac{2}{3}\sin^2\theta_W$	$-\frac{1}{2}$

Using the expressions for $A^\pm(\lambda)$ and $B^\pm(\lambda)$ given above, one may now write down expressions for the coefficients corresponding to scattering with left- and right-handed beam leptons $L, \lambda = -1, R, \lambda = +1$. Starting with electrons $e^-_{L,R}$

$$\begin{aligned}
A_i^-(L) &= e_i^2 - 2e_i v_i(v_e + a_e)P_Z + (v_e + a_e)^2(v_i^2 + a_i^2)P_Z^2 \\
A_i^-(R) &= e_i^2 - 2e_i v_i(v_e - a_e)P_Z + (v_e - a_e)^2(v_i^2 + a_i^2)P_Z^2 \\
B_i^-(L) &= -2e_i a_i(v_e + a_e)P_Z + 2a_i v_i(v_e + a_e)^2 P_Z^2 \\
B_i^-(R) &= 2e_i a_i(v_e - a_e)P_Z - 2a_i v_i(v_e - a_e)^2 P_Z^2
\end{aligned} \quad (8.35)$$

for positrons $e^+_{L,R}$ on finds the relations

$$\begin{aligned}
A_i^+(L) &= +A_i^-(R), & A_i^+(R) &= +A_i^-(L), \\
B_i^+(L) &= -B_i^-(R), & B_i^+(R) &= -B_i^-(L),
\end{aligned} \quad (8.36)$$

It is very important to understand that in these definitions, when the electron is changed to the positron, the electron couplings, v_e, a_e specified in Table 8.2 are not changed. In principle, the antifermion charges and weak couplings are the opposite of those of the fermions, but it has become conventional to use the fermion values and modify the formulae as specified in Eq. (8.36). Note that in some of the sources for these formulae in the literature this convention is not applied and the signs of the fermion couplings v_l, a_l must be changed when the sign of the fermion charge is changed, as for example in Cooper-Sarkar et al. (1998).

Using $A_i^\pm(L,R), B_i^\pm(L,R)$ and Eq. (8.32) one can now write reduced cross-sections for $e^\pm_{L,R}$ scattering

$$\sigma^\pm_{L,R} = Y_+ f_2^\pm(L,R) + Y_- x f_3^\pm(L,R), \qquad (8.37)$$

where

$$\begin{aligned}
f_2^-(L,R) &= \sum_i x(q_i + \bar{q}_i) A_i^-(L,R), \\
f_2^+(L,R) &= \sum_i x(q_i + \bar{q}_i) A_i^-(R,L), \\
x f_3^-(L,R) &= \sum_i x(q_i - \bar{q}_i) B_i^-(L,R), \\
x f_3^+(L,R) &= -\sum_i x(q_i - \bar{q}_i) B_i^-(R,L),
\end{aligned}$$

where all the couplings have been written in terms of those appropriate for the electron.

Finally it is convenient to define the following combinations of coefficients from $A_i^-(L,R), B_i^-(L,R)$: first those for unpolarized lepton beams

$$\begin{aligned}
A_i^0 &= \tfrac{1}{2}\left(A_i^-(R) + A_i^-(L)\right) \\
&= e_i^2 - 2 e_i v_i v_e P_Z + (v_e^2 + a_e^2)(v_i^2 + a_i^2) P_Z^2 \qquad (8.38) \\
B_i^0 &= \tfrac{1}{2}\left(B_i^-(R) + B_i^-(L)\right) \\
&= -2 e_i a_i a_e P_Z + 4 a_i v_i v_e a_e P_Z^2; \qquad (8.39)
\end{aligned}$$

then those associated with the polarization of the lepton

$$\begin{aligned}
A_i^P &= \tfrac{1}{2}\left(A_i^-(R) - A_i^-(L)\right) \\
&= 2 e_i a_e v_i P_Z - 2 a_e v_e (v_i^2 + a_i^2) P_Z^2 \qquad (8.40) \\
B_i^P &= \tfrac{1}{2}\left(B_i^-(R) - B_i^-(L)\right) \\
&= 2 e_i a_i v_e P_Z - 2 a_i v_i (v_e^2 + a_e^2) P_Z^2. \qquad (8.41)
\end{aligned}$$

The polarization of the lepton beam is defined to be

$$P = \frac{N_R - N_L}{N_R + N_L},$$

where N_R and N_L are the number of right- and left-handed leptons in the beam respectively. The reduced cross-section for scattering may now be written in terms of the L and R cross-sections of Eq. (8.37), for example for an electron beam with polarization P

$$\sigma^-(P) = \frac{1}{2}(1-P)\sigma^-(L) + \frac{1}{2}(1+P)\sigma^-(R).$$

Using the relationships between the coefficients for e^- and e^+ given in Eq. (8.36) and the forms for the structure functions in Eq. (8.37), one finds

$$\sigma^\pm(P) = H_0^\pm + PH_P^\pm, \qquad (8.42)$$

$$H_{0,P}^\pm = Y_+ F_2^{0,P} \mp Y_- x F_3^{0,P}, \qquad (8.43)$$

$$F_2^{0,P} = \sum_i x(q_i + \bar{q}_i) A_i^{0,P}, \qquad (8.44)$$

$$x F_3^{0,P} = \sum_i x(q_i - \bar{q}_i) B_i^{0,P}. \qquad (8.45)$$

In these expressions all the sign dependence on the beam charge is removed from the coefficients and it appears in the sign of the xF_3 term and the definition of P.

9
DIS at low x

This chapter is concerned with physics at low values of Bjorken x, below 0.01. In this region, the $q\bar{q}$ sea and gluon dynamics are dominant and their study has been made possible by the advent of data from the HERA collider (see Fig. 5.1). The principal new feature of the data is the dramatic rise of F_2 at low x, the rise becoming more rapid as Q^2 increases (see Fig. 5.11). There are hints, as explained in Chapter 6, that standard DGLAP NLO QCD may not be adequate to describe the data at low x. The problem may be stated in the following way: sub-leading terms in $\ln(Q^2/\mu^2)$ involve powers of $\alpha_s \ln(1/x)$ which become large as x decreases (e.g. for $Q^2 \approx 10\,\text{GeV}^2$ and $x \approx 0.01$, $\alpha_s \ln(1/x) \approx 1$). The question which then arises is should the $\alpha_s \ln(1/x)$ terms be be resummed and if so how? This chapter will cover in some detail the various approaches within and beyond pQCD that have been developed for this extension to the traditional DIS phase space.

The chapter starts with a general overview of the approaches to be considered and their regions of validity in the $\ln(1/x) - \ln Q^2$ plane, relating them where appropriate to different perturbative summation schemes. Sections 9.2 and 9.3 cover what might be regarded as the 'standard' approaches, the limits of the DGLAP equations at low x and the Regge approach to total cross-sections, extended to encompass both soft and hard Pomerons (i.e. both slowly and steeply rising behaviours with energy). The new ideas begin in Section 9.4 with a description of the BFKL equation, which can either be seen as an attempt to calculate the Pomeron perturbatively or to predict the low x behaviour of F_2. A related and important aspect of QCD is angular ordering of gluon radiation and the related CCFM approach to DIS, these topics are covered in Section 9.5. All the approaches mentioned so far are linear and lead to an undamped increase in F_2 as x decreases. This must violate unitarity eventually. How this is avoided, by including non-linear effects that occur when the gluon density becomes large enough, is outlined in Section 9.6. This topic is important for both low-x DIS and the high density nuclear matter produced in heavy ion collisions. Sections 9.7 and 9.8 cover dipole models which provide a powerful framework for studying non-linear models respecting the constraint of unitarity. Section 9.9 gives a brief summary of the comparison of the various pQCD

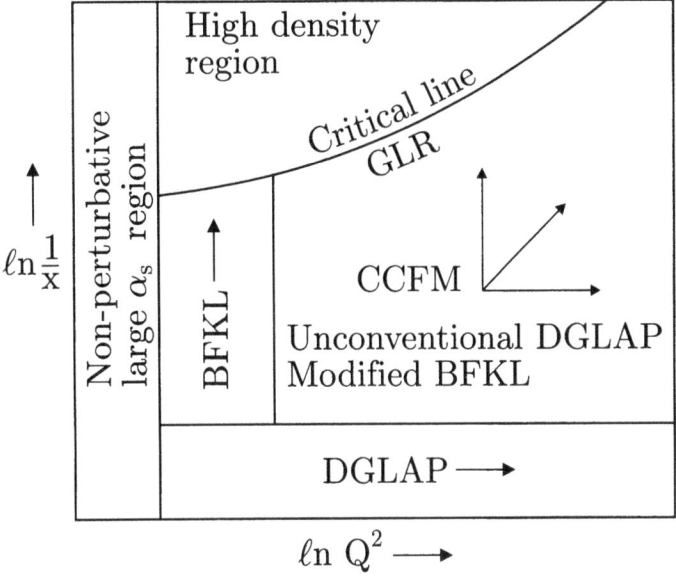

Fig. 9.1 Approaches to physics at low-x. (Courtesy of A D Martin.)

approaches with F_2 data and the need for precision measurements of other observables, particularly F_L. The chapter ends with an overall summary and problems.

Some sections in this chapter are quite technical in places, but it is to be hoped that the essential ideas can be extracted from a superficial reading. Many of the topics discussed in this chapter are the subject of active research, some of the formalism is not yet stable and conclusions are inevitably somewhat tentative.

9.1 Approaches at low x

Before embarking on more technical discussions, it is worth summarizing the possible approaches with the help of Fig. 9.1 which shows a 'map' of the $\ln Q^2 - \ln(1/x)$ plane. The familiar DGLAP evolution in Q^2, or summation of leading $\alpha_s \ln(Q^2/\mu^2)$ terms, is shown near the bottom as being valid for x not too small and Q^2 above a minimum value. Since small x is equivalent to a large $\gamma^* p$ CM energy W, the Regge framework which describes hadronic and photoproduction cross-sections at high energies may be applicable for large $1/x$ and small Q^2, the region marked 'non-perturbative large α_s'. At the top of the diagram, the 'high density region' shows the area where the gluon density becomes so large that non-linear effects must be taken into account. Below the 'critical line' and above the DGLAP region non-linear effects become negligible but $\ln(1/x)$ is still large. Here BFKL refers to an equation that results from summing $\alpha_s \ln(1/x)$ terms. CCFM is a different approach based on angular ordering of gluon emission, which in a sense interpolates between the results of the DGLAP and BFKL summations. All these ideas will be elaborated in due course, but first a more precise description of the various summations in $\ln Q^2$ and $\ln(1/x)$ is presented.

9.1.1 Summation schemes

In the linear regime and for Q^2 large enough for perturbative techniques to be applicable, the behaviour of the parton densities at small x is dominated by the behaviour of the splitting functions $P_{ij}(z)$ as $z \to 0$ (in the convolution integrals $z < x$). At LO one has

$$P_{qq}(z) \to \frac{4}{3}, \quad P_{qg}(z) \to \frac{1}{2}, \quad P_{gq}(z) \to \frac{8}{3z}, \quad P_{gg}(z) \to \frac{6}{z}. \tag{9.1}$$

The gluon splitting functions are the most singular and P_{gg} has the larger coefficient. Hence the behaviour of the DGLAP equations becomes dominated by the behaviour of the gluon splitting functions. The form of the DGLAP evolution equation for the gluon distribution at low x becomes dominated by the term containing P_{gg}

$$\frac{\partial g(x, Q^2)}{\partial \ln Q^2} = \int_x^1 \frac{d\xi}{\xi} P(x/\xi, \alpha_s) g(\xi, Q^2),$$

where the overall factor of $\alpha_s/(2\pi)$ has been absorbed into the splitting function and the label gg has been dropped. For small x the leading term in $P(x, \alpha_s)$ is then $(3\alpha_s)/(\pi x)$. This equation can be solved to give a prediction of a steeply rising behaviour of the gluon distribution at low x. The exact behaviour of the solution depends on the boundary conditions, and the two possible classes of solution are explored in Section 9.2.1 and 9.2.2

In general, any splitting function $P(x, \alpha_s)$ may be expanded in powers of $\alpha_s/(2\pi)$

$$P(x, \alpha_s) = \sum_{j=1}^{\infty} \left(\frac{\alpha_s}{2\pi}\right)^j P^{(j)}(x)$$

at low x the functions $xP^{(j)}(x)$ have a leading behaviour of $\ln^{j-1}(1/x)$ and may be written as

$$xP^{(j)}(x) = \sum_{k=1}^{j} A_k^{(j)} \ln^{k-1}(1/x) + x\overline{P}^{(j)}(x),$$

where $\overline{P}^{(j)}$ is finite as $x \to 0$. So the full expansion for the splitting function is

$$xP(x, \alpha_s) = \sum_{j=1}^{\infty} \left(\frac{\alpha_s}{2\pi}\right)^j \left[\sum_{k=1}^{j} A_k^{(j)} \ln^{k-1}(1/x) + x\overline{P}^{(j)}(x)\right]$$

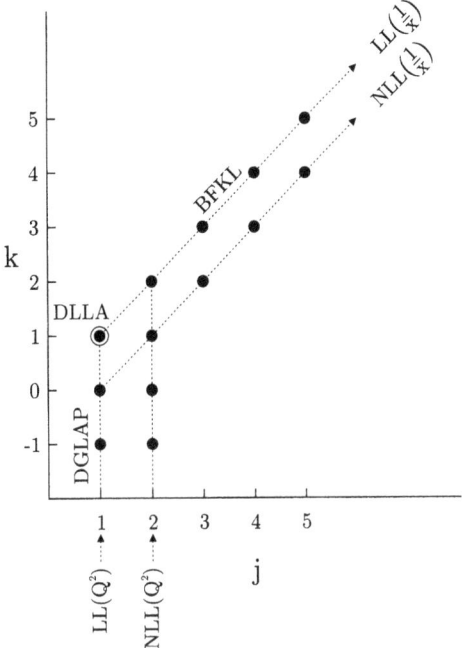

Fig. 9.2 Various resummation schemes. (Figure courtesy of R G Roberts.)

Such an expression holds for all splitting functions. Considering only the singular terms as $x \to 0$ and absorbing the factors of 2π in the coefficients, the expansion becomes a double sum over terms of the form $\alpha_s^p \ln^q(1/x)$

$$xP(x,\alpha_s) = \sum_{j=1}^{\infty}\sum_{k=1}^{j} \tilde{A}_k^{(j)} \alpha_s^j \ln^{k-1}(1/x). \qquad (9.2)$$

To connect with more formal OPE approaches it is useful to write the above equation in terms of its Mellin moments (anomalous dimensions)[1]

$$\gamma(n,\alpha_s) = \sum_{j=1}^{\infty}\sum_{k=1}^{j} a_{jk} \alpha_s^j (n-1)^{-k},$$

where the Mellin transform $x^{-1}\ln^{k-1}(1/x) \to (k-1)!(n-1)^{-k}$ has been used. The leading coefficient for γ_{gg} is $a_{11} = 3\alpha_s/\pi$. The DGLAP splitting functions correspond to a particular form of the double sum which is valid for $\ln Q^2$ large and x not too small. Referring to Fig. 9.2 various approximations to the splitting functions are defined as:

---

[1] See Section 4.1.2 and Eq. (4.31).

DLLA leading double asymptotic scaling limit with $j = 1, k = 1$;

LL(Q^2) LO DGLAP, $j = 1$ and the sum over k is extended to run from negative powers up to $k = 1$ (in order to include the expansion of the non-singular part);

NLL(Q^2) NLO DGLAP, $j = 1, 2$ and the sum over k from negative powers as in the previous case;

LL($1/x$) LO BFKL, $j \geq 1, k = j$;

NLL($1/x$) NLO BFKL, $j \geq 1, k = j, j - 1$.

BFKL refers to Balitsky, Fadin, Kuraev and Lipatov and their approach will be described in detail later in the chapter. Note the the DLLA limit is common to both the BFKL and DGLAP summations at LO. The challenge is to find techniques for performing the summations and to determine the regions of validity of a given approximation.

9.1.2 Low x and Mellin moments

Useful insights and powerful techniques are provided by the language and formalism of the operator product expansion. In particular, the use of moments (Mellin transforms). The nth moment of a parton density or structure function is defined by

$$\tilde{q}(n, Q^2) = \int_0^1 dx\, x^{n-1} q(x, Q^2).$$

For splitting functions the moments are the anomalous dimension functions $\gamma(n)$ (see Section 4.1.2). The inverse Mellin transform is a contour integral

$$q(x, Q^2) = \frac{1}{2\pi i} \int_C dn\, x^{-n} \tilde{q}(n, Q^2)$$

where the contour C runs parallel to the imaginary axis and is to the right of the right-most singularity of $\tilde{q}(n, Q^2)$ in complex n-space. The behaviour at low x of a PDF or structure function is controlled by the position of the rightmost singularity. When the PDF is given by a convolution in x-space, or equivalently a product of moments in n-space, the singularity could arise from either the PDF at the starting scale or the anomalous dimension.

Although extensive mathematical derivations are beyond the scope of this book, the above ideas will aid understanding at various points. For a full and lucid account of the OPE moment method, see Chapter 4 of Yndurán (1999).

9.2 DGLAP at low x

In Chapters 4 and 6, the QCD-enhanced parton model was described and its use in extracting parton density functions from global fits explained. While the DGLAP approach gives the evolution of parton densities and structure functions in Q^2, it requires other input for the x dependence at the starting scale. Usually simple parameterized functions of x are used with the parameters determined by the global fit. The question now is, in the more limited region of low x is it possible to get a more specific prediction, or at least one dependent on fewer parameters?

It is informative to look at the structure of the splitting functions in moment space. The LO expressions for the anomalous dimensions are given in Section 4.9 and they have poles in n at the negative integers and $n = 0, 1$. The poles at $n = 1$ occur in $\gamma_{gq}^{(0)}$ and $\gamma_{gg}^{(0)}$ for the gluon and they correspond to the $1/z$ terms in the gluon splitting functions since

$$\int_0^1 dz\, z^{n-1}(1/z) = 1(n-1).$$

Suppose that the gluon density at the starting scale μ^2 and for small x has the behaviour

$$xg(x,\mu^2) = Ax^{-\lambda_g}$$

then

$$\tilde{g}(n,\mu^2) = A \int_0^1 dx\, x^{n-1} x^{-\lambda_g - 1} = A/(n - \lambda_g - 1),$$

showing that $\tilde{g}(n, \mu^2)$ has a pole at $n = 1 + \lambda_g$. Considering the grossly simplified case of LO gluon evolution in which the singlet quark terms are ignored, it can be seen that the rightmost pole in n-space is at $n = 1$ from the anomalous dimension unless $\lambda_g > 0$. For $\lambda_g \leq 0$ ('valence like', or flat, gluon density) the asymptotic behaviour as $x \to 0$ is given by the evolution kernel. By contrast for $\lambda_g > 0$ the singular power behaviour of the input distribution is the dominant behaviour for all Q^2.

The calculation is more complicated if both quarks and gluons are included at LO or NLO as the singlet-gluon anomalous dimension matrix has to be diagonalized. The critical value of λ_g (or the equivalent power for the structure function) maybe somewhat larger than 0 (e.g. $\lambda_g \sim 0.2$), with this value also depending on the starting scale μ^2. However the essential feature of two possible behaviours at low x for DGLAP evolution remains. If μ^2 is not large and $xg(x, \mu^2)$ is valence like, flat or only mildly singular then the low x behaviour is given by that of the splitting function. If $xg(x, \mu^2)$ is more singular at low x then one obtains $xg(x, Q^2) \sim x^{-\lambda_g}$ with the behaviour in x now independent of Q^2.

Both types of behaviour are considered in turn and compared with early HERA data at low x. Both approaches were able to describe the data in the mid 1990's. Although global fits across the complete kinematic plane are now the norm for the analysis of the much larger complete HERA-I data, it is useful to understand the features of the DGLAP low-x limiting behaviours.

9.2.1 Double asymptotic scaling

The DGLAP limit at low x with a non-singular input is known as the double leading log approximation (DLLA) or double asymptotic scaling (DAS). The result was an early prediction of QCD (De Rujula et al. 1974), which was revived by the work of Ball and Forte (1994). A simplified account, at LO, following the approach of the latter authors is given here.

The first step is to simplify the gluon evolution equation by dropping the singlet term[2] to give

$$\frac{\partial g(x,t)}{\partial t} \approx \frac{\alpha_s(t)}{2\pi} \int_0^{\ln(1/x)} d(\ln(1/\xi)) g(\xi,t) P_{gg}(x/\xi)$$

where a change of integration variable has been made and $t = \ln(Q^2/\Lambda^2)$ introduced (the appearance of the constant Λ makes no difference to the derivative with respect to $\ln Q^2$). Next approximate P_{gg} by its most singular term and substitute the LO expression $\alpha_s(Q^2) = 1/(b_0 t)$ to give

$$t\frac{\partial x g(x,t)}{\partial t} \approx \frac{3}{\pi b_0} \int_0^{\ln(1/x)} d(\ln(1/\xi)) \xi g(y\xi, t).$$

This equation is now differentiated with respect to $\ln(1/x)$. First introduce the two variables

$$u = \ln\left(\frac{t}{t_o}\right), \quad v = \ln\left(\frac{x_0}{x}\right), \tag{9.3}$$

then differentiating with respect to v and using u in place of $\ln t$ gives

$$\frac{\partial^2 G(u,v)}{\partial u \partial v} = \gamma^2 G(u,v) \quad (\gamma^2 = \frac{3}{\pi b_0}), \tag{9.4}$$

where $G(u,v) = xg(x, Q^2)$. (Again the extra constants introduced, t_0 and x_0, make no difference to the derivative.) Equation (9.4) is of the form of a wave equation. To find the solution for the double asymptotic limit, $\ln Q^2 \to \infty$, $\ln(1/x) \to \infty$, it is convenient to make another change of variables

$$\rho = \left(\frac{v}{u}\right)^{1/2}, \quad \sigma = (uv)^{1/2}. \tag{9.5}$$

The limit required corresponds to $\sigma \to \infty$ with ρ fixed (at a value $O(1)$). Changing the variables and keeping only the leading terms for large σ, Eq. (9.4) becomes

$$\frac{\partial^2 G(\sigma, \rho)}{\partial \sigma^2} = 4\gamma^2 G(\sigma, \rho). \tag{9.6}$$

This equation may now be solved immediately

---

[2] The numerical studies of Section 4.3.2 also show that the gluon contribution is by far the most important at low x.

$$G(\sigma, \rho) = A \exp(2\gamma\sigma), \tag{9.7}$$

where A is a constant, or

$$G(u, v) = A \exp(2\gamma\sqrt{uv}), \tag{9.8}$$

or

$$xg(x, Q^2) = A \exp\left\{\sqrt{\frac{12}{\pi b_0} \ln\left(\frac{t}{t_0}\right) \ln\left(\frac{x_0}{x}\right)}\right\} \tag{9.9}$$

which are different forms of the double asymptotic scaling (DAS) limit for the gluon density. This corresponds to a rise of the gluon at small x, which becomes steeper as Q^2 increases. Note that, in addition to the asymptotic dependence on σ, the gluon density, xg, is predicted to be independent of ρ.

Having got an expression for xg at small x, the LO relationship between F_2 and xg may be approximated to give

$$F_2(\rho, \sigma) = \frac{A\bar{e}^2}{9} \frac{\gamma}{\rho} \exp(2\gamma\sigma)[1 + O(1/\sigma)] \tag{9.10}$$

for $\sigma \to \infty$ and ρ fixed (see Problem 1 at the end of the chapter). Thus the steep rise in the gluon distribution at low x translates into a steep rise of F_2 at low x, which also becomes steeper as Q^2 increases (see Problem 2 at the end of the chapter). Ball and Forte found that with appropriate boundary conditions the asymptotic region could be reached for values of σ and ρ well within the range of the HERA experiments. The constant A has to be determined from the data and the choice of x_0 and Q_0^2 will define the range of data points included. Figure 9.3 shows the results of a study by H1 (1995) of the LO DAS prediction, using their 1993 F_2 data[3] with $x_0 = 0.1$, $Q_0^2 = 1\,\text{GeV}^2$ and $\Lambda = 185\,\text{MeV}$. The upper plot of the figure shows $\ln(R'_F F_2)$ as a function of σ, where $R'_F \sim (\rho/\gamma)^4$, resulting in a quantity that should rise linearly with σ with slope 2γ. The lower plot shows $\ln(R_F F_2)$ as a function of ρ, where $R_F = \exp(-2\gamma\sigma)R'_F$. This tests the second prediction of DAS that this quantity should be independent of ρ. It can be seen that 'ρ-scaling' is reasonable for $\rho > 1.2$ or so. Applying this cut to the data and fitting the slope of the first plot gives $2.22 \pm 0.04(\text{stat.}) \pm 0.10(\text{sys.})$, to be compared with 2.4, the LO prediction for 2γ with four flavours.

9.2.2 Singular input distribution

The DGLAP limit with singular input distributions has been explored in detail by Yndurá in and co-workers (Barreiro et al. 1996). F_2 is split into its singlet (F_S) and non-singlet (F_{NS}) components respectively. The Q^2 behaviours of both pieces are calculated at LO and NLO in terms of α_s

---

[3] The data covered the range $4.5 < Q^2 < 1600\,\text{GeV}^2$ and $1.8 \times 10^{-4} < x < 0.13$.
[4] Up to some smaller sub-leading terms that have been ignored here

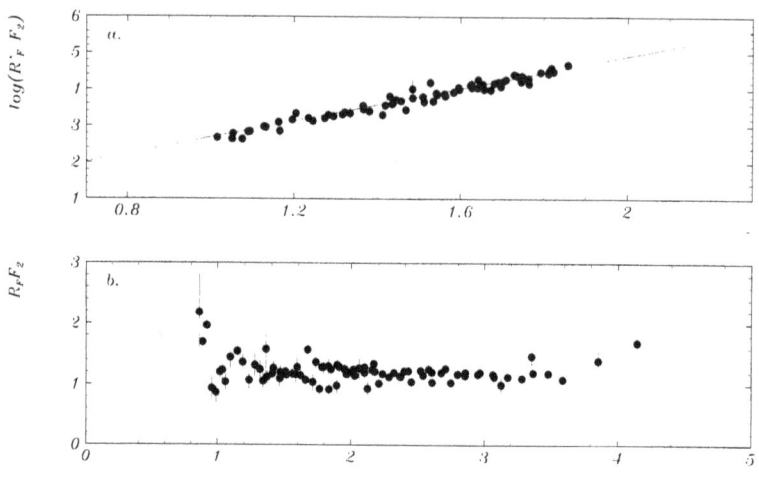

Fig. 9.3 Evidence for double asymptotic scaling, early data from the H1 experiment compared with the LO DAS prediction: (a) $\ln(R'_F F_2)$ as a function of σ; (b) $R_F F_2$ as a function of ρ. R_F, R'_F and the variables σ and ρ are defined in the text. (From H1 1995).

and anomalous dimensions. The x dependence is constructed from powers of x and $(1-x)$ with a small number of parameters. At small x, F_2 is dominated by the singular term in F_S, which at LO takes the form

$$F_S(x, Q^2) \approx B_S[\alpha_s(Q^2)]^{-d_+(1+\lambda_S)} x^{-\lambda_S},$$

where d_+ is the largest eigenvalue of the singlet anomalous dimension matrix and λ_S is a parameter to be determined from the data. Fitting to HERA 1994 data (H1 and ZEUS separately), good descriptions of the data are obtained with vales of λ_S in the range 0.33–0.36. An example of the fit to the ZEUS data is shown in Fig. 9.4.

Before the significance of these results for the structure function at low x can be fully appreciated, it is necessary to review the description of the energy dependence of hadronic total cross-sections given by Regge Theory.

9.3 Regge Theory

Regge theory, which predates QCD, is concerned with the high energy behaviour of hadron–hadron and photon–hadron cross-sections. This is relevant for DIS at low x because W, the $\gamma^* p$ CM energy, is given by $W^2 = Q^2(1/x - 1) \approx Q^2/x$. At large W^2, the structure function F_2 is simply related to the $\gamma^* p$ total cross-section (Eq. (C.16) of Appendix C)

$$\sigma^{tot}_{\gamma^* p}(W^2, Q^2) \equiv \sigma_T + \sigma_L \approx \frac{4\pi^2 \alpha}{Q^2} F_2(x, Q^2). \tag{9.11}$$

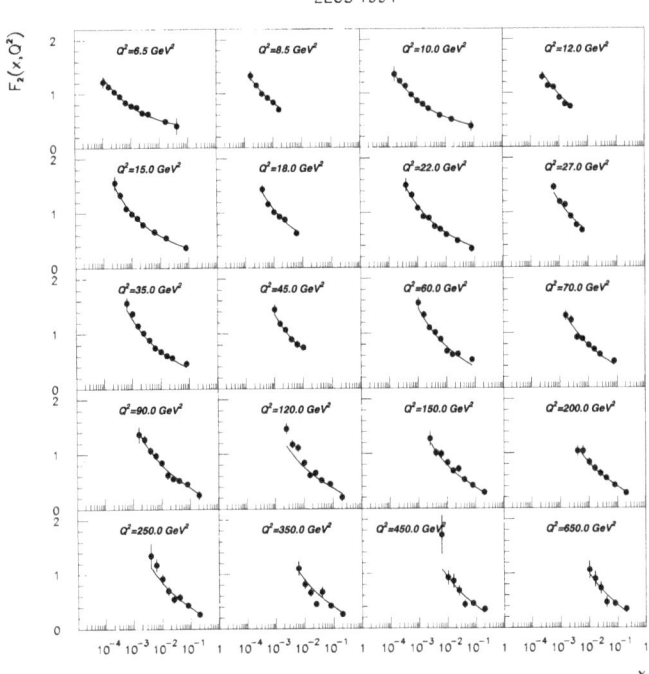

Fig. 9.4 The fit to the ZEUS 1994 F_2 data by Barreiro et al. (1996) using the singular solution to the DGLAP equations at low x.

Regge theory grew out of the study of the analytic properties of the scattering amplitudes for strong interaction processes.[5] It provides a systematic framework for describing the high energy behaviour of hadronic total cross-sections and forward differential cross-sections. Consider a two-body scattering process $a + b \to c + d$ with the usual Mandelstam variables s, t, u ($s = (p_a + p_b)^2$, $t = (p_a - p_c)^2$, $u = (p_a - p_d)^2$) and ignore spin and masses for simplicity. The key physical idea is to relate the high energy behaviour of the s-channel amplitude to the quantum numbers that are exchanged in the t-channel. But why is Regge theory necessary for this? Expand the s-channel amplitude in terms of t-channel partial wave amplitudes $a_l(t)$

$$A_{ab:cd}(s,t) = \sum_{l=o}^{\infty}(2l+1)a_l(t)P_l(\cos\theta_t),$$

where θ_t is the t-channel scattering angle and $\cos\theta_t = 1 + 2s/t$. Suppose now that a single resonance with mass M_J and spin J dominates the t-

---

[5] A comprehensive modern account is given in the book by Donnachie et al. (2002a) and a brief introduction in Forshaw & Ross (1997).

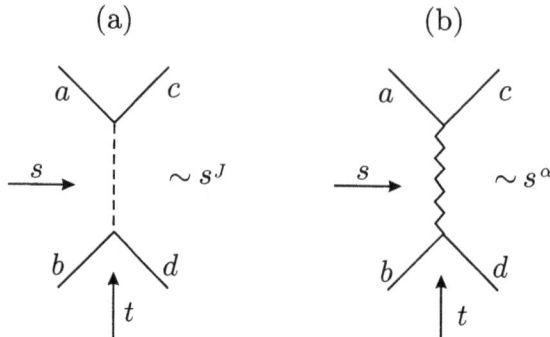

Fig. 9.5 Contributions in the t-channel for $ab \to cd$: (a) a resonance with spin J; (b) a Reggeon R with trajectory α.

channel (Fig. 9.5(a)), then at high energy the t-channel amplitude has the form

$$A_{ab:cd}(s,t) \sim \frac{C}{t - M_j^2}\left(\frac{2s}{t}\right)^J$$

where C is a constant and $P_l(z) \sim z^l$ for z large has been used. There is evidence to support this idea for states with low values of J and small masses that are then 'near' the s-channel physical region. For example, single pion exchange dominates $ep \to e\pi^+ n$ at low energy and small $|t|$, providing a way of measuring the EM form factor of the pion. However, this fails in general for large J as the amplitude grows too large at large s and a single t-channel exchange is unlikely to be dominant. Regge theory comes to the rescue by exploiting the properties of analytic functions. First, the t-channel partial wave expansion is written as a contour integral

$$A_{ab:cd}(s,t) = \frac{1}{2i}\oint_{C_1} dl(2l+1)\frac{a(l,t)}{\sin \pi l}P(l, \cos\theta_t),$$

where the contour C_1 surrounds the positive real axis as shown in Fig. 9.6. The integrand has poles at the positive integers from the $\sin \pi l$ and thus reproduces the original sum. The contour C_1 is deformed to C_2, along the vertical at $\mathrm{Re}\, l = -1/2$. Suppose the partial wave amplitude has a pole

$$a(l,t) = \frac{\tilde{\beta}(t)}{l - \alpha(t)}$$

then the contribution of that pole to A is picked up by the closed contour C_3 to give

$$A(s,t) = \frac{(2\alpha(t) + 1)\pi\tilde{\beta}(t)}{\sin \pi\alpha(t)}P(\alpha(t), \cos\theta_t).$$

In the high energy limit the contribution of the contour C_2 is small and may be neglected. Thus the complete t-channel partial wave sum has been

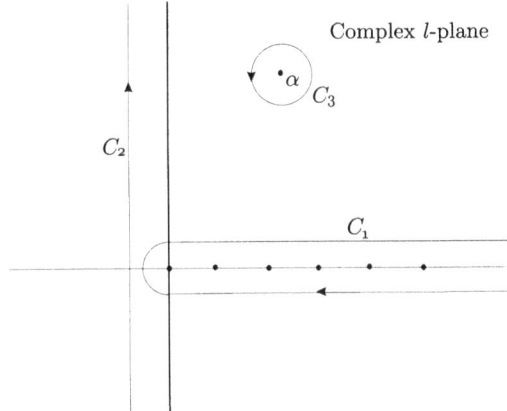

Fig. 9.6 The complex l-plane showing the contours for manipulating the partial wave sum.

replaced by that of a single pole — the Regge pole — or possibly a finite sum of such terms. Using again the large $|z|$ limit for $P(\alpha, z)$, one has the expression for the contribution of a single Regge pole (for $s \gg |t|$)

$$A_{ab:cd}(s,t) \sim \beta(t) \left(\frac{s}{s_0}\right)^{\alpha(t)} \qquad (9.12)$$

where s_0 is a scale factor, usually taken to be $1\,\text{GeV}^2$ (and often omitted). The function $\beta(t)$ is related to $\tilde{\beta}$ but has absorbed various other factors, some t dependent. The function $\alpha(t)$ is known as the *trajectory* of the Regge pole and the one with the largest real part will control the high energy behaviour of the amplitude.

What has been gained? For t-channel ($a\bar{c} \to \bar{b}d$) scattering $\alpha(t)$ passes through integer spin values J at physical states with mass m_J such that $\alpha(m_J^2) = J$ giving rise to resonance contributions (for technical reasons J actually jumps by two units). However, when the trajectory is continued to the s-channel with $t < 0$, $\alpha(t)$ is usually less than 1, thus no longer causing a problem at high energy. In a sense the Regge pole is equivalent to exchanging a 'tower' of angular momentum states and if the trajectory can be determined the energy dependence of the amplitude is given.. For a simple Regge pole, R, the residue function factorizes $\beta(t) = \gamma_{acR}(t)\gamma_{bdR}(t)$ (see Fig. 9.5(b)) but the forms of these functions are not specified by the theory. A Regge pole has definite quantum numbers, such as isospin, charge conjugation and 'natural or unnatural parity'. Natural parity means that the parity of the states on the trajectory satisfy $P = (-1)^J$, unnatural parity corresponds to $P = -(-1)^J$. The contribution of a single Regge pole to the differential cross-section takes the form

240 *DIS at low x*

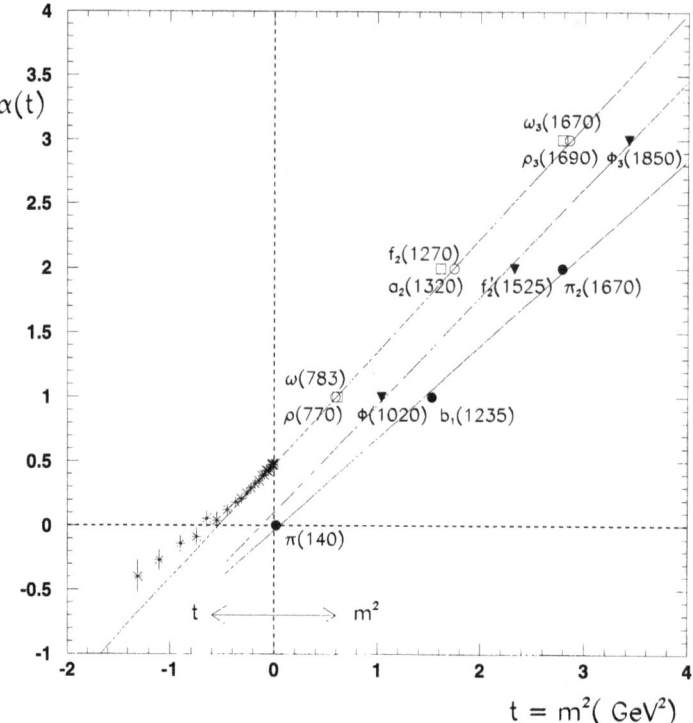

Fig. 9.7 Chew–Frautschi plot of spin versus mass-squared for low-lying meson states and associated Regge trajectories. (From Abramowicz and Caldwell 1999).

$$\frac{d\sigma}{dt} \sim f(t) s^{2\alpha(t)-2},$$

and if there is more than one pole there will be interference effects. Of more interest here is the contribution of a Regge pole to the total cross-section for $ab \to all$. Using the optical theorem this is related to the contribution of the pole to the forward elastic amplitude for $ab \to ab$, giving

$$\sigma_{ab}^{tot} \approx \text{const.} \times s^{\alpha(0)-1},$$

with an obvious generalisation if more than one pole contributes.

Figure 9.7 shows some of the low-lying meson states plotted on a Chew–Frautschi diagram (spin versus mass-squared) with associated Regge trajectories, which are approximately linear ($\alpha(t) = \alpha(0) + \alpha' t$). The trajectories are built from valence $q\bar{q}$ states and follow the pattern of their quantum numbers. The $\rho - a_2$ trajectory ($I = 1$, C odd, natural parity) and $\omega - f_2$ ($I = 0$, C odd, natural parity) are essentially degenerate with $\alpha(0) = 0.55$ and $\alpha' = 0.86 \, \text{GeV}^{-2}$. The $\phi - f_2'$ ($I = 0$, C odd, natural parity) is from states predominantly $s\bar{s}$ and has a lower intercept. The $\pi - b_1$ ($I = 1$, C even, unnatural parity) has a still lower intercept. Assuming that the linear form of the trajectory may be extrapolated to negative t values (at least for small $|t|$) one can predict the energy dependence in the s-channel. All these trajectories correspond to the exchange of quantum numbers in the t-channel and give contributions to total cross-sections decreasing as $s^{-0.45}$ or faster. This is in line with general results proved by Pomeranchuk and Okun. However total cross-sections are mostly constant or slowly increasing with increasing energy. Foldy and Peierls showed that such behaviour must be associated with the exchange of states with vacuum quantum numbers. A new Regge trajectory, with $I = 0$, C even and natural parity is required. It is called the Pomeron (after Pomeranchuk) and must have an intercept $\alpha_P(0) \approx 1$, its slope (from diffractive scattering which also involves vacuum exchanges) is around $0.25\,\text{GeV}^{-2}$. To meet the requirement of another high energy theorem of Pomeranchuk, namely that $\sigma_{ab}^{\text{tot}} \to \sigma_{a\bar{b}}^{\text{tot}}$ as $s \to \infty$, the Pomeron must couple equally to ab and $a\bar{b}$. To date it has not been possible to relate the Pomeron trajectory unambiguously to meson states, but it may be related to a glueball candidate. In QCD, the Pomeron would be a colourless object with vacuum quantum numbers composed of at least two gluons.

9.3.1 Hard and soft Pomerons

Regge theory provides a compact framework for describing the energy dependence of hadronic total cross-sections above the resonance region (Donnachie and Landshoff 1992, Cuddell *et al.* 1997). In particular, pp and p$\bar{\text{p}}$ total cross-section data may be fit by the form

$$\sigma_i^{\text{tot}}(s) = A_i^P s^{\alpha_P(0)-1} + A_i^R s^{\alpha_R(0)-1} \qquad (9.13)$$

where P, R refer to the Pomeron and other Regge exchanges respectively and the A_i are process dependent constants. The intercepts of the Pomeron and Regge trajectories are universal. By fitting to cross-section data, it is found that $\alpha_R(0) \approx 0.5$ in line with the $\rho - a_2$ and $\omega - f_2$ trajectory intercepts. For the Pomeron intercept, $\alpha_P(0)$, values in the range 1.081–1.094 are found and for reasons which will become apparent below this mild high energy behaviour is now qualified as that of the 'soft' Pomeron.

It has also been shown that Eq. (9.13) (with the intercepts fixed) describes other hadronic total cross-section data including that of real photoproduction on protons. A number of authors have extended the Regge approach in a variety of ways to accommodate virtual photon–proton scattering, for both the region at small Q^2 and the deep inelastic region at small x. To do both requires features not present in the Regge framework as applied to hadronic cross-sections. There are many models, some combine features of Regge theory and DGLAP evolution, others allow the Regge trajectory and intercept to have Q^2 dependence. Here the approach of Donnachie and Landshoff is followed, both for its simplicity and because it is closest to Regge paradigm of describing the high energy behaviour of

F_2 in terms of a limited number of Regge exchanges with Q^2 independent intercepts. The early work of Donnachie and Landshoff was also important in that it provided a model independent criterion for judging whether the low-x behaviour of F_2 was exceptional. In Donnachie and Landshoff (1993), they showed that the Pomeron plus Regge exchange model could describe successfully the then existing data on $\sigma_{\gamma^* p}^{tot}$ (or F_2) up to Q^2 values of about $10\,\text{GeV}^2$. They used a simple parameterization in which the intercepts controlling the energy dependence remained universal and all Q^2 dependence was contained in the Regge pole residue functions. Their prediction for F_2 (which relates to $\sigma_{\gamma^* p}^{tot}$ as given in Eq. (9.11)) is

$$F_2(x, Q^2) = A x^{1-\alpha_P(0)} \left(\frac{Q^2}{Q^2+a}\right)^{\alpha_P(0)} + B x^{1-\alpha_R(0)} \left(\frac{Q^2}{Q^2+b}\right)^{\alpha_R(0)}, \tag{9.14}$$

with

$$4\pi^2 \alpha A a^{-\alpha_P(0)} = A_{\gamma p}^P \quad \text{and} \quad 4\pi^2 \alpha B b^{-\alpha_R(0)} = A_{\gamma p}^R.$$

The latter constraints ensure that the fit to real photoproduction data (given by Eq. (9.13)) is recovered as $Q^2 \to 0$. The important result from this approach is that the structure function F_2 is expected to rise only very slowly as $x \to 0$, roughly as $F_2 \sim x^{1-\alpha_P(0)}$ or $\sim x^{-0.08}$.

Given the success of Regge theory in correlating the high energy behaviour of hadronic cross-sections, it might seem safe to extend it to the small x behaviour of structure functions. However, as Donnachie and Landshoff remark at the end of their 1993 paper: *'If the HERA experiments find results for F_2 significantly larger at small x than our extrapolation, we claim that this will be a clear signal that they have discovered new physics.'* What they had in mind, if this happened, was possible evidence for the 'Lipatov' (BFKL) or 'hard' Pomeron.

The striking result from HERA is that F_2 does rise very much more steeply than the 'soft' Pomeron of hadronic physics would predict. A nice way of summarizing the behaviour of the data at low x is to plot the effective power of F_2 as function of Q^2, $F_2 \sim x^{-\lambda_{\text{eff}}}$ or $\lambda_{\text{eff}} = \partial \ln F_2 / \partial \ln(1/x)$. Figure 9.8 shows a compilation of λ_{eff} values versus Q^2 from H1 and ZEUS. In Regge language $\lambda_{\text{eff}} = \alpha_P(0) - 1$. The plot shows that below $Q^2 \sim 1\,\text{GeV}^2$ the data are compatible with a conventional hadronic Pomeron exponent but that for large Q^2 values λ_{eff} increases steadily to a value in the range 0.3–0.4 for $Q^2 \sim 100\,\text{GeV}^2$ and above. Interestingly, this larger value is close to the exponent λ_S found in the analysis by Ynduráin of the 'singular' DGLAP limit discussed in Section 9.2.2 above. This raises the possibility that the energy dependence of $\gamma^* p$ scattering (and γp scattering) is not given by a single Regge pole, but rather a sum of terms of the form $a(Q^2)s^0 + b(Q^2)s^\varepsilon$, where $\varepsilon \sim 0.35$. For $Q^2 \approx 0$, $a \gg b$ whereas for large Q^2, $a \ll b$. Only in DIS is the true nature of the high energy behaviour revealed.

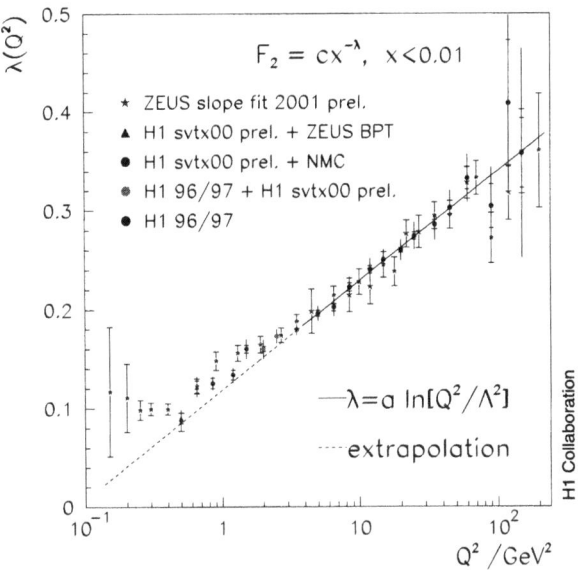

Fig. 9.8 The effective slope of F_2 ($\lambda_{\text{eff}} = \lambda(Q^2) = \partial \ln F_2 / \partial \ln(1/x)$) at low x from H1 and ZEUS as a function of Q^2. (From Gayler 2002).

Following these ideas, Donnachie and Landshoff have added a 'hard' Pomeron to their Regge ansatz for F_2. In a series of papers[6] the three component model is refined and gives a good description of F_2 proton data with $x < 0.07$, the charm structure function F_2^c and $\sigma(\gamma p \to J/\psi p)$. In more detail, the structure function is written as

$$F_2(x, Q^2) = \sum_i f_i(Q^2) x^{-\epsilon_i},$$

where $\epsilon_0 + 1$ is the intercept of the hard Pomeron to be determined from the data. The parameters $\epsilon_1 = \alpha_P(0) - 1 = 0.0808$ and $\epsilon_2 = \alpha_R(0) - 1 = -0.4525$ are fixed by the intercepts of the conventional 'soft' Pomeron and $f - a$ Regge poles as determined from hadron–hadron data. The functions $f_i(Q^2)$ take the forms

$$f_0(Q^2) = A_0 \left(Q^2\right)^{1+\epsilon_0} \left(1 + Q^2/Q_0^2\right)^{-1-\epsilon_0/2}$$

$$f_i(Q^2) = A_i \left(Q^2\right)^{1+\epsilon_i} \left(1 + Q^2/Q_i^2\right)^{-1-\epsilon_i}, \quad i = 1, 2$$

The model contains a total of seven parameters, two determined from the energy dependence of the photoproduction total cross-section and the remaining five from structure function data with $x < 0.001$. The value of ϵ_0 is

---

[6] Donnachie and Landshoff (2001, 2002b) and references therein.

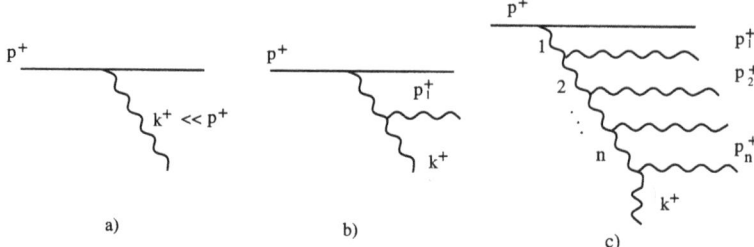

Fig. 9.9 Development of a gluon 'ladder' from a fast moving parton with longitudinal momentum $p^+ = x_0 P^+$, where P^+ is the momentum of the parent hadron: (a) single gluon emitted with momentum $k^+ = xP^+$; (b) radiative correction; (c) multiple emissions. (From Iancu 2002).

found to be 0.4372. A good description of F_2 is obtained and the behaviour of F_2^c can be described with the hard Pomeron contribution alone. The value of ϵ_0, if translated to the behaviour of the gluon at small x, gives a steeper rise than is found in the recent global fits using the HERA-I data (MRST2001 and CTEQ6M for example).

9.4 The BFKL equation

Although Regge theory provides a successful framework for parameterizing the high energy behaviour of total cross-sections, it does not provide predictions for key ingredients such as the Pomeron trajectory or intercept. Does the greater understanding of the strong interaction given by QCD offer any insights here? Large parts of a total cross-section are build up by soft processes which may not be susceptible to perturbative techniques, but the hard Pomeron which may describe F_2 at low x might be. Building on earlier work using model field theories Balitsky, Fadin, Kuraev and Lipatov (BFKL) tried to do just this. This section will concentrate on the LO BFKL equation with a brief mention of the complications of allowing α_s to run and the still unresolved issue of meaning of the NLO BFKL results. Whether the BFKL equation can be used to describe data directly is a hotly debated topic of active research. Whatever the outcome, the BFKL calculation has been seminal in leading to new thinking about the gluon dominance of physics underlying the behaviour of scattering amplitudes at low x. The approach to the BFKL equation outlined here follows closely that of the book by Forshaw and Ross (1997), which should be consulted for more details and full references.

9.4.1 Multiple gluon emission at small x

The BFKL equation provides a mechanism for summing the contributions of multiple gluon emission at small x. Before starting on the discussion consider the following simple argument (due to Iancu (2002)) as to why this may lead to a power law growth of a cross-section with energy. Figure 9.9 shows the development of a gluon 'ladder' by successive radiation from a gluon emitted by a fast moving parton with longitudinal momentum $p^+ = x_0 P^+$, where P^+ is the momentum of the parent hadron.[7] The first emission, shown in (a) gives a gluon with $k^+ = xP^+$, where $x \ll x_0$, this gluon then radiates a second gluon with momentum p_1^+ (shown in (b)). Since gluons have a bremsstrahlung-like spectrum, the probability for this is of order

$$\frac{3\alpha_s}{\pi} \int_{k^+}^{p^+} \frac{dp_1^+}{p_1^+} = \bar{\alpha}_s \ln \frac{p^+}{k^+} = \bar{\alpha}_s \ln \frac{x_0}{x},$$

with the probability enhanced by the large logarithm, $\ln(x_0/x)$. For the cascade of gluons shown in (c), with strongly ordered longitudinal momenta $p^+ \gg p_1^+ \gg p_2^+ \gg \cdots \gg p_n^+ \gg k^+$, each emission will have a similarly enhanced probability, giving a contribution $(\bar{\alpha}_s \ln(x_0/x))^n/n!$. For small enough x, $\ln(x_0/x) \sim 1/\bar{\alpha}_s$ and the multiple gluon corrections must be summed. The sum gives an overall probability depending exponentially on $\ln(x_0/x)$, $\sim \exp(\omega \bar{\alpha}_s \ln(x_0/x))$ or $\sim (x/x_0)^{-\omega \bar{\alpha}_s}$, where ω is a numerical factor. One is picking out and summing the leading logs in $1/x$, the $LL(1/x)$ series of Fig. 9.2.

9.4.2 The reggeized gluon and the LO BFKL equation

In the 1960s, it was established that sums of ladder diagrams in scalar field theories could give Regge behaviour (i.e. the scattering amplitude behaving like $s^{\alpha(t)}$ for $s \gg |t|$). If the momentum of the ith vertical leg in the ladder is k_i, then the Regge limit is given by strong ordering of the longitudinal components (with respect to the external momenta p_1, p_2) of the k_i.

BFKL applied these ideas to gluon ladders in QCD. The strong ordering of the longitudinal components of the ladder momenta is a strong ordering in the x_i of the internal momenta in DGLAP language. In the DGLAP limit, the internal loop transverse momenta are strongly ordered as well. Together this means that only simple gluon ladder diagrams have to be considered in the DGLAP case. For the BFKL Pomeron, the internal transverse momenta are not ordered, which means that many more diagrams have to be considered. In addition to the 'straight' rungs of the DGLAP ladders, rungs may cross over or the gluon rung may split — ladders are built using the three basic types of half section shown from the left in Fig. 9.10. In the Regge limit, the relevant diagrams have a ladder like structure in which the vertical links are themselves constructed from ladders (the 'reggeized' gluons) coupled to simple gluon 'rungs' using the effective vertex of Fig. 9.10. Other diagrams involving more complicated crossed rungs or quark line insertions give sub-leading behaviour and are

---

[7] Formally, p^+ denotes the light cone coordinate (see Section 3.7.1), but the crucial physics is that the transverse momentum is assumed to be approximately the same for all gluons.

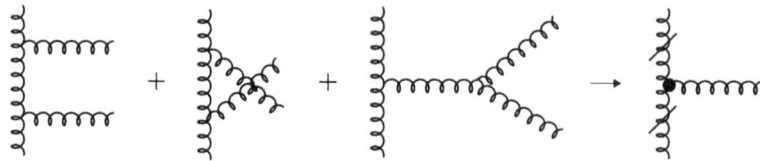

Fig. 9.10 The basic idea of the 'reggeized' gluon and effective vertex.

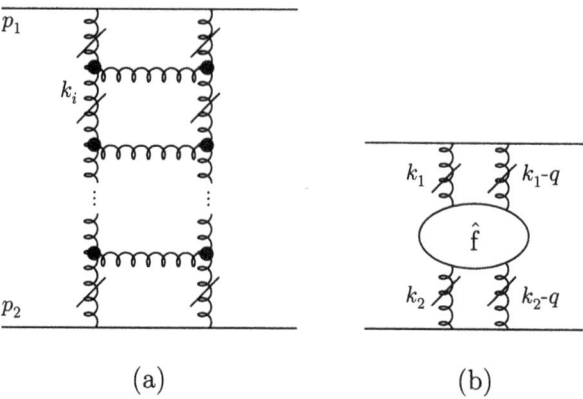

Fig. 9.11 Diagram showing reggeized two-gluon exchange between two quarks: (a) the lowest order contribution to the singlet state, (b) the general term.

ignored. To match the vacuum quantum numbers of the Pomeron, one must have colour singlet exchange and this implies a minimum of two gluon exchange in the amplitude. Here the gluons get 'reggeized' to give diagrams of the form shown in Fig. 9.11(a), with (b) representing the summed contribution of all such diagrams. BFKL found and solved an integral equation for this reggeized gluon four-point function, essentially the diagram (b) above with the external quark lines removed.

The BFKL amplitude $\hat{f}(\omega, \mathbf{k}_1, \mathbf{k}_2, \mathbf{q})$ is related to the colour singlet quark scattering amplitude, $A^S(s,t)$, by a Mellin transform[8]

$$\tilde{A}^S(\omega, t) = \int_1^\infty dz z^{-\omega-1} \frac{A^S(s,t)}{s}$$
$$= 4i\alpha_s^2 G^S \int \frac{d^2\mathbf{k}_1 d^2\mathbf{k}_2}{\mathbf{k}_2^2 (\mathbf{k}_1 - \mathbf{q})^2} \hat{f}(\omega, \mathbf{k}_1, \mathbf{k}_2, \mathbf{q}), \quad (9.15)$$

where $z = s/\mathbf{k}^2$ ($\mathbf{k}^2$ is a scale factor typical of the external transverse momenta) and G^S projects out the colour singlet term. Referring to Fig. 9.11(b),

---
[8]Note that the Mellin transform variable ω here corresponds to $n-1$ used earlier in the chapter — low x corresponds to large s.

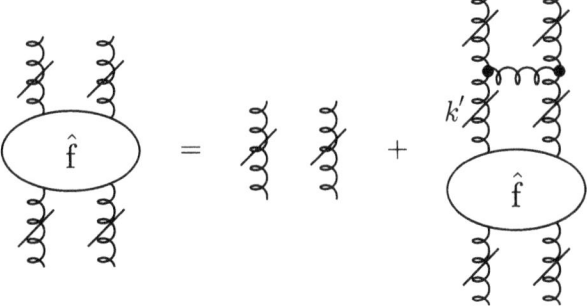

Fig. 9.12 Diagrammatic form of the BFKL equation.

k_1, k_2 are the transverse momenta of the gluons at the top and bottom of the BFKL amplitude. It will also be convenient to define $F(s, \mathbf{k}_1, \mathbf{k}_2, \mathbf{q})$, the Mellin inverse of $\hat{f}(\omega, \mathbf{k}_1, \mathbf{k}_2, \mathbf{q})$. The simplest term in $A^S(s, t)$ is given by the exchange of two reggeized gluons and further terms have the form of ladder diagrams as in Fig. 9.11(a). The BFKL amplitude satisfies a linear integral equation of the form

$$\omega \hat{f}(\omega, \mathbf{k}_1, \mathbf{k}_2.\mathbf{q}) = \delta^2(\mathbf{k}_1 - \mathbf{k}_2) + K_0 \otimes \hat{f}(\omega, \mathbf{k}_1, \mathbf{k}_2, \mathbf{q}) \tag{9.16}$$

which is shown diagrammatically in Fig. 9.12. The BFKL equation provides a way of summing an infinite number of ladder diagrams, a term with n rungs behaving like $(\alpha_s \ln s)^n$.

Since the forward amplitude $A^S(s, 0)$ is required for the total cross-section, only the forward BFKL amplitude with $\mathbf{q} = 0$ will be considered here. The convolution on the RHS with the LO BFKL kernel K_0 is

$$K_0 \otimes \hat{f}(\omega, \mathbf{k}_1, \mathbf{k}_2, 0) = \frac{\bar{\alpha}_s}{\pi} \int \frac{d^2 \mathbf{k}'}{(\mathbf{k}_1 - \mathbf{k}')^2}$$

$$[\hat{f}(\omega, \mathbf{k}', \mathbf{k}_2, 0) - \frac{\mathbf{k}_1^2}{\mathbf{k}'^2 + (\mathbf{k}_1 - \mathbf{k}')^2} \hat{f}(\omega, \mathbf{k}_1, \mathbf{k}_2, 0)], \tag{9.17}$$

where $\bar{\alpha}_s = 3\alpha_s/\pi$. Note that in the LO BFKL equation α_s does not run but is fixed at a scale typical of the external transverse momentum. The BFKL amplitude is infrared finite as the terms in the [] brackets in the above expression vanish as $\mathbf{k}' \to \mathbf{k}_1$. The solution to the BFKL equation may be written schematically in terms of the eigenfunctions $\phi_i(\mathbf{k})$ of the kernel

$$\hat{f}(\omega, \mathbf{k}_1, \mathbf{k}_2, 0) = \sum_i \frac{\phi_i(\mathbf{k}_1) \phi_i^*(\mathbf{k}_2)}{\omega - \lambda_i} \tag{9.18}$$

where

$$K_0 \otimes \phi_i(\mathbf{k}) = \lambda_i \phi_i(\mathbf{k}).$$

In 2-dimensional $\mathbf{k}'$ space, $\mathbf{k}' = (k', \theta')$, so that the eigenvalue, $\lambda_i \equiv \omega_n(\nu) = \bar{\alpha}_s \chi_n(\nu)$, depends on two variables, ν and n, corresponding to

k' and θ', respectively. The function $\chi_n(\nu)$ is the characteristic function of the BFKL kernel. The eigenfunctions have the form

$$\phi_\nu^n(\mathbf{k}) = \frac{1}{\pi\sqrt{2}} (k^2)^{-1/2+i\nu} e^{in\theta} \qquad (9.19)$$

(so in Eq. (9.18) $\sum_i$ is shorthand for $\sum_{n=0}^{\infty} \int_{-\infty}^{\infty} d\nu$). $A^S(s,0)$ is given by the Mellin inversion of Eq. (9.18) and thus the leading behaviour in s is given by rightmost singularity in ω which means the eigenvalue with the largest real part. The function $\chi_n(\nu)$ decreases as n increases from 0 and decreases as $|\nu|$ increases. Keeping only the leading $n = 0$ term

$$\hat{f}(\omega, \mathbf{k}_1, \mathbf{k}_2, 0) = \frac{1}{\pi k_1 k_2} \int_{-\infty}^{\infty} \frac{d\nu}{2\pi} \left(\frac{k_1^2}{k_2^2}\right)^{i\nu} \frac{1}{\omega - \bar{\alpha}_s \chi_0(\nu)}. \qquad (9.20)$$

It is convenient to rewrite this result in terms of the variable $\gamma = \frac{1}{2} + i\nu$, since it takes the form of a Mellin integral in γ-space (conjugate to transverse momentum)

$$\hat{f}(\omega, \mathbf{k}_1, \mathbf{k}_2, 0) = \frac{1}{\pi k_1^2} \int_{\frac{1}{2}-\infty}^{\frac{1}{2}+\infty} \frac{d\gamma}{2\pi i} \left(\frac{k_1^2}{k_2^2}\right)^{\gamma} \frac{1}{\omega - \bar{\alpha}_s \chi(\gamma)}, \qquad (9.21)$$

where the subscript 0 has been dropped from χ and

$$\chi(\gamma) = 2\psi(1) - \psi(\gamma) - \psi(1-\gamma),$$

where $\psi(x)$ is the digamma function $\frac{d}{dx} \ln \Gamma(x)$. Using Eq. (9.21), the amplitude $F(s, \mathbf{k}_1, \mathbf{k}_2)$ may be expressed as a double Mellin integral

$$F(s, \mathbf{k}_1, \mathbf{k}_2) = \int \frac{d\omega}{2\pi i} \left(\frac{s}{k^2}\right)^\omega \frac{1}{\pi k_1^2} \int \frac{d\gamma}{2\pi i} \left(\frac{k_1^2}{k_2^2}\right)^\gamma \frac{1}{\omega - \bar{\alpha}_s \chi(\gamma)}, \qquad (9.22)$$

where the contour in ω runs parallel to the imaginary axis and to the right of all singularities in the ω-plane. The integrand has a simple pole at $\omega = \bar{\alpha}_s \chi(\gamma)$ and thus the ω integration gives

$$F(s, \mathbf{k}_1, \mathbf{k}_2) = \int \frac{d\gamma}{2\pi i} \left(\frac{s}{k^2}\right)^{\bar{\alpha}_s \chi} \frac{1}{\pi k_1^2} \left(\frac{k_1^2}{k_2^2}\right)^\gamma, \qquad (9.23)$$

where the contour runs parallel to the imaginary γ axis. For large s/k^2, the integral is controlled by χ, Fig. 9.13 shows Re(χ) in the complex γ plane and the integration contour for the final Mellin inversion. Concentrating on the leading behaviour in s, the integral may be evaluated by a saddle point approximation[9] about Re(γ) = 1/2 giving,

---

[9]The details can be followed in Problem 3 at the end of the chapter.

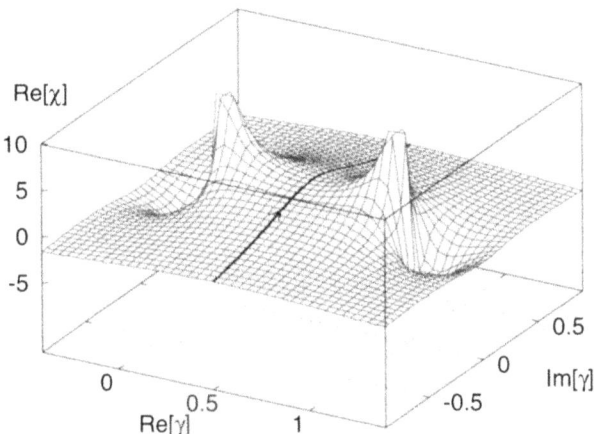

Fig. 9.13 Real part of the LO BFKL characteristic function χ in the complex γ plane and the contour for the Mellin inversion using the saddle point method at $\text{Re}(\gamma) = 1/2$. (From Salam 1999.)

$$F(s, \mathbf{k}_1, \mathbf{k}_2) = \frac{1}{\pi k_1 k_2} \frac{(s/\mathbf{k}^2)^{\bar{\alpha}_s \chi(1/2)}}{\sqrt{2\pi \bar{\alpha}_s |\chi''(1/2)| \ln(s/\mathbf{k}^2)}}. \quad (9.24)$$

Finally $A(s, 0)$ is obtained by integrating over $\mathbf{k}_1^2$ and $\mathbf{k}_2^2$

$$\frac{A(s, 0)}{s} = 4i\alpha_s^2 G^S \int \frac{d^1\mathbf{k}_1}{\mathbf{k}_1^2} \frac{d^2\mathbf{k}_2}{\mathbf{k}_2^2} F(s, \mathbf{k}_1, \mathbf{k}_2),$$

giving a high energy behaviour

$$A(s, 0) \sim \frac{(s/\mathbf{k}^2)^{1+\omega_0}}{\sqrt{\ln(s/\mathbf{k}^2)}},$$

where $\omega_0 = \bar{\alpha}_s \chi(1/2) = 4\bar{\alpha}_s \ln 2$. If α_s is taken to be 0.2, $\omega_0 = 0.53$ giving a very strongly rising cross-section for quark–quark scattering, $\sigma_{qq} \sim s^{0.53}$.

9.4.3 BFKL for hadronic processes

To connect the BFKL result to hadron–hadron scattering or DIS requires replacing the quark lines by *impact factors* which are, in general, non-perturbative objects relating the hadrons to their quark content. For the case of DIS, or more precisely, the $\gamma^* p$ total cross-section this is shown in Fig. 9.14 and is

$$\sigma_\lambda(W^2, Q^2) = \frac{1}{(2\pi)^4} \int \frac{d^2\mathbf{k}_1}{\mathbf{k}_1^2} \frac{d^2\mathbf{k}_2}{\mathbf{k}_2^2} \Phi_{\gamma,\lambda}(\mathbf{k}_1) \Phi_p(\mathbf{k}_2) F(W^2, \mathbf{k}_1, \mathbf{k}_2), \quad (9.25)$$

where $\lambda = T, L$ designates the polarization state of the γ^* and $\Phi_{\gamma,\lambda}$, Φ_p are the virtual photon and proton impact factors, respectively. The colour

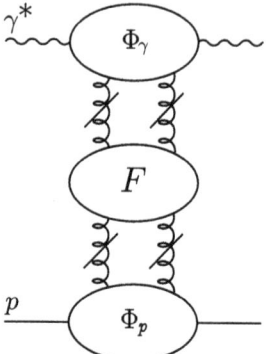

Fig. 9.14 The BFKL contribution to the γ^*p total cross-section.

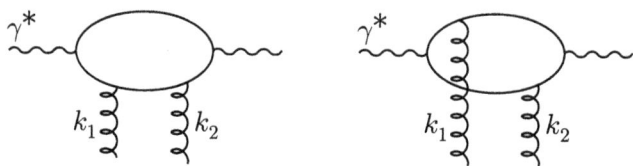

Fig. 9.15 The BFKL impact factor Φ_λ for the virtual photon.

factor G^S has been absorbed into Φ_p. Φ_γ can be calculated in terms of $\gamma^* \to q\bar{q}$ as shown in Fig. 9.15.[10] Equation (9.25) is an example of k_T factorization, which is analogous to the collinear factorization of the DGLAP approach.

Depending on the process under consideration Φ_p may be calculable perturbatively, as for example in the case of forward jets or forward particle production in DIS (both with transverse energy above some suitable minimum). The BFKL amplitude may also be used to calculate the total cross-section for $\gamma^* + \gamma^* \to X$ at high energies by attaching Φ_γ at both ends. In all these cases one has hard scales in both impact factors, and these scales may be chosen to be comparable giving the so-called 'single scale' process. On the other hand, for inclusive DIS Φ_p is non-perturbative and the scales in the two impact factors differ — giving the so-called 'two scale' process.

For inclusive DIS, Φ_p when convoluted with $F(x, \mathbf{k}_1, \mathbf{k}_2)$ gives the unintegrated gluon density function $F_g(x, \mathbf{k})$, where the relation $x \approx Q^2/W^2$ has been used to replace the energy variable W^2 by x. More precisely

$$F_g(x, \mathbf{k}) \equiv \frac{1}{(2\pi)^3} \int \frac{d^2\mathbf{k}'}{\mathbf{k}'^2} \Phi_p(\mathbf{k}') \mathbf{k}^2 F(x, \mathbf{k}, \mathbf{k}'), \quad (9.26)$$

[10] Details are given in Forshaw and Ross (1997), pp169–173.

and $F_g(x, \mathbf{k})$ is related to the gluon density that appears in the DGLAP equations by

$$xg(x, Q^2) = \frac{1}{\pi} \int_0^{Q^2} \frac{d^2\mathbf{k}}{\mathbf{k}^2} \Theta(Q^2 - \mathbf{k}^2) F_g(x, \mathbf{k}). \tag{9.27}$$

Although a complete calculation of xg requires a model for Φ_p, it can be seen that the leading behaviour at low x is given by $xg(x, Q^2) \sim x^{-\omega_0}/\sqrt{\ln(1/x)}$. Similarly $\sigma_\lambda(W^2, Q^2) \sim (W^2)^{\omega_0}/\sqrt{\ln(W^2)}$. If the BFKL exponent is re-expressed as the Pomeron intercept, then $\alpha_P^{\text{BFKL}}(0) = 1 + \omega_0 \approx 1.53$ which is much larger than the values of about 1.08–1.09 that are found from hadron–hadron data. However it is much closer to the hard Pomeron intercept in the Donnachie–Landshoff fit or the x-exponent λ_S of the DGLAP singular input found by Ynduráin, and this is one of the reasons why the result has generated so much interest.

9.4.4 Beyond the LO BFKL with fixed α_s

Although the fixed α_s LO BFKL result for F_2 is suggestive, the running of α_s and the size of NLO terms need to be addressed before one can use the calculations with confidence to describe data. There has been much discussion since the publication of the NLO calculation in the fixed-α_s framework by Fadin and Lipatov, and Camici and Ciafaloni in 1998. Only a brief summary will be attempted here, following in part the Cracow lectures of Salam (1999) where more details and references may be found. First a brief comment on diffusion in transverse momentum.

Diffusion

The calculation above concentrated on establishing the leading power behaviour in s of the BFKL amplitude. Returning to Eq. (9.20) and expanding $\chi(\nu)$ in powers of ν, $\bar{\alpha}_s \chi(\nu) = \omega_0 - a^2 \nu^2 + \cdots$ where $a^2 = 16.83 \bar{\alpha}_s$, the Mellin inversion gives an additional factor multiplying Eq. (9.24)

$$\exp\left(-\frac{\ln^2(\mathbf{k}_1^2/\mathbf{k}_2^2)}{4a^2 \ln(s/\mathbf{k}^2)}\right).$$

This term controls the diffusion or spread in transverse momentum along the BFKL ladder, the amount depends on both the 'mismatch' in transverse momenta at the ends $(\mathbf{k}_1, \mathbf{k}_2)$ and the energy, since the width of the gaussian increases with energy. The physical basis of this spread is the lack of ordering in k_T which allows the k_T down the BFKL ladder to follow a random walk, rather than the strict k_T ordering of the DGLAP limit. Although the fixed α_s BFKL equation giving rise to this result is infrared safe, one now has a potential inconsistency. Unless the diffusion is limited it is not clear that the assumption of a fixed value of $\alpha_s(\langle \mathbf{k}^2 \rangle)$ at some appropriate average value of $\mathbf{k}^2$ will be adequate. Early attempts to produce a realistic phenomenology of the BFKL equation have involved the

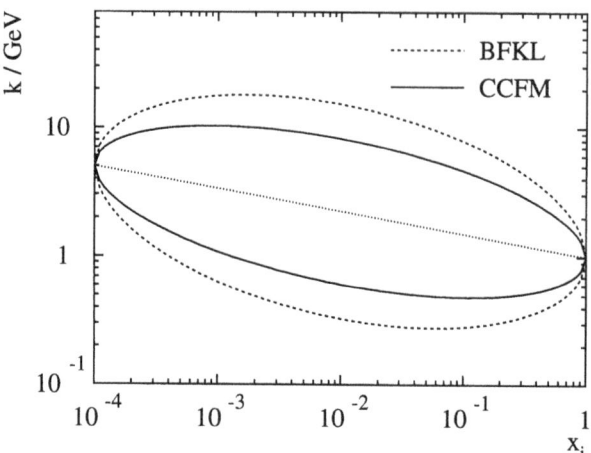

Fig. 9.16 The range over which the transverse momentum in a gluon ladder diffuses as a function of the intermediate x_i for the CCFM and BFKL equations. This plot, from Salam (1997), is for $x = 10^{-4}$, $k = 5\,\text{GeV}$ and $\alpha_s = 0.2$.

imposition of kinematic constraints to control this diffusion (Kwiecinski et al. 1997). A typical example of the scale of the diffusion is shown in Fig. 9.16 and this raises the next and more serious problem, that the diffusion takes **k** to values small enough for α_s to be outside the perturbative region.

Running α_s LO BFKL

Once α_s is allowed to run the BFKL kernel is no longer infrared finite because $\mathbf{k}^2$ can diffuse into the non-perturbative region. This changes the nature of the LO BFKL equation considerably. Various approximate methods, both analytical and numerical, have been used to try to solve the LO BFKL equation with running α_s. In some approaches it is argued that α_s should be allowed to run down to some scale $\mathbf{k}_0^2$ then frozen. This does not lead to a unique prediction without further assumptions but it can be shown that the BFKL exponent ω_0 is bounded by

$$1.2\bar{\alpha}_s(\mathbf{k}_0^2) < \omega_0 < 4\ln 2\bar{\alpha}_s(\mathbf{k}_0^2).$$

Another approach is to consider the modified equation in double Mellin transform space (see Eq. (9.22)). The fixed α_s equation is an algebraic equation. When $\bar{\alpha}_s$ is replaced by $\bar{\alpha}_s(\mathbf{k}^2) = [\bar{b}_0 \ln(\mathbf{k}^2/\Lambda_{QCD}^2)]^{-1}$, the equation becomes a first order differential equation in the Mellin variable γ (from the extra $\ln(\mathbf{k}^2/\Lambda^2)$ factors). Although the equation can be solved by use of an integrating factor, the singularity structure in complex γ spaced has changed. This leads to a procedure analogous to regularizing the infrared

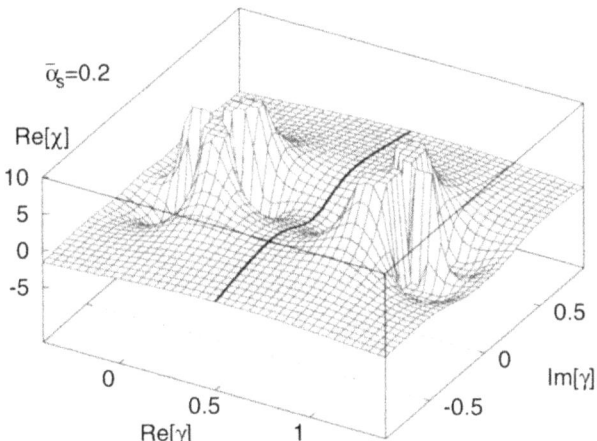

Fig. 9.17 Real part of the LO+NLO BFKL characteristic function χ in the complex γ plane and the contour for the Mellin inversion using the saddle point method at $\text{Re}(\gamma) = 1/2$. (From Salam 1999.)

singularities of parton densities, which isolates the non-perturbative terms but requires more assumptions to fix them.

Fixed α_s NLO BFKL

The result of the NLO calculation may be considered as the expansion of the characteristic function of the BFKL kernel as a power series in α_s

$$\chi(\gamma) = \chi^{(0)}(\gamma) + \bar{\alpha}_s \chi^{(1)}(\gamma) + \cdots$$

where $\chi^{(0)} = \chi_0$ of the LO kernel. The surprising result is that the behaviour of χ near $\gamma = 1/2$ changes radically with

$$\chi(1/2) = \chi^{(0)}(1/2)[1 - 6.47\bar{\alpha}_s] \tag{9.28}$$

which in turn means that the exponent ω_0 of F_2 at small x may even change sign (e.g. for $\alpha_s = 0.2$ one finds the LO result of -0.53 becomes 0.13). There has been much discussion about the meaning of this result for the validity of the BFKL approach. The essence of the problem is to be seen by comparing Fig. 9.17, which shows $\text{Re}(\chi)$ at LO+NLO, with Fig. 9.13 showing only the LO contribution. The shape of the function has changed completely and it calls into question the validity of the saddle point approximation for evaluating the Mellin inversion at NLO. The single saddle point along the $\text{Re}(\gamma) = 1/2$ integration contour at LO has become a complex conjugate pair, thus giving rise to oscillations in the BFKL cross-section as a function of $\ln(\mathbf{k}_1^2/\mathbf{k}_2^2)$, where $\mathbf{k}_1^2/\mathbf{k}_2^2$ is the ratio of the transverse momenta at top and bottom of the BFKL amplitude.

9.4.5 BFKL — discussion

A number of approaches have been put forward to ameliorate the situation. They fall into two classes, those concentrating on the implications for BFKL as applied to DIS and those for the BFKL kernel more generally. For DIS what matters at low x is the behaviour of the kernel near $\text{Re}(\gamma) = 0$.

Thorne (2000) and Altarelli et al. (2000) concentrate on running α_s arguing that for 'two-scale' processes, such as DIS, that this will be the most important effect. In effect what both calculations attempt to do is to improve the DGLAP splitting functions at low x by using the BFKL equation with running α_s to sum $\alpha_s \ln(1/x)$ terms. One is recasting the leading twist part of the BFKL k_T factorization into a collinear form with $\ln(1/x)$ terms included. The calculation of the gluon anomalous dimensions rederives the BFKL result at $LL(1/x)$, but the approach to steep asymptotic behaviour is rather slow because the gluon anomalous dimension is given by the series

$$\gamma_{gg}(n, \alpha_s) = \sum_{j=1}^{\infty} A_j^{gg} \alpha_s^j (n-1)^{-j}$$

and the coefficients A_j^{gg} are zero for $j = 2, 3, 5$ due to strong cancellations coming from colour coherence. However, the quark anomalous dimensions can also be improved by summing $\ln(1/x)$ terms. These anomalous dimensions are zero at $LL(1/x)$ (recall that the quark splitting functions are not singular as $z \to 0$, from Eq. (9.1)) but the $NLL(1/x)$ contributions are quite significant since all the coefficients A_j^{qg} in the series

$$\gamma_{qg}(n, \alpha_s) = \sum_{j=1}^{\infty} A_j^{qg} \alpha_s \alpha_s^j (n-1)^{-j}$$

are positive and large. Thus the splitting function P_{qg} is much steeper than in conventional DGLAP. However an absolute prediction of the low-x behaviour of F_2 is lost, since non-perturbative effects from the diffusion in transverse momentum have to be fixed by additional assumptions. In essence, although the parton splitting functions have been improved, the parton densities at the starting scale, or at least their leading power in x, still have to be fixed by fitting to data. However, these approaches do give some improvement in χ^2 over the conventional DGLAP approach, when fits are made to low-x F_2 data.

Turning to more general studies of the kernel Ciafaloni et al. (1999) have investigated the structure of the NLO contributions to χ. They argue that $\chi^{(1)}$ is dominated, to a good approximation, by a small number of 'collinear' contributions from the running of α_s, the non-singular piece of the splitting function and the choice of energy scale. In this collinear approximation $\chi^{(0)}(\gamma) = 1/\gamma + 1/(1-\gamma)$ and $\chi^{(1)}(\gamma) = A/\gamma^2 + B/(1-\gamma)^2 - 1/(2\gamma^3) - 1/(2(1-\gamma)^3)$, where A, B are known coefficients. The origin of the inverse cubic terms is from the choice of energy scale. In the double

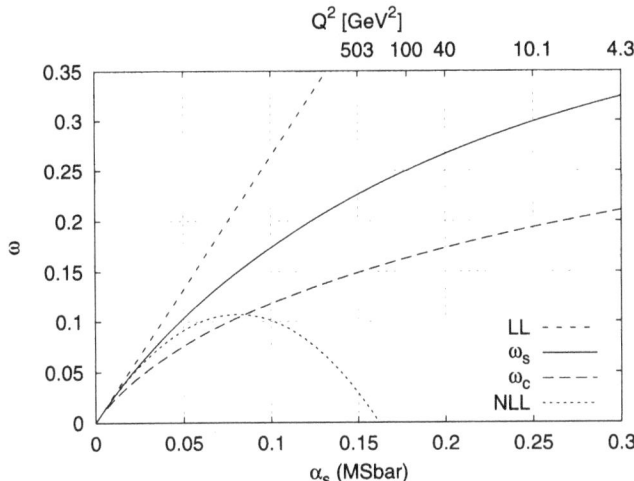

Fig. 9.18 BFKL exponent ω for various calculations: LO, NLO, resummed collinear approximation of Ciafaloni et al. . (From Salam 1999.)

Mellin integral of Eq. (9.22) for $F(s, \mathbf{k}_1, \mathbf{k}_2)$ the scale $\mathbf{k}^2$ was used. However, this choice does not respect the symmetry of the BFKL kernel between $\mathbf{k}_1$ and $\mathbf{k}_2$ and a better choice might be $k_1 k_2$.[11] At LO, one is summing terms of the form $(\bar{\alpha}_s \ln(s/s_0))^n$ and a change of scale s_0 can be absorbed in discarded higher order terms in $\bar{\alpha}_s$. This is not possible at NLO and above. Consider the scales as they appear in the Mellin integral

$$\left(\frac{s}{k_1 k_2}\right)^\omega \left(\frac{k_1^2}{k_2^2}\right)^\gamma = \left(\frac{s}{k_2^2}\right)^\omega \left(\frac{k_1^2}{k_2^2}\right)^{\gamma+\omega/2},$$

the change in energy scale is equivalent to a shift $\gamma \to \gamma - \omega/2$ in the second Mellin integration variable. The scale on the LHS respects the symmetry of the BFKL kernel whereas the choice on the RHS is relevant for DIS 'two-scale' processes in which $k_1^2 \gg k_2^2$. This gives a technique for resumming these 'energy scale' terms to all orders. The resulting function $\chi(\gamma, \omega)$ can then be used to estimate the high energy LO+NLO (or small x) exponent as shown in Fig. 9.18. More precisely, ω_s is the minimum value of $\bar{\alpha}_s \chi(\gamma, \omega)$ and is the power expected for $F(s, \mathbf{k}_1, \mathbf{k}_2)$ controlling high energy $\gamma^* \gamma^*$ cross-sections or DIS forward jets. ω_c is the position of the leading singularity in the anomalous dimension and gives the power growth at small x of the splitting functions. Both ω's vary slowly with α_s and lie between the LO result and the NLO result derived from Eq. (9.28).

---

[11]The BFKL amplitude is symmetric with respect to the transverse momenta at either end of the gluon ladder ($\mathbf{k}_1$, $\mathbf{k}_2$) and this is reflected in the symmetry of $\chi(\gamma)$ about the line $\mathrm{Re}(\gamma) = 1/2$.

To summarize the BFKL calculations, a lot of progress has been made in understanding why the LO fixed α_s calculation cannot be expected to describe inclusive DIS (or the 'single scale' hard processes) directly. It is too early to say which approach to the NLO BFKL will be the most successful, but it is clear that controlling transverse momentum diffusion and the running of α_s will be vital.

9.5 Angular ordering and the CCFM equation

Angular ordering, or coherence of soft radiation, is a property of both QED and QCD. It is a powerful tool that may be used in many areas of QCD in addition to providing an alternative approach to low-x physics. Many of the ideas and applications are to be found in the book by Dokshitzer et al. (1991). The basic idea may be understood by following the heuristic discussion of photon radiation in pair creation, given in Chapter 4 of that book. Referring to Fig. 9.19(a), a photon creates an e^+e^- pair with momenta p^+, p^- respectively, and opening angle θ_{ee}. After a time t_γ^-, the electron radiates a photon with momentum k at an angle $\theta_{\gamma e}$. The lifetime of the virtual e^- state may be estimated using $\Delta E \Delta t \sim 1$ and allowing for time dilation.

$$t_\gamma^- \sim \frac{1}{M_{\text{virt}}} \frac{E}{M_{\text{virt}}} = \frac{E}{(k+p^-)^2} \approx \frac{1}{k\theta_{\gamma e}^2},$$

where M_{virt} is the mass of the virtual state and E its energy. In the last step the electron mass has been neglected. During this time the e^+e^- pair will have separated a transverse distance $r_T^{ee} = \theta_{ee} t_\gamma^-$. Now $k\theta_{\gamma e} = k_T = \lambda_T^{-1}$, where λ_T is the transverse wavelength of the emitted photon. Using this one has

$$r_T^{ee} = \lambda_T \frac{\theta_{ee}}{\theta_{\gamma e}}$$

and for $\theta_{\gamma e} \gg \theta_{ee}$ the charge separation at the time of emission is very much less than λ_T. Under these circumstances the emitted photon cannot resolve the separate charges only their sum (which is zero) and the radiation is strongly suppressed. Independent emission of photons only occurs for $\theta_{\gamma e} < \theta_{ee}$ for which $r_T^{ee} > \lambda_T$. This phenomena, under the name of the Chudakov effect was observed in 1950s in large cosmic ray air showers. Essentially the same phenomenon occurs for g → q$\bar{\text{q}}$ in QCD with the modification that the wide angle soft gluon radiation is not zero (because the initial gluon carries a non-zero colour charge) but 'as if it were emitted from the parent gluon'. The details of the QCD calculation are given in Ellis et al. (1996), Section 5.5. The angular ordering constraint applies to all QCD branching processes involving both quarks and gluons and in particular g → gg. Thus in a cascade the angles of the emitted particles must decrease as one proceeds down any one branch. This has important consequences for the emission of final state hadrons as well as providing the basis for an alternative approach to inclusive DIS.

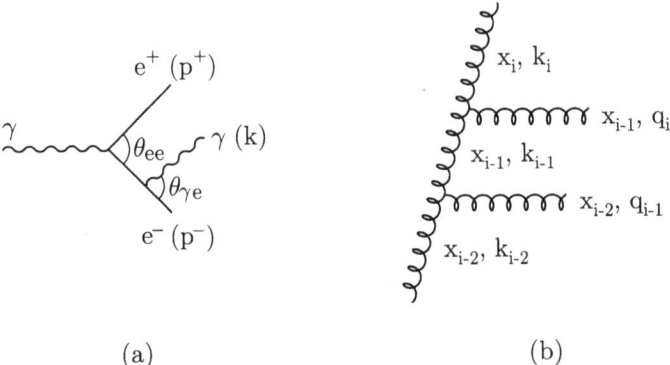

Fig. 9.19 Angular ordering: (a) soft photon emission in $\gamma \to e^+e^-$; (b) details of momenta in the CCFM gluon ladder.

The CCFM (Ciafaloni, Catani, Fiorani & Marchesini) equation is the application of angular ordering to the calculation of the gluon ladder. Fig. 9.19(b) shows a section of the ladder with transverse momenta k_i of gluons between emissions and q_i of the emitted gluon. The x_i variables are the usual fractions of the initial longitudinal momentum. For a small angle emission of gluon i, $\theta_i \approx q_i/(x_{i-1}p)$, angular ordering $\theta_i > \theta_{i-1}$ gives the condition that $q_i > z_{i-1}q_{i-1}$, where $z_i = x_i/x_{i-1}$. The soft gluon emission probability takes the form

$$dP = \frac{d^2 q_i}{\pi q_i^2} dz_i \frac{\bar{\alpha}_s}{z_i} \Delta(z_i, q_i, k_i) \Theta(q_i - z_{i-1}q_{i-1}),$$

where $z_i \ll 1$, the theta function reflects the angular ordering condition and the form factor Δ (which screens the $1/z$ singularity) is given by

$$\Delta(z_i, q_i, k_i) = \exp\left\{-\int_{z_i}^1 dz' \frac{\bar{\alpha}_s}{z'} \int \frac{dq'^2}{q'^2} \Theta(k_i - q')\Theta(q' - z'q_i)\right\}.$$

By considering successive emissions along the chain, an integral equation for $F(x, k, Q)$, the CCFM unintegrated gluon density, may be derived

$$F(x, k, Q) = F^{(0)}(x, k, Q) + \bar{\alpha}_s \int_x^1 \frac{dz}{z} \int \frac{d^2 q}{\pi q^2} \Delta(z, q, k) \Theta(Q - zq) F(x/z, k', Q),$$

where $k' = |\mathbf{k} + \mathbf{q}|$ and Q is the hard scale of the probe which sets the maximum angle for gluon emission. For larger x, the singular part of the gluon splitting function $\bar{\alpha}_s/z$ is replaced by the full P_{gg} splitting function. One of the interesting features of the CCFM equation is that it 'interpolates' between the BFKL equation at small x and the DGLAP equation at moderate x. For small x, small z will dominate the integral and the

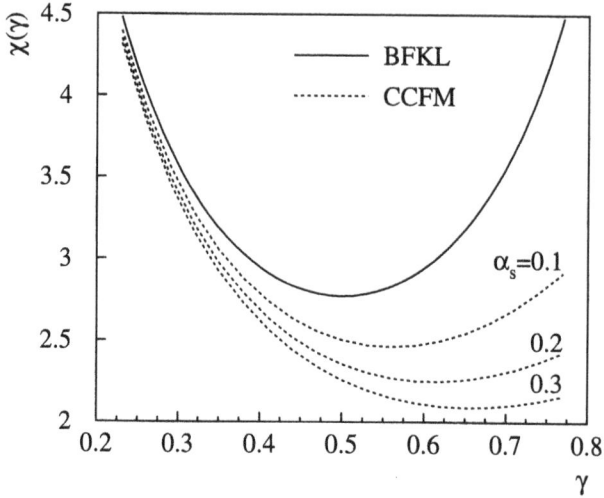

Fig. 9.20 The CCFM characteristic function in Mellin space for different values of α_s together with that for the LO BFKL equation. (From Salam 1997.)

$\Theta(Q - zq) \to 1$ which gives $F(x, k, Q)$ independent of Q as in the BFKL equation. For moderate to large x, $\Delta \to 1$, the variables $z_i \approx 1$ and the condition $q_i > z_{i-1} q_{i-1}$ becomes $q_i > q_{i-1}$, which is the strong ordering in transverse momentum of the DGLAP equation.

The CCFM equation cannot be solved analytically but it has been studied numerically. In the context of the present discussion, it is useful to compare the characteristic function for the CCFM equation with that from the BFKL equation. Figure 9.20, from Salam (1997), shows the characteristic function of the CCFM equation in Mellin space for different values of α_s together with that for the LO BFKL equation. The close similarity of all the curves for small values of γ demonstrates that solutions of the CCFM equation in this region approximate those of BFKL. At low x the CCFM equation exhibits BFKL like diffusion of transverse momentum along the gluon ladder, but the extent of the diffusion is not so large, as Fig. 9.16 shows.

The DIS final state may provide more sensitive measures for comparing the differences between the CCFM/BFKL approach and conventional DGLAP. The key difference to be tested is the more restricted forward energy flow that follows from the strict ordering of transverse momentum in DGLAP compared to the more randomly ordered k_T final state in BFKL and related approaches. Experimental observables that have been studied are forward jet production and forward particle production in DIS. Figure 9.21 shows the p_T spectra of charged particles at fixed x in the hadronic CMS of DIS, as measured in the H1 detector (H1 1997b) at $\sqrt{s} = 300 \, \text{GeV}$

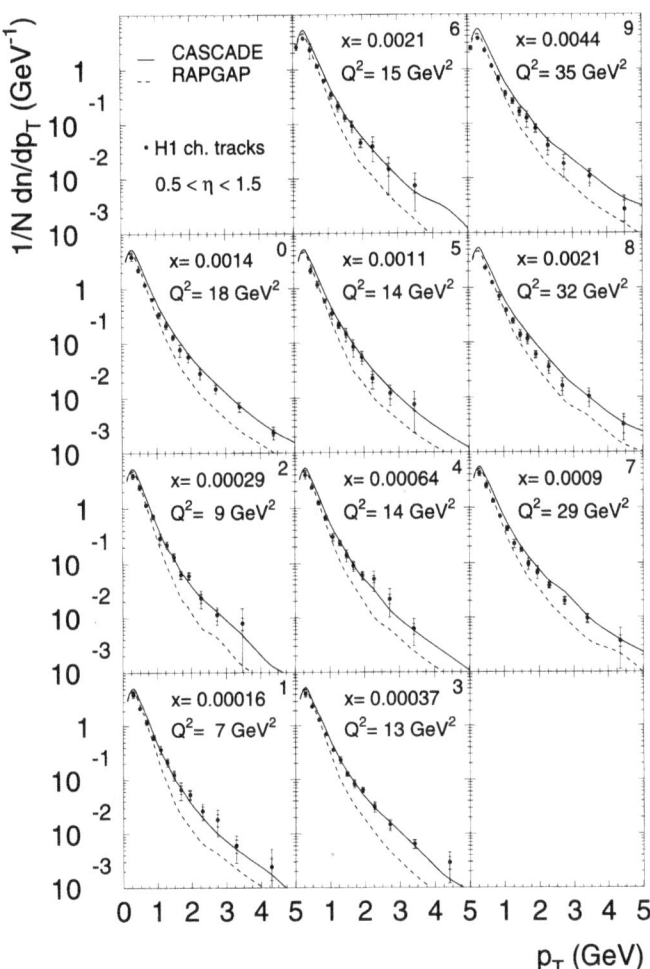

Fig. 9.21 The p_T spectra of charged particles in the hadronic CMS at fixed x in DIS at $\sqrt{s} = 300\,\text{GeV}$, measured by H1, with $5 < Q^2 < 50\,\text{GeV}^2$ and $1.5 < \eta < 2.5$. The data are compared to expectations from the CCFM code CASCADE (full curves) and the DGLAP code RAPGAP (dashed curves). (From Jung and Salam 2001.)

260 *DIS at low x*

with $5 < Q^2 < 50\,\mathrm{GeV}^2$ and $1.5 < \eta < 2.5$. The full curve shows the expectation of the CCFM CASCADE Monte Carlo simulation of Jung and Salam (2001) and the dashed curve that from the DGLAP based code RAPGAP (both codes use the JETSET realisation of the Lund string model for fragmentation[12]). The spectrum in the smallest x bins is clearly harder than the RAPGAP expectation but is well described by CASCADE. It should be noted that the CASCADE calculation shown is based only on gluon dynamics, with the unintegrated gluon distribution function fitted to F_2 data with $x < 0.01$ and $Q^2 > 5\,\mathrm{GeV}^2$. While this is encouraging, it cannot be taken as unambiguous evidence for BFKL as there are other approaches, such as modifications to the fragmentation model that can also describe the data. What can be concluded is that the data do point to a deficiency of the standard DGLAP approach at low x. Another example of the comparison of conventional DGLAP and CCFM predictions for $F_2^{c\bar{c}}$ is shown in Fig. 5.17 of Chapter 5. Here also there are indications that the CASCADE code provides a better description of the data at the lowest x values for low Q^2.

9.6 Unitarity and saturation

Cross-sections are limited by unitarity which gives a non-linear constraint on the scattering amplitude. At very high energies this leads to the Froissart–Martin bound on the total cross-section

$$\sigma_{\mathrm{tot}}(s) < A \ln^2 s,$$

where s is the centre of mass energy squared and $A \sim 60\,\mathrm{mb}$ is a constant determined from the pion mass. All the models and approximations considered so far for the high energy (low x) behaviour of the structure function, or $\sigma_{\gamma^* p}^{\mathrm{tot}}$, have given high energy behaviours growing like powers of W^2 (or $1/x$) and thus will eventually violate the Froissart bound. For the Regge pole description of hadronic total cross-sections given by Donnachie and Landshoff, the rate of increase with s is very gentle, $s^{0.08}$, which does not violate the Froissart bound until astronomically large energies are reached. The much larger effective exponents given by the DGLAP or BFKL equations are a different matter. However, as we have already intimated, non-linear effects at high parton density can modify this steep behaviour. Some simple physical arguments give an idea of how and when non-linear effects might affect deep inelastic scattering.

In the language of the QCD improved parton model, there are two effects that need to be considered as the density of gluons increases at low x: the first is 'shadowing' in which the effective total cross-section is reduced because the target partons overlap each other; the second is 'saturation' in which the growth of low x gluons is tamed by $gg \to g$ absorption (the

---

[12] Brief details and references to Monte Carlo codes are given in Appendix E.

inverse of $g \to gg$ splitting). Shadowing is a term taken from DIS on large nuclei in which it is known that $\sigma(\gamma^*A) = A^\alpha \sigma(\gamma^*N)$, with $\alpha < 1$ from this 'overlap' effect. However, it is saturation that is of particular interest here.

To get a semi-quantitative feel for the problem work in the infinite momentum frame, where $xg(x,Q^2)$ gives the number of gluons per unit of rapidity ($\ln(1/x)$ or dx/x). The gluon–gluon recombination or absorption cross-section is of order $\alpha_s(Q^2)/Q^2$ at a transverse scale Q^2, thus one might expect saturation effects when the total effective gluon–gluon recombination cross-section in the proton approaches that of the size of the proton disk,

$$\frac{\alpha_s(Q^2)}{Q^2} xg(x,Q^2) \sim \pi R^2. \tag{9.29}$$

If R is taken to be 0.8 fm then at $Q^2 \sim 10\,\text{GeV}^2$ the above relation gives $xg \sim 2000$. Such densities are not likely to be reached within the low-x range of HERA. However there is no compelling reason why the gluon density should be uniform across the nucleon, an upward fluctuation in a restricted region could lead to saturation occurring at much lower average densities.

9.6.1 High density gluon dynamics

Gribov, Levin and Ryskin (1983) and later Mueller and Qiu (1986) extended the low-x DLLA approximation to include the effect of saturation with the equation

$$\frac{\partial^2 xg(x,Q^2)}{\partial \ln Q^2 \partial \ln(1/x)} = \frac{3\alpha_s}{\pi} xg(x,Q^2) - \frac{81\alpha_s^2}{16Q^2 R^2}(xg(x,Q^2))^2. \tag{9.30}$$

In this equation, the first term on the RHS gives essentially Eq. 9.4 of DLLA or DAS, while the second term is an estimate of gluon recombination. Note that this term will damp the growth of xg at low x, in fact giving zero growth when $xg(x,Q^2) = 48\pi Q^2 R^2/(81\alpha_s)$, in agreement with the crude estimate above.

Since this early work the subject has developed considerably. Although a detailed exposition is beyond the scope of this book, a brief account of the key ideas is given here showing how the evidence for gluon dominated dynamics at low x revealed by HERA data may be linked to the physics of the quark–gluon plasma.

As demonstrated by the discussion above, an essential outcome of all models of saturation is the existence of an x-dependent *saturation scale*, $Q_s^2(x)$. For $Q^2 \gg Q_s^2$, saturation effects may be ignored and the linear BFKL or DGLAP equations are valid; for $Q^2 \ll Q_s^2$, the non-linear effects of saturation must be taken into account. At high enough energies $\alpha_s(Q_s^2)$ may be small enough to allow weak coupling techniques to be used. The problem is not perturbative because of the very large gluon density, but it may be susceptible to a semi-classical description. Note that saturation effects are likely to be enhanced in nuclei since $xg_A \propto A$, $\pi R_A^2 \propto A^{2/3}$ and from Eq. (9.29)

$$Q_s^2(x,A) \sim \alpha_s x g_A(x,Q_s^2)/\pi R_A^2 \sim A^{1/3} Q_s^2(x).$$

The model that has emerged from the work of many authors is known as the *Colour Glass Condensate* (CGC), for a pedagogic account see the lectures by Iancu, Leonidov and McLerran (2002) and for a short qualitative introduction with references to the original literature, the article by Iancu (2002). The CGC is a model for the wave-function of a hadron interacting in a situation where many soft gluons may be exchanged — such as DIS at small x with $Q^2 < Q_s^2$. The essential physical ideas, expressed in the infinite momentum frame, are that: the multiple soft gluons may be described by a classical Yang–Mills field the sources for which are the fast moving partons; at a given x the internal dynamics of the these partons are 'frozen' through Lorentz time-dilation for the duration of the interaction; QCD provides a non-linear equation for evolution in x. The effective theory has similarities with techniques for spin glasses, one is dealing with interacting colour charges and, as gluons are bosons, one may have multiple occupancy of quantum states, as in a boson condensate, hence the name colour glass condensate.

In the infinite momentum frame for a $\gamma^* p$ scattering at a given low x value, the fast partons are taken to be moving at almost the speed of light in the positive z direction (x^+ using light cone variables) generating a colour current, $J_a^\mu = \delta^{\mu+} \rho_a$, concentrated around $z = t$ or $x^- = 0$. The classical equation for the colour field is then

$$(D_\nu F^{\mu\nu})_a(x) = \delta^{\mu+} \rho_a(x^-, x_\perp), \quad \text{or in the covariant gauge} \quad \nabla_\perp^2 A^+ = -\rho.$$

Field theory techniques are then used to construct physical quantities from A^+ by averaging over appropriately weighted configurations of ρ. Referring to Fig. 9.9(c), the classical field A^+ represents the 'last' gluon in the ladder with momentum $k^+ = xP^+$ and encodes the non-linear effects at this scale. The sources contributing to ρ_a are the quarks and gluons with momenta $p^+ \gg k^+$. Moving to a new, lower, scale in x, $x' < x$, partons with momenta p'^+, where $x'P^+ < p'^+ < k^+$, will no longer be described by A^+, but now contribute to the source ρ_a, so a new field A'^+ must be constructed. It has been possible, under certain conditions, to derive evolution equations for physical quantities in x, or more precisely in $Y = \ln(1/x)$.

As will be explained below, one of the key physical objects in the connection of the CGC to measurements is the colour-dipole hadron scattering amplitude $N(x_\perp, Y)$ or its transverse momentum representation $\phi(k_\perp, Y)$. In the large N_c limit, Balitsky and Kovchegov derived the non-linear evolution equation

$$\frac{\partial \phi}{\partial Y} = K_{\text{BFKL}} \otimes \phi - \bar{\alpha}_s \phi \otimes \phi, \tag{9.31}$$

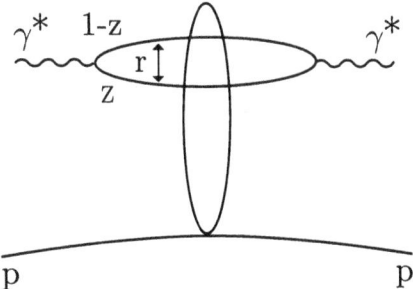

Fig. 9.22 The qq̄ dipole model for the elastic $\gamma^* p$ amplitude and hence for deep inelastic scattering.

where $\bar{\alpha}_s = N_c \alpha_s / \pi$ and K_{BFKL} is the kernel of the (linear) BFKL equation. This equation, which has close similarities to the GLR equation above, describes both the production and merging of soft gluons through the $g \to gg$ and $gg \to g$ interactions, respectively. Through the latter, the growth of the cross-section is curbed and unitarity respected.

An important question, to be discussed below, is whether there is experimental evidence for any such effect within the x range accessible at HERA. Another prediction of the CGC concerns the rapidity distribution of particles produced at central rapidities in nucleus-nucleus collisions. In the A–A CM frame at rapidities close to those of the projectiles, η_{beam}, the shape of the rapidity distribution should be roughly universal and independent of the energy — controlled by the 'fast partons'. As the energy increases, particle production should increase at central rapidities, far from η_{beam}, controlled by gluons with small x or large $Y = \eta - \eta_{\text{beam}}$. There is evidence that this so-called limiting fragmentation is indeed seen in data from heavy-ion collisions at RHIC (the relativistic heavy ion collider operating at Brookhaven).

9.7 Dipole models — general formalism

Colour dipole models provide a rather general framework for studying the constraints of unitarity and the approach to saturation at low x. They are closely related to an alternative approach to the BFKL equation and provide a natural link between inclusive DIS and diffraction. They have been studied and developed by many authors, useful references relevant to the discussion here and containing references to the original literature are Forshaw et al. (1999) and Golec-Biernat (2002).

The dipole model can be given a very direct physical interpretation in the rest frame of the proton. Consider Fig. 9.22, the virtual photon, with 4-momentum q, splits into a qq̄ pair in which the q takes a fraction z of the initial momentum and the splitting occurs a 'long' time $\sim 1/(mx)$ before the pair interacts with the proton (see Problem 3 at the end of the chapter). The $q\bar{q}$ pair then scatters coherently from the proton in a time which is short in comparison to $\sim 1/(mx)$. The $\gamma^* \to q\bar{q}$ process is described by QED and the strong interaction physics is then contained in the cross-section for the dipole–proton interaction. During the interaction, it is assumed that the transverse size of the pair, r, remains constant.

Although the resulting formalism will have greater generality, it is convenient to start from the BFKL $\mathbf{k}_T$ factorized form of Eqs (9.25) and (9.26), in which impact factors[13] were introduced. Using these two equations, the $\gamma^* p$ total cross-section for polarization λ may be written as

$$\sigma_\lambda(x, Q^2) = \frac{1}{2\pi} \int \frac{d^2\mathbf{k}}{\mathbf{k}^4} \Phi_{\gamma,\lambda}(\mathbf{k}) F_g(x, \mathbf{k}). \tag{9.32}$$

For the longitudinal polarization state $\lambda = L$, the $\gamma^* \to q\bar{q}$ impact factor is

$$\Phi_{\gamma,L}(\mathbf{k}) = 32\alpha\alpha_s \sum_{i=1}^{n_f} e_i^2 \int_0^1 dz \int d^2\mathbf{r}(1 - e^{i\mathbf{k}\cdot\mathbf{r}}) Q^2 z^2 (1-z)^2 K_0^2(\epsilon r), \tag{9.33}$$

where $\epsilon = z(1-z)Q^2 + m_i^2$, n_f if the number of active quark flavours of mass m_i and charge e_i and K_0 is a Macdonald (or modified Bessel) function. Referring to Fig. 9.15, there are two types of coupling between the $q\bar{q}$ pair and the two gluons: in the first both gluons couple to the same quark (or antiquark); in the second one gluon couples to the quark and the other to the antiquark. In the above expression for $\Phi_{\gamma,L}$ the former coupling gives rise to the 1 and the latter to the $e^{i\mathbf{k}\cdot\mathbf{r}}$ in the factor $(1 - e^{i\mathbf{k}\cdot\mathbf{r}})$. Inserting the expression for $\Phi_{\gamma,L}$ into Eq. (9.32) and re-arranging gives

$$\sigma_L(x, Q^2) = \int_0^1 dz \int d^2\mathbf{r} |\Psi_L(z, \mathbf{r})|^2 \hat{\sigma}(x, r^2), \tag{9.34}$$

where

$$|\Psi_L(z, \mathbf{r})|^2 = \frac{6\alpha}{\pi^2} \sum_{i=1}^{n_f} e_i^2 Q^2 z^2 (1-z)^2 K_0^2(\epsilon r) \tag{9.35}$$

and the *dipole cross-section*, $\hat{\sigma}$, is given by

$$\hat{\sigma}(x, r^2) = \frac{8\pi\alpha_s}{3} \int \frac{d^2\mathbf{k}}{\mathbf{k}^4} F_g(x, \mathbf{k})(1 - e^{i\mathbf{k}\cdot\mathbf{r}}). \tag{9.36}$$

Starting from the impact factor $\Phi_{\gamma,T}$ for the transversely polarized γ^* states an equation for σ_T, of identical form to (9.34), may be deduced with

$$|\Psi_T(z, \mathbf{r})|^2 = \frac{3\alpha}{2\pi^2} \sum_{i=1}^{n_f} e_i^2 \left\{ [z^2 + (1-z)^2]\epsilon^2 K_1^2(\epsilon r) + m_i^2 K_0^2(\epsilon r) \right\}. \tag{9.37}$$

A number useful physical insights may already be gained from the forms of these expressions. The functions $K_0(x)$ and $K_1(x)$ are exponentially

---

[13] The required γ^* impact factors are calculated in Forshaw and Ross (1997), Chapter 6.

damped for large positive values of x and for purposes of approximating the integrals they may be replaced by

$$K_0(x) = c_0 \Theta(1-x), \quad K_1(x) = \frac{c_1}{x}\Theta(1-x), \tag{9.38}$$

where c_i are constants. Thus the integrals from which the cross-sections are calculated are dominated by small values of the argument $x = \epsilon r$ and since $\epsilon = z(1-z)Q^2 + m_i^2$, this may be satisfied in various different ways. Note that there are two transverse scales with which r may be compared: $1/Q$, from the virtuality of the γ^* and R, a typical hadron radius of about 1 fm at which the cross-section saturates. The following qualitative results may be obtained (see Problems 5–7 at the end of the chapter):

- At small transverse separation, $\hat{\sigma}(x, r^2) \sim r^2$. This behaviour is an example of 'colour transparency', at small separations the $q\bar{q}$ dipole will appear to be an almost colourless object and thus have a much reduced cross-section with other hadronic matter.
- Contribution of 'small' sized $q\bar{q}$ pairs $r < 1/Q$. The constraint $\epsilon r < 1$ may be satisfied without any restriction on z and both transverse and longitudinal cross-sections 'scale' with Q^2, i.e.

$$\sigma_T(x, Q^2) \sim 1/Q^2 \quad \sigma_L(x, Q^2) \sim 1/Q^2,$$

which is equivalent to Bjorken scaling of the structure functions.
- Contribution of 'large' sized $q\bar{q}$ pairs $r \sim R \gg 1/Q$. The integrals are dominated by values of $\epsilon r < 1$ giving z or $1-z \ll 1/(Q^2 R^2)$ and from the forms of $|\Psi_{T,L}(z,\mathbf{r})|^2$ it follows that

$$\sigma_T(x, Q^2) \sim 1/Q^2 \quad \sigma_L(x, Q^2) \sim m_i^2/Q^4.$$

With either one or other of the $q\bar{q}$ pair taking almost all the momentum of the incoming γ^*, these contributions are known as *aligned jet* configurations. Note that the contribution to σ_T is leading twist ('scales') but that to σ_L is higher twist (one power down in Q^2).
- At large $r \sim R$ (the radius of the hadron), some form of saturation mechanism must damp the cross-section to respect the unitarity limit.

Dipole models provide a natural framework for the description of F_2 (or $\sigma(\gamma^* p)$) through the transition region from $Q^2 = 0$ to deep inelastic scattering.

Although not followed here in any detail, the colour dipole formalism also describes diffractive processes. For example, the forward diffractive differential cross-section is given by

$$\frac{d\sigma_{T,L}^D}{dt}\bigg|_{t=0} = \frac{1}{16\pi} \int_0^1 dz \int d^2\mathbf{r}\, |\Psi_{T,L}(z,\mathbf{r})|^2 |\hat{\sigma}(x, r^2)|^2. \tag{9.39}$$

Note that the dipole cross-section enters the above expression squared, so diffractive scattering will be more sensitive to dipoles with large separation.

More details of applications to diffraction are covered in the review by Wüsthoff and Martin (1999).

9.8 Dipole models — examples

Many authors have developed dipole models for a range of phenomena from $\sigma^{\text{tot}}(\gamma^*p)$ to vector meson production. Here a brief summary is given of two contrasting examples of phenomenological dipole models for the $\sigma^{\text{tot}}(\gamma^*p)$, followed by a discussion of attempts to build models incorporating linear and non-linear QCD effects.

9.8.1 The Forshaw–Kerley–Shaw model

The aim of these authors (Forshaw et al. 1999) is to extract the dipole cross-section by fitting a flexible parameterization to photo- and electro-production total cross-section data with the minimum of assumptions. The dipole cross-section is constructed from two components both with Regge type energy dependence, a 'soft' term with which dominates at large r (the dipole separation) and a 'hard' term which dominates at small r but vanishes for r large.

$$\sigma(s,r) = \sigma_{\text{soft}}(s,r) + \sigma_{\text{hard}}(s,r)$$

where

$$\sigma_{\text{soft}}(s,r) = a_0^S \left(1 - \frac{1}{1 + (a_1^S r + a_2^S r^2)^2}\right)(r^2 s)^{\lambda_S}$$

$$\sigma_{\text{hard}}(s,r) = (a_1^H r + a_2^H r^2 + a_3^H r^3)^2 \exp(-\nu_H^2 r)(r^2 s)^{\lambda_H}, \quad (9.40)$$

with a_i^S, a_i^H λ^S, λ^H and ν_H^2 are parameters to be determined by the fit. The data used in the fit were F_2 from HERA 1994/5 and E665 with $s \geq 100\,\text{GeV}^2$, $Q^2 < 60\,\text{GeV}^2$, $0 \geq x \geq 0.01$ and real photoproduction data at intermediate energies with two high energy points from HERA. The parameter controlling the energy dependence of the soft term was fixed at $\lambda_S = 0.06$ not far from the intercept of the soft Pomeron. Generally a good description of the fitted data is obtained with λ_H around 0.38, consistent with the energy dependence of the hard Pomeron in the Donnachie–Landshoff Regge model. The model also predicts very little Q^2 and s dependence of the ratio $\sigma^D(\gamma^*p)/\sigma(\gamma^*p)$, the diffractive to inclusive cross-sections in DIS, in accord with observation. Two other features are: the behaviour of large dipoles is constrained by accurate photoproduction data; although saturation at large r is incorporated into the parameterization, there is no evidence for for an energy dependent saturation scale.

9.8.2 The Golec-Biernat–Wüsthoff model

In contrast, the Golec-Biernat–Wüsthoff (GBW) model (Golec-Biernat and Wüsthoff 1999) explicitly includes an x dependent saturation scale, the 'saturation radius', R_0, where

$$R_0(x) = \frac{1}{Q_0}\left(\frac{x}{x_0}\right)^{\lambda/2}, \tag{9.41}$$

with parameters λ, x_0 and $Q_0 = 1\,\text{GeV}$ to fix the dimensions. $R_0(x)$ then scales the q$\bar{\text{q}}$ transverse separation r in the dipole cross-section

$$\hat{\sigma}(x, r^2) = \sigma_0 g(\hat{r}^2), \qquad \hat{r} = \frac{r}{2R_0(x)}, \tag{9.42}$$

where σ_0 is a constant. The function $g(\hat{r}^2)$ is constrained to vanish like $\hat{r}^2$ as $\hat{r} \to 0$ and it should tend to 1 for $\hat{r} \geq 1$. The latter condition guarantees that saturation occurs and the total cross-sections calculated from $\hat{\sigma}$ will tend to a constant for small Q^2. A simple functional form is chosen for g

$$g(\hat{r}^2) = 1 - \exp(-\hat{r}^2). \tag{9.43}$$

In addition to the three parameters of the model (λ, x_0, σ_0), values have to be chosen for the quark masses. For the light quark masses, $m_q = 140\,\text{MeV}$ is chosen as this gives a good description of data at very low Q^2. To ensure a smooth transition as $Q^2 \to 0$, and to have the correct threshold behaviour, the Bjorken variable is also modified by $\bar{x} = x(1 + 4m_q^2/Q^2)$. The model parameters were determined to be $\sigma_0 = 23.03\,\text{mb}$, $\lambda = 0.288$, $x_0 = 3.04 \times 10^{-4}$ by fitting low-x DIS data ($x < 0.01$)[14] with statistical and systematic errors added in quadrature. The model is very compact and gives a good description of F_2 data at low x from $Q^2 \sim 100\,\text{GeV}^2$ through the transition region to photoproduction, as Fig. 9.23 shows. Contrast this with Fig. 5.18 of Chapter 5 which shows the same data with the non-perturbative Regge model at low Q^2 and a ZEUS NLO QCD fit at larger Q^2. The GBW model also gives very little Q^2 and s dependence of the ratio $\sigma^D(\gamma^*p)/\sigma(\gamma^*p)$ in agreement with observation.

The parameterization explicitly gives a constant value of σ_0 for the dipole cross-section, $\hat{\sigma}$ at saturation (large r), but because of the x dependence of R_0 $\hat{\sigma}$ saturates for smaller dipoles sizes as x decreases. In terms of the earlier discussion of saturation scales, $Q_s^2(x) = 1/R_0^2(x)$. This may be understood qualitatively as follows (Golec-Biernat and Wüsthoff 1999). The photon virtuality gives a scale, $1/Q$, for the $q\bar{q}$ pair. For small dipoles, $r < 1/Q$, the γ^*p cross-section behaves as

$$\sigma_T(r\,\text{small}) \sim \sigma_0/(Q^2 R_0^2).$$

For large dipoles, $r > 1/Q$, one finds

$$\sigma_T(r\,\text{large}) \sim \sigma_0 + \sigma_0 \ln(1/(Q^2 R_0^2)).$$

Rough equality between $\sigma_T(r\,\text{small})$ and $\sigma_T(r\,\text{large})$ occurs for $Q^2 R_0^2 \approx 1$. Looking at Fig. 9.23 and remembering that $F_2 \sim Q^2 \sigma_T$ at fixed x, one sees

---

[14] The data set used was essentially the HERA 1994 F_2 data from H1 and ZEUS together with shifted vertex and ZEUS BPC data at low Q^2.

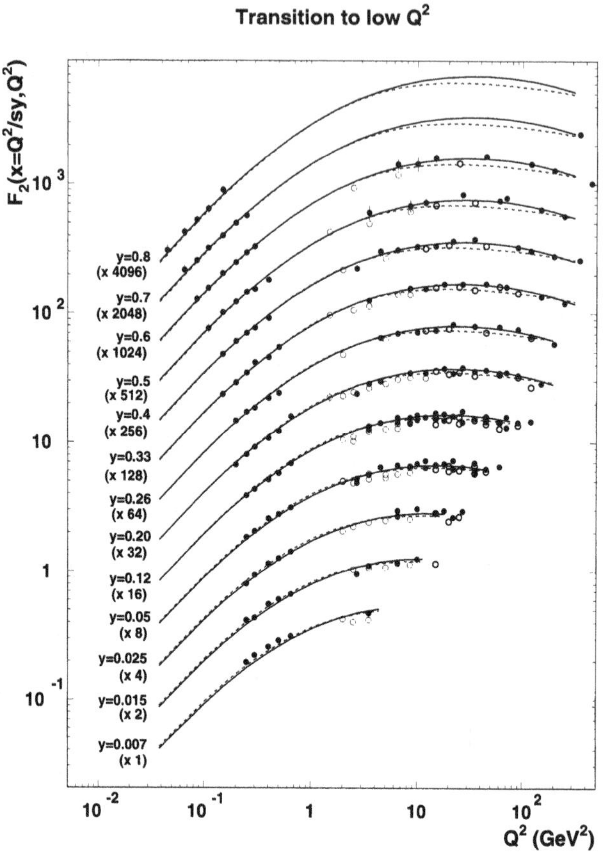

Fig. 9.23 The GBW dipole model fit to F_2 data through the transition region to photoproduction. The data are plotted versus Q^2 in bins of y. The dashed curves show the original model, the full curve the model with DGLAP evolution. (From Bartels et al. 2002.)

a clear change in behaviour around $Q^2 \approx 1\,\mathrm{GeV}^2$, which might then be taken as a rough estimate of the saturation scale for HERA data.

One needs to be somewhat careful before jumping to the conclusion that the GBW model shows that saturation effects have been seen at HERA. As discussed at the beginning of this section, on fairly general grounds, one may not have the 'kinematic reach' at HERA to probe densities large enough to require saturation. Also alternative models, such as the Forshaw–Kerley–Shaw model, do not require this feature to fit much the same data.

9.8.3 Geometrical scaling

Another general feature of saturation models that is incorporated in the GBW model is the dependence of the dipole cross-section on a scaled variable, here $r/R_0(x)$, as shown by Eqs (9.42) and (9.43). Stasto et al. (2001) have made the interesting observation that this leads to a new scaling property of $\sigma(\gamma^* p)$. At low x $\sigma(\gamma^* p)$ should depend only on the dimensionless variable $\tau = Q^2 R_0^2$. What is more, such a scaling property seems to be satisfied by HERA data. Qualitatively, using the arguments outlined at the end of the last section, one expects

$$\sigma(\gamma^* p) \sim \sigma_0 \ (\tau \text{ small}) \rightarrow \sigma(\gamma^* p) \sim \sigma_0/\tau \ (\tau \text{ large}).$$

Since $R_0(x)$ changes slowly with x, small τ corresponds to small Q^2 and large dipole separation, whereas large τ corresponds to large Q^2 and the DIS region dominated by dipole scattering with small separation. The authors use a parameterization of $R_0(x)$ that is correlated with the energy dependence of $\sigma(\gamma^* p)$, but less model dependent than that of the GBW model. Figure 9.24 shows HERA data with $x < 0.01$ as a function of τ. Not only do the data show scaling very clearly, but also the expected change in behaviour discussed above seem to occur around values of $\tau \sim 1$. The authors also show that this new scaling — geometrical scaling — is a property of low x data. It does not hold for data with $x > 0.01$. It is interesting to note that the data shown following geometric scaling have Q^2 values up to the order of $400\,\text{GeV}^2$, well above the $O(1)\,\text{GeV}^2$ estimates that emerge from the GBW model.

9.8.4 Dipole models and Q^2 evolution

Both the models considered above concentrate on the providing simple parameterizations that allow the main features of the dipole framework to be tested against DIS data at low x. Neither allows for the logarithmic scaling violations expected from QCD as Q^2 increases, so it is not surprising that the models are either limited to $Q^2 < 100\,\text{GeV}^2$ or do not provide a good description of data at these scales and above. Bartels et al. (2002) have attempted to extend the GBW model to include LO DGLAP evolution. The essence of their idea is to use the relationship between the dipole cross-section and the gluon density, valid for small r,[15]

$$\hat{\sigma}(x,r) \approx \frac{\pi^2}{3} r^2 \alpha_s x g(x, \mu^2), \tag{9.44}$$

where the scale $\mu^2 \sim C/r^2$ for small r^2. The key feature of the GBW model, that $\hat{\sigma} \rightarrow \sigma_0$ for large r must also be preserved. This leads to the following modification

$$\hat{\sigma}(x,r) = \sigma_0 \left\{ 1 - \exp\left(-\frac{\pi^2 r^2 \alpha_s(\mu^2) x g(x, \mu^2)}{3\sigma_0}\right) \right\},$$

---
[15]See Problem 8.

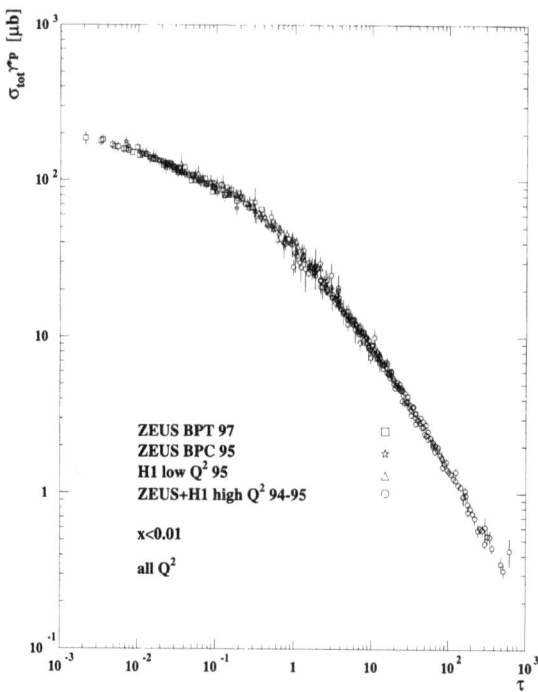

Fig. 9.24 Geometrical scaling. Data on $\sigma(\gamma^*p)$ with $x < 0.01$ plotted versus the scaling variable $\tau = Q^2 R_0^2(x)$. (From Stasto et al. 2001.)

where $\mu^2 = \mu_0^2 + C/r^2$ with C and μ_0^2 parameters to be determined from a fit to data. Since the model is to be applied to data at low x, only the gluon density is included in the DGLAP LO evolution. The model is applied to DIS data (H1, ZEUS and E665) with $x < 0.01$ and $0.1 < Q^2 < 500\,\text{GeV}^2$ and a satisfactory fit obtained. This is shown as the full line in Fig. 9.23. As expected, the differences between the original GBW model and the dipole with evolved gluon are most noticeable at the largest Q^2 values. However, on the scale of the plot the improvement looks quite modest. A more pronounced difference is in the description of the logarithmic slope of $F_2 \sim x^{-\lambda}$ at low x. The data on this slope is shown in Fig. 9.8. The prediction from the original GBW model already falls below the data for Q^2 as low as $10\,\text{GeV}^2$, tending to a value of $\lambda \sim 0.25$ by $Q^2 \sim 100\,\text{GeV}^2$. However the extended model with gluon evolution follows the trend of the data accurately over the full range, $0.1 < Q^2 < 100\,\text{GeV}^2$.

9.8.5 Dipole models and the colour glass condensate

The previous section has shown how linear QCD effects may be incorporated into a dipole model, but is it possible to go further and relate the

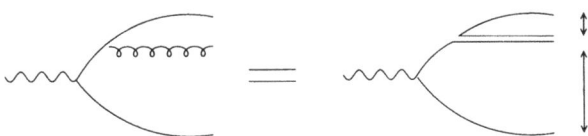

Fig. 9.25 Gluon emission from a dipole and its equivalence as two dipoles in the large N_c limit. (From Golec-Biernat 2002.)

saturation aspect to non-linear high gluon density QCD dynamics? The first step is to reformulate multiple gluon emission in dipole language. This is valid in the large N_c limit and is most conveniently explained by referring to Fig. 9.25. On the left-hand side one has gluon emission from the $q\bar{q}$ pair coupled to the virtual photon, on the right-hand side is the same process but now drawn as two colour dipoles by replacing the gluon's colour charge with that of a new (coloured) $q\bar{q}$ pair. In this way multiple gluon emission may be replaced by a cascade of dipoles, which is what the Balitsky–Kovchegov equation describes. More formally, the dipole cross-section $\hat{\sigma}$ is related to the forward $q\bar{q}$–hadron scattering amplitude N by

$$\hat{\sigma}(\mathbf{r},x) = 2\int d^2\mathbf{b}\, N(\mathbf{r},\mathbf{b},Y),$$

where $\mathbf{b}$ is the impact parameter of the $q\bar{q}$ with respect to the centre of the hadron and $Y = \ln(1/x)$. The BK equation (Eq. 9.31), which is valid for small r^2, applies to the Fourier transform of N

$$\phi(\mathbf{k},\mathbf{b},Y) = \int \frac{d^2\mathbf{r}}{2\pi} e^{-i\mathbf{k}\cdot\mathbf{r}} \frac{N(\mathbf{r},\mathbf{b},Y)}{r^2}.$$

In the GBW model, N has the form

$$N_{\mathrm{GBW}}(\mathbf{r},\mathbf{b},Y) = \left[1 - \exp\left(-\frac{r^2}{4R_0^2(x)}\right)\right]\Theta(b-b_0),$$

with $\sigma_0 = 2\pi b_0^2$. In other words the hadron is modelled in impact parameter space by a disc with a sharp edge. Note that at saturation $N \to \Theta(b-b_0)$, there are no edge effects and saturation is a uniform blackening of the disc. Solving the BK equation with an initial condition $N(\mathbf{r},\mathbf{b},Y_0) = N(\mathbf{r})S(\mathbf{b})$, with $S(\mathbf{b})$ a profile function giving a less sharp edge, gives an approach to saturation as $x \to 0$, which is illustrated in Fig. 9.26. There are two different regimes, in the inner black disc one has full non-linearity and saturation, but in the outer grey rim one has a region in which the gluon density is lower and the linear term of the BK equation dominates, i.e. BFKL growth. Note the size of both discs is growing as $x \to 0$. This corresponds to a cross-section growing slowly with energy, but respecting the Froissart bound. Another very interesting observation is that the non-linear term in

272 DIS at low x

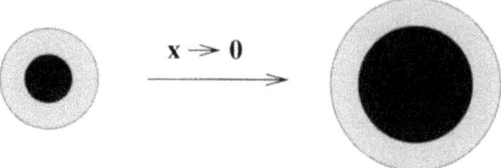

Fig. 9.26 Black disc expansion and saturation in a non-linear QCD dipole model. (From Golec-Biernat 2002.)

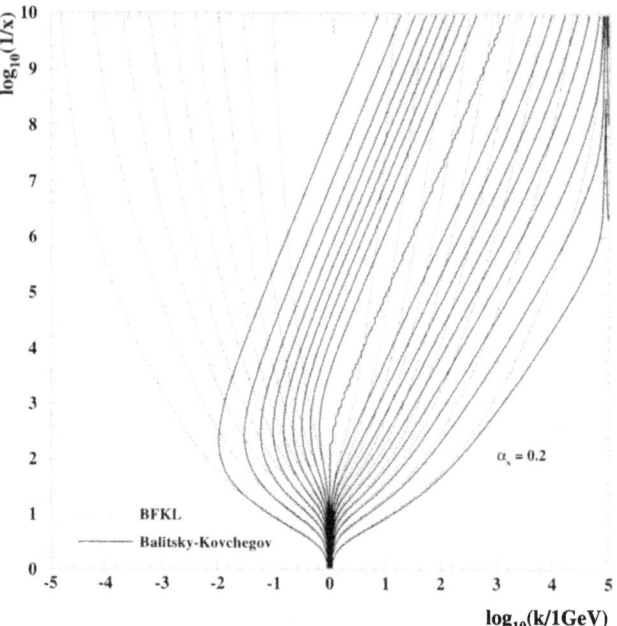

Fig. 9.27 Comparison of transverse momentum diffusion in the BFKL and BK equations. (From Golec-Biernat 2002.)

the BK equation constrains the diffusion in transverse momentum which is a problem in the linear BFKL equation. Figure 9.27 shows the dipole amplitudes $\phi(k, Y)$ projected onto the (k, Y) plane from the BFKL kernel alone (dotted lines) and from the full BK equation (full lines). In both cases, the initial condition in transverse momentum is at $k = k_0 = 1\,\text{GeV}$. The BFKL solutions spread out to both larger and smaller vales of k, symmetrical in $\log k$ as $\log(1/x)$ increases, as expected from the gaussian profile of transverse momentum. However, for the BK solutions, although there is some initial diffusion to values of $k < k_0$, as $\log(1/x)$ increases all solutions are 'pulled' across to the 'safe' region with $k > k_0$. These results are very promising, but there is much work still to be done before this type of model can be applied quantitatively to precision data.

9.9 The description of low–x inclusive data

In this chapter many approaches to the description of F_2 at low x have been described, some are still at a stage where a full description of data would not be realistic, but many do give a very fair description of the precision F_2 data at low x from HERA. Those that attempt to describe the data through the transition region to photoproduction have, of necessity, to include some non-perturbative effects. Confining the discussion to $Q^2 > 1\,\text{GeV}^2$, where α_s should allow either pQCD methods or weak coupling techniques to have some validity, can one distinguish between the various approaches? Is there clear evidence for the limiting values of x and Q^2 where the standard NLO DGLAP approach fails? Is NNLO DGLAP better, or some form of 'BFKL enhanced' splitting functions, or is there a need for non-linear terms?

These last points have been discussed at some length in Chapter 6, where the conclusion is reached that for Q^2 less than $\sim 2\,\text{GeV}^2$ and $x < 0.001$, the parton densities (and F_L) are starting to look distinctly unphysical. It is worth recalling that at low x

$$F_2(x, Q^2) \sim x\Sigma(x, Q^2) \quad \text{and} \quad \frac{\partial F_2}{\partial \ln Q^2} \sim P_{qg} x g(x, Q^2)$$

F_2 measures the $q\bar{q}$ sea distributions directly, but the gluon distribution is determined via a convolution with the splitting function. Within the DGLAP formalism, the splitting functions are well known, so that the behaviour of the F_2 scaling violations at low Q^2 translates into the flattening of the gluon distribution observed in Fig. 6.1. However, an extension to the formalism would alter the splitting factors (or even add an extra term to the evolution equation in the case of non-linear effects), and thus change the shape of the fitted gluon distribution. To distinguish the different approaches one needs another measured quantity which is dependent on the gluon shape at low x in a different way from $\partial F_2/\partial \ln Q^2$. The longitudinal structure function F_L is the obvious candidate. However, data on F_L are not of sufficient accuracy to be helpful.

Thorne has made a comparison of NLO, NNLO DGLAP and a fit including $\ln(1/x)$ resummation terms ('BFKL enhanced' splitting functions) in the context of the MRST global fit. He finds that all these approaches describe the low x data quite well. The fit χ^2's do point towards marginal improvement for the $\ln(1/x)$ resummation fit but the DGLAP fits are still acceptable. The gluon shape for the NNLO fit is still flattening and turning over at low Q^2, see Fig. 6.9. The $\ln(1/x)$ resummation fit does not predict parton densities directly, it predicts measurable quantities like F_2 and F_L.

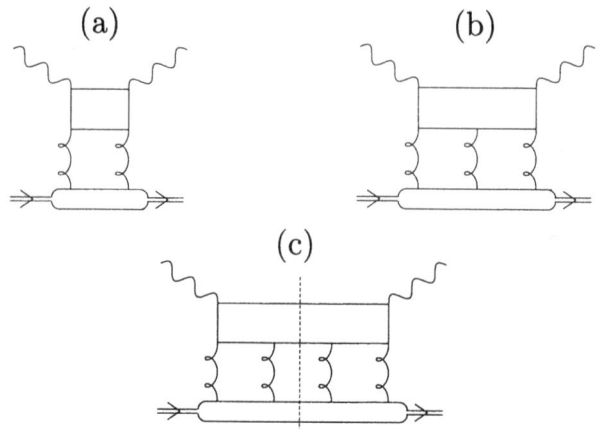

Fig. 9.28 The simplest QCD diagrams contributing to higher twist with (a) two, (b) three or (c) four gluons in the t-channel.

The predictions for $F_L(x, Q^2)$ as a function of x at four values of Q^2 are shown in Fig. 6.10. The curves show calculations from approaches using LO, NLO and NNLO DGLAP and the $\ln(1/x)$ resummation fit. At large Q^2 all approaches give the same rapidly increasing behaviour as x decreases, but for low Q^2 there are large differences in shape. The shape of F_L predicted by the resummation fit seems more 'reasonable' than that of the DGLAP fits. At the time of writing, some authors are investigating fits including non-linear terms, which also give marginal χ^2 improvements and predict more 'reasonable' gluon and F_L shapes. However, this does not constitute hard evidence, clearly, a precision measurement of F_L would transform our understanding of low x physics.

There is another reason why F_L is a very interesting quantity for distinguishing the appropriate theoretical approach at low x. This is made apparent by using the operator product twist expansion (i.e. expanding the operators in inverse powers of Q^2 beyond the leading DGLAP terms, see Section 4.4). Diagrams, shown in Fig. 9.28, with two, three and four gluons in the t-channel have been studied by Bartels and coworkers using the GBW model to provide parton distributions at the starting scale. They have studied contributions to F_2 and F_L up to twist-8, as functions of a variable $\xi = 1/Q^2 R_0^2(x)$, depending on the GBW function $R_0(x)$. It is found that the higher twist terms for both σ_T and σ_L are large around $\xi = 1$, but have opposite signs and hence tend to cancel out in their sum, which is proportional to F_2. Thus the fact that the leading twist DGLAP formalism works down to $Q^2 \sim 2\,\text{GeV}^2$, maybe an artefact of this cancellation. However, F_L is proportional to σ_L alone and for that there is no cancellation. Thus a measurement of F_L could indicate if there is a need to consider higher twist terms in the formalism.

9.10 Summary

Possible approaches to DIS at low x have been explored from 'classical' hadronic Regge theory, through the linear pQCD approaches of DGLAP, BFKL and CCFM to the non-linear regime of high density gluon dynamics. Although the last is hinted at in a tantalizing way by the low x data from HERA, it is not proven that it is strictly necessary. However, the rise of F_2 at low x is surely a necessary precursor for non-linearity and it has stimulated much new work in this area. Much of this focuses around dipole models which provide a flexible framework for studying the approach to saturation in DIS and related processes. Although good progress is being made, there is a need for precision DIS data at low x on F_L as well as F_2 and also for data from different sources, such as high energy hadron–hadron and heavy-ion colliders.

9.11 Problems

1. Double asymptotic scaling.
   (i) Making the change of variable $(u, v) \to (\rho, \sigma)$ defined by Eq. (9.5) show that
   $$4\frac{\partial^2}{\partial u \partial v} = \frac{\partial^2}{\partial \sigma^2} - \frac{\rho}{\sigma^2}\frac{\partial}{\partial \rho} + \frac{1}{\sigma}\frac{\partial}{\partial \sigma} - \frac{\rho^2}{\sigma^2}\frac{\partial^2}{\partial \rho^2},$$
   and hence justify Eq. (9.6).
   (ii) Expression for F_2. The method is a modification of the Prytz approximation explained in Section 4.2.2. By expanding $G(x/(1-z))$ about $z = 0$ and keeping only the leading term, derive the approximate equation
   $$t\frac{\partial F_2}{\partial t} \approx \frac{\bar{e}^2}{3\pi b_0} G(x, Q^2).$$
   Writing this in terms of variables (u, v) and substituting the DAS result $G(u, v) = A \exp(2\gamma\sqrt{uv})$, show that the resulting equation may be integrated to give Eq. (9.10).

2. Writing Eq. (9.10) for the DAS limit in terms of x and Q^2, find an expression for the corresponding effective slope at small x, $\lambda_{\text{eff}} = \partial \ln F_2 / \partial \ln(1/x)$.

3. The saddle point approximation to the integral
   $$I = \int_{-\infty}^{+\infty} dx\, g(x) e^{-f(x)},$$
   may be used when $g(x)$ is slowly varying around the minimum of the function $f(x)$. By expanding $f(x)$ around its minimum at x_0

$$f(x) = f(x_0) + \frac{1}{2}f''(x_0)(x-x_0)^2 + \ldots$$

show that

$$I \approx g(x_0)e^{-f(x_0)}\sqrt{\frac{2\pi}{|f''(x_0)|}}.$$

Using this result evaluate the integral Eq. (9.23) to give the LO BFKL result (9.24) for $F(s, \mathbf{k}_1, \mathbf{k}_2)$.

4. Justify the statement made at the start of Section 9.7 that in the proton rest frame the $q\bar{q}$ pair is produced a long time $\sim 1/(mx)$ before the interaction. Use light cone vectors introduced in Section 3.7.1:

$$q = (q^+, q^-, \mathbf{0}), \quad Q^2 = -2q^+q^-, \quad p = (m/\sqrt{2}, m/\sqrt{2}, \mathbf{0})$$

$$q_1 = \left(zq^+, \frac{k_T^2}{2zq^+}, \mathbf{k}_T\right) \quad q_2 = \left((1-z)q^+, \frac{k_T^2}{2(1-z)q^+}, -\mathbf{k}_T\right),$$

where m is the proton mass and q_1 and q_2 are the momenta of the q and $\bar{q}$, respectively. Consider the energy difference, $\Delta E = q_1^0 + q_2^0 - q^0$, between the $q\bar{q}$ pair and the γ^*. Show that

$$\Delta E \approx \frac{mx}{2}\left[1 + \frac{k_T^2}{z(1-z)Q^2}\right],$$

and hence that the required result follows.

5. This and following two questions refer to the general considerations on dipole models in Section 9.7. By expanding the exponential factor $e^{i\mathbf{k}\cdot\mathbf{r}}$ in Eq. (9.36) and performing the angular integral, show that at small transverse separation, $\hat{\sigma}(x, r^2) \sim r^2$.

6. Verify the results that for 'large' sized $q\bar{q}$ pairs, that the integrals are dominated only by highly asymmetric configurations and

$$\sigma_T(x, Q^2) \sim 1/Q^2 \quad \sigma_L(x, Q^2) \sim m_i^2/Q^4.$$

7. Verify the result for 'small' sized $q\bar{q}$ pairs, that

$$\sigma_T(x, Q^2) \sim 1/Q^2, \quad \sigma_L(x, Q^2) \sim 1/Q^2.$$

8. Verifying Eq. (9.44). Start from the definition of the dipole cross-section given by Eq. (9.36). Show that for small r^2 this may be approximated to give

$$\hat{\sigma}(x, r^2) \approx \frac{4\pi\alpha_s r^2}{3}\int \frac{d^2\mathbf{k}}{k^2}F_g(x, \mathbf{k}).$$

Then use the relationship between F_g and xg given by Eq. (9.27) to show that

$$\hat{\sigma}(x, r^2) \approx \frac{4\pi^2\alpha_s r^2}{3}xg(x, Q^2),$$

with $Q^2 \approx 1/r^2$.

[For Problems 6 and 7, either use appropriate approximation of the $\int dz$ integrals first, or by use the result that $\int d^2\mathbf{r}\sigma(r)K_i^2(\epsilon r) \propto \epsilon^{-4}$ on dimensional grounds. Further details are to be found in Forshaw and Ross (1997), Chapter 7.]

10
Hadron induced DIS

The description of hard scattering in hadron–hadron interactions using the QCD enhanced parton model is a large subject. This chapter concentrates on those areas which are important for the determination of parton densities, or which complement lepton induced DIS. Because of the very large momentum transfers involved, high E_T jet production in pp or p$\bar{\text{p}}$ interactions probes hadronic structure to very small distance scales. At the Tevatron with $\sqrt{s} = 1.8\,\text{TeV}$, one reaches a scale of around $10^{-4}\,\text{fm}$, comparable or even exceeding the scale reached in the highest Q^2 events at HERA. At the LHC, one will be able to probe a couple of orders of magnitude smaller. It is in this sense that one may speak of hadron induced DIS.

The chapter starts with the basic idea and essential assumptions. The crucial result is factorization, as this enables hard processes to be calculated using pQCD matrix elements and PDFs determined from other processes. Much of the initial discussion of detail is focused on Drell–Yan lepton-pair production. This then leads naturally to W and Z production at p$\bar{\text{p}}$ colliders. The core of the chapter concerns high E_T jet production in hadron–hadron collisions - a pure QCD process. High E_T isolated photon production is treated briefly and the chapter ends with a discussion of the importance of hadron induced DIS for LHC physics.

10.1 Rapidity

The majority of high energy hadron–hadron interactions, whether in fixed target or collider modes, produce final states with limited transverse momentum (with respect to the initial beam direction). In the parton model, the transverse momentum of the partons is also assumed small in comparison to the longitudinal momentum. In these circumstances it is very convenient to use *rapidity*[1], $y = \dfrac{1}{2}\ln\left(\dfrac{E+p_z}{E-p_z}\right)$, together with transverse momentum, p_T and an azimuthal angle, ϕ to describe a particle's momentum. If
$$p = (E, p_T\cos\phi, p_T\sin\phi, p_z)$$
then, in terms of y,

---
[1] Not to be confused with the y variable of conventional DIS.

$$p = (m_T \cosh y, p_T \cos \phi, p_T \sin \phi, m_T \sinh y), \tag{10.1}$$

where $m_T = \sqrt{m^2 + p_T^2}$ is the *transverse mass*. Rapidities are additive under Lorentz boosts along the beam direction and a rapidity difference is an invariant under the same transformation. At high energies, when masses can be ignored, the rapidity is often replaced by the pseudo-rapidity $\eta = -\ln \tan(\theta/2)$ (θ is the polar angle), as it is much easier to measure. Note that η has an infinite range $(-\infty, \infty)$ corresponding to $(-1 < \cos\theta < 1)$, whereas y always has a finite range, approximately $(\ln(m/2E), \ln(2E/m))$. The difference is usually only important in the regions close to the beam line where it is difficult, if not impossible, to measure particle momenta.

A useful relation following from the Jacobian of the transformation from Cartesian to rapidity momentum components is

$$\frac{d^3 p}{E} = p_T dp_T d\phi\, dy \equiv d^2 p_T dy \to \pi dp_T^2 dy \approx \pi dp_T^2 d\eta, \tag{10.2}$$

where for the last expressions the azimuthal angle has been integrated out.

10.2 The cross-section for a hard hadronic process

The application of the QCD improved parton model to hard processes in hadron–hadron scattering is summarised in Fig. 10.1. It is a natural extension of the approach used for lepton–induced DIS. The hadron–hadron cross-section is constructed from a convolution of the calculable parton-level cross-section for the hard process with the parton momentum densities for the incident hadrons. The new features are the presence of two parton densities in the convolution and parton fragmentation or vector boson decay in the final state, depending on the process considered (neither is shown in the figure).

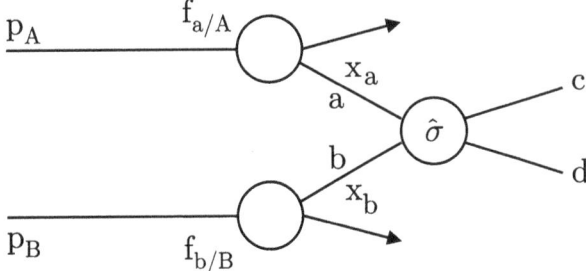

Fig. 10.1 The basic idea of the QCD improved parton model for a hard hadron–hadron interactions.

One may write

$$d\sigma_{\text{hard}}(p_A, p_B, Q^2) = \sum_{ab} \int dx_a dx_b f_{a/A}(x_a, \mu^2) f_{b/B}(x_b, \mu^2)$$

$$\times d\hat{\sigma}_{ab \to cd}(\alpha_s(\mu^2), Q^2/\mu^2)(1 + \delta_{\text{had}}) \quad (10.3)$$

where $d\hat{\sigma}_{ab \to cd}$ is the parton–parton cross-section at a hard scale Q^2 and $f_{a/A}$ is the parton momentum density of parton a in hadron A at a factorization scale μ^2. The initial parton momenta are $p_a = x_a p_A$, $p_b = x_b p_B$ and the factor $(1 + \delta_{\text{had}})$ represents hadronization corrections, which may be necessary if the final state of the hard process contains quarks or gluons fragmenting into jets. The hard scale will be provided by jet E_T or lepton-pair mass, for example. Strictly, the scale involved in the definition of α_s in the cross-section (the renormalization scale) could be different from the factorization scale, but it is usual to set the two to be equal and indeed the choice $\mu^2 = Q^2$ is often made as well.[2] As before in any calculation at finite order in α_s, it will be important to ensure that the parton densities have been extracted at the same order as that used for the calculation of $d\hat{\sigma}_{ab \to cd}$.

The formal proof of factorization is a difficult technical task[3] involving the detailed analysis of the short and long distance singularities of classes of Feynman diagrams in the limit in which all scalar products of the external momenta are at scales $\sim Q^2$ very much larger than the external hadronic masses involved. In this limit one may treat the partons as massless and the paradigm gauge theory QED provides useful analogies and insights. A few qualitative remarks may help towards the underlying physics. At the simplest level, one has the assumption of the parton model shown in the cartoon of Fig. 10.2(a), in which the two hadrons approaching each other at large equal and opposite momenta in the CM frame are Lorentz contracted. Time dilation ensures that during the instant of overlap, in which the partons a, b with momentum fractions x_a, x_b experience the hard scatter, the spectator partons in the two hadrons are 'frozen'.

The crucial difference between this process and that of leptonic DIS is the possibility that soft interactions between the colour fields of one hadron with the partons of the other may modify the calculation of the hard scattering cross-section $\hat{\sigma}_{ab \to cd}$, either before or after the hard interaction. In the equivalent problem in classical electrodynamics a relativistic charged particle passes close to a stationary charge and it is well known (see for example, Jackson 1975 p552) that the electric field of the moving particle is Lorentz contracted from spherical to disk with axis along the direction of motion. This means that the stationary test particle only feels a sizeable field around the instant of closest approach. The fact that partons are confined into colourless hadrons in QCD on the scale of 1 fm implies that the colour fields will decrease even more rapidly at large distances. In QED gauge invariance is crucial in ensuring the cancellation of many apparent singularities that arise from collinear or infrared divergences. For example, consider a hard scattering between two electrons, as shown in Fig. 10.2(b),

---

[2] The choice $\mu^2 \approx Q^2$ avoids the appearance of large $\ln(\mu^2/Q^2)$ terms.
[3] For an accessible summary see Collins *et al.* 1989.

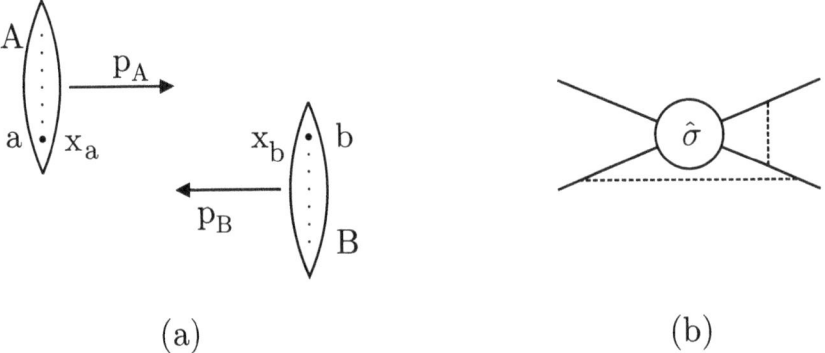

Fig. 10.2 (a) A cartoon of the Lorentz contracted hadrons in the CM frame at high energies showing the two partons with momentum fractions x_a, x_b that will be involved in the hard collision.(b) A two-body hard scattering diagram with soft photon or gluon insertions soft as the dashed lines.

with two soft photon insertions. The QED Bloch–Nordsieck theorem ensures that when all diagrams required by gauge invariance are included the hard cross-section is not modified by the soft insertions. In the similar QCD diagram for the hard scattering between two coloured partons, the crucial insertions for factorization are those connecting the initial and final states. Although QCD is a gauge theory, it has a more complicated non-Abelian structure, and this means that not all soft cancellations can be guaranteed. In hadron–hadron scattering factorization has been proved at leading twist (i.e. for the leading terms in the expansion of the cross-section in inverse powers of Q^2), but it has also been shown that factorization is not exact beyond the single loop for higher twist terms. Explicit calculations for Drell–Yan production at two loops show factorization breaking of the order m^4/Q^4, where m is the scale of the external hadron masses.[4] For the rest of this chapter factorization will be assumed for leading twist QCD improved parton model calculations.

10.3 The Drell–Yan process

The Drell–Yan (D–Y) process (Drell and Yan 1970) of lepton pair production ($q\bar{q} \to \gamma^* \to \ell^+\ell^-$) was the first application of the parton model to a hard hadronic interaction. As it is the simplest such process and is also the basis for the calculation of W and Z boson production in hadron–hadron interactions, it will be discussed in some detail. The leading order diagram is shown in Fig. 10.3. The hard scale is set by the invariant mass of the lepton pair.

[4]See Collins et al. 1989 for references.

282 Hadron induced DIS

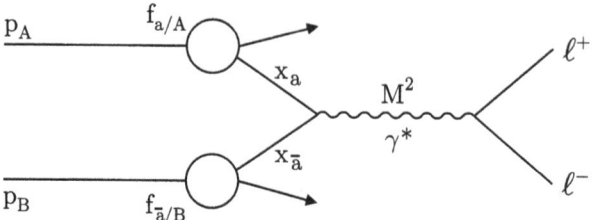

Fig. 10.3 Leading order (parton model) diagram for Drell–Yan lepton pair pair production in hadron–hadron scattering.

Although the D–Y cross-section is a tiny fraction of the total hadron–hadron cross-section, experimentally the presence of an energetic e^+e^- or $\mu^+\mu^-$ pair is a clean signature for the process. The early success of the Drell–Yan model gave an important boost to the parton model generally and it allowed direct confirmation of the assumed limited parton transverse momentum. Because the D–Y process may be initiated by any pair of hadrons, it opens the possibility of studying parton momentum densities in hadrons other than the proton, e.g. the pion.

10.3.1 Kinematics and LO cross-section

Experimentally, for a fixed hadronic CM energy $\sqrt{s}$, three quantities may be reconstructed for the D–Y pair: the invariant mass (M), the rapidity (y) and the transverse momentum (p_T with respect to the incident hadron beam line). Ignoring p_T for the moment and referring to Fig. 10.3 the 4-momenta of the incident partons in the hadron–hadron CM frame are

$$p_a = \frac{\sqrt{s}}{2}(x_a, 0, 0, x_a), \qquad p_b = \frac{\sqrt{s}}{2}(x_b, 0, 0, -x_b),$$

where $s = (p_A + p_B)^2$ and parton masses have been ignored. Then

$$M^2 = (p_a + p_b)^2 = x_a x_b s, \qquad y = \frac{1}{2}\ln\left(\frac{E+p_z}{E-p_z}\right) = \frac{1}{2}\ln(x_a/x_b),$$

where E, p_z are the energy and longitudinal momentum of the lepton pair.

For D–Y, it is conventional to define $\tau = x_a x_b$ and as M is also the CM energy, $\sqrt{\hat{s}}$, of the $q\bar{q} \to \ell^+\ell^-$ sub-process $\hat{s} = \tau s$ and $\tau = M^2/s$. The parton momentum fractions, x_a, x_b, may also be written in terms of τ and y as:

$$x_a = \sqrt{\tau}\,e^y, \qquad x_b = \sqrt{\tau}\,e^{-y}. \tag{10.4}$$

An alternative to the variable y is the normalized longitudinal momentum of the lepton pair — Feynman x or x_F.

$$x_F = \frac{2}{\sqrt{s}}(p_{\ell+} + p_{\ell-}) \approx x_a - x_b \qquad \text{ignoring } p_T,$$

where an obvious notation has been used for the lepton momenta. Sometimes, the definition $x_F = x_a - x_b$ is used. In terms of τ and x_F one has

$$x_a = \frac{\sqrt{x_F^2 + 4\tau} + x_F}{2}, \qquad x_b = \frac{\sqrt{x_F^2 + 4\tau} - x_F}{2}. \qquad (10.5)$$

The cross-section for the sub-process is the same as that for $e^+e^- \to q\bar{q}$ via a virtual photon

$$\hat{\sigma}(q_a\bar{q}_a \to \ell^+\ell^-) = \frac{Q_a^2}{3}\sigma_0, \quad \text{where} \quad \sigma_0 = \frac{4\pi\alpha^2}{3\hat{s}} \qquad (10.6)$$

where Q_a is the electric charge of quark q_a and the angular dependence of the final state lepton pair has been integrated out. Note that both the quark flavour and colour have to match for the annihilation. This gives a factor $1/9$ which is multiplied by 3 when summed over all colours, giving the overall colour factor of $1/3$. Using the above expression in Eq. (10.3) gives the parton model expression for D–Y pair production

$$\frac{d\sigma_{\text{DY}}}{dM^2} = \frac{\sigma_0}{3}\int_0^1 dx_a dx_b \delta(x_a x_b s - M^2) \qquad (10.7)$$
$$\times \sum_a Q_a^2 \left[f_{a/A}(x_a)f_{\bar{a}/B}(x_b) + f_{\bar{a}/A}(x_a)f_{a/B}(x_b) \right],$$

where the sum is only over quark flavours as the $\bar{q}$ contributions are shown explicitly. Using the definitions of τ and σ_0 above the last equation may be rewritten as

$$M^4 \frac{d\sigma_{\text{DY}}}{dM^2} = \frac{4\pi\alpha^2}{9}\tau F_0(\tau), \qquad (10.8)$$

where

$$F_0(\tau) = \int_0^1 dx_a dx_b \delta(x_a x_b - \tau) \qquad (10.9)$$
$$\times \sum_a Q_a^2 \left[f_{a/A}(x_a)f_{\bar{a}/B}(x_b) + f_{\bar{a}/A}(x_a)f_{a/B}(x_b) \right].$$

At the parton level, the PDFs $f_{a/A}(x_a)$ do not depend on M^2 and the above equations show that $M^4 \frac{d\sigma_{\text{DY}}}{dM^2}$ is a function of τ only — it 'scales' in the same sense that $F_2(x, Q^2)$ in DIS scales at the parton level. The data do show approximate scaling and this was an important result for the model. The D–Y cross-section is also predicted to fall as M^{-4} at fixed τ.

Starting from Eq. (10.7) and using Eq. (10.4) for x_a, x_b one finds

$$s\frac{d^2\sigma_{\text{DY}}}{d\tau\, dy} = \frac{4\pi\alpha^2}{9\tau}\sum_a Q_a^2 \left[f_{a/A}(x_a)f_{\bar{a}/B}(x_b) + f_{\bar{a}/A}(x_a)f_{a/B}(x_b) \right] \quad (10.10)$$

for the double differential cross-section. At $y = 0$, the right-hand side of the above equation is also a function of τ only. Data is often presented with respect to M^2 and x_F rather than τ and y

$$M^4 \frac{d^2\sigma_{\text{DY}}}{dM^2 \, dx_F} = \frac{4\pi\alpha^2}{9} \frac{\tau}{(x_a + x_b)} \times \sum_a Q_a^2 \left[f_{a/A}(x_a) f_{\bar{a}/B}(x_b) + f_{\bar{a}/A}(x_a) f_{a/B}(x_b) \right], \qquad (10.11)$$

where x_a, x_b are given by Eq. (10.5) and the extra factor of $(x_a + x_b)$ comes from the Jacobian $dx_a dx_b \to dM^2 dx_F$. For small x_F, one has $x_a \approx x_b \approx \sqrt{\tau}$. Figure 10.4 shows the approximate scaling of $M^4 \frac{d^2\sigma_{\text{DY}}}{dM^2 dx_F}$ (actually $M^3 \frac{d^2\sigma_{\text{DY}}}{dM \, dx_F}$) from fixed target experiments on $pA \to \mu^+\mu^- X$ at 400 and 800 GeV incident proton momenta. At a fixed value of x_F, the cross-section is a slowly varying function of τ and data from different CM energies lie on the same curve. Note that the apparently large scaling violation of the E439 data in the range $0.3 < \sqrt{\tau} < 0.4$ is caused by the production of the Upsilon family of narrow $b\bar{b}$ resonances.

By measuring the double differential cross-section over a wide range of values of M^2 and y (or x_F), one has a direct constraint on the PDFs of the parent hadrons A and B. For example, the ratio of the D-Y rates for pp and pd scattering at high energy is sensitive to the $\bar{d}/\bar{u}$ ratio in the valence region. The E866 fixed target experiment (E866 1998) measured the pd/pp ratio with 800 GeV incident protons. Since $x_F > 0$, the PDF factor will be dominated by the first terms which are products of beam quark with target antiquark momentum distribution functions so that, using Eq. (10.11),

$$\frac{\sigma_{\text{DY}}(pd)}{2\sigma_{\text{DY}}(pp)} \approx \frac{[4u(x_a) + d(x_a)][\bar{u}(x_b) + \bar{d}(x_b)]}{2[4u(x_a)\bar{u}(x_b) + d(x_a)\bar{d}(x_b)]}, \qquad (10.12)$$

where strong isospin symmetry has been used to write the deuteron PDFs in terms of those for the proton. This ratio may be written as

$$\frac{\sigma_{\text{DY}}(pd)}{2\sigma_{\text{DY}}(pp)} \approx \frac{\left(1 + \frac{1}{4}\frac{d_A}{u_A}\right)\left(1 + \frac{\bar{d}_B}{\bar{u}_B}\right)}{2\left(1 + \frac{1}{4}\frac{d_A}{u_A}\frac{\bar{d}_B}{\bar{u}_B}\right)} \approx \frac{1}{2}\left(1 + \frac{\bar{d}_B}{\bar{u}_B}\right), \qquad (10.13)$$

(using an obvious change of notation) where the last approximation follows from the fact that $d < u$ in the valence region and $\bar{d} \approx \bar{u}$. If $\bar{d} = \bar{u}$, as was often assumed for the $q\bar{q}$ nucleon sea, the measured ratio would be approximately independent of x_b. As Fig. 10.5 shows, this is not the case and the experiment has provided important data for refining global PDF determination (as discussed in §6.5.1).

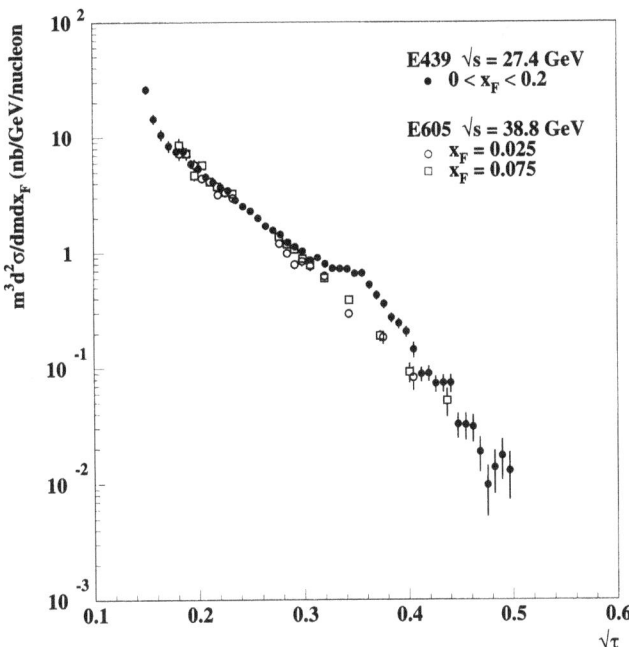

Fig. 10.4 Approximate scaling of the D–Y $pA \to \mu^+\mu^- X$ double differential cross-section at fixed x_F. Data are from E439 at 400 GeV (Smith *et al.* 1981) and E605 at 800 GeV (E605 1991).

10.3.2 Transverse momentum and QCD corrections

There are two sources of transverse momentum corrections for the D–Y process. Firstly, the intrinsic transverse momentum of the q and q̄ in the initial state hadrons. In the parton model, this is assumed to be small in comparison with both the longitudinal momentum scale and the dilepton mass scale. Secondly, the potentially larger transverse momenta introduced by QCD corrections such as gluon emission. The effect of the first source cannot be calculated perturbatively, but it can be measured and the parton model assumption of limited p_T is verified. Simple models for the intrinsic parton p_T, for example, a Gaussian distribution, may be used to describe this part of the dilepton p_T spectrum. Measurements from fixed–target hadron–hadron D–Y production give a mean value, $\langle k_T \rangle \sim 700 \,\text{MeV}$. At larger p_T there is a non-gaussian tail and this is taken to be evidence for the QCD corrections. The other evidence of the need for corrections is that at fixed target energies in particular, although the parton model explains the approximate scaling seen in the data, it predicts a cross-section that is

286 *Hadron induced DIS*

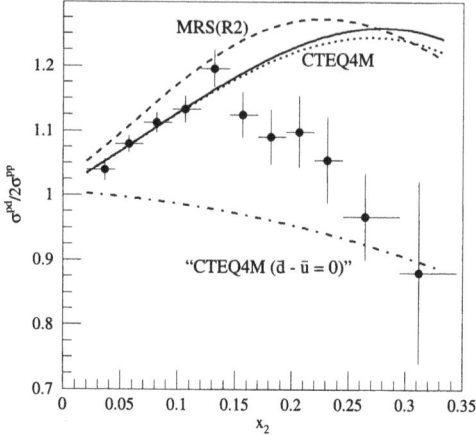

Fig. 10.5 Ratio of Drell–Yan rates $\sigma_{DY}(pd)/(2\sigma_{DY}(pp))$ as measured by the E866 experiment (E866 1998). The curves show a number of predictions calculated using PDFs available prior to this experiment.

too low by as much as 50%.

The QCD diagrams for order α_s corrections to the D-Y process are shown in Fig. 10.6. There are three types: the two diagrams on the left are

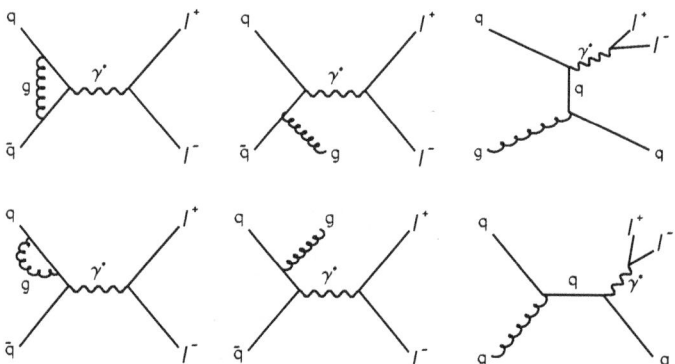

Fig. 10.6 Order α_s QCD corrections to Drell–Yan.

virtual gluon vertex and loop insertions; the next two come from $q\bar{q} \to \gamma^* g$ (inverse of photon–gluon fusion) and finally the two on the right are from $qg \to \gamma^* q$ (inverse QCD Compton scattering). It is no coincidence that the same diagrams occur in DIS (as can be seen by referring back to Figs 4.2 and 4.3 of Chapter 4.) The net effect of the QCD corrections is similar to that explained in detail in Chapter 4. The various divergences that arise from the loop integrals, collinear parton configurations and infrared parton emission have to be regularized by the introduction of 'renormalized' parton densities at a renormalization scale μ^2. This introduces $\ln(M^2/\mu^2)$ scaling violations into the PDFs and explicit correction terms to the parton model cross-sections. Because the diagrams are intrinsically the same as in DIS, the renormalized parton densities are those governed by the DGLAP equations. At order α_s and in the $\overline{\text{MS}}$ scheme with scale $\mu^2 = M^2$ the D–Y cross-section of Eq. (10.8) becomes

$$M^4 \frac{d\sigma_{\text{DY}}}{dM^2} = \frac{4\pi\alpha^2}{9}\tau \left[F_1(\tau, M^2) + G_1(\tau, M^2)\right], \quad (10.14)$$

where

$$F_1(\tau, M^2) = \int_0^1 dx_a dx_b dz \delta(x_a x_b z - \tau)\left(\delta(1-z) + \frac{\alpha_s(M^2)}{2\pi}D_q(z)\right)$$
$$\times \sum_a Q_a^2 \left[f_{a/A}(x_a, M^2)f_{\bar{a}/B}(x_b, M^2) + (a \leftrightarrow \bar{a})\right]$$

$$G_1(\tau, M^2) = \int_0^1 dx_a dx_b dz \delta(x_a x_b z - \tau)\frac{\alpha_s(M^2)}{2\pi}D_g(z)$$
$$\times \sum_a Q_a^2 \left[f_{g/A}(x_a, M^2)\left(f_{a/B}(x_b, M^2) + f_{\bar{a}/B}(x_b, M^2)\right) + (a, \bar{a} \leftrightarrow g)\right]$$

and D_q and D_g are the D–Y coefficient functions

$$D_q(z) = C_F \left[4(1+z^2)\left(\frac{\ln(1-z)}{1-z}\right)_+ - 2\frac{1+z^2}{1-z}\ln z + \frac{2\pi^2 - 24}{3}\delta(1-z)\right]$$

$$D_g(z) = T_R \left[(z^2 + (1-z)^2)\ln\frac{(1-z)^2}{z} + \frac{1}{2} + 3z - \frac{7}{2}z^2\right].$$

The first term, F_1, in the [ ] on the RHS of Eq. (10.14) is the modified $q\bar{q}$ term corresponding to the LO F_0 of Eq. (10.8). The second term, G_1, is a new contribution which arises from the two diagrams on the right of Fig. 10.6 and depends explicitly on the gluon density. At fixed-target energies, the first term dominates and gives rise to a large positive correction. At the higher energies of collider experiments, corresponding to smaller values of τ, the qg terms (which are negative) are larger and the overall correction to the parton model result is smaller. With the order α_s corrections the calculation of the D–Y cross-section is in good agreement with the data in both shape and normalization. $O(\alpha_s^2)$ corrections have also been calculated (Hamberg et al. 1991) and are generally smaller than the α_s corrections.

The quality of the NLO description of D-Y pair production is shown in Fig. 10.7, in which data from the E866/NuSea Collaboration (E866 2003) for $M^3 \frac{d^2\sigma}{dM\,dx_F}$ at fixed x_F values between 0.05 and 0.8 for $pp \to \mu^+\mu^- X$

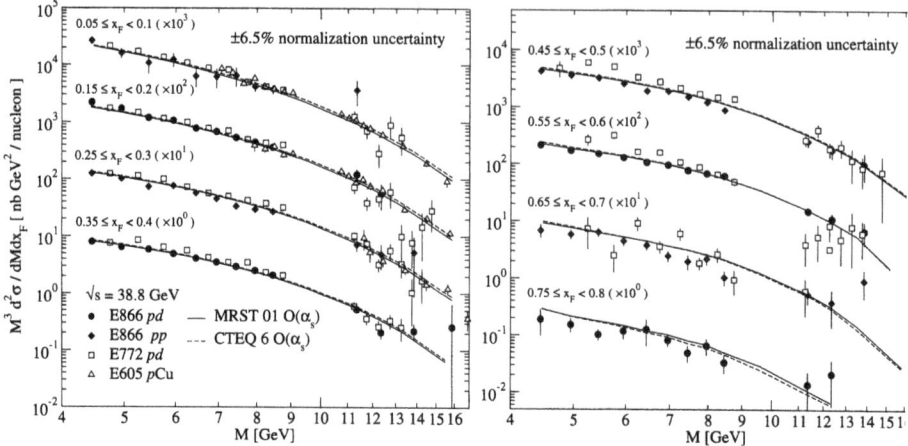

Fig. 10.7 Comparison of the invariant cross-section $M^3 \dfrac{d^2\sigma}{dM\, dx_F}$ at fixed x_F values for D–Y mu-pair production in pp and pd fixed–target scattering at proton beam momenta of 800 GeV with $O(\alpha_s)$ calculations using the CTEQ6 and MRST01 PDFs. (From E866 2003.)

and $pd \to \mu^+\mu^- X$ fixed–target scattering at a proton beam momentum of 800 GeV are compared with NLO calculations using the CTEQ6 and MRST01 PDFs. Also shown are some earlier data on various targets at the same beam momentum. The E866/NuSea experiment was optimized for high precision measurement of the absolute D–Y cross-section at $x_F > 0$. Overall there is very good agreement between data and calculation, and this is reflected in values of $K' = \sigma^{exp}/\sigma^{NLO}$ factors obtained using the statistical and point-to-point systematic uncertainties of the data, but not the overall normalization uncertainty of 6.5%. For pp, $K' = 1.016$, 0.980 for CTEQ6 and MRST01 with $\chi^2/\text{dof} = 1.39$, 1.45, respectively. For the pd, the corresponding figures are $K' = 1.001$, 0.966 with $\chi^2/\text{dof} = 2.56$ and 2.44. Both the CTEQ6 amd MRST01 PDFs take into account the E866 ratio data shown in Fig. 10.5. The somewhat larger χ^2 values for the pd K' fits may be traced to discrepancies in the description of the data at large $x_a > 0.6$ (proton beam valence distributions) which are present in both pp and pd data, but larger in pd. Unlike large-x DIS off deuterium targets, these data are in a kinematic region where nuclear corrections are known to be small. Thus these D–Y data imply that the u- and d-quark valence distributions may be too large as $x \to 1$. Although the proton valence–quark distributions are generally well determined, there is lack of high precision data at large x. This emphasizes the importance of accurate D–Y data and the potential of high luminosity HERA-II NC and CC cross-section measurements to further constrain PDFs in this region — which is important for many searches for evidence of effects beyond the Standard

Model.

10.4 W & Z boson production

The Drell–Yan process was one of the discovery channels for the J/ψ in 1974 and the discovery channel for the Υ family of heavy $Q\bar{Q}$ vector mesons in 1977. Provided the detector components have sufficient resolution, the D–Y process enables a very wide range of dilepton invariant mass to be covered in a single experiment. At high enough energies, the W and Z gauge bosons can be produced and once more the dilepton mode was crucial for their discovery in $p\bar{p}$ collisions. To $O(\alpha_s)$ the hadronic part of the lepton-pair production cross-section for the W and Z is the same as that for a virtual photon, since the gluon coupling is flavour blind. At $O(\alpha_s^2)$ there is a numerically small difference for Z production.

10.4.1 Z boson production

The situation here is similar to that in high energy e^+e^- annihilation — diagrams for $q\bar{q} \to Z \to \ell^+\ell^-$ must be added to those for $q\bar{q} \to \gamma^* \to \ell^+\ell^-$. Including the angular dependence of the final state lepton pair, the sub-process cross-section becomes

$$\frac{d\hat{\sigma}}{d\cos\theta^*}(q_i\bar{q}_i \to \ell^+\ell^-) = \frac{\pi\alpha^2}{6\hat{s}}\left[(1+\cos^2\theta^*)A + \cos\theta^* B\right] \quad (10.15)$$

where
$$A = e_i^2 - 2e_i v_i v_\ell \chi_1(\hat{s}) + (a_i^2 + v_i^2)(a_\ell^2 + v_\ell^2)\chi_2(\hat{s}),$$
$$B = -4e_i a_\ell a_i \chi_1(\hat{s}) + 8a_\ell v_\ell a_i v_i \chi_2(\hat{s}),$$

θ^* is the angle of ℓ^+, in the Z^0 rest frame, with respect to the incident proton beam direction (in $p\bar{p}$ interactions), and[5]

$$\chi_1(\hat{s}) = \kappa \frac{\hat{s}(\hat{s}-M_Z^2)}{(\hat{s}-M_Z^2)^2 + \Gamma_Z^2 M_Z^2}, \quad \chi_2(\hat{s}) = \kappa^2 \frac{\hat{s}^2}{(\hat{s}-M_Z^2)^2 + \Gamma_Z^2 M_Z^2}.$$

e_i is the electric charge and v_i, a_i are the electroweak vector and axial couplings of quark q_i (defined in Section 8.7, see also Appendix D) and

$$\kappa = \frac{\sqrt{2}G_F M_Z^2}{4\pi\alpha}.$$

As in e^+e^- annihilation the angular distribution of the final state leptons contains sensitive information on the vector boson couplings through the vector-axial interference term, given by B in Eq. (10.15) above. A direct

---

[5]The functions χ_1, χ_2 are the equivalent of P_Z, P_Z^2 in Eq. (8.3) (Section 8.1.1), making due allowance for the time-like nature of the Z^0 exchange here and using $\sin^2 2\theta_W = 1/\kappa$.

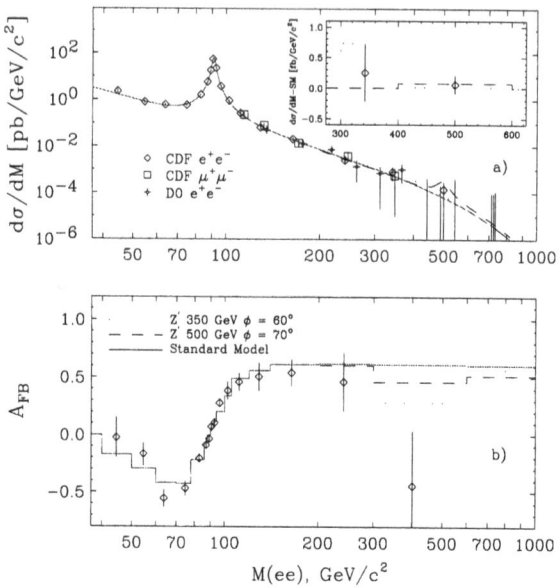

Fig. 10.8 Drell–Yan lepton pair production around the mass of the Z^0 boson from the CDF experiment (CDF 2001b). (a) $d\sigma/dM$ together with an $O(\alpha_s^2)$ using MRST99 PDFs (full curve); (b) $A_{\rm FB}$ together with an $O(\alpha_s)$ standard model calculation.

measure of the interference term is given by the lepton forward-backward asymmetry

$$A_{\rm FB} = \frac{\sigma(\theta^* > \pi/2) - \sigma(\theta^* < \pi/2)}{\sigma(\theta^* > \pi/2) + \sigma(\theta^* < \pi/2)} = \frac{3B}{8A},$$

in terms of the coefficients defined in Eq. (10.15. Most detectors cannot measure over the full solid angle, so $A_{\rm FB}$ must either be corrected for acceptance or the theory calculated over the measured angular range.

Figure 10.8 shows the cross-section $d\sigma/dM$ and $A_{\rm FB}$ for $p\bar{p} \to \ell^+\ell^- X$ at $\sqrt{s} = 1.8\,{\rm TeV}$ with $M > 40\,{\rm GeV}$. The data are from the CDF experiment (CDF 2001b) and correspond to 108 pb^{-1}. Electrons are measured in the range $|\eta| < 4.2$ and muons in $|\eta| < 1.1$. The curve shown in plot (a) is an $O(\alpha_s^2)$ calculation using the MRST99 PDFs normalized to the Z peak cross-section (factor 1.11). For $A_{\rm FB}$, plot (b), the data are compared with an $O(\alpha_s)$ calculation. The calculations describe the shapes of both distributions around the position of the Z peak very well. The dashed curves show the sensitivity to a model with an additional Z' boson and indicate the continuing potential of D–Y for discovering new physics.

For studies of the properties of the Z^0 itself, one integrates the data over the region of the resonance ($|M - M_Z| < \Delta$ where $\Gamma_Z \ll \Delta \ll M_Z$). The expression for the cross-section also simplifies as only the term involving

χ_2 is important. To calculate the integrated cross-section it is enough to use the 'narrow resonance approximation'

$$\frac{1}{(\hat{s}-M_Z^2)^2+\Gamma_Z^2 M_Z^2} \to \frac{\pi}{M_Z \Gamma_Z}\delta(\hat{s}-M_Z^2),$$

where the normalization is given by the full area under the Breit–Wigner line shape. Using this gives

$$\hat{\sigma}(q_i\bar{q}_i \to Z \to \ell^+\ell^-) = \frac{4\pi^2\alpha^2}{9}\frac{M_Z}{\Gamma_Z}\kappa^2(a_i^2+v_i^2)(a_\ell^2+v_\ell^2)\delta(\hat{s}-M_Z^2). \quad (10.16)$$

Using the standard model expression,

$$B(Z \to \ell^+\ell^-) = \frac{G_F M_Z^3}{6\sqrt{2}\pi\Gamma_Z}(v_\ell^2+a_\ell^2),$$

one may write Eq. (10.16) as

$$\hat{\sigma}(q_i\bar{q}_i \to Z \to \ell^+\ell^-) = \hat{\sigma}(q\bar{q} \to Z)B(Z \to \ell^+\ell^-),$$

where

$$\hat{\sigma}(q\bar{q} \to Z) = \frac{\pi}{3}\sqrt{2}G_F M_Z^2(a_i^2+v_i^2)\delta(\hat{s}-M_Z^2). \quad (10.17)$$

10.4.2 W boson production

Since the width of the W is also small in comparison with its mass, one may use the narrow resonance approximation to calculate the sub-process production cross-section. The form is that of Eq. (10.17) for the Z, but with the electroweak coupling factor replaced by the appropriate CKM matrix element $|V_{ij}|$

$$\hat{\sigma}(q_i\bar{q}_j \to W) = \frac{\pi}{3}\sqrt{2}G_F M_W^2 |V_{ij}|^2 \delta(\hat{s}-M_W^2) \quad (i \neq j). \quad (10.18)$$

To calculate the rate for a leptonic final state, one also needs

$$B(W \to \ell\nu_\ell) = \frac{G_F M_W^3}{6\sqrt{2}\pi\Gamma_W}.$$

Until the advent of high energy running at LEP, above the $e^+e^- \to W^+W^-$ threshold, precise measurement of the mass and width of the W was the preserve of $p\bar{p}$ collider experiments via the D–Y process. Because of much higher rates at hadron machines such measurements will remain competitive.

Note that many experimental and theoretical uncertainties can be reduced by measuring the ratio of lepton pair production at the W and Z

$$R_{\text{DY}} = \frac{N(W \to \ell\nu_\ell)}{N(Z \to \ell^+\ell^-)} = \frac{\sigma_W}{\sigma_Z}\frac{B(W \to \ell\nu_\ell)}{B(Z \to \ell^+\ell^-)}$$

$$= \frac{\sigma_W \Gamma_Z}{\sigma_Z \Gamma_W} \frac{\Gamma(W \to \ell \nu_\ell)}{\Gamma(Z \to \ell^+ \ell^-)}. \tag{10.19}$$

These questions will not be followed any further here, but the possible use of W or Z production as a 'luminosity monitor' for next generation hadron colliders will be considered in Section 10.7.2.

10.4.3 W decay asymmetry

The technique for calculating the rapidity distributions for W and Z production in hadron–hadron collisions follows that for D–Y production. For both the photon and Z, the rapidity distribution must be symmetric about $y = 0$ (this follows from CP invariance since they are both neutral particles). The same does not hold for $W^\pm$ and, as the charge of the W may be determined from its leptonic decay, one has a handle on the hadronic parton densities at very large mass scales. Define the positive z axis, and hence $y > 0$, to be along the direction of the incoming proton in a $p\bar{p}$ collision. Then W^+ will tend to be produced forward, $u\bar{d} \to W^+$, as u quarks carry more momentum than d quarks at a given x value. Conversely $\bar{u}d \to W^-$ will tend to produce W^- at $y < 0$. What is measured is the angular asymmetry of the decay $\ell^\pm$ (where ℓ is μ or e). Consider the production and decay of the W in its rest frame. Figure 10.9(a) shows W^+ production and

Fig. 10.9 Production and decay of W in its rest frame: (a) W^+; (b) W^-.

decay, with the polar axis defined to be along the incoming proton beam direction and θ^* the decay angle of ℓ^+ in the W rest frame. Neglecting parton transverse momenta the u and $\bar{d}$ are aligned along the z-axis and neglecting all masses, the alignment of the initial and final spins are shown as the double arrows. The total spin is $J = 1$, with $J_z = -1$ in the initial state and $J_z = +1$ in the final state. The angular dependence of the matrix element is given by the rotation matrix $d^1_{1,-1}(\theta^*) = (1 - \cos\theta^*)/2$. This shows that ℓ^+ is produced preferentially backwards. Similarly, diagram (b) shows the situation for W^-, now both initial and final state have $J_z = -1$ and the angular dependence of the matrix element is given by $d^1_{-1,-1}(\theta^*) = (1 + \cos\theta^*)/2$. For W^- decay, ℓ^- is produced preferentially forwards. h The lepton decay asymmetry will be somewhat smaller than the W production asymmetry but it is nonetheless a useful quantity to measure. A typical example from the CDF experiment (CDF 1998) is shown in Fig. 10.10, together with predictions using a number of PDFs.

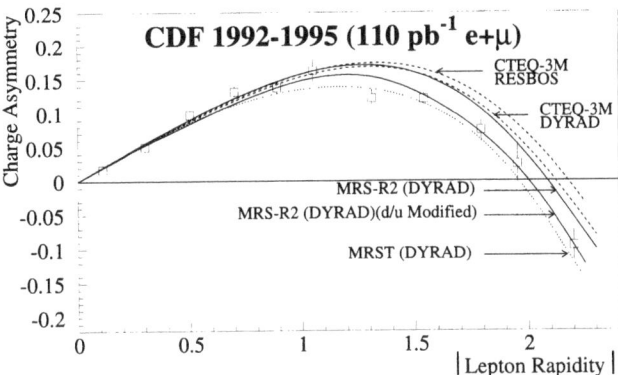

Fig. 10.10 W decay asymmetry as measured by CDF in p$\bar{\text{p}}$ at $\sqrt{s} = 1.8\,\text{TeV}$. (From CDF 1998.)

In the W rest frame, $E_\ell = |p_\ell| = M_W/2$, so

$$\cos^2\theta^* = 1 - \frac{4p_T^2}{M_W^2},$$

where p_T refers to the decay $\ell^\pm$ and is invariant under boosts along the z-axis. Averaging the W decay angular distribution over both charge states gives

$$\frac{1}{\sigma}\frac{d\sigma}{d\cos\theta^*} \propto (1 + \cos^2\theta^*).$$

Changing variables from $\cos\theta^*$ to p_T^2 gives

$$\frac{1}{\sigma}\frac{d\sigma}{dp_T^2} \propto \frac{1}{M_W^2}\left(1 - \frac{4p_T^2}{M_W^2}\right)^{-1/2}\left(1 - \frac{2p_T^2}{M_W^2}\right).$$

When smeared over the finite width of the W, the cross-section shows a peak at $p_T \sim M_W/2$ — the Jacobian peak. This is one of the methods of measuring the W mass. The p_T distribution is also smeared by W transverse motion. Another method, which is less sensitive to this last effect, is based on the transverse mass

$$M_T^2 = |p_{T\ell}||p_{T\nu}|(1 - \cos\Delta\phi),$$

where $p_{T\nu}$ is identified with the missing transverse momentum in the event and $\Delta\phi$ is the difference in azimuthal angles between the charged lepton track and the missing momentum vector. Since M_T depends on $p_{T\ell}$, it will also exhibit a Jacobian peak and thus sensitivity to M_W. Measurement

294 *Hadron induced DIS*

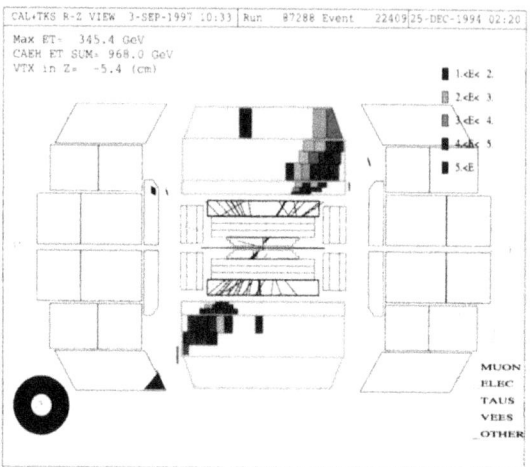

Fig. 10.11 A high E_T dijet event from p$\bar{\text{p}}$ interactions at $\sqrt{s} = 1.8\,\text{TeV}$ as measured by the DØ experiment.

from Tevatron Run-I data achieved values of M_W with an error of less than 230 MeV. For the most precise value of M_W, quoted in the Particle Data Group tables, data from p$\bar{\text{p}}$ at the Tevatron is combined with data from LEPII on $e^+e^- \to W^+W^-$.

As in classic D–Y, the W and Z bosons produced at large transverse momenta correspond to QCD processes beyond the LO parton model, specifically q$\bar{\text{q}} \to Vg$ and $qg \to Vq$. There is a difference, however, in that there are now two scales involved. For pQCD to give valid results, $p_T \ll \langle k_T \rangle$ (the typical intrinsic parton transverse momentum of $\sim 700\,\text{MeV}$), but the vector boson mass M_V also sets a scale giving rise to terms involving $\ln(M_V^2/p_T^2)$. For $\langle k_T \rangle \ll p_T \ll M_V$, it may be necessary to sum multiple 'soft' gluon emission. For more detail on this and other aspects of M_W measurement see Ellis et al. 1996, Chapter 9.

10.5 High p_T jet production

So far, the hard parton processes that have been discussed involve electroweak couplings at the leading, parton model, order and QCD effects only enter as $O(\alpha_s)$ corrections. High p_T jet production in hadron–hadron scattering is a manifestation of an underlying hard two-parton interaction involving gluon or quark exchange at the leading order, giving cross-sections of $O(\alpha_s^2)$. More details on the leading order diagrams and matrix elements are given in the appendix to this chapter. At the high energies of p$\bar{\text{p}}$ colliders, particularly the Tevatron, these processes give rise to spectacular events with two back-to-back high-p_T jets standing out from the remaining hadronic remnants nearer the beam line — an example is shown in Fig. 10.11. Higher order processes give rise to events with more than two jets.

Most of the published jet data from hadron–hadron colliders have been produced using a cone jet definition[6] in η, ϕ space with cone radius, $R = \sqrt{(\Delta\eta)^2 + (\Delta\phi)^2}$, typically of about 1 unit. However, it has been argued that cone algorithms are not 'infrared' safe when measuring multi-jet events. For the Tevatron Run-II a k_T algorithm is being used and some Run-I data has been re-analysed using this method. Such concerns do not affect the present discussion which concentrates on the lowest order two-jet events. For a proper comparison of theory with data, the efficiency of the jet finder must be taken into account, but 100% efficiency will be assumed here. At Tevatron energies and for simple jet topologies, this is a reasonable approximation to the truth.

10.5.1 Parton–parton kinematics

Figure 10.12(a) shows the parton–parton hard sub-process $ab \to cd$, where the initial partons a, b are contained in the colliding hadrons A, B. The Mandelstam variables for the sub-process are

$$\hat{s} = (p_a + p_b)^2, \quad \hat{t} = (p_a - p_c)^2, \quad \hat{u} = (p_a - p_d)^2.$$

Assuming that all parton masses may be neglected, these may be approx-

Fig. 10.12 Two body parton–parton kinematics: (a) definition of of sub-process variables $\hat{s}$ and $\hat{t}$; (b) the parton–parton CM frame and the definition of the CM scattering angle θ^*.

imated as

$$\hat{s} \approx 2p_a \cdot p_b, \quad \hat{t} \approx -2p_a \cdot p_c, \quad \hat{u} \approx -2p_a \cdot p_d.$$

In the same approximation the following relations also hold

$$\hat{s} + \hat{t} + \hat{u} = 2E_a(E_a + E_b - E_c - E_d) = 0, \quad (10.20)$$

as there is overall conservation of 4-momentum in the hard sub-process. Diagram (b) shows the scattering angle, θ^*, in the parton–parton CM frame.

---
[6] See Section 7.5 for a discussion of jet measures.

Since masses are neglected all four partons have the same energy in the CM, $E^* = \sqrt{\hat{s}}/2$.

In terms of θ^*

$$\hat{t} = -\frac{\hat{s}}{2}(1 - \cos\theta^*), \qquad \hat{u} = -\frac{\hat{s}}{2}(1 + \cos\theta^*).$$

The initial state sub-process momenta are given in terms of the hadron–hadron CM energy, $\sqrt{s}$ by

$$p_a = \frac{\sqrt{s}}{2}(x_a, 0, 0, x_a), \qquad p_b = \frac{\sqrt{s}}{2}(x_b, 0, 0, -x_b),$$

where x_a, x_b are the fractions of the initial-state hadron beam momenta involved in the hard collision. As with other kinematic quantities of interest these have to be inferred from the final-state parton momenta which also cannot be measured directly, but are assumed to be those of the two jets. Ignoring any initial parton transverse momentum, one has

$$\begin{aligned} p_c &= p_T(\cosh y_c, \cos\phi, \sin\phi, \sinh y_c), \\ p_d &= p_T(\cosh y_d, -\cos\phi, -\sin\phi, \sinh y_d). \end{aligned}$$

where y_c and y_d are the rapidities of the final state partons (jets). Here the transverse mass $m_T = p_T$ as parton masses are being neglected. From these expressions, and using energy and longitudinal momentum conservation, one finds

$$x_a = \frac{x_T}{2}\left(e^{y_c} + e^{y_d}\right), \qquad x_b = \frac{x_T}{2}\left(e^{-y_c} + e^{-y_d}\right), \qquad (10.21)$$

where $x_T = 2p_T/\sqrt{s}$. It is also convenient to have expressions in terms of parton CM quantities. Defining

$$y_{CM} = \frac{1}{2}(y_c + y_d), \qquad y^* = \frac{1}{2}(y_c - y_d)$$

one finds

$$x_a = x_T e^{y_{CM}} \cosh y^*, \qquad x_b = x_T e^{-y_{CM}} \cosh y^*.$$

Note that y^* is invariant under boosts along the beam direction and that $\cos\theta^* = \tanh y^*$. The sub-process CM energy, $\sqrt{\hat{s}}$, is also the invariant jet-jet mass, M_{JJ}. From the above relations one has

$$\hat{s} \equiv M_{JJ}^2 = x_a x_b s = 4p_T^2 \cosh^2 y^*. \qquad (10.22)$$

10.5.2 Single jet inclusive cross-section

For the two-body parton–parton scattering sub-process shown in Fig. 10.12, the cross-section is given by the standard expression

$$d\hat{\sigma} = \frac{1}{32\pi^2 \hat{s}} \sum |\mathcal{M}|^2 \frac{d^3 p_c}{E_c} \frac{d^3 p_d}{E_d} \delta^4(p_a + p_b - p_c - p_d), \qquad (10.23)$$

where $\hat{\sum}|M|^2$ is the final spin summed and initial spin averaged squared matrix element (details given in the appendix at the end of the chapter). Using one of final state momentum integrals to eliminate $\delta^3(\mathbf{p})$ gives

$$d\hat{\sigma} = \frac{1}{16\pi^2 \hat{s}} \hat{\sum}|\mathcal{M}|^2 \frac{d^3 p_c}{E_c} \delta(\hat{s} + \hat{t} + \hat{u}), \qquad (10.24)$$

where Eq. (10.20) has been used. Inserting the above into the master equation (10.3) gives an expression for the single-jet inclusive cross-section

$$E\frac{d^3\sigma}{d^3p} = \frac{1}{16\pi^2 s} \sum_{a,b,c,d} \int_0^1 \frac{dx_a}{x_a} \frac{dx_b}{x_b} f_{a/A}(x_a, \mu^2) f_{b/B}(x_b, \mu^2)$$

$$\times \hat{\sum}|\mathcal{M}(ab \to cd)|^2 \frac{1}{1+\delta_{cd}} \delta(\hat{s} + \hat{t} + \hat{u}), \qquad (10.25)$$

where E and p are the total energy and momentum of the jet and the sum on the RHS is over all valid quark and gluon two-body processes. The $1/(1+\delta_{cd})$ term accounts for cases that have identical partons in the final state. For comparison with experimental data it is usual to use the transformation given in Eq. (10.2) to get the inclusive jet cross-section in terms of jet E_T and η. Figure 10.13 shows the inclusive jet E_T spectrum from $p\bar{p}$ interactions at $\sqrt{s} = 1.8\,\text{TeV}$ as measured by the DØ experiment (DØ 1999a). Jets were defined by an iterative $\Delta\eta, \Delta\phi$ cone algorithm with radius $R = 0.7$. The cross-section is averaged over $|\eta| < 0.5$ and is for $E_T > 60\,\text{GeV}$. The two curves indicate the $\pm 1\sigma$ systematic error band, the largest contribution to which is the energy scale uncertainty. The data are compared with the NLO calculation from JETRAD (Giele et al. 1994) with scale $\mu = E_T/2$ and CTEQ3M PDFs. There is good agreement over seven orders of magnitude in cross-section. The DØ experiment has also measured inclusive jet cross-sections using a k_T algorithm for events with $|\eta| < 0.5$ and $p_T > 60\,\text{GeV}$ (DØ 2002). Reasonable agreement is found with the NLO QCD prediction using JETRAD except at the lowest p_T values. While the measurements using the k_T algorithm are consistent with those using a cone algorithm, the former are consistently above the latter. Some of the difference may be understood in the details of how the hadronization process affects jet reconstruction by the two algorithms.

How reliable are the NLO calculations? The agreement between data and theory just demonstrated for the inclusive jet cross-section is impressive, but a number of choices have been made. The calculations are performed at NLO using the $\overline{\text{MS}}$ renormalization scheme and with a 'standard choice' of $\mu = E_T/2$ for the factorization scale. Ultimately one would like to have the calculations at NNLO, but this is a very large project and although progress is being made it will be some years before it is complete. Within a given renormalization scheme, one way to estimate the effect of higher orders is to vary the renormalization and factorization scales — giving the so-called scale dependence. In an extensive investigation Anandam

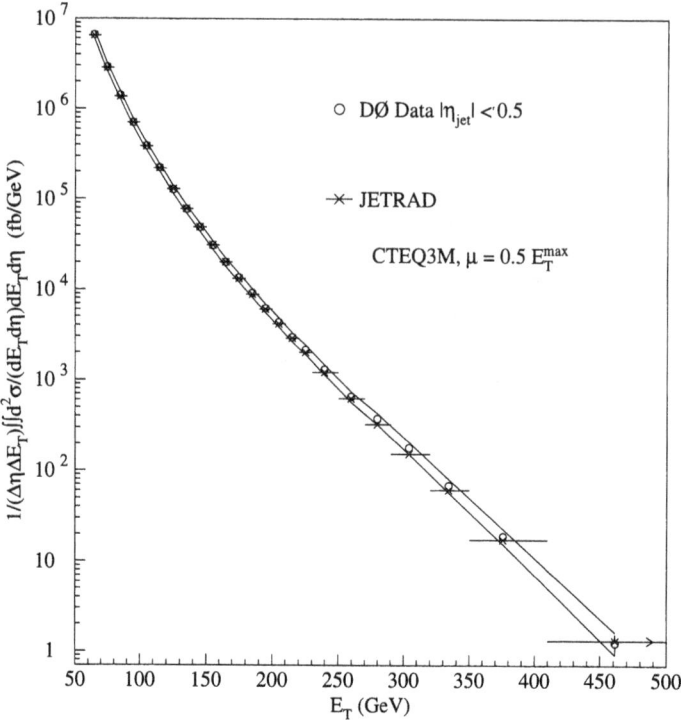

Fig. 10.13 Inclusive jet E_T spectrum in p$\bar{\text{p}}$ collisions at $\sqrt{s} = 1.8\,\text{TeV}$ as measured by DØ (1999a). The data are shown as open circles and the two lines indicate the $\pm 1\sigma$ systematic error band. The NLO QCD calculations from JETRAD are shown as crosses with a horizontal uncertainty bar.

and Soper (2000) have studied the simultaneous dependence on scale and renormalization scheme at NLO. Two parameters are introduced, λ which allows the renormalization scheme to be changed ($\lambda = 0, 1$ for the $\overline{\text{MS}}$ and DIS schemes, respectively) and μ, which varies the factorization scale. The renormalization scale is fixed at $\mu_{UV} = E_T/2$ and CTEQ4 PDFs are used. The changes in the inclusive jet cross-section $d\sigma/dE_T$ are then calculated as μ and λ vary over the ranges $E_T/8 < \mu < 2E_T$ and $-2 < \lambda < 2$. Results for $E_T = 50, 400\,\text{GeV}$ are shown in Fig. 10.14, the contours representing 1% change in $d\sigma/dE_T$. The plots show that $\mu = E_T/2$ is indeed a good choice of factorization scale, as the dependence on λ is then weak. The authors use these calculations to estimate the overall uncertainty in the NLO calculation as a function of E_T for two choices of parameter change: 'minimum' [$\Delta \log_2(\mu) = \pm 1, \Delta \lambda = \pm 1$]; 'conservative' [$\Delta \log_2(\mu) = \pm 2, \Delta \lambda = \pm 2$]. For the minimum change, the errors vary from 3 to 7% as E_T varies from 50 to 400 GeV and for the conservative case the variation is from 9 to 32%.

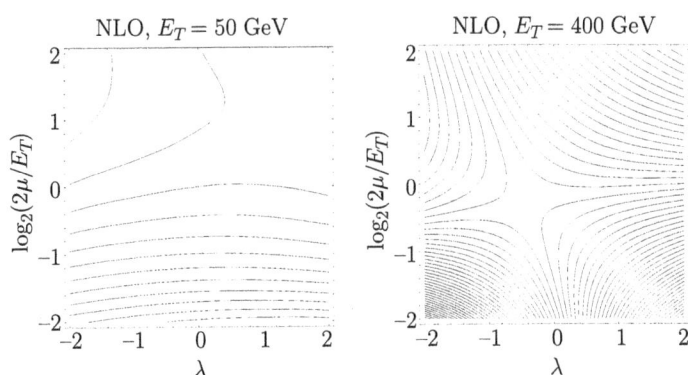

Fig. 10.14 Contour plots showing 1% changes in the inclusive jet cross-section $d\sigma/dE_T$ as the scale (μ) and scheme (λ) parameters are varied. (From Anandam and Soper 2000.)

10.5.3 Dijet cross-sections

With enough data, more information on the hard processes is available from measurement of various dijet cross-sections. Start from the differential form of the hadron–hadron master equation, Eq. (10.3),

$$\frac{d\sigma(AB \to jjX)}{dx_a dx_b d\cos\theta^*} = \sum_{ab} f_{a/A}(x_a, \mu^2) f_{b/B}(x_b, \mu^2) \frac{d\hat{\sigma}_{ab \to cd}}{d\cos\theta^*},$$

where the parton–parton CM scattering angle θ^* is defined in Fig. 10.12(b). In the CM frame, Eq. (10.23) for the sub-process cross-section gives

$$\frac{d\hat{\sigma}}{d\cos\theta^*} = \frac{1}{32\pi \hat{s}} \sum |\mathcal{M}|^2.$$

Inserting this into the previous equation and using the Jacobian,

$$dp_T^2 dy_c dy_d \to \frac{s}{2} dx_a dx_b d\cos\theta^*,$$

the dijet cross-section is given by

$$\frac{d^3\sigma}{dy_c dy_d dp_T^2} = \frac{1}{16\pi s^2} \sum_{a,b,c,d} \frac{f_{a/A}(x_a, \mu^2)}{x_a} \frac{f_{b/B}(x_b, \mu^2)}{x_b}$$
$$\times \sum |\mathcal{M}(ab \to cd)|^2 \frac{1}{1 + \delta_{cd}}. \qquad (10.26)$$

Usually for comparison with experimental data

$$dy_c dy_d dp_T^2 \to d\eta_1 d\eta_2 dE_T^2,$$

where E_T is the transverse energy of the leading (highest E_T) jet and $\eta_{1,2}$ are the pseudo-rapidities of the two jets, now taken to represent those of

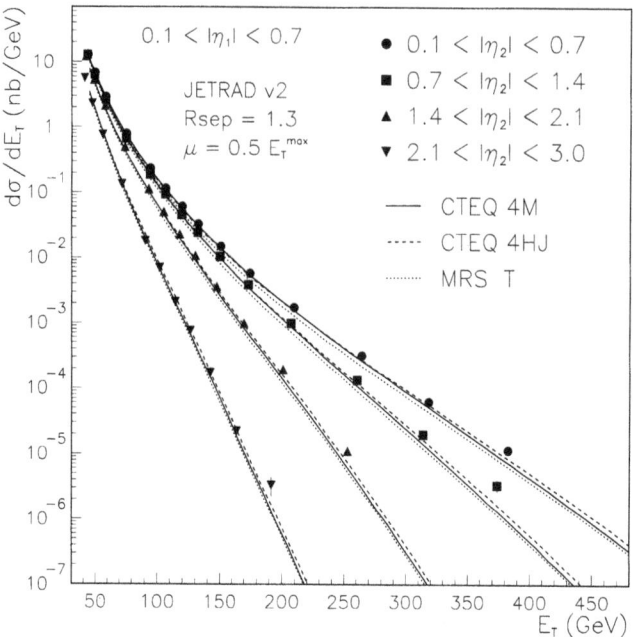

Fig. 10.15 Dijet $d^3\sigma/d\eta_1 d\eta_2 dE_T$ as measured by CDF (2001c) is plotted for four bins of 2nd jet pseudo-rapidity η_2 as a function of the leading jet E_T with $0.1 < |\eta_1| < 0.7$. The QCD predictions are from JETRAD with three choices of PDFs.

the scattered partons c and d. Hence, the initial momentum fractions are given in terms of the jet variables by

$$x_a = \frac{E_T}{\sqrt{s}}(e^{\eta_1} + e^{\eta_2}) \qquad x_b = \frac{E_T}{\sqrt{s}}(e^{-\eta_1} + e^{-\eta_2}).$$

The CDF collaboration (2001c), see Fig. 10.15, have used data from $86\,\text{pb}^{-1}$ of $p\bar{p}$ collisions at $\sqrt{s} = 1.8\,\text{TeV}$ to measure the dijet cross-section as a function of the leading jet E_T with $0.1 < |\eta_1| < 0.7$ for four bins of pseudo-rapidity of the second jet in the range $0.1 < |\eta_2| < 3.0$. Jets are defined using a cone algorithm in $\Delta\eta$, $\Delta\phi$ with radius 0.7. The NLO QCD calculation describes both the E_T and rapidity dependence of the data.

Apart from establishing the underlying quark and gluon dynamics, dijet measurements give sensitive measures for new phenomena. At the highest E_T, the single inclusive jet cross-section measured by the CDF experiment showed a small excess over the standard NLO QCD prediction (CDF 1996,

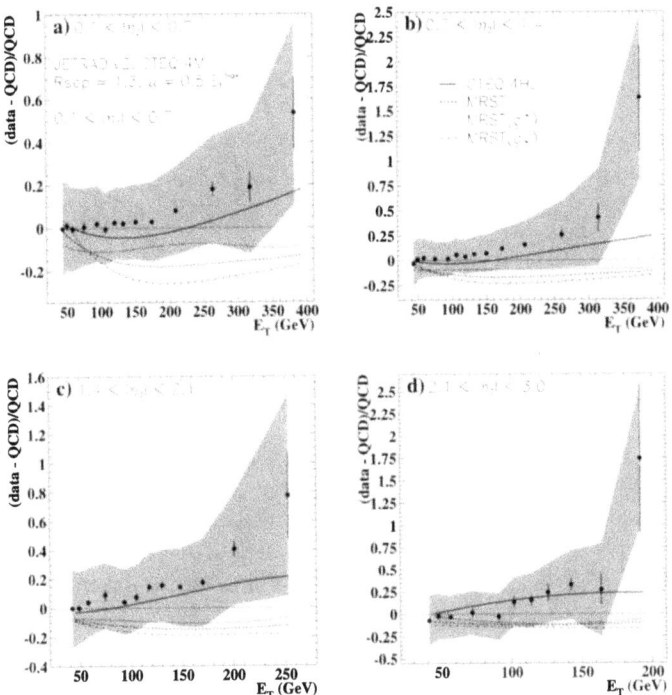

Fig. 10.16 The dijet data shown in the previous figure are plotted as (Data-QCD)/QCD are plotted using the same binning. The reference QCD prediction is from JETRAD with CTEQ4M PDFs and predictions using different PDFs are shown as curves. The error bars show statistical errors and the shaded band the experimental correlated systematic errors. (From CDF 2001c).

2001a).[7] The CDF excess may be studied in more detail by using dijet data. Figure 10.16 shows the comparison of the dijet data and NLO QCD in more detail. The excess of events over the standard model prediction is clearly seen in all rapidity bins at the highest E_T values. The shaded band shows the correlated systematic errors, which are dominated by the experimental energy scale uncertainty.

Rather than characterising the dijet events in terms of rapidities, another useful choice of variables is the jet–jet invariant mass M_{JJ} and CM scattering angle θ^*. Either by a change of variables from Eq. (10.26) or by analogy with the D–Y cross-section, Eq. (10.7), the double differential cross-section takes the form

---

[7] The excess was not seen by the DØ experiment, but given the quite large systematic errors at large E_T, the data from the two experiments are not inconsistent.

$$\frac{d^2\sigma}{dM_{JJ}^2 d\cos\theta^*} = \frac{1}{32\pi M_{JJ}^2} \sum_{a,b,c,d} \int_0^1 dx_a dx_b f_{a/A}(x_a,\mu^2) f_{b/B}(x_b,\mu^2)$$

$$\times \delta(x_a x_b s - M_{JJ}^2) \sum |\mathcal{M}(ab \to cd)|^2 \frac{1}{1+\delta_{cd}}, \quad (10.27)$$

where $\cos\theta^* = \tanh(|\Delta\eta|/2)$ and $M_{JJ}^2 = 2E_1^T E_2^T (\cosh\Delta\eta - \cos\Delta\phi)$. Massless jets are assumed where $E_{1,2}^T$ are the jet transverse energies and $\Delta\eta = \eta_1 - \eta_2$, $\Delta\phi$, are the differences in jet pseudo-rapidities and azimuthal angles, respectively.

The angular dependence of the squared matrix elements given in the appendix may be expressed conveniently in terms of

$$\chi = \frac{\hat{u}}{\hat{t}} = \frac{(1+\cos\theta^*)}{(1-\cos\theta^*)}.$$

For example the angular factor for $qq' \to qq'$ is $4(2\chi^2 + 2\chi + 1)/9$. The dominant terms for dijet production are $gg \to gg$, $gq \to gq$ and for $p\bar{p}$, $q\bar{q} \to q\bar{q}$ (see Problem 6). All of these processes involve gluon exchange in the $\hat{t}$-channel, giving rise to the characteristic $1/\sin^4(\theta^*/2)$ angular behaviour at small θ^*. It is often convenient to remove this rapid angular variation by plotting $d\sigma/d\chi$ rather than $d\sigma/d\cos\theta^*$, since $d\sigma/d\chi \sim$ constant for small angles (large χ). For two-jet final states, there is usually no way of identifying quark or gluon jets so the cross-sections are symmetrized over $\hat{t}$ and $\hat{u}$ exchanges, which means that the data are presented in terms of

$$\chi' = \frac{(1+|\cos\theta^*|)}{(1-|\cos\theta^*|)}.$$

Dijet invariant mass and angular distributions give sensitive tests of standard QCD and competitive limits on new phenomena such as composite quarks and new contact interactions.

Figure 10.17 shows $d^3\sigma/dM_{JJ}d\eta_1 d\eta_2$ with $|\eta_i| < 1.0$ as measured by DØ in $p\bar{p}$ collisions at $\sqrt{s} = 1.8\,\text{TeV}$ (DØ 1999b). The upper plot shows the jet mass spectrum compared to the NLO QCD prediction from JETRAD with CTEQ3M PDFs, the agreement between theory and data is good over the whole range of M_{JJ} shown from 300 to 850 GeV. The lower plot shows (Data-QCD)/QCD with a number of alternative choices of PDFs and scales. A change of renormalization/factorization scale from $\mu = 0.5E_T^{max}$ to $\mu = 2.0E_T^{max}$ produces a roughly 10% change in the prediction independent of jet mass. The shaded region shows the $\pm 1\sigma$ experimental systematic uncertainty on the data, the largest contribution comes from the jet energy scale uncertainty which rises from 7% at 200 GeV to around 30% at the highest energies.

As discussed above, $d\sigma/d\chi'$ should be fairly flat as a function of χ' if gluon exchange diagrams dominate the hard parton-parton interaction. Deviations from this behaviour could signal new effects. The dijet angular

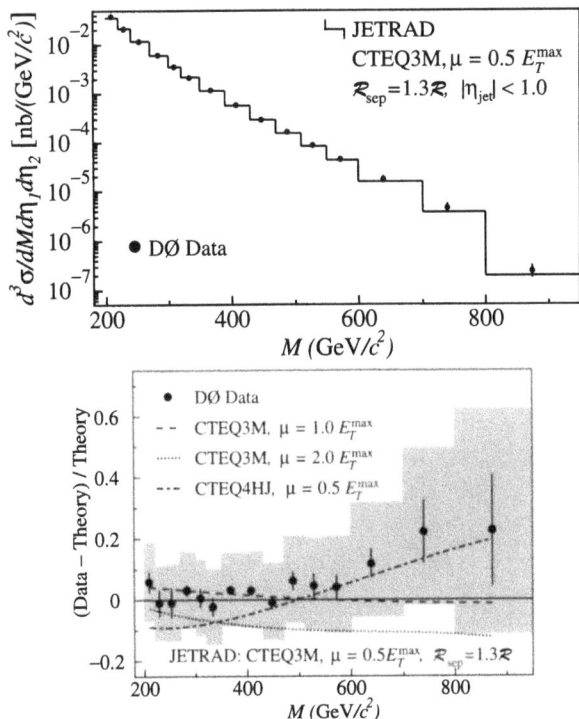

Fig. 10.17 Dijet invariant mass distribution as measured by the DØ experiment in p$\bar{\text{p}}$ at $\sqrt{s} = 1.8\,\text{TeV}$ (DØ 1999b). The upper plot shows $d^3\sigma/dM_{JJ}d\eta_1 d\eta_2$ with $|\eta_i| < 1.0$; the lower plot shows (Data-QCD)/QCD, with a number of alternative choices of PDFs and scales. The standard QCD prediction is from JETRAD with CTEQ3M PDFs and renormalization/factorization scale $\mu = 0.5 E_T^{max}$.

distribution $d\sigma/d\chi'$ is indeed found to be flat in the range $2 < \chi' < 5$ rising by about 30% between 1 and 2. An isotropic angular distribution (in θ^*) would give a very rapidly rising distribution at small χ' values. The ratio of events $R_\chi = N(\chi' < 2.5)/N(2.5 < \chi' < 5.0)$ is therefore a useful number to quantify the shape of the dijet angular distribution. Figure 10.18 shows R_χ as a function of M_{JJ} as measured by the CDF experiment from dijet events with $E_T > 50\,\text{GeV}$, $|\eta_{1,2}| < 2$ and $\chi' < 5$. The ratio shows little variation with M_{JJ} and the level is reasonably well described by NLO QCD with either $\mu = M_{JJ}$ or $\mu = p_T$ as choice of renormalization scale. The other curves refer to models with additional contact interactions which are discussed briefly in Chapter 12 (Section 12.4).

The overall conclusion on inclusive and dijet physics from p$\bar{\text{p}}$ interactions at high energies (particularly the Tevatron data) is that NLO QCD

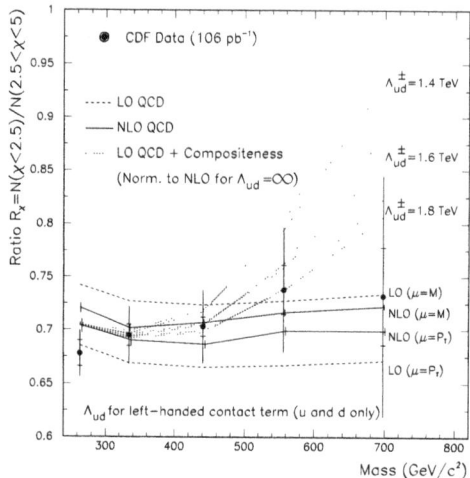

Fig. 10.18 Dijet angular ratio R_χ as a function of dijet mass, M_{JJ}. $R_\chi = N(\chi' < 2.5)/N(2.5 < \chi' < 5.0)$ where χ' is defined in the text. (From CDF 1996.)

gives a very good description. Apart from direct support for the QCD enhanced parton model and the dominance of gluon exchange diagrams, the jet data can provide useful constraints on parton momentum densities as discussed in Section 6.5.2. The limitations come primarily from the large systematic uncertainty in the jet energy scale at large E_T and to a lesser extent from the renormalization and factorization scale uncertainties.

10.6 Isolated photon production

The production of isolated[8], high E_T, photons in hadron–hadron collisions offers, in principle, a clean approach to parton dynamics. The leading order processes are $q\bar{q} \to \gamma^* g$ (annihilation) and $qg \to \gamma^* q$ (QCD Compton), diagrams for which are shown in Fig. 10.19(a) and (b), respectively. Expressions for the squared matrix elements are given in the appendix. Similar diagrams are also relevant for the production of high E_T vector bosons.

The advantages of direct photons over high E_T jets are: no complications from parton fragmentation and jet algorithms; generally much better energy resolution for EM calorimeters. The disadvantages are: the smaller rate, $O(\alpha\alpha_s)$ rather than $O(\alpha_s^2)$; the potentially large background from π^0 and η decays; the problems of intrinsic k_T and soft gluon radiation. Using

---

[8] Also known as direct or prompt photon production.

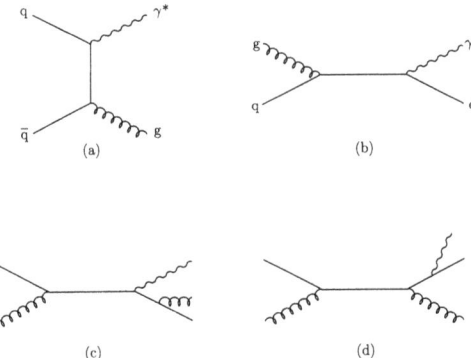

Fig. 10.19 Diagrams for high E_T isolated photon (or vector boson) production: (a) annihilation; (b) QCD Compton; (c),(d) NLO diagrams.

Eq. (10.21), for photons produced centrally ($y \approx 0$), the initial state partons are probed at $x \approx 2E_T^\gamma/\sqrt{s}$. At very large E_T^γ in p$\bar{\text{p}}$ the annihilation diagram is most important, but generally for hadron–hadron scattering with moderate E_T^γ the Compton process dominates, which implies a sensitivity to the gluon density. In principle the prompt photon process allows study of the gluon density at larger x values than those accessible through scaling violations in DIS (see Section 6.5.2).

QCD calculations have been performed to NLO. Fig. 10.19(c) and (d) show a couple of NLO diagrams and (d) — with a photon radiated from the final quark line — requires the introduction of a non-perturbative photon fragmentation function, D_q^γ, for its complete evaluation. D_q^γ obeys an evolution equation analogous to the DGLAP equations and is similar to hadron fragmentation functions familiar from e^+e^- applications (see e.g. Ellis et al. 1996, Chapter 6 for details). The collinear singularity which arises when the momenta of the quark and photon are parallel is absorbed into D_q^γ in a similar way to the renormalization of parton densities. However, there is an important difference due to the point-like coupling of the photon which gives an inhomogeneous but calculable addition to the evolution equations. The effect of this addition is to enhance the numerical importance of the contribution of D_q^γ in the NLO corrections.

For p$\bar{\text{p}}$ at Tevatron energies of $\sqrt{s} = 1.8\,\text{TeV}$, direct photon production has been measured to E_T^γ of well over 100 GeV. Figure 10.20 shows the cross-section $d^2\sigma/dE_T^\gamma d\eta$ for two ranges of pseudo-rapidity, central $|\eta| < 0.9$ and forward $1.6 < |\eta| < 2.5$. In both cases, the NLO calculation, with a scale E_T^{max} (the larger of the photon or leading jet E_T) and CTEQ4M PDFs, agrees well with the data for $E_T^\gamma > 40\,\text{GeV}$. At lower E_T^γ, the measured data points lie above the NLO calculation. A similar excess of data over the NLO calculation has been seen in lower energy pp and pA direct

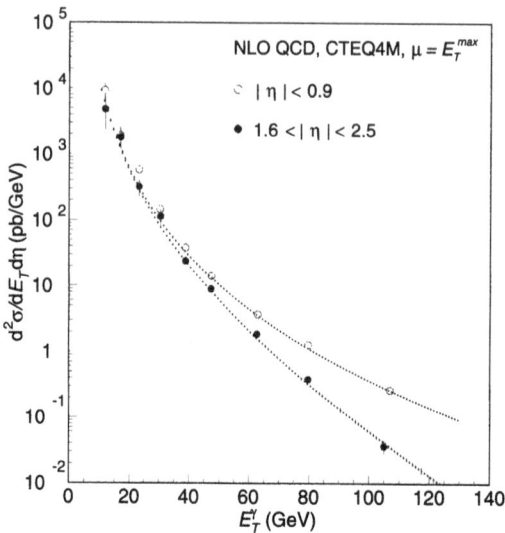

Fig. 10.20 The cross-section for isolated photon production in two rapidity ranges as measured by the DØ experiment (2000). The curves show an NLO QCD calculation using CTEQ4M PDFs.

photon fixed–target experiments. The reason for the excess is not completely clear and various suggestions have been put forward:

- an incorrect experimental determination of the background from the $\gamma\gamma$ decays of π^0 and η;
- intrinsic parton k_T;
- soft gluon emissions.

Some authors have attempted to include the effect of intrinsic k_T by allowing a Gaussian smearing with an additional fit parameter. The mean value $\langle k_T \rangle$ has been estimated from a number of processes, fixed target D–Y muon pair production from pion and proton beams, direct gamma production from $pp(\bar{p})$ interactions at both fixed target and collider energies and direct gamma production in high energy γp interactions at HERA. It is found that the measured $\langle k_T \rangle$ increases with CM energy W, as shown in Fig. 10.21 from ZEUS (2001b). The introduction of intrinsic partonic k_T does improve the description of prompt photon data at low E_T^γ and allowing $\langle k_T \rangle$ to increase with W does provide a reasonably consistent picture across a wide range of energies and experiments. However, at the relatively low values of E_T^γ where the discrepancy between data and theory (without explicit addition of intrinsic k_T) is worse, other factors may be as important and these are not yet included in the calculations. Furthermore, it seems

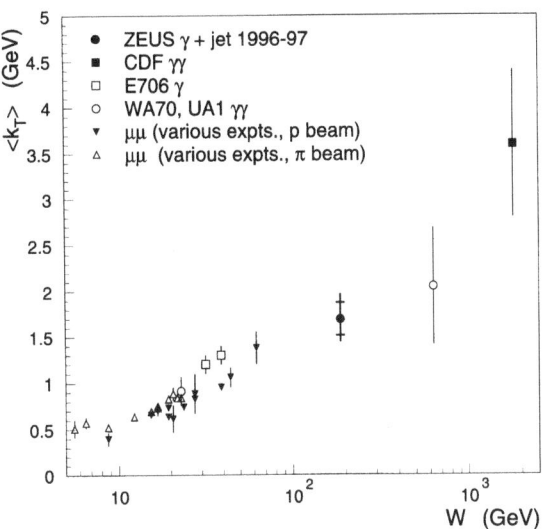

Fig. 10.21 Mean intrinsic parton transverse momentum $\langle k_T \rangle$ from D–Y and direct photon processes as a function of CM energy W. (Summary plot from ZEUS 2001b.)

that not all prompt photon experiments give completely consistent results. More details are given in the critical study by Aurenche et al. (1999). Work is underway to include the effect of soft gluon emissions by resummation techniques and there is some optimism that this may lead to a resolution of the problem. Until this problem is understood, the use of low E_T^γ direct photon data as a constraint on the gluon density is suspect.

10.7 Hadronic DIS at the LHC

The methods described so far in this chapter will be essential for calculating many standard model and beyond the standard model processes at the LHC. Accurate prediction of the former will often be a prerequisite to claims of a discovery of the latter. To produce a state of mass M with rapidity y requires[9]

$$x_a = \frac{M}{\sqrt{s}} e^{+y}, \quad x_b = \frac{M}{\sqrt{s}} e^{-y} \qquad (10.28)$$

[9] The kinematics are those of the D–Y process discussed in Section 10.3.1.

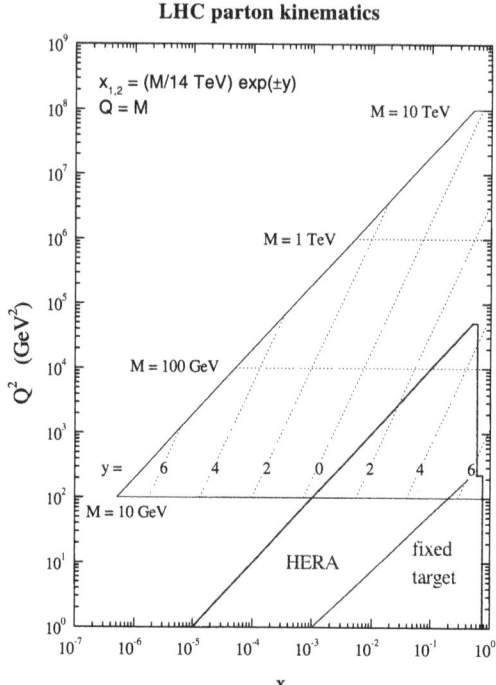

Fig. 10.22 The LHC kinematic plane for producing a state of mass M and rapidity y in terms of DIS variables x, Q^2. The kinematic regions covered by HERA and the fixed target experiments are shown for comparison. Figure courtesy of W J Stirling.

where $\sqrt{s}$ is the pp CM energy (14 TeV for the LHC) and x_a, x_b are the fractions of the incident beam momenta involved in the hard collision. For states with masses less than 500 GeV, this means that at least one of the interacting partons will have an $x < 10^{-3}$ and thus be well into the $q\bar{q}$ sea region. This is shown graphically in Fig. 10.22 which shows the LHC kinematic plane in terms of DIS variables x, Q^2 with the HERA and fixed target regions for comparison. It can be seen from the figure that the HERA x range is well matched to the regions required for 'low mass' LHC physics, but the parton densities will have to be evolved over two or three orders of magnitude in Q^2. The precise determination of the $q\bar{q}$ sea and gluon densities from global analyses using HERA, Tevatron and fixed target data will be very important.

10.7.1 Jet physics at the LHC

With an increase in CM energy and luminosity of an order of magnitude between the Tevatron and the LHC, jet physics will be be systematics limited

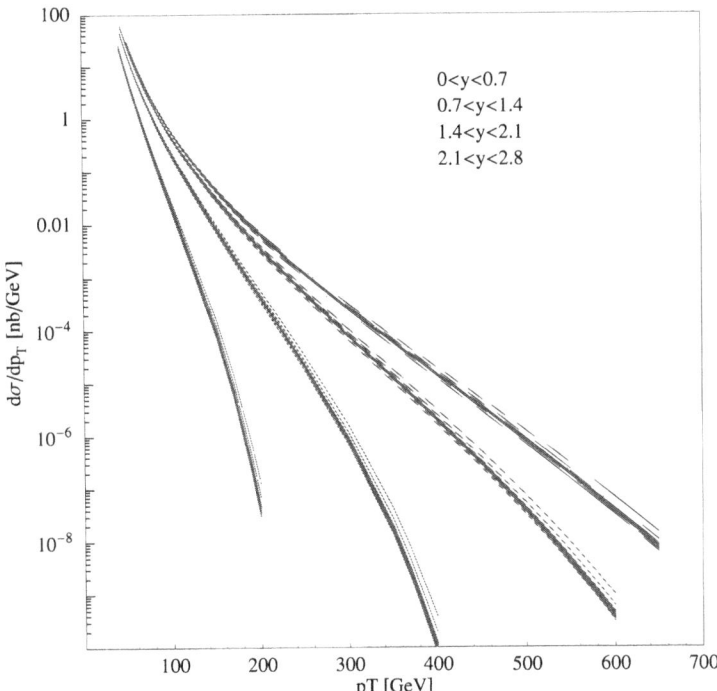

Fig. 10.23 Jet cross-sections as function of p_T for the Tevatron Run-II in the four η bins used by CDF. The spread in the curves is given by the ensemble of PDFs compatible with a tolerance of $T = 10$ around the global minimum. From a study by the CTEQ group, (Stump *et al.* 2003).

at E_T scales in excess of 3 TeV. At 'low' luminosity ($10^{37}\,\mathrm{m^{-2}s^{-1}}$), the integrated luminosity for one year is estimated to be $10\,\mathrm{fb^{-1}}$, at the full design luminosity it is a factor of ten larger. To get an idea of the increase in physics reach that this implies, consider the corresponding planned increase in energy and luminosity for Run-II at the Tevatron. The CM energy increase is modest, from 1.8 to 1.96 TeV, but the integrated luminosity is hoped to be around $15\,\mathrm{fb^{-1}}$ by the end of Run-IIb, compared with $\sim 100\,\mathrm{pb^{-1}}$ for a typical Run-I analysis. The consequences for inclusive jet physics are shown in Figs 10.23 and 10.24 for the Tevatron and LHC, respectively. The plots are from a detailed study by the CTEQ group (Stump *et al.* 2003) of high–energy jets and searches for new physics. It uses the CTEQ6 PDFs which are based on fits using most of the published HERA-I structure functions and Tevatron Run-I jet data. In addition the group have used the Hessian method to estimate the uncertainty in the resulting predictions from the experimental systematic errors of the input data. The Hessian method is explained in Section 6.7.2 of Chapter 6. The Hessian matrix is diagonalized

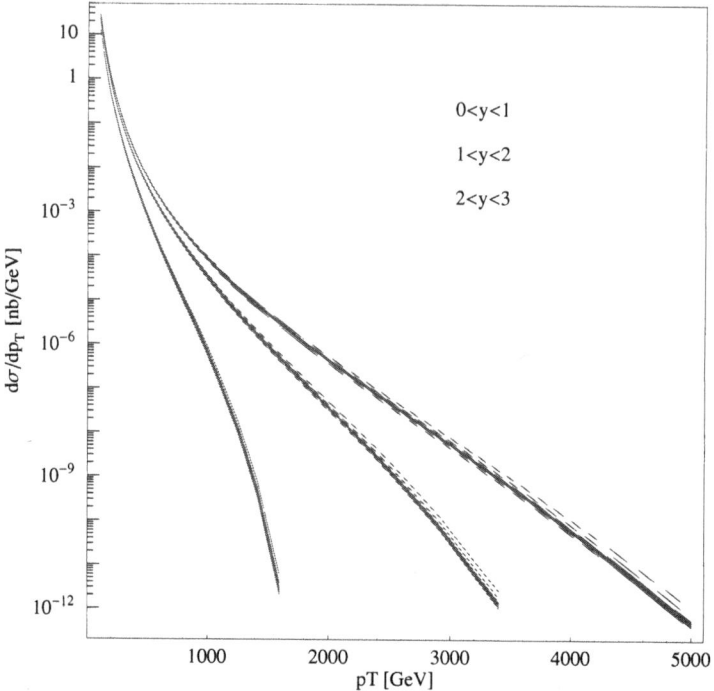

Fig. 10.24 Jet cross-sections as function of p_T for the LHC in three η bins. The spread in the curves is given by the ensemble of PDFs compatible with a tolerance of $T = 10$ around the global minimum. From a study by the CTEQ group (Stump et al. 2003).

and the positive and negative distances along each eigenvector direction that give an increase in overall χ^2 from the global minimum by an amount T^2 determined. The tolerance, T, is set equal to 10 (for more discussion of this point see Section 6.7.6). This procedure gives a set of PDFs corresponding to the extremes of the eigenvectors compatible with the global minimum. These PDFs can then be used to give an estimate of the uncertainty in any physical quantity derived from the PDF set. Figure 10.23 shows the predictions for inclusive jet cross-sections as a function of p_T for the four bins in rapidity used by the CDF experiment. The increase in luminosity will almost double the maximum jet p_T accessible by the end of Run-II (compare Fig. 10.15).

Figure 10.24 shows the corresponding plot for the LHC. The almost order of magnitude increase in the CM energy from 1.95 to 14 TeV means that the jet cross-sections will be measurable to almost 5 TeV at central rapidities and to 1.5 TeV in the forward direction. The uncertainty on the predictions at the highest p_T are comparable for the Tevatron and LHC experiments. To exploit the physics reach implied by this large increase in sensitivity at very large p_T and correspondingly small distance scales, the LHC experiments have set themselves ambiitious goals of controlling the experimental systematic uncertainty on the detector energy scales, for example ATLAS are aiming for 1% at 1 TeV and about 10% at 3 TeV. A further challenge for the design of both the LHC machine and detectors, is that the inclusive jet cross-section at the LHC runs over more than 10 orders of magnitude, with the cross-section at the largest p_T at least three orders of magnitude smaller than the smallest jet cross-section at the Tevatron.

10.7.2 Collider luminosity

Apart from needing to know the parton densities accurately for the prediction of standard model rates, there is also the vital question of how the collider luminosity will be determined. The traditional approach to luminosity determination has been to measure a process with high rate and accurately known cross-section. In colliders with an electron beam (e^+e^-, ep) this means a QED process. In p$\bar{\text{p}}$ or pp scattering QED processes tend to be too small or too difficult to measure in a collider geometry, so an alternative is to measure the total pp rate N_{tot} and the forward elastic rate N_{el}. Using the optical theorem and assuming that the real part of the elastic amplitude is small,

$$\left.\frac{d\sigma_{\text{el}}}{dt}\right|_{t=0} \approx \frac{\sigma_{\text{tot}}^2}{16\pi}.$$

Assuming that acceptances can be determined accurately, one has $N_{\text{tot}} = L\sigma_{\text{tot}}$ and $N_{\text{el}} = L\sigma_{\text{tot}}^2$, from which both the luminosity L and σ_{tot} can be determined. Apart from the very high rates at small angles to the beam line, one has the problem of extrapolating the elastic cross-section or rate from the measured region at small t to $t = 0$.

Because of the potential problems with small-angle pp scattering, alternative methods of measuring the collider luminosity are being considered. The most promising processes with accurately known cross-sections and clear experimental signatures are W and Z production. The parton subprocesses and the DGLAP coefficient functions are known to NNLO and there are reasonable approximations to the full NNLO splitting functions. The MRST team have investigated the predictions for $\sigma \cdot B$ for W and Z production (branching to leptonic deacy modes) at the Tevatron and LHC at different orders but with the same input data determining the parton densities. The results are shown in Fig. 10.25. The data points are Run-I measurements from the DØ and CDF Tevatron experiments. The large jump from LO to NLO is the large $O(\alpha_s)$ correction to the parton model D-Y prediction that was discussed in Section 10.3.2. The convergence between NLO and the NNLO estimate is good.

As has been discussed in detail in Chapter 6, particularly Section 6.7, much progress has been made in developing methods to include proper esti-

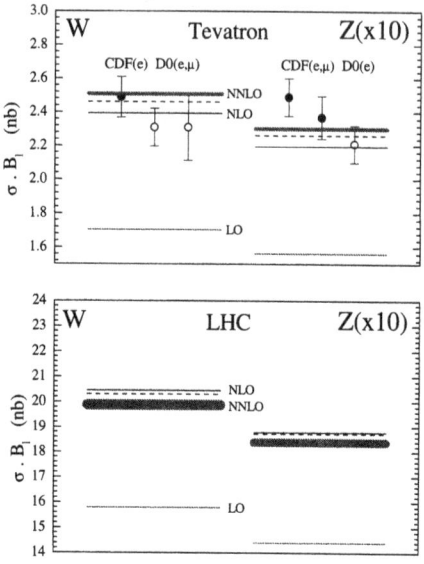

Fig. 10.25 Predictions for $\sigma \cdot B$ for W and Z production (branching to leptonic decay modes) at the Tevatron and LHC colliders as a function of the order of calculation with fixed parton input. The band and dashed line indicate the uncertainty in present NNLO calculations. (From Khoze et al. 2001.)

mates of the correlated and uncorrelated systematic errors on experimental data in global fits. The MRST team (Martin et al. 2002) have investigated both the Hessian and Lagrange multiplier approaches. They take a tolerance of $T^2 = 50$ as a reasonable measure of the overall experimental uncertainty on a quantity derived from their MRST01 PDF determination. Both methods agree reasonably well that σ_W is determined to within $^{+2.5}_{-2.0}\%$ and σ_H to about $\pm 3\%$. The two quantities are positively correlated and the effect of fixing α_s is modest, reducing the error on σ_H to around $\pm 2\%$.

A further advantage of using W and Z production is that the ratio of W to Z production ($R_{\rm DY}$ defined in Eq. (10.19)) is independent of the collider luminosity and thus gives an observable that may be used to check how well a particular PDF set actually describes W and Z production.

10.7.3 Parton-parton luminosity

Dittmar et al. (1997) make the point that for many hard processes at the LHC it is actually the parton-parton luminosity[10] that is needed, rather than that for the complete pp interaction. For W and Z production, the basic processes are

---

[10]The concept of parton-parton luminosity is intuitively obvious but more details and examples are given in Ellis et al. 1996, Chapter 7.

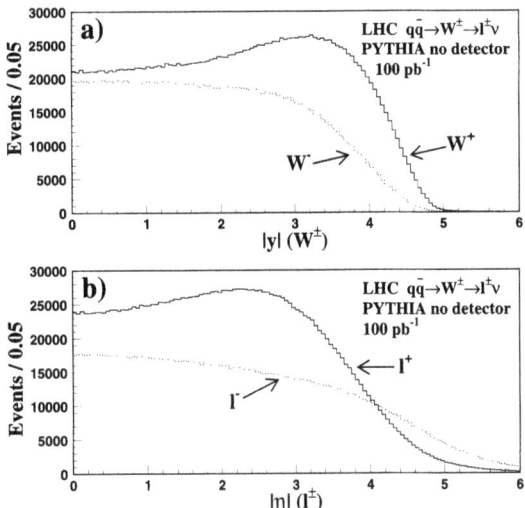

Fig. 10.26 Predictions for the $W^\pm$ asymmetry and the $W^\pm \to \ell^\pm \nu$ decay asymmetry at the LHC using MRSA PDFs. (From Dittmar et al. 1997.)

$$u\bar{d} \to W^+ \to \ell^+\nu; \quad d\bar{u} \to W^- \to \ell^-\nu; \quad u\bar{u}(d\bar{d}) \to Z^0 \to \ell^+\ell^-.$$

So by measuring the rapidity distribution (through that of the decay $\ell^\pm$), one may constrain the luminosity of the u and d quarks and anti-quarks. At the LHC, one is dealing with pp collisions rather than the p$\bar{\text{p}}$ process and furthermore the CM energy is an order of magnitude larger. For small y values, $x_a \approx x_b \sim 10^{-3}$, both well into the region of the q$\bar{\text{q}}$ sea. For medium to large rapidities, x_a will be in the valence region with $x_b \ll 10^{-3}$. So only in the latter case will there be a substantial difference in $W^\pm$ rates, with W^+ larger because of the two u-valence quarks in the proton. For an actual observable process, one must fold in the decay kinematics (as discussed in Section 10.4.3). The event distributions in rapidity for $W^\pm$ production and decay corresponding to $100\,\text{pb}^{-1}$ of pp interactions at the LHC energy of $\sqrt{s} = 14\,\text{TeV}$ are shown in Fig. 10.26. The idea is that by measuring the rate of decay leptons in small fixed intervals Δy_i one will eventually be able to constrain the corresponding q and $\bar{\text{q}}$ densities over an x range of something like $(5 \times 10^{-4}, 0.1)$ at a $Q^2 \approx 10^4\,\text{GeV}^2$. Statistical precision will not be a problem. Other processes that are initiated by the same q$\bar{\text{q}}'$ combinations may now be calculated accurately, without the uncertainty from the extrapolation of parton densities from lower energies.

10.8 Summary

This chapter has shown the power of the QCD enhanced parton model, which relies on pQCD calculations at NLO and the universality of parton momentum density functions. At least at leading twist, one has a well defined framework, supported by proofs of factorization, in which to perform the calculations. The early success of the parton model predictions of scaling in the Drell–Yan process helped consolidate the parton approach. The later QCD calculation of corrections to D–Y gave impressive agreement between data and theory in both shape and normalization. The D–Y process is central to understanding how to use the QCD enhanced parton model in hadron–hadron processes, because of its simplicity and its importance as a discovery channel for high mass states such as the W and Z. Jet production in high energy $p\bar{p}$ and pp collisions is a 'pure QCD' process and the success of both experimental groups in measuring its properties with precision over a wide range of energies and theoretical calculations based on the the parton idea and pQCD, shows the enormous advances in understanding of hard processes since the early 1970s. Generally the overall agreement at NLO between jet data and theory is good. There is a need for NNLO calculations to reduce the scale uncertainties at the highest E_T and there is a problem with a complete understanding of soft gluon emission effects at low E_T. The latter affects, in particular, the reliability of predictions for direct photon production. Despite these problems, data from hadron–hadron DIS are an important and complementary ingredient to global fits for determining hadron PDFs. Accurate parton densities will be crucial for standard model and 'new' physics at the LHC.

10.9 Problems

1. Calculate the Jacobian of the transformation from Cartesian to rapidity momentum components given in Eq. (10.2).
2. Check the kinematics of D-Y lepton pair production at the start of Section 10.3.1 and, in particular, Eqs (10.4,10.5).
3. Consider the Drell–Yan process for π C scattering. The carbon nucleus has an equal number of protons and neutrons and thus also an equal number of u and d quarks. Using Eq. (10.10) with $y \approx 0$, show that in the quark-parton model, the ratio

$$r = \frac{\sigma(\pi^+ C \to \mu^+\mu^- X)}{\sigma(\pi^- C \to \mu^+\mu^- X)}$$

equals 1/4 when τ approaches 1. What value does r have for small τ?
4. Derive the expression given in Eq. (10.16) for $\hat{\sigma}(q_i\bar{q}_i \to Z \to \ell^+\ell^-)$ using the narrow resonance approximation.

5. Check the parton–parton kinematics of Section 10.5.1 in the hadron–hadron and parton–parton frames. Show that the Jacobian $\dfrac{\partial(x_a, x_b, \cos\theta^*)}{\partial(y_c, y_d, p_T^2)}$ has the value $2/s$.

6. Estimate the relative sizes of the parton–parton squared matrix elements of the appendix as follows. Express each in terms of the variable χ (defined in Section 10.5.3), for example the gg term is

$$\frac{9}{2}\left(3 - \frac{\chi}{(\chi+1)^2} + \chi + \chi^2 + \frac{1}{\chi^2} + \frac{1}{\chi}\right).$$

Then evaluate each term at $\chi = 1$ ($\theta^* = 90°$).

7. Use Eq. (10.26) and the squared matrix element given in the appendix for gluon–gluon scattering to estimate the dijet cross-section at central rapidities and jet $E_T = 200\,\text{GeV}$ at $\sqrt{s} = 1.8\,\text{TeV}$. Repeat the calculation for the LHC energy of $\sqrt{s} = 14\,\text{TeV}$. Explain why the LHC cross-section is larger.

[Parton densities may be found at http://durpdg.dur.ac.uk/HEPDATA/.]

10.10 Appendix: $\sum|\mathcal{M}|^2$ for tree-level sub-processes

Process	$\sum|\mathcal{M}|^2/(4\pi\alpha_s)^2$	Fig. 10.27
$qq' \to qq'$	$\dfrac{4}{9}\dfrac{\hat{s}^2+\hat{u}^2}{\hat{t}^2}$	a
$q\bar{q}' \to q\bar{q}'$	$\dfrac{4}{9}\dfrac{\hat{s}^2+\hat{u}^2}{\hat{t}^2}$	a
$qq \to qq$	$\dfrac{4}{9}\left(\dfrac{\hat{s}^2+\hat{u}^2}{\hat{t}^2}+\dfrac{\hat{s}^2+\hat{t}^2}{\hat{u}^2}\right)-\dfrac{8}{27}\dfrac{\hat{s}^2}{\hat{u}\hat{t}}$	a, b
$q\bar{q} \to q'\bar{q}'$	$\dfrac{4}{9}\dfrac{\hat{t}^2+\hat{u}^2}{\hat{s}^2}$	c
$q\bar{q} \to q\bar{q}$	$\dfrac{4}{9}\left(\dfrac{\hat{s}^2+\hat{u}^2}{\hat{t}^2}+\dfrac{\hat{t}^2+\hat{u}^2}{\hat{s}^2}\right)-\dfrac{8}{27}\dfrac{\hat{u}^2}{\hat{s}\hat{t}}$	a, c
$q\bar{q} \to gg$	$\dfrac{32}{27}\dfrac{\hat{t}^2+\hat{u}^2}{\hat{t}\hat{u}}-\dfrac{8}{3}\dfrac{\hat{t}^2+\hat{u}^2}{\hat{s}^2}$	d, e, f
$gg \to q\bar{q}$	$\dfrac{1}{6}\dfrac{\hat{t}^2+\hat{u}^2}{\hat{t}\hat{u}}-\dfrac{8}{3}\dfrac{\hat{t}^2+\hat{u}^2}{\hat{s}^2}$	d, e, f reversed
$gq \to gq$	$-\dfrac{4}{9}\dfrac{\hat{s}^2+\hat{u}^2}{\hat{s}\hat{u}}+\dfrac{\hat{s}^2+\hat{u}^2}{\hat{t}^2}$	g, h, i
$gg \to gg$	$\dfrac{9}{2}\left(3-\dfrac{\hat{t}\hat{u}}{\hat{s}^2}-\dfrac{\hat{s}\hat{u}}{\hat{t}^2}-\dfrac{\hat{t}\hat{s}}{\hat{u}^2}\right)$	j, k, l
	$\sum|\mathcal{M}|^2/(16\pi^2\alpha_s\alpha e_i^2)$	Fig. 10.19
$q\bar{q} \to \gamma^*g$	$\dfrac{8}{9}\dfrac{\hat{t}^2+\hat{u}^2+2\hat{s}(\hat{s}+\hat{t}+\hat{u})}{\hat{t}\hat{u}}$	a
$qg \to \gamma^*q$	$-\dfrac{1}{3}\dfrac{\hat{s}^2+\hat{u}^2+2\hat{t}(\hat{s}+\hat{t}+\hat{u})}{\hat{s}\hat{u}}$	b

$\sum|\mathcal{M}|^2$ is summed over final state and averaged over the initial state spins and colours. The Mandelstam variables for the sub-processes, $\hat{s},\hat{t},\hat{u}$, are defined at the start of Section 10.5.1. A useful summary together with plots of angular dependences is given by Tymieniecka and Zarnecki (1992).

Appendix: $\sum |\mathcal{M}|^2$ for tree-level sub-processes

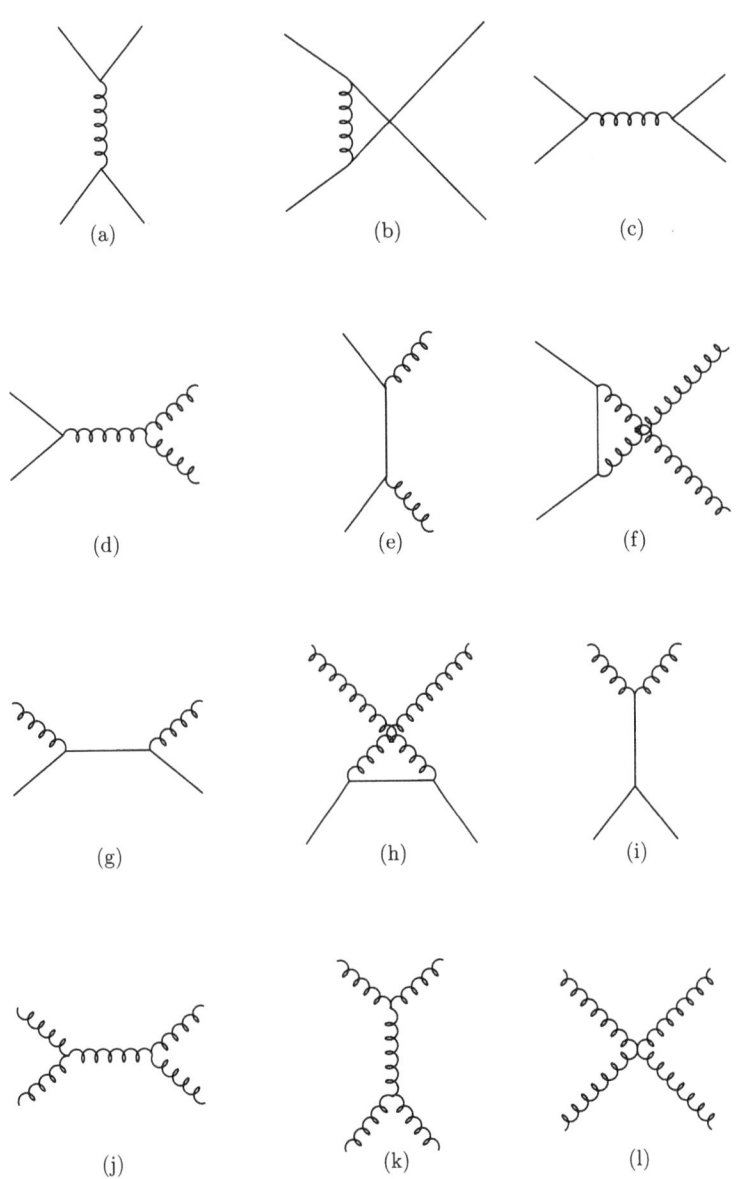

Fig. 10.27 Leading order QCD diagrams for quark and gluon scattering sub-processes as indicated in the table.

11
Polarized DIS

This chapter concentrates on the classic measurements of nucleon spin structure made using inclusive polarized deep inelastic scattering. A brief outline is given of how the unpolarized formalism is extended to describe polarized scattering. The subject is put into context by a short account of the 'spin crisis' — how a crucial early measurement did not support the simple quark model prediction that the valence quarks would carry most of the nucleon's spin. A summary is given of the sources and quality of the data on polarized structure functions. A section on theory outlines the operator product expansion for the polarized case, spin sum rules and the axial anomaly. This last turns out to be the key to a resolution of the problems raised by the 'spin crisis'. The polarized analogues of the DGLAP evolution equations are then outlined and how they are being used to analyse data and provide information on the polarized quark and gluon densities. The remaining sections cover: semi-inclusive polarized DIS; how better information might be gained in the future on the polarized gluon density and a short mention of transverse spin. Experiments on spin and polarization require subtle and delicate techniques which it is not appropriate to cover here but which are well worth following up for the range of topics covered from nuclear physics to cryogenic plumbing.

11.1 Formalism

A full derivation of the formalism for polarized deep inelastic scattering is quite complicated and beyond what is needed here. A brief summary is given in the appendix to this chapter, here the key formulae relating measured quantities to the spin dependent scaling functions or structure functions are presented. Since all polarized DIS experiments to date are fixed target with centre of mass energies well below the mass of the Z^0, only terms arising from virtual photon exchange are considered.

The primary measured quantities are the asymmetry for longitudinally polarized leptons scattering off longitudinally polarized nucleons

$$A_\parallel = \frac{d\sigma^{+-} - d\sigma^{++}}{d\sigma^{+-} + d\sigma^{++}}, \tag{11.1}$$

and the asymmetry for longitudinally polarized leptons scattering off transversely polarized nucleons

$$A_\perp = \frac{d\sigma^{+\uparrow} - d\sigma^{+\downarrow}}{d\sigma^{+\uparrow} + d\sigma^{+\downarrow}}. \tag{11.2}$$

In the above expressions $d\sigma^{+-}$ is the differential cross-section for the scattering of a lepton with spin parallel to its momentum off a nucleon with spin antiparallel to the lepton momentum; $d\sigma^{++}$ has both spins aligned parallel to the lepton momentum; $d\sigma^{+\uparrow}, d\sigma^{+\downarrow}$ have the nucleon spin aligned along directions perpendicular to the lepton momentum, the precise definitions are given in the appendix.

The asymmetries $A_\parallel$ and $A_\perp$ are related to the virtual photon asymmetries A_1, A_2 and to the structure functions g_1, g_2 by

$$A_\parallel = D(A_1 + \eta A_2), \qquad A_\perp = d(A_2 - \xi A_1),$$

$$A_1 = (g_1 - \gamma^2 g_2)/F_1, \qquad A_2 = \gamma(g_1 + g_2)/F_1,$$

where D, d, η, ξ and γ are kinematic factors and F_1 is the unpolarized structure function. The virtual photon asymmetries are bounded: $|A_1| \leq 1$ and $|A_2| \leq \sqrt{R}$, where $R = \sigma_L/\sigma_T$. Expressions for the kinematic factors are given in the appendix and η, ξ, γ are all small at high energies. Given that R is also small, $A_2 \approx 0$ and $A_1 \approx A_\parallel/D \approx g_1/F_1$. If both $A_\parallel$ and $A_\perp$ are measured, then g_1 and g_2 may be determined independently.

In the quark–parton model

$$F_1(x) = \frac{1}{2}\sum_i e_i^2 \left(q_i^+(x) + q_i^-(x)\right) = \frac{1}{2}\sum_i e_i^2 q_i(x)$$

$$g_1(x) = \frac{1}{2}\sum_i e_i^2 \left(q_i^+(x) - q_i^-(x)\right) = \frac{1}{2}\sum_i e_i^2 \Delta q_i(x), \tag{11.3}$$

where $q_i^+(x), q_i^-(x)$ are the quark distribution functions with momentum fraction x and helicities parallel, antiparallel to the nucleon helicity respectively. The sums run over active quark and antiquark flavours. These equations both define the quantities $\Delta q_i(x)$ and show the relationship between the polarized and unpolarized parton densities. In the QPM, $g_2(x) = 0$. One of the essential assumptions of the QPM is that transverse momentum of the initial state quarks is ignored. If this is relaxed, then it can be shown that g_2 is non-zero, however to make realistic estimates it is necessary to go beyond the QPM.

11.2 The 'spin crisis'

One of the key measurements in polarized DIS is the integral of g_1 over x,

$$\Gamma_1^N = \int_0^1 g_1^N(x, Q^2) dx = \frac{1}{2}\sum_i e_i^2 \Delta q_i, \tag{11.4}$$

where $\Delta q_i = \int_0^1 \Delta q_i(x)dx$. So for the proton, taking $n_f = 3$,

$$\Gamma_1^p = \frac{1}{2}\left[\frac{4}{9}\Delta u + \frac{1}{9}\Delta d + \frac{1}{9}\Delta s\right] = \frac{1}{9}a_0 + \frac{1}{12}a_3 + \frac{1}{36}a_8, \qquad (11.5)$$

where

$$a_0 \equiv \Delta\Sigma = \Delta u + \Delta d + \Delta s; \quad a_3 = \Delta u - \Delta d; \quad a_8 = \Delta u + \Delta d - 2\Delta s.$$

The rightmost part of Eq. (11.5) is a rearrangement to give terms with definite symmetry properties under $SU(3)_F$. As will be outlined later, this enables the combinations a_3, a_8 to be identified with the matrix elements responsible for neutron and hyperon beta decay, specifically $a_3 = F + D$, $a_8 = 3F - D$, where F, D are the two constants that occur in the matrix elements for the $SU(3)_F$ octet of weak axial currents.[1] Assuming that these are fixed from other experiments, one may write

$$a_0 = 9\Gamma_1^p - \frac{3}{4}a_3 - \frac{1}{4}a_8,$$

which then means that the net spin of the quarks, a_0, may be determined from the measurement of Γ_1^p.

The first polarized DIS experiment to have sufficient reach in x to measure Γ_1^p was the EMC experiment at CERN. The data, published in 1988, gave $0.126 \pm 0.01(\text{stat}) \pm 0.015(\text{sys})$ and using the values of a_3 and a_8 available at the time ($a_3 \approx 1.25$, $a_8 \approx 0.58$) gave $a_0 \approx 0.05$. Later experiments have refined this result but not changed the fact that a_0 is very small. Using the expressions given below Eq. (11.5) for the a_i in terms of their quark content gives $\Delta u \approx 0.79, \Delta d \approx -0.46, \Delta s \approx -0.18$. The large negative value for Δs was also a surprise. Simple application of quark model ideas give $a_0 \approx 0.7$, with the valence quarks carrying most of the proton's spin and other constituents giving small contributions. With hindsight, it is not surprising that this approach is too naive, given the importance of the $q\bar{q}$ sea and gluons at small x for unpolarized DIS. Formally, one may write for the spin of the proton

$$\frac{1}{2} = \frac{1}{2}\Delta\Sigma + \Delta g + \langle L_z \rangle, \qquad (11.6)$$

where $\Delta\Sigma$ is the contribution from the quarks (and antiquarks), Δg from the gluons and the last term the mean contribution of any orbital angular momentum of the constituents. The EMC result initiated intense activity on polarized DIS, both theoretical and experimental. It emerged that the relation between a_0 and $\Delta\Sigma$ is not as simple as assumed. It involves a contribution from gluons through the 'axial anomaly' as will be explained below. The challenge remains to find experimental handles on the three components, particularly Δg.

---

[1] See, for example, Section 6.4 of Commins and Buckbaum (1983).

Table 11.1 Key parameters of polarized DIS experiments.

Exp.	Beam	E (GeV)	x	Target	P_B	P_T	f
E154/5 (SLAC)	e^-	48	0.014 – 0.7	NH_3	0.81	0.80	0.15
				^3He	0.83	0.38	0.55
				LiD	0.81	0.22	0.36
EMC	μ^+	200	0.01 – 0.7	NH_3	0.79	0.78	0.16
SMC (CERN)	μ^+	190	0.003 – 0.7	NH_3	0.80	0.89	0.16
				H-butanol	0.80	0.86	0.12
				D-butanol	0.80	0.50	0.20
HERMES (DESY)	$e^\pm$	28	0.023 – 0.6	^3He	0.55	0.46	1.0
				H	0.55	0.88	1.0
				D	0.55	0.85	1.0
COMPASS (CERN)	μ^+	160		NH_3	0.80	0.90	0.16
				LiD	0.80	0.50	0.50

11.3 Experiments and data

All the existing data on polarized DIS comes from fixed target experiments at SLAC, CERN and DESY. At SLAC and CERN solid or liquid targets are used, for example H- or D-butanol, NH_3, LiD or ^{3}He. At DESY, the electron or positron beam of HERA is scattered off an internal gas jet target (H, D, ^{3}He). The measured asymmetry is reduced compared to the ideal asymmetry by the beam and target polarization and the fraction of the target nucleons that are polarized (dilution factor). Thus

$$A_{\text{measured}} = P_B P_T f A_{\text{ideal}},$$

where P_B, P_T are the beam and target polarizations and f the fraction of polarizable nucleons in the target. Typical values of these and other parameters for representative experiments are given in Table 11.1. It can be seen that for the CERN and SLAC solid targets polarizations vary over a considerable range and dilution factors are generally quite low. In the case of HERMES, the beam polarization is lower but this is compensated by a dilution factor of 1.0. The factors P_B, P_T, f must be monitored with great care, otherwise they will introduce unwanted systematic errors. Typically the polarizations have errors in the range 2.5–5%. Overall, these factors give rise to a normalization error of around 4% on the measured values of A_1 and hence g_1. In addition, to get from A_1 to g_1 requires knowledge of the unpolarized structure function F_1. This is calculated from the precisely measured F_2 and the less well known ratio $R = \sigma_L/\sigma_T$ using $F_1 = F_2(1 + \gamma^2)/(2x(1 + R))$. This adds a further uncertainty of around 2–4%.

The left-hand plots of Fig. 11.1 show a summary of data on xg_1 for the proton, deuteron and neutron (extracted from ^{3}He). As the Q^2 range of all experiments is quite restricted and g_1 varies slowly with Q^2, the data are plotted as functions of x with Q^2 at the measured mean values. For the

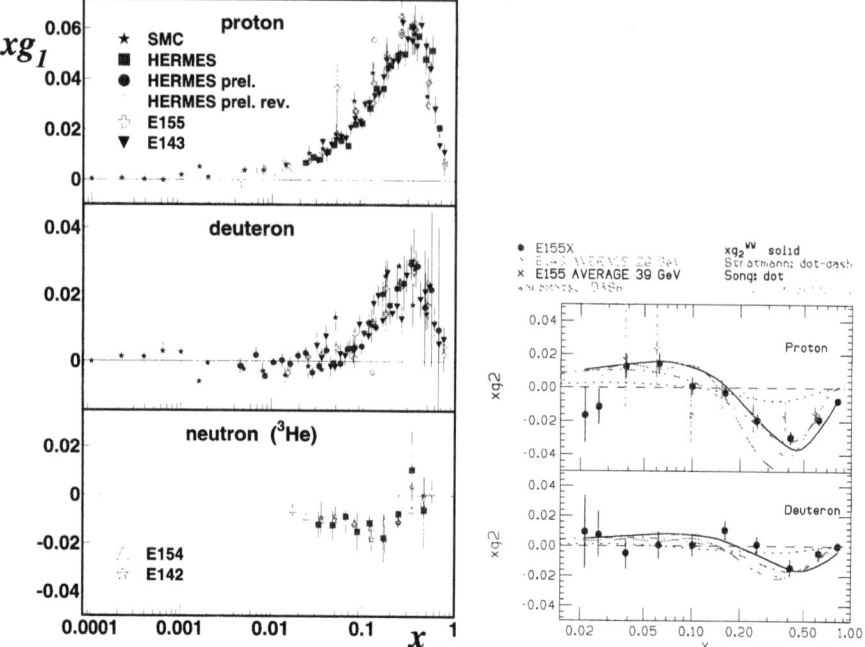

Fig. 11.1 Compilation of data on xg_1 (LH plots) and xg_2 (RH plots). The data (at measured mean values of Q^2) are plotted as functions of x. The solid curves on the RH plots show a calculation based on the Wandzura-Wilczek relation. (From Stösslein 2002.)

proton and deuteron, g_1 is directly related to measured quantities, but g_1^n must be extracted from ^{2}H or ^{3}He data using nuclear corrections. For ^{3}He, the polarization of the nucleus is largely that of the neutron so $g_1^n \approx g_1^{^3\text{He}}$. More precisely

$$g_1^n = \left(g_1^{^3\text{He}} - 2\rho_p g_1^p\right)/\rho_n,$$

where ρ_n, ρ_p are calculated parameters with values $\rho_n = 0.86 \pm 0.02$, $\rho_p = -0.028 \pm 0.004$ and other nuclear corrections are small. Although it is clear[2] from the figure that $g_1^d \approx g_1^p + g_1^n$, extracting g_1^n from g_1^d is not so precise experimentally as it requires subtracting the much larger g_1^p. The data for g_1^p and g_1^n are peaked in the valence region but with opposite signs.

For smaller x, below 0.01, only the SMC experiment has a large enough centre-of-mass energy ($\sqrt{s} \approx 18.9\,\text{GeV}$) to provide data, g_1^p is positive and g_1^n (extracted from g_1^d) negative. Although with large errors, both g_1^p and g_1^n are increasing in magnitude for $x < 0.01$. The data do not allow one to determine whether g_1 will show a large increase at very small x, as

---

[2] The more exact relation includes a factor allowing for the D-state contribution to the deuteron wave-function.

found for the unpolarized structure functions. (This would indicate a strong polarization of the $q\bar{q}$ sea.)

The right-hand plots of Fig. 11.1 show the world data on xg_2 with the high statistics experiment E155X establishing that for both the proton and deuteron xg_2 is non-zero and negative at large x. These results show that more than the simple QPM is required to describe the data on polarized structure functions. The solid curves show a calculation based on the Wandzura–Wilczek relation which is covered in Section 11.4.2 below.

11.4 Theoretical framework

As for unpolarized DIS, the theoretical underpinning of the QCD enhanced parton model is provided by the Operator Product Expansion and the polarized analogues of the DGLAP evolution equations. The latter will be covered in the next section. Here a brief outline is given of the OPE for $g_{1,2}$ and the relationship to various sum rules that have played an important role in spin physics. The aim is to give an understanding of the connection between polarized structure functions and weak matrix elements, concentrating on g_1. For full details, one should refer to the original literature and a good starting point is Section 5 of the review by Anselmino et al. (1995).

For simplicity, it will be assumed that $n_f = 3$ and the u, d, s quarks are massless with an SU(3)$_F$ symmetry. The OPE provides expressions for the odd moments of the structure functions in terms of products of calculable coefficient functions with matrix elements of derivative operators. The flavour structure of the matrix elements is an SU(3)$_F$ octet plus two singlet terms, labelled 0 and g and associated with quark and gluon terms, respectively. The matrix elements are Dirac axial vector currents because the g_i are related to *differences* of nucleon or quark helicity states, i.e. $\gamma^\mu(1+\gamma_5) - \gamma^\mu(1-\gamma_5)$. For g_1, at leading twist 2, the OPE gives

$$\int_0^1 x^{n-1} g_1(x, Q^2) dx = \frac{1}{2} \sum_i \delta_i a_n^i E_{1,i}^n(Q^2, \alpha_s) \quad n = 1, 3, 5\ldots \quad (11.7)$$

where a_n^i are the matrix elements, $E_{1,i}^n$ the coefficient functions, and δ_i SU(3)$_F$ numerical factors. The sum over i runs over the flavour indices, but only the terms corresponding to $i = 3, 8$ and the two singlet terms are non-zero. The coefficient functions are calculated perturbatively as power series in α_s and will depend on the renormalization scheme and scale. The matrix elements are given by

$$\frac{2m_N a_n^i}{n} \{s^\sigma p^{\mu_1} \ldots p^{\mu_{n-1}}\}_{\text{sym}} = -\langle p, s | \hat{R}_{1,i}^{\sigma\mu_1\ldots\mu_{n-1}} | p, s \rangle$$

where s is the spin 4-vector and p the 4-momentum, with the operators $\hat{R}$

$$\hat{R}_{1,i}^{\sigma\mu_1\ldots\mu_{n-1}} = (i)^{n-1} \left\{ \bar{\psi}\gamma_5\gamma^\sigma D^{\mu_1} \ldots D^{\mu_{n-1}} \left(\frac{\lambda_i}{2}\right) \psi \right\}_{\text{sym}} \quad n \geq 1$$

$$\hat{R}^{\sigma\mu_1...\mu_{n-1}}_{1,0} = (i)^{n-1}\left\{\overline{\psi}\gamma_5\gamma^\sigma D^{\mu_1}...D^{\mu_{n-1}}\psi\right\}_{sym} \quad n \geq 1$$

$$\hat{R}^{\sigma\mu_1...\mu_{n-1}}_{1,g} = (i)^{n-1}Tr\left\{\varepsilon^{\sigma\alpha\beta\gamma}\mathbf{G}_{\beta\gamma}D^{\mu_1}...D^{\mu_{n-2}}\mathbf{G}^{\mu_{n-1}}_\alpha\right\}_{sym} \quad n \geq 2$$

where ψ are the quark fields, $\mathbf{G}_{\mu\nu}$ the gluon field tensor, D^μ the QCD covariant derivative and $\boldsymbol{\lambda}_i$ the Gell-Mann SU(3)$_F$ matrices. Note that the gluon operators do not contribute directly to lowest moment $n = 1$. Similar expressions may be constructed for g_2 from operators of twist 3, with $n = 3$ the lowest non-zero moment.

11.4.1 Sum rules

Of particular importance is the $n = 1$ moment of g_1, which takes the form

$$\Gamma^{p(n)}_1(Q^2) \equiv \int_0^1 g^{p(n)}_1(x, Q^2)dx = \frac{C^{NS}_1(Q^2)}{12}\left[\frac{a_8}{3} \pm a_3\right] + \frac{C^S_1(Q^2)}{9}a_0(Q^2), \tag{11.8}$$

where $+,-$ in the square brackets refer to proton and neutron, respectively and $C^{NS,S}_1$ are the SU(3)$_F$ non-singlet and singlet coefficient functions known to order α_s^3. From this result, one gets immediately

$$\Gamma^p_1(Q^2) - \Gamma^n_1(Q^2) = \frac{C^{NS}_1(Q^2)}{6}a_3 \tag{11.9}$$

Eqs (11.8) and (11.9) are the QCD forms of the Ellis–Jaffe and Bjorken sum rules respectively. Note that only a_0 depends on Q^2, this is because a_3, a_8 are matrix elements of conserved currents in the limit of massless quarks. Conservation of the related axial current operators is a physical statement independent of scale. The reason why this is not the case for a_0 will be discussed in connection with the axial anomaly below.

Going beyond the OPE, and assuming that $g_2(x, Q^2) \to 0$ less singularly than $1/x^2$, two sum rules involving g_2 have been conjectured:

$$\Gamma_2(Q^2) \equiv \int_0^1 g_2(x, Q^2)dx = 0 \quad \text{Burkhardt} - \text{Cottingham},$$

$$\int_0^1 x[g_1(x, Q^2) + 2g_2(x, Q^2)]dx = 0 \quad \text{Efremov} - \text{Leader} - \text{Teryaev},$$

details are given in Anselmino et al. (1995).

11.4.2 Wandzura–Wilczek relation

The Wandzura–Wilczek relation is a relation between g_1 and g_2 requiring the same assumption on the low x behaviour of g_2 as in the previous section and the further assumption that twist-3 terms can be neglected:

$$g_1(x, Q^2) + g_2(x, Q^2) = \int_x^1 \frac{dy}{y}g_1(y, Q^2).$$

Taking this to define g_2^{WW} one may write $g_2(x,Q^2) = g_2^{WW}(x,Q^2) + \overline{g}_2(x,Q^2)$, where

$$\overline{g}_2(x,Q^2) = -\int_x^1 \frac{\partial}{\partial y}\left(\frac{m_q}{m_N}h_T(y,Q^2) + \xi(y,Q^2)\right)\frac{dy}{y}.$$

The function h_T represents a twist-2 contribution depending on the transverse polarization of the quark in the nucleon, but for light quarks it is suppressed by the mass ratio and its contribution will be ignored. The other term, ξ, is the twist-3 part arising from quark–gluon correlations. The OPE then gives the relations

$$d_n = 2\int_0^1 dx\, x^n \left[\frac{n+1}{n}g_2(x,Q^2) + g_1(x,Q^2)\right] = 2\frac{n+1}{n}\int_0^1 dx\, x^n\, \overline{g}_2(x,Q^2),$$

where d_n are the moments of the twist-3 matrix elements. In particular

$$d_2 = 3\int_0^1 dx\, x^2\, \overline{g}_2(x,Q^2).$$

Assuming that any Q^2 dependence may be ignored in the region measured, the E155 collaboration (E155 2003) have extracted values for $d_2^{p,d}$ at a mean Q^2 of 5 GeV^2 from the difference between their measured data for g_2 and g_2^{WW} calculated from their accurate data on g_1. The twist-3 contribution is very small as may be seen from the right-hand plots of Fig. 11.1. More precisely $d_2^p = 0.0025 \pm 0.0019$, $d_2^d = 0.0054 \pm 0.0024$, which shows that the twist-3 contributions are consistent with zero within two standard deviations.

11.4.3 Axial anomaly

If QCD interactions are ignored, the normalization of the coefficients in Eq. (11.8) is such that C_1^{NS} and C_1^S both reduce to 1 and the simple quark model relation of Eq. (11.5) is recovered, with the quark content of the a_i as given there. In particular, a_0 is identified with $\Delta\Sigma$ — the singlet quark term.

Consider the axial current operator for quarks of flavour i

$$j_{5\mu}^i = \overline{\psi}_i(x)\gamma_\mu\gamma_5\psi_i(x).$$

Using the free-field Dirac equation for ψ_i one finds that

$$\partial^\mu j_{5\mu}^i = 2im_i\overline{\psi}_i(x)\gamma_5\psi_i(x),$$

where m_i is the mass of the quark. In the massless, or chiral limit, $m_i \to 0$ and the current $j_{5\mu}^i$ would appear to be conserved. If this were true, even approximately, then there would be symmetry between LH and RH quark helicity states leading to a parity degeneracy of the hadron spectrum. As there is no evidence for this (no chiral partners for the proton or neutron for example), the symmetry must be broken, or the axial current is not conserved through other contributions. The apparent conservation, derived above, is based on the free–field equations of motion. When quark–gluon interactions are allowed, there is a contribution from the quark 'triangle' diagram connecting the $\gamma_\mu \gamma_5$ current to two gluon fields. This is an example of the Adler–Bell–Jackiw anomaly and, for QCD, the diagram gives a non-zero contribution to $\partial^\mu j^i_{5\mu}$

$$\partial^\mu j^i_{5\mu} = \frac{\alpha_s}{8\pi} G^a_{\mu\nu} \varepsilon^{\mu\nu\rho\sigma} G^a_{\rho\sigma}$$

where G is the gluon field tensor and a the colour label. The anomaly introduces an effectively pointlike interaction between the quark singlet axial current and gluons and this in turn gives a contribution to the hadronic matrix element a_0, which for three flavours is

$$a_0^g(Q^2) = -3\frac{\alpha_s}{2\pi} \int_0^1 \Delta g(x, Q^2) dx.$$

Δg is the difference between the density of gluons with the same helicity as that of the nucleon and those with the opposite helicity. The complete expression for a_0 becomes

$$a_0 = \Delta\Sigma - 3\frac{\alpha_s}{2\pi}\Delta g. \qquad (11.10)$$

This result shows that a_0 cannot be equated directly with the mean quark spin and that a small value for a_0 does not have to imply that $\Delta\Sigma$ is small. It also means that the 'missing' spin contributions to the valence u and d quarks could be carried by the strange quark component or gluons or a combination of the two.

11.5 QCD analysis

The application of QCD factorization theorem and the resulting structure of the evolution equations for polarized DIS are very similar to the unpolarized case. The equations have been derived up to NLO and many groups have performed quite detailed analyses using the full polarized structure function data sample. Here, for definiteness, the formalism of Blümlein and Böttcher (2002) is followed. They analyse all available data and give the 1σ correlated error bands on the resulting polarized parton densities from propagation of statistical errors. Their paper contains full references to original calculations and other QCD analyses. The SMC collaboration (SMC 1998) has also performed a QCD analysis of polarized structure functions, taking account of statistical, systematic and theoretical uncertainties.

The twist 2 structure function $g_1(x, Q^2)$ is written as convolution of coefficient functions and polarized parton densities.

$$g_1(x, Q^2) = \frac{1}{2}\sum_{i=1}^{n_f} e_i^2 \int_x^1 \frac{dz}{z}\left[\frac{1}{n_f}\Delta\Sigma\left(\frac{x}{z},\mu_F^2\right)\Delta C_q^S\left(z,\frac{Q^2}{\mu_F^2}\right) + \Delta G\left(\frac{x}{z},\mu_F^2\right)\right.$$
$$\left.\times \Delta C^G\left(z,\frac{Q^2}{\mu_F^2}\right) + \Delta q_i^{NS}\left(\frac{x}{z},\mu_F^2\right)\Delta C_q^{NS}\left(z,\frac{Q^2}{\mu_F^2}\right)\right], \quad (11.11)$$

where e_i denotes the charge of the ith quark flavor and n_f is the number of flavors. The factorization scale μ_F is introduced to remove the collinear singularities from the parton densities. In addition, there will be dependence on the renormalization scale of α_s. As always with factorization dependent perturbative expansions of physical quantities it is essential to match the scheme and order dependence of the parton densities and coefficient functions. The complete, all orders, expression for $g_1(x, Q^2)$ will be independent of both scales. The singlet and non-singlet parton densities are given by

$$\Delta\Sigma\left(x,\mu_F^2\right) = \sum_{i=1}^{n_f}\left[\Delta q_i(x,\mu_F^2) + \Delta\bar{q}_i(z,\mu_F^2)\right] \quad \text{and}$$

$$\Delta q_i^{NS}\left(z,\mu_F^2\right) = \Delta q_i(z,\mu_F^2) + \Delta\bar{q}_i(z,\mu_F^2) - \frac{1}{n_f}\Delta\Sigma\left(z,\mu_F^2\right).$$

The evolution of the parton densities with the scale $\mu_F^2 = Q^2$ is given by the polarized equivalents of the DGLAP equations

$$\frac{\partial \Delta q_i^{NS}(x, Q^2)}{\partial \ln Q^2} = P_{NS}^-(x,\alpha_s) \otimes \Delta q_i^{NS}(x, Q^2), \quad (11.12)$$

$$\frac{\partial}{\partial \ln Q^2}\begin{pmatrix}\Delta\Sigma(x,Q^2)\\ \Delta G(x,Q^2)\end{pmatrix} = \boldsymbol{P}(x,\alpha_s) \otimes \begin{pmatrix}\Delta\Sigma(x,Q^2)\\ \Delta G(x,Q^2)\end{pmatrix}, \quad (11.13)$$

where P_{ij} are the splitting functions

$$P_{NS}^-(x,\alpha_s) = \alpha_s P_{NS}^{(0)}(x) + \alpha_s^2 P_{NS}^{-(1)}(x) + O(\alpha_s^3), \quad (11.14)$$

$$\boldsymbol{P}(x,\alpha_s) \equiv \begin{pmatrix}P_{qq}(x,Q^2) & P_{qg}(x,Q^2)\\ P_{gq}(x,Q^2) & P_{gg}(x,Q^2)\end{pmatrix}$$

$$= \alpha_s \boldsymbol{P}^{(0)}(x) + \alpha_s^2 \boldsymbol{P}^{(1)}(x) + O(\alpha_s^3). \quad (11.15)$$

The polarized coefficient functions and splitting functions have been calculated to next-to-leading order.

The equations given here are for the $\overline{\text{MS}}$ renormalization scheme. Blümlein and Böttcher solve the evolution equations using Mellin n-space (moments) techniques and calculate g_1 from the three light flavours only ($n_f = 3$). The running coupling, α_s, is calculated with $n_f = 4$. The structure of the equations described here follows very closely the structure used for unpolarized

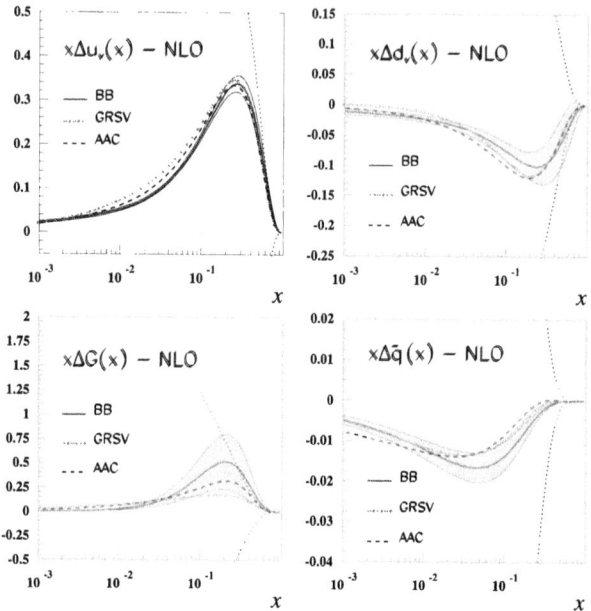

Fig. 11.2 NLO polarized parton densities at the starting scale $Q_0^2 = 4\,\text{GeV}^2$, with 1σ error bands, from the fit by Blümlein and Böttcher (2002). Curves from the GRSV and AAC NLO fits are shown for comparison.

DIS. For polarized DIS, other choices of renormalization scheme are often used, for example those of Adler and Bardeen and the so-called JET scheme (for details see Anselmino et al. 1995). For these two schemes, the quark singlet term $\Delta\Sigma$ is defined to be a conserved quantity and is then scale independent. The singlet terms in the different schemes are related by

$$\Delta\Sigma^{\overline{\text{MS}}} = \Delta\Sigma^{\text{AB/JET}} - n_f \frac{\alpha_s}{2\pi} \Delta g,$$

where Δg is the same in all three schemes.

The polarized quark densities are parameterized at the starting scale, Q_0^2 by functions of the form

$$x\Delta q_i(x, Q_0^2) = \eta_i A_i x^{a_i} (1-x)^{b_i} \left(1 + \gamma_i x + \rho_i \sqrt{x}\right),$$

with a similar expression for the gluon density Δg. The normalization constants are chosen such that η_i give the first moments of the parton densities at the starting scale $\eta_i = \int_0^1 \Delta q_i(x, Q_0^2) dx$. This then allows the constraints on the u and d valence distributions from neutron and hyperon decay to be implemented easily:

$$\eta_{u_v} - \eta_{d_v} = F + D, \qquad \eta_{u_v} + \eta_{d_v} = 3F - D.$$

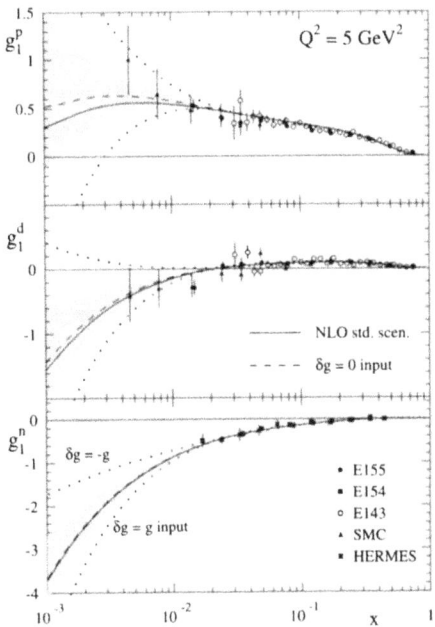

Fig. 11.3 The x dependence of g_1^N from the GRSV NLO fit. The full and dashed curves show the NLO fit with Δg included and set to zero at the input scale, respectively. The shaded band shows the 1σ error band on the best fit. (From Glück et al. (2001).)

SU(3)$_F$ symmetry is assumed for the light (u, d, s) quarks and a flavour symmetric q$\bar{\text{q}}$ sea is assumed ($\Delta\bar{q}(x) = \Delta\bar{u}(x) = \Delta\bar{d}(x) = \Delta s(x) = \Delta\bar{s}(x)$). The extent and precision of the polarized data, in comparison to the unpolarized case, does not allow nearly as much detail to be determined on the shape and symmetry properties of the polarized parton densities, so that even after these assumptions are applied there are still too many free parameters in the general form given above. Hence, further restrictions are necessary: the ρ parameters are set to zero and γ allowed to be non-zero only for u_v and d_v densities. Finally, 12 parameters are used to describe the parton densities at the starting scale. In addition, $\Lambda_{\rm QCD}^{(4)}$ is an additional fit parameter determined through the dependence of the evolution equations on α_s. Even with this reduced set of parameters, it is found that the data does not significantly constrain the parameters, $\gamma_{u_v}, \gamma_{d_v}, b_{\bar{q}}, b_g$, so they were set to values determined from an initial fit, and do not enter into the error analysis. The small x behaviours of the sea and gluon distributions are strongly correlated, so this is built in by assuming $a_g = a_{\bar{q}} + c$, with c a constant in the range 0.5–1.0. In the end, seven PDF parameters and $\Lambda_{QCD}^{(4)}$ are determined from a total 435 data points on the asymmetry $A_1(\approx g_1/F_1)$ for p, n and d targets. Only by restricting the number of parameters allowed to vary in the fit to those that can be determined reasonably well, can reliable error estimates for the parton densities and other quantities be obtained from the parameter covariance matrix. Fits have been performed at both LO and NLO and the quality of the overall description of the data is good with χ^2/ndf in the range 0.9–1.0. The parton densities at the starting scale from the NLO fit with $a_g = a_{\bar{q}} + 1.0$ are shown in Fig. 11.2, together with their 1σ error bands. Also shown for comparison are curves from two other NLO fits: GRSV from Glück et al (2001) and from the AAC group, Goto et al. (2000).

The results show a pattern rather reminiscent of that from unpolarized NLO QCD fits: the best determined of the polarized parton densities is Δu_v, followed by Δd_v and $\Delta \bar{q}$, with Δg the least well determined. This is not surprising as the data is concentrated in the valence region, with the only handle on the gluon density through the rather weak scaling violations of g_1. The uncertainty at low x is shown in Fig. 11.3 from the GRSV fit. This figure shows the x dependence of g_1^N at $Q^2 = 5\,\text{GeV}^2$, rather than the more usual xg_1^N. The structure function is tightly constrained in the valence region, but the 1σ uncertainty grows rapidly for $x < 0.01$. Also shown is the result of putting $\Delta g = 0$ at the input scale (dashed curves). The dotted curves show the result of allowing the gluon component to take on the maximum values allowed by the positivity constraint $\Delta g(x, Q^2) = \pm g(x, Q^2)$ — where g is the unpolarized density. This is becoming disfavoured by the data.

The value of α_s derived from the Blümlein–Böttcher NLO fit is $\alpha_s(M_Z^2) = 0.113^{+0.010}_{-0.008}$ (with statistical and systematic errors combined), somewhat low compared to the world average of 0.118 ± 0.002, but compatible within the errors. Apart from the clear need for more experimental information on Δg, a major source of uncertainty in the present analyses is the need to know the unpolarized R which is used in the calculation of A_1 from g_1 and F_2.

Despite all the caveats, the present NLO QCD analyses of polarized data do represent a major advance in understanding of nucleon spin physics. They also provide an unambiguous answer to an earlier problem encountered in testing sum rules — particularly those for g_1 — namely how to extrapolate the measured values to a common value of Q^2 and to values of x below the region of experimental measurements. This is particularly important for the Bjorken sum rule (Eq. 11.9). This sum rule has the smallest theoretical uncertainty since a_3 does not evolve with Q^2 and is fixed from neutron decay. For $n_f = 3$ massless quarks (u, d, s)

$$\Gamma_1^p(Q^2) - \Gamma_1^n(Q^2) = \frac{a_3}{6}\left[1 - \frac{\alpha_s}{\pi} - 3.58\frac{\alpha_s}{\pi}^2 - 20.22\frac{\alpha_s}{\pi}^3 \cdots\right],$$

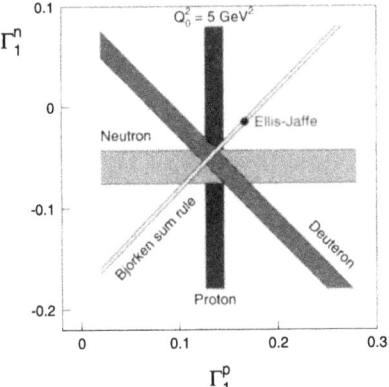

Fig. 11.4 Summary of the data on $\Gamma_1^{p,n}$ using NLO QCD for extrapolation. From the SMC experiment, together with predictions for the Bjorken and Ellis–Jaffe sum rules. (Data from SMC 1998.)

which gives a numerical value of 0.182 ± 0.005 at $Q^2 = 5\,\text{GeV}^2$. A number of experimental groups have used NLO QCD fits to extrapolate their data and make accurate measurements of the Bjorken sum rule at $Q^2 = 5\,\text{GeV}^2$, for example $0.176 \pm 0.003 \pm 0.007$ from E155 (2000) and $0.174^{+0.024}_{-0.012}$ from SMC (1998). The experimental results are well compatible with each other and both agree with the theoretical prediction to within 10%. A nice graphical summary of the status of data on first moments of g_1 from the SMC experiment is shown in Fig. 11.4. The point labelled Ellis–Jaffe is the quark model prediction for $\Gamma_1^{p,n}$ separately and in the light of earlier discussion it is not surprising that it does not agree that well with the data.

11.6 Semi-inclusive asymmetries

Another route to separating the polarized quark distributions is to measure semi-inclusive asymmetries, that is, to measure a specific hadron h in the final state in coincidence with the scattered lepton. If the chosen hadron is strongly correlated with an underlying quark flavour then one has access to the associated Δq distribution. The additional information needed is the appropriate $q \to h$, fragmentation function, $D_q^h(z,Q^2)$, where $z = E_h/(E - E')$ is the normalized hadronic energy in the lab frame. Fragmentation functions are non-perturbative objects assumed to be universal and independent of spin orientation. Assuming further that the fragmentation process occurs on a long timescale in comparison to the hard polarized lepton–quark interaction, the hadronic asymmetry may be written

$$A_1^h(x,Q^2,z) = \frac{\sum_i e_i^2 \Delta q_i(x,Q^2) D_i^h(z,Q^2)}{\sum_i e_i^2 q_i(x,Q^2) D_i^h(z,Q^2)} \frac{(1+R(x,Q^2))}{(1+\gamma^2)}. \qquad (11.16)$$

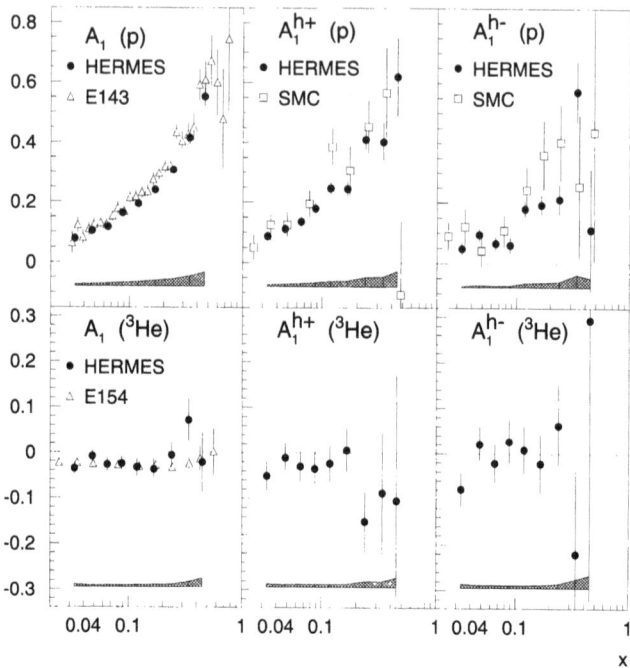

Fig. 11.5 Examples of semi-inclusive asymmetries A_1^h for positive and negative hadrons measured on proton and ^{3}He targets, together with the fully inclusive asymmetries. (From HERMES 1999).

The last factor $(1+R)/(1+\gamma^2)$ is close to unity as both R and γ are small. To extract the polarized quark densities, the measured A_1^h, as defined above, are first integrated over Q^2 and $0.2 < z < 1.0$ giving

$$A_1^h(x) = \sum_i P_i^h(x) \frac{\Delta q_i(x)}{q_i(x)} \frac{(1 + R(x, Q^2))}{(1 + \gamma^2)},$$

where P_i^h — the purity of finding quark flavour i from hadron h — is

$$P_i^h(x) = \frac{e_i^2 q_i(x) \int_{0.2}^1 D_i^h(z) dz}{\sum_i e_i^2 q_i(x) \int_{0.2}^1 D_i^h(z) dz},$$

with the integration over Q^2 implied. Examples from proton and ^{3}He targets are shown in Fig. 11.5. The Δq may now be unfolded using purities determined from a Monte Carlo simulation of the fragmentation process (using the LUND model, for example) and including detector effects (see Appendix E). Given that the measured data is concentrated in the valence region in x, simplifying assumptions have to be made on the number of quark densities that can be extracted from the data. Assuming an SU(3)$_F$

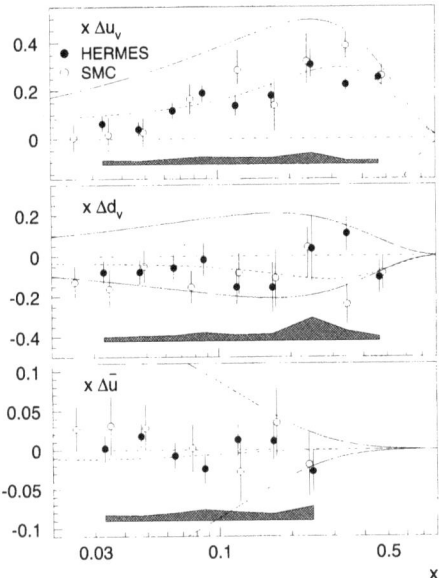

Fig. 11.6 Polarized quark distributions $x\Delta q$ extracted from semi-inclusive asymmetries. The solid lines show positivity limits and the dashed lines are from the calculation of Gehrmann & Stirling, based on a LO QCD fit. (From HERMES 1999.)

symmetric sea allows u_v, d_v and $\bar{q}$ distributions to be extracted as shown in Fig. 11.6. Within the quite large errors, the data agree with the corresponding distributions extracted from the Gehrmann and Stirling (1996) LO QCD fit to the fully inclusive data. The dominant distribution in the valence region is $x\Delta u_v$, with both $x\Delta d_v$ and $x\Delta \bar{q}$ small.

An important goal of semi-inclusive asymmetry measurements is to give a direct estimate of the strange quark distribution Δs. For this to be possible, particle identification is needed and, in particular, $K^\pm/\pi^\pm$ separation.

11.7 Δg

One of the biggest challenges in polarized nucleon physics is to find experimental handles on the polarized gluon distribution Δg. An obvious, but perhaps remote, possibility is to extend the reach of polarized DIS by modifying HERA to operate with polarized protons as well as polarized $e^\pm$. This would allow measurements at low x where the $q\bar{q}$ sea and gluon dominate, giving Δg through scaling violations of g_1. What of the possibilities at existing accelerators: HERMES at HERA ($e^\pm$ fixed target), COMPASS at CERN (combined muon and hadron beam fixed–target experiment) and fully polarized pp scattering at RHIC? Each will be considered in turn.

11.7.1 HERMES

The HERMES collaboration has attempted to make a direct measurement of Δg via the boson or photon–gluon fusion (BGF) process, in which a $q\bar{q}$ pair is produced by the interaction of the virtual photon with a gluon in the target. This process has already been considered in some detail in Sections 3.2 and 4.5 as it is the primary mechanism for the production of heavy flavours and particularly $c\bar{c}$ pairs in DIS. It is also one of the mechanisms by which high energy jets are produced. Given the low CM energy of the HERMES experiment, measurement of charm production or of jet production is not a realistic possibility. However, the production of high p_T hadrons via the BGF process is a possible channel from which information on Δg can be obtained (HERMES 2000). Events with at least two oppositely charged hadrons with large p_T from DIS off a polarized hydrogen target are selected. Requiring $p_T^{h_1} > 1.5\,\text{GeV}$ and $p_T^{h_2} > 1.0\,\text{GeV}$, the double spin asymmetry is found to be $A_\| = -0.28 \pm 0.12(\text{stat}) \pm 0.02(\text{sys})$. In addition to the BGF process, the QCD Compton (QCDC) process and vector meson production could contribute to the measured asymmetry. Vector mesons give negligible contributions to $A_\|$ but do dilute it. One then expects

$$A_\| \approx \left(\hat{a}_{\text{BGF}} \frac{\Delta g}{g} f_{\text{BGF}} + \hat{a}_{\text{QCDC}} \frac{\Delta q}{q} f_{\text{QCDC}} \right) D$$

where f_i are the unpolarized fractions of events from each process with $f_{\text{BGF}} + f_{\text{QCDC}} + f_{\text{VM}} = 1$, $\hat{a}_i$ are the asymmetries for the two hard processes and D is the dilution factor. For LO QCD with massless quarks one finds $\langle \hat{a}_{\text{BGF}} \rangle = -1.0$ and $\langle \hat{a}_{\text{QCDC}} \rangle = -0.5$, where the average is taken over the kinematics of the events (using Monte-Carlo techniques). The quark contribution $\Delta q/q$ is a weighted combination of the $\Delta u/u$ and $\Delta d/d$ distributions measured in semi-inclusive DIS. The process fractions f_i are estimated using the PYTHIA Monte Carlo. Finally a mean value of $\Delta g/g = 0.41 \pm 0.18(\text{stat}) \pm 0.03(\text{sys})$ is extracted at $\langle x \rangle = 0.17$ and $\langle p_T^2 \rangle = 2.1\,\text{GeV}^2$ (the hard scale here is set by p_T^2 rather than Q^2). That Δg is non-zero and positive at this x value agrees with the distribution extracted from NLO QCD fits (see Fig. 11.2).

11.7.2 COMPASS

The COMPASS experiment at CERN is a facility for spectroscopy and spin physics using hadron and muon beams from the SpS with a variety of targets and a range of sophisticated detectors for analysing the final state. The spin programme, in which muon beams of typical energies 100–200 GeV will be used with a range of cyrogenic polarized targets is a continuation of the CERN SMC programme, but with the aim of using final state properties to complement the earlier inclusive measurements. With a wide range of particle identification devices, the measurement of Δg through the BGF

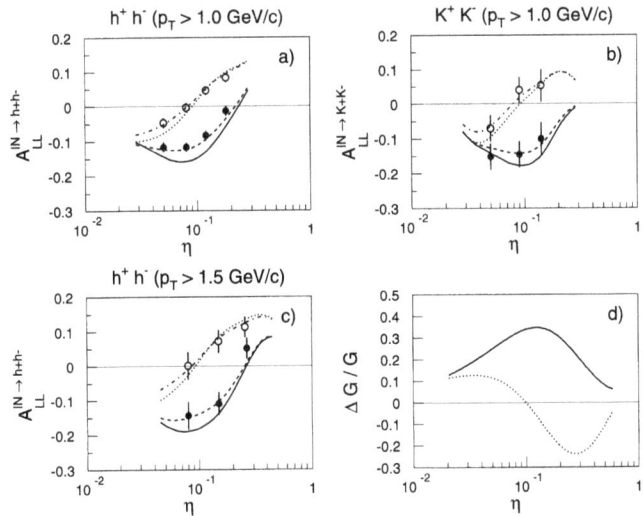

Fig. 11.7 $A_{\parallel}$ for $\mu N \to h^+h^- X$ from simulated data from the COMPASS experiment as a function of rapidity. The closed points and full lines correspond to Δg from Gehrmann–Stirling fit A and the open points and dotted lines to fit C. (From Bravar et al. 1998.)

production of both $c\bar{c}$ pairs and high p_T hadrons will be a primary aim. For high p_T hadrons, the measurement and analysis will follow the procedure outlined for HERMES, except that with the higher CM energy (17 GeV as opposed to 7 GeV for HERMES) only the BGF and QCDC processes need to be considered. Bravar et al. (1998) have made estimates of a typical year of running with a 200 GeV μ^+ beam incident on a deuteron target and with cuts to select two oppositely charged hadrons in central rapidities. They find the numbers of events expected are of the order of 700k h^+h^- pairs with $p_T > 1.0$ GeV, 70k K^+K^- with $p_T > 1.0$ GeV and 80k h^+h^- pairs with $p_T > 1.5$ GeV. Fig. 11.7 shows the expectation for various channels for two possible gluon distributions from the Gehrmann and Stirling analysis. It should be possible to distinguish between the two possibilities shown. Compass will also have a high enough CM energy to collect a reasonable number of $c\bar{c}$ events and will identify the events by D^0 and D^* decays. For example, with the first data collected using 160 GeV muons on a LiD target, 16k charm events have been reconstructed with an expectation that $\Delta g/g$ can be measured to about 18% at a mean rapidity of 0.1.

11.7.3 RHIC $\vec{p}\vec{p}$

In addition to its operation as a heavy ion collider, the RHIC facility at Brookhaven will also run with polarized proton beams giving $\vec{p}\vec{p}$ collisions at CM energies of up to $\sqrt{s} = 500\,\text{GeV}$. Typical polarizations of the proton beams will be around 70%. A first run with polarized protons has taken place at $\sqrt{s} = 200\,\text{GeV}$ and $\vec{p}\vec{p}$ operations will be interleaved on a regular basis with heavy ion running. Since both longitudinal and transverse polarizations states can be prepared for the protons the number of possible spin asymmetry measurements is large. Concentrating on longitudinal spin measurements first, the polarized quark distributions are written $\Delta q = q_+^+ - q_+^-$ where the upper and lower indices indicate the quark and hadron helicities respectively. The observed double longitudinal spin asymmetry is given by

$$A_{LL} = \frac{1}{P_1 P_2} \frac{(N_{++} + N_{--}) - (N_{+-} + N_{-+})}{(N_{++} + N_{--}) + (N_{+-} + N_{-+})},$$

where N_{ij} is the number of events with $\vec{p}\vec{p}$ helicities i and j and P_1, P_2 are the beam polarizations. As such asymmetries are expected to be quite small, the accuracy will be approximately

$$\Delta A_{LL} \sim \frac{1}{P_1 P_2} \times \frac{1}{\sqrt{N}},$$

where N is the total number of events summed over all spin states, for example, for $P_1 \approx P_2 \approx 0.7$ and 10^4 events $\Delta A_{LL} \sim 0.02$. A primary goal of the RHIC spin programme is to make direct measurements of Δg. Processes such as high p_T prompt photon production $\vec{p}\vec{p} \to \gamma + X$, high p_T jet production $\vec{p}\vec{p} \to j_1 j_2 + X$ and heavy flavour production $\vec{p}\vec{p} \to c\bar{c}\,(b\bar{b}) + X$ will all be considered. The underlying physics and the factorization of the hadronic cross-section into a convolution over a partonic cross-section and two parton densities is almost identical to that discussed in Chapter 10 on hadronic DIS, with the exception that many of the calculations are only available at LO. As an example consider the prospects for measuring Δg with direct photons. The longitudinal asymmetry for this process may be written as

$$A_{LL}^\gamma = \frac{\Delta g}{g}(x_1) A_1^p(x_2) \hat{a}(gq \to \gamma q) + (1 \leftrightarrow 2), \qquad (11.17)$$

where A_1^p is the proton asymmetry already measured in DIS and $\hat{a}(gq \to \gamma q)$ is the asymmetry for the hard process. Figure 11.8 shows the results of a simulated measurement of Δg in the STAR detector using $320\,\text{pb}^{-1}$ at $\sqrt{s} = 200\,\text{GeV}$ (open circles) and $800\,\text{pb}^{-1}$ at $\sqrt{s} = 500\,\text{GeV}$ (filled circles) with three different input gluon distributions from the Gehrmann–Stirling LO set.

Another interesting possibility at RHIC is to use $W^\pm$ to measure the polarized u and d distributions directly. The physics and formalism follows

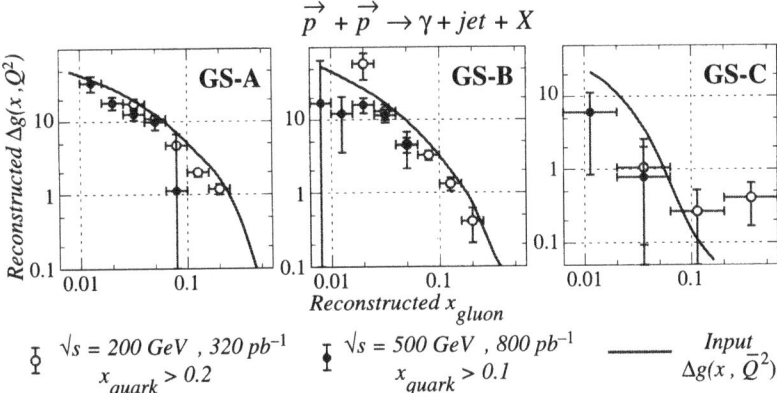

Fig. 11.8 Simulated measurements of Δg in the STAR detector at RHIC from $\vec{p}\vec{p} \to \gamma + X$ for the energies and luminosities shown and Gehrmann–Stirling gluons A, B, C. (From Bunce et al. 2000.)

very closely that given in Section 10.4.3 making allowance for the difference between $p\bar{p}$ and pp colliders. The primary measurement is the parity violating single spin asymmetry

$$A_L^W = \frac{1}{P} \frac{N_-(W) - N_+(W)}{N_-(W) + N_+(W)},$$

where the helicity labels refer to those of one of the protons, the helicities of the other being summed over, and P is the beam polarization. W^+ is produced from $u\bar{d}$ so

$$A_L^{W^+} = \frac{\Delta u(x_1)\bar{d}(x_2) - \Delta\bar{d}(x_1)u(x_2)}{u(x_1)\bar{d}(x_2) + \bar{d}(x_1)u(x_2)}. \tag{11.18}$$

For W^- production one interchanges u and d. At RHIC energies the quark densities probed by the W production are in the valence region. The x_i are related to the rapidity of the W, y_W, by

$$x_1 = \frac{m_W}{\sqrt{s}} e^{y_W}, \quad x_2 = \frac{m_W}{\sqrt{s}} e^{-y_W},$$

provided $p_T^W \approx 0$. If y_W is defined relative to the polarized proton used in A_L^W, then for $y_W \gg 0$, $A_L^{W^+} \to \Delta(u)/u$ and for $y_W \ll 0$, $A_L^{W^+} \to -\Delta(d)/d$, and correspondingly for W^- interchanging u and d. The $W^\pm$ events will be identified by the presence of a high p_T $e^\pm$ or $\mu^\pm$ from leptonic decays. The W kinematics can be reconstructed from the the lepton variables y_ℓ and p_T^ℓ measured in the lab

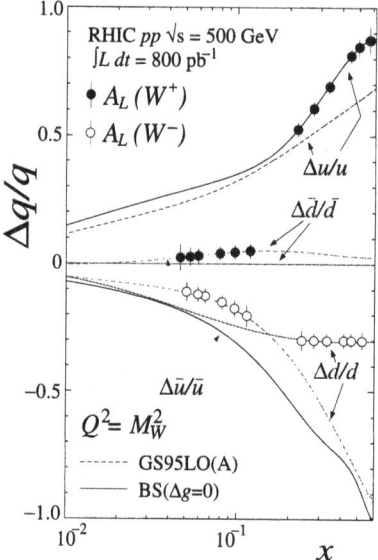

Fig. 11.9 Expected sensitivity to polarized u and d quark and antiquark distributions from measurements of $\vec{p}\vec{p} \to W^{\pm} + X$ at RHIC. The darker(lighter) points refer to W^+ (W^-) respectively and the lines are from analyses of polarized DIS data. (From Bunce et al. 2000.)

$$y_W = y_\ell - \frac{1}{2}\ln\left[\frac{1+\cos\theta^*}{1-\cos\theta^*}\right], \qquad \sin\theta^* = \frac{2p_T^\ell}{m_W},$$

where θ^* is the charged lepton decay angle in the W rest frame and p_T^W has been ignored. To give an example of the sort of measurements that could be made, Fig. 11.9 shows results for $\Delta q/q$ based on about 8000 W's of each sign identified using muons with $p_T^\ell > 20\,\text{GeV}$ in the PHENIX detector from $800\,\text{pb}^{-1}$. Many more details of the RHIC $\vec{p}\vec{p}$ spin programme are given in the review article by Bunce et al. (2000).

11.8 Transverse spin asymmetries

So far this chapter has been concerned mainly with the analysis of nucleon and quark longitudinal spin components. This fits very well within the framework of the QCD enhanced parton model in the nucleon's infinite momentum frame. Just as one can learn more by not integrating out the transverse momentum components of parton densities, so the behaviour of the nucleon and quark transverse spin components should be studied. A transversely polarized parton density may be defined in analogy to Δq by $\delta q = q_\uparrow^\uparrow - q_\uparrow^\downarrow$, i.e. the difference in probabilities of the quark with transverse spin aligned parallel and antiparallel to the nucleon's transverse spin (with respect to the nucleon momentum direction). The distributions δq cannot be studied in inclusive polarized DIS for quarks assumed massless. This is because helicity is conserved at the $\gamma - q$ vertex. The normal polarized density Δq involves the difference of quark distributions with spins parallel or antiparallel to the nucleon's longitudinal momentum direction, but the helicities of the individual (polarized) densities do not change. A non-zero contribution to δq requires a spin change between the initial and final state at the amplitude level — it is thus an 'off-diagonal' process in comparison to the usual 'diagonal' amplitudes studied in inclusive DIS. Interestingly, there is no corresponding quantity for the gluon because this would require a spin change of 2 units, which cannot be produced with a spin-$\frac{1}{2}$ nucleon target. The δq may be studied using semi-inclusive DIS in which a spin property of a final state particle is measured, or in fully polarized hadron–hadron scattering such as RHIC $\vec{p}\vec{p}$. In both cases, soft physics plays a crucial role and spin information is passed from the initial to the final state via a quark line. Examples of suitable semi-inclusive processes are $ep^\uparrow \to e\Lambda^\uparrow X$ or $ep^\uparrow \to e\pi^+ X$, in the first case, the Λ polarization is measured from its decay angular distribution and in the second, the azimuthal dependence of the single spin asymmetry is measured. At RHIC, the Drell–Yan process $p^\uparrow p^\uparrow \to \ell^+\ell^- X$ is sensitive to the δq through the double spin transverse asymmetry.

11.9 Summary

Experiments on spin and polarization have a habit of disrupting simple ideas based on unpolarized data and deep inelastic scattering is no exception. The measurement of Γ_1^p by the EMC experiment which caused the spin crisis should, perhaps, have been anticipated but it wasn't. That event led to a re-examination of the theoretical framework and more accurate measurements, both of which contributed to a deeper understanding of the problem. The central question remains of how the nucleon spin is distributed amongst its constituents $\frac{1}{2} = \frac{1}{2}\Delta\Sigma + \Delta g + \langle L_z \rangle$. The use of NLO QCD analyses and better data give relatively good information on Δu and Δd in the valence region and some constraint on Δq_{sea} and Δg. New measurements are needed to disentangle the gluon and strange sea components. Encouragingly these are on the horizon. The final challenge, perhaps the hardest, and one not touched on here is to find a way to measure $\langle L_z \rangle$.

11.10 Appendix: formalism for polarized DIS

A full derivation of the formalism for polarized deep inelastic scattering may be found in the review article by Anselmino et al. (1995). Here the procedure is outlined and the more important formulae summarised.

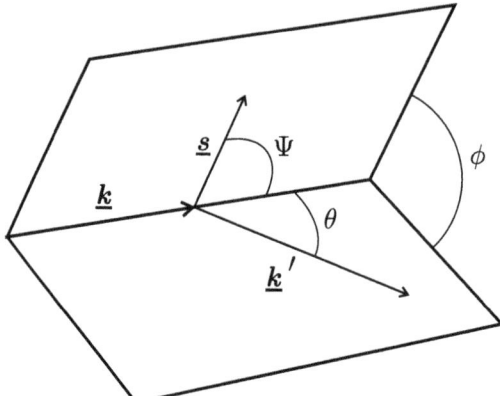

Fig. 11.10 Lepton scattering and polarization planes and the azimuthal angle ϕ.

As in the unpolarized case, the differential cross-section for polarized scattering via a virtual photon is proportional to $\frac{\alpha^2}{Q^4} L_{\mu\nu} W^{\mu\nu}$, where $L_{\mu\nu}$ and $W_{\mu\nu}$ are the leptonic and hadronic tensors (see Appendix C). Now, in addition to the lepton and nucleon 4-momenta, there are also the covariant spin vectors with which to construct contributions to the tensors. For a spin-$\frac{1}{2}$ particle with mass m and momentum p, the covariant spin vector s is given by

$$\bar{u}(p,s)\gamma^\mu \gamma^5 u(p,s) = -\bar{v}(p,s)\gamma^\mu \gamma^5 v(p,s) = 2ms^\mu, \quad \text{and} \quad s_\mu s^\mu = -1$$

The presence of $\gamma^\mu \gamma^5$ in the spin vectors shows how axial vector terms enter into the expressions for polarized DIS. The extra terms involving the spin vectors give rise to contributions both symmetric and antisymmetric under the interchange of the Lorentz indices μ, ν. In particular, the hadronic tensor, $W_{\mu\nu}$, now contains two contributions which are antisymmetric under the interchange of indices. For polarized scattering, these terms do not disappear in the contraction since $L_{\mu\nu}$ also contains antisymmetric terms. These terms give rise to two new structure functions g_1 and g_2. The cross-section for arbitrary polarized scattering then depends on four structure functions: F_1, F_2, g_1 and g_2.

To connect to experimental observables one needs to define two planes as shown in Fig. 11.10 for the fixed target frame. The first is the usual lepton scattering plane defined by the initial and final 3-momenta $\mathbf{k}, \mathbf{k}'$ with the scattering angle θ between them. The second is the polarization plane defined by $\mathbf{k}$ and $\mathbf{s}$, the polarization direction of the initial state nucleon (at rest), with angle Ψ between them. The two planes intersect along the direction of $\mathbf{k}$ with angle ϕ between them. The cross-sections

Appendix: formalism for polarized DIS

σ^{++}, σ^{+-} are those for which the initial lepton spin direction is parallel to **k**, with the nucleon spin either parallel ($\Psi = 0$) or antiparallel ($\Psi = \pi$) to **k**, respectively. The cross-sections $\sigma^{+\uparrow}, \sigma^{+\downarrow}$ are those for which the initial lepton spin direction is parallel to **k**, with the nucleon spin perpendicular to **k** either $\Psi = \pi/2$ ($\uparrow$) or $\Psi = 3\pi/2$ ($\downarrow$). Then using the cross-section differences

$$\Delta d\sigma_{||} = d\sigma^{+-} - d\sigma^{++}, \qquad \Delta d\sigma_\perp = d\sigma^{+\uparrow} - d\sigma^{+\downarrow},$$

the polarized part of the cross-section for arbitrary nucleon spin **s** is given by

$$\Delta\sigma = \cos\Psi \Delta\sigma_{||} + \sin\Psi \cos\phi \Delta\sigma_\perp,$$

with $\Delta\sigma_{||}, \Delta\sigma_\perp$ related to the polarized structure functions by

$$\frac{d^2\Delta\sigma_{||}}{dxdQ^2} = \frac{16\pi\alpha^2 y}{Q^4}\left[\left(1 - \frac{y}{2} - \gamma^2\frac{y^2}{4}\right)g_1 - \gamma^2\frac{y}{2}g_2\right], \quad (11.19)$$

$$\frac{d^3\Delta\sigma_\perp}{dxdQ^2 d\phi} = -\cos\phi \frac{8\alpha^2 y}{Q^4}\gamma\sqrt{1-y-\gamma^2\frac{y^2}{4}}\left[\frac{y}{2}g_1 + g_2\right], \quad (11.20)$$

where $\gamma^2 = 4m^2x^2/Q^2$, with x, y, Q^2 the usual DIS kinematic invariants. Just as in the unpolarized case, it is useful to relate the polarized cross-sections and asymmetries for $ep \to e'X$ to those for $\gamma^*p \to X$ and through the optical theorem to the amplitudes for virtual Compton scattering. In particular the two virtual photon asymmetries are defined by

$$A_1 = \frac{\sigma_{1/2} - \sigma_{3/2}}{\sigma_{1/2} + \sigma_{3/2}}, \qquad A_2 = \frac{2\sigma_{TL}}{\sigma_{1/2} + \sigma_{3/2}}, \quad (11.21)$$

where $\sigma_{1/2, 3/2}$ are the γ^*p cross-sections for total spins $\frac{1}{2}, \frac{3}{2}$, respectively and σ_{TL} is the transverse–longitudinal interference term. $A_{1,2}$ are related to the asymmetries $A_{||,\perp}$ (defined in Eqs (11.1) and (11.2)) by

$$A_{||} = D(A_1 + \eta A_2), \qquad A_\perp = d(A_2 - \xi A_1),$$

$$A_1 = (g_1 - \gamma^2 g_2)/F_1, \qquad A_2 = \gamma(g_1 + g_2)/F_1,$$

where D, d, η, ξ and γ are kinematic factors and F_1 is the unpolarized structure function. In the fixed target frame, with the nucleon at rest, the kinematic factors may be written in terms of the initial and final lepton energies, E, E' and the virtual photon polarization ε (see Eq. 5.25)

$$D = \frac{E - \varepsilon E'}{E(1 + \varepsilon R)} \qquad \eta = \frac{\varepsilon\sqrt{Q^2}}{E - \varepsilon E'}$$

$$d = D\sqrt{\frac{2\varepsilon}{1 + \varepsilon}} \qquad \xi = \eta\frac{1 + \varepsilon}{2\varepsilon}.$$

Although the use of both sets of asymmetries may seem an unnecessary complication it does allow a clear view of the relative importance of the various terms, as outlined in Section 11.1. Also, given the definitions of $A_{1,2}$ above, the bounds $|A_1| < 1$ and $|A_2| < \sqrt{R}$ are highly plausible without any formal derivation.

At high energies (with $Q^2 \approx M_Z^2$ and above[3]) the contribution from Z^0 exchange must be included in the cross-section expressions for polarized NC scattering. As in the unpolarized case, the contributions are separated into γ^*, Z^0 and γ–Z interference. In addition to the unpolarized xF_3, three more parity violating polarized structure functions have to be introduced: g_3, g_4 and g_5. The additional structure functions are related to the polarized parton density differences Δq_i by the electroweak couplings of the Z^0, g_V and g_A. For fully polarized CC scattering, the cross-sections have a similar set of structure functions and cross-section expressions given in terms of W exchange. The details are given in Anselmino et al. 1995.

---

[3]Such energies could be achieved with a fully polarized HERA collider.

12
Beyond the Standard Model

Beyond the Standard Model (BSM) is a vast subject and it is not appropriate to attempt to cover it in any depth here. The aim of this chapter is to explain, with the help of some specific examples, the sort of techniques used in deep inelastic scattering to search for BSM signals and to set limits. To limit the scope further, the discussion will concentrate on DIS at HERA, that is, high energy $e^{\pm}p$ scattering. The topics covered are: setting limits on possible compositeness scales using the classic form factor technique; direct searches for leptoquark resonances; the use of the contact interaction approach for indirect searches. All these depend on accurate measurement of the NC and CC cross-sections at large Q^2, the mass and coupling scales that can be probed are limited by the statistical and systematic errors of the measurements and ultimately by the CM energy of the HERA collider.

At the time of writing, no positive signals for BSM effects have been confirmed at HERA or indeed any existing collider. In 1997, there was a brief flurry of excitement when both H1 and ZEUS saw an excess of events at very large Q^2 and large x, but these anomalies disappeared with the increase in data recorded since then. The one area where there might be a signal is in events with a high p_T $e^{\pm}$ or $\mu^{\pm}$ and large missing energy in the final state. The SM source for such events is W production, ZEUS see a rate consistent with the SM whereas H1 have an excess of events. This is a topic to be continued at HERA-II. It is covered at the end of the chapter.

12.1 General remarks

Many of the BSM processes lead to final states that are identical to those of a standard model process, thus searching for new effects is looking for deviations from the SM expectation for that process. To squeeze the maximum information from the experimental data, one needs to take proper account of normalization errors and correlated systematic errors. To make precise tests and set limits, χ^2 or log likelihood methods may be used.[1]

---

[1]For more details on statistical methods for limit calculations, see, for example, the book by Cowan (1998), or the proceedings of the Durham Workshop 2002 (Durham 2002).

The χ^2 method is applied as follows. Consider a binned experimental distribution, such as the NC cross-section $d\sigma/dQ^2$ shown in Fig. 12.1. Define χ^2, as outlined in Section 6.7 by

$$\chi^2 = \sum_i \left(\frac{fA_i^{exp} - A_i^{th}(\lambda_{\text{BSM}})(1 + \sum_k \Delta_{ik}(s_k))}{f\sigma_i} \right)^2 + \left(\frac{f-1}{\sigma_f} \right)^2 + \sum_k s_k^2,$$

where A_i^{exp} represents the experimental quantity in the ith bin, $A_i^{th}(\lambda_{\text{BSM}})$ the corresponding theoretical quantity with λ_{BSM} representing an unknown parameter of a BSM extension of the theory. The overall normalization of the data is given by f with uncertainty σ_f, σ_i is the statistical and uncorrelated systematic uncertainties for the ith bin added in quadrature and $\Delta_{ik}(s_k)$ are functions describing the effect of the correlated systematic uncertainty from source k on data in bin i. The parameters s_k can be determined during the fitting process, but in a sense have been measured by the experiment to be $s_k = 0 \pm 1$ from the 1σ estimates of the systematic uncertainties.[2] Using this χ^2 definition, the first step is to find the minimum value of χ^2 for the best SM fit to the data, i.e. the fit with $\lambda_{\text{BSM}} = 0$. Then λ_{BSM} is varied successively and the fit repeated until the relevant increase in χ^2 is reached, for example, $\chi^2 - \chi^2_{\text{SM}} = 3.84$ for a 95% confidence level (CL) limit. The likelihood method will be outlined in the context of a contact interaction study in Section 12.4.1 below. For the sort of cross-section data used here, the dominant experimental uncertainties will typically be those to do with the absolute energy scales and the error on the luminosity. For the SM calculations, dominant errors are those on α_s and the parton densities.

12.2 Quark compositeness and form factors

Using the technique of inelastic scattering to probe the spatial extent of a particle's charge distribution is to return to the roots of the subject and the basic idea is outlined in the first chapter of the book. Here the charge distributions are those of the electroweak couplings of the quarks and leptons. In principle, the high Q^2 probes are γ^*, Z^0 for NC scattering and $W^\pm$ for CC processes. The Fourier or momentum space transform of the spatial distribution of a charge is the form factor $F_i(Q^2)$, where i labels the particle and charge. The normalization of the form factor and corresponding charge distribution is chosen so that $F_i(0) = 1$. In Section 1.1.1, it was shown that under certain circumstances

$$F_i(Q^2) \approx 1 - \frac{1}{6} R_i^2 Q^2,$$

where $R_i = \sqrt{\langle r_i^2 \rangle}$ is the root mean square of the charge distribution. For simplicity it is usually assumed that R_i is the same for all EW charges for

---

[2]Different treatments of these systematic uncertainty parameters are considered in Section 6.7.

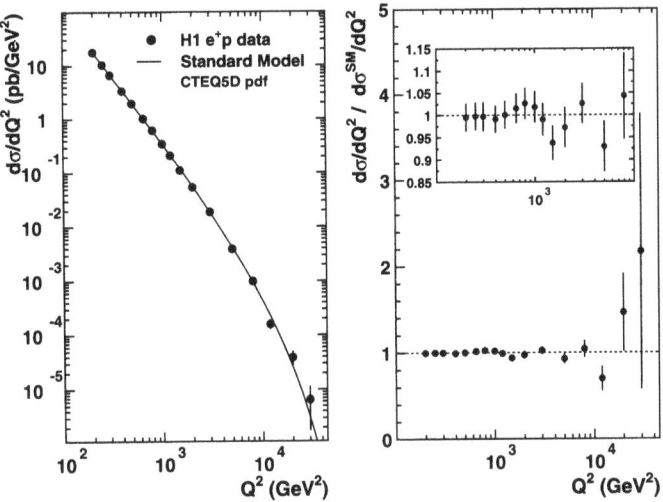

Fig. 12.1 An example of the quality of the SM description of the Q^2 dependence of the NC cross-section. (From H1 2000.)

a given particle and it may be taken to give an estimate of the 'size' of the particle. For high Q^2 NC scattering the cross-section is modified by the form factors of the electron and quark

$$\frac{d\sigma}{dQ^2} = \frac{d\sigma^{SM}}{dQ^2} F_e^2(Q^2) F_q^2(Q^2),$$

where σ^{SM} is the cross-section for the e and q assumed point-like. Measurements on the anomalous magnetic moment of the electron give the electron's charge radius as bounded by $R_e < 2 \times 10^{-23}$ m, so F_e will be taken as unity.[3] Using the data for high Q^2 NC e^+p scattering shown in Fig. 12.1 (which corresponds to 35.6 pb^{-1}) and using CTEQ5D parton densities for the SM cross-section, H1 find a limit on R_q at 95% CL of $R_q < 1.7 \times 10^{-18}$ m. A similar result has been reported by ZEUS. These results are comparable to limits from $p\bar{p}$ scattering. For more details on the method and limits from other process see Köpp et al. 1995.

12.3 Direct leptoquark searches

Leptoquarks are particles that are formed by the fusion of a lepton and a quark, so they can be produced as direct resonant excitations in $e^\pm p$ scattering at HERA provided the mass $m_{LQ} < \sqrt{s}$. Leptoquarks exist in many

---

[3]The very stringent limit on R_e from $(g-2)_e$ assumes that non-standard contributions scale linearly with mass. If the mass dependence is quadratic, the bound may be weakened significantly — see the discussion and references in Köpp et al. 1995.

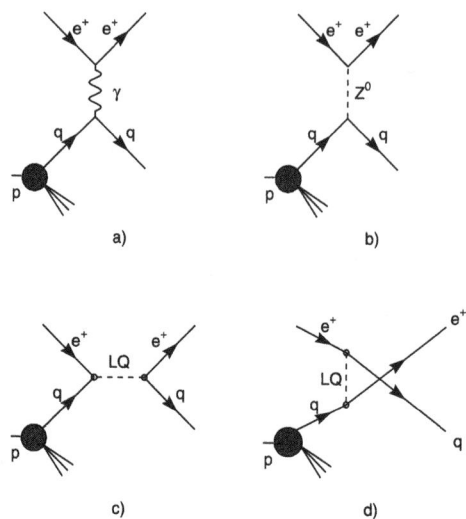

Fig. 12.2 Leptoquark production in e^+p NC scattering (c) s-channel $F = 0$ states; (d) u-channel $F = 2$ states. Also shown are the diagrams for the SM γ^* (a) and Z^0 (b) exchanges. (From ZEUS 2000c.)

BSM theories: grand unified theories; composite models; and supersymmetric theories with R-parity violation. It has become standard in many searches to use the general classification of leptoquarks (LQs) introduced by Buchmüller, Rückl and Wyler (BRW) (1987). This is based on the assumption that any new interaction will: respect the $SU(3)_F \times SU(2)_L \times U(1)_Y$ symmetry of the SM; have family diagonal couplings to avoid flavour changing neutral currents; preserve baryon and lepton number to avoid rapid proton decay and couple to lepton states of definite helicity. The leptoquarks are characterised by a fermion number $F = L + 3B = 0, 2$ where L and B are lepton and baryon number, respectively. With these restrictions there are 14 leptoquark states of scalar (S) and vector (V) particles. LQs decay to either an eq final state with branching ratio β, or to νq with BR $1 - \beta$. Details of the quantum numbers and coupling modes for the LQ states are given in Table 12.2 in the appendix to this chapter.

Only LQs coupling to the first generation of SM leptons and quarks are considered here. In addition to being produced in the ep s-channel, LQs may also be exchanged in the u-channel. Both possibilities are shown in Fig. 12.2 for the e^+p NC. For e^-p NC the s- and u-channel quantum numbers are interchanged, so that $F = 2$ LQs are formed in the s-channel and $F = 0$

in the u-channel. With the appropriate changes in quantum numbers, LQs can also couple to $e\bar{q}$ and $\nu\bar{q}$ channels. The LQ events will tend to have the same final state characteristics as the corresponding SM NC or CC event (the SM diagrams are also given in Fig. 12.2). As the LQ is expected to have a relatively large mass, the s-channel process will dominate with the quark in the valence region at $x_0 = m_{LQ}^2/s$. For a LQ with a small coupling λ_{LQ}, it is appropriate to estimate the contribution to the total cross-section in the s-channel using the narrow resonance approximation

$$\sigma_{LQ}^{NRA} = (J+1)\frac{\pi}{4s}\lambda_{LQ}^2 q(x_0, m_{LQ}^2), \qquad (12.1)$$

where J is the LQ spin and the quark density is evaluated at $x = x_0$ with the scale given by the leptoquark mass. This simple estimate is modified by radiative corrections. QED ISR radiation decreases the cross-section by 5–25% as m_{LQ} increases from 100 to 290 GeV, depending only weakly on LQ type. NLO QCD corrections (K-factors) have only been evaluated for scalar LQs, giving values 1.17(1.17) to 1.15(1.35) for $eu(ed)$ combinations as m_{LQ} varies from 100 to 300 GeV. For consistency, no K-factors are applied to either scalar or vector resonances.

Let θ^* give the angle of the final state $e^{\pm}$ with respect to the incoming lepton beam direction in the LQ rest frame. The distribution of θ^* contains useful information. For e^+p scattering, a scalar LQ in the s-channel and a vector LQ in the u-channel will decay isotropically, while a vector LQ in the s-channel and a scalar LQ in the u-channel follow a $(1 + \cos\theta^*)^2$ distribution. For e^-p scattering the assignments are reversed. The angle θ^* is related to the DIS y variable by $\cos\theta^* = 1 - 2y$.

The two HERA experiments have followed different procedures for reconstructing the mass variable for LQ searches in the NC channel. H1 use $m_{LQ} \equiv M_e = \sqrt{xs}$ with x reconstructed using the electron method, whereas ZEUS calculate m_{LQ} from the final state using the scattered positron and the high p_T jet: $m_{LQ}^2 \equiv M_{ej}^2 = 2E'E_j(1 - \cos\theta_{ej})$, where θ_{ej} is the angle between the positron and the jet. For the νq CC channel only the final state method is available, with the neutrino energy and momentum reconstruction using energy–momentum conservation assuming that the ν is the only missing particle. For the NC channels and for $e^+p \to \bar{\nu}q$ the SM cross-sections peak at small values of y, so in these cases additional cuts on y or $\cos\theta^*$ may be applied to enhance the sensitivity.

A typical M_{ej} distribution from the ZEUS experiment is shown in the top two plots of Fig. 12.3. The data correspond to 47.7 pb^{-1} of data collected between 1994 and 1997 at a CM energy of $\sqrt{s} = 300$ GeV, with event selection cuts requiring: a final state positron with energy at least 25 GeV, a well reconstructed primary vertex and a hadronic jet with $p_T^{jet} > 15$ GeV. The figure shows the distribution of events with $M_{ej} > 100$ GeV on a log scale with an inset showing the distribution for events with masses greater than 180 GeV on a linear scale, both compared with the SM expectation.

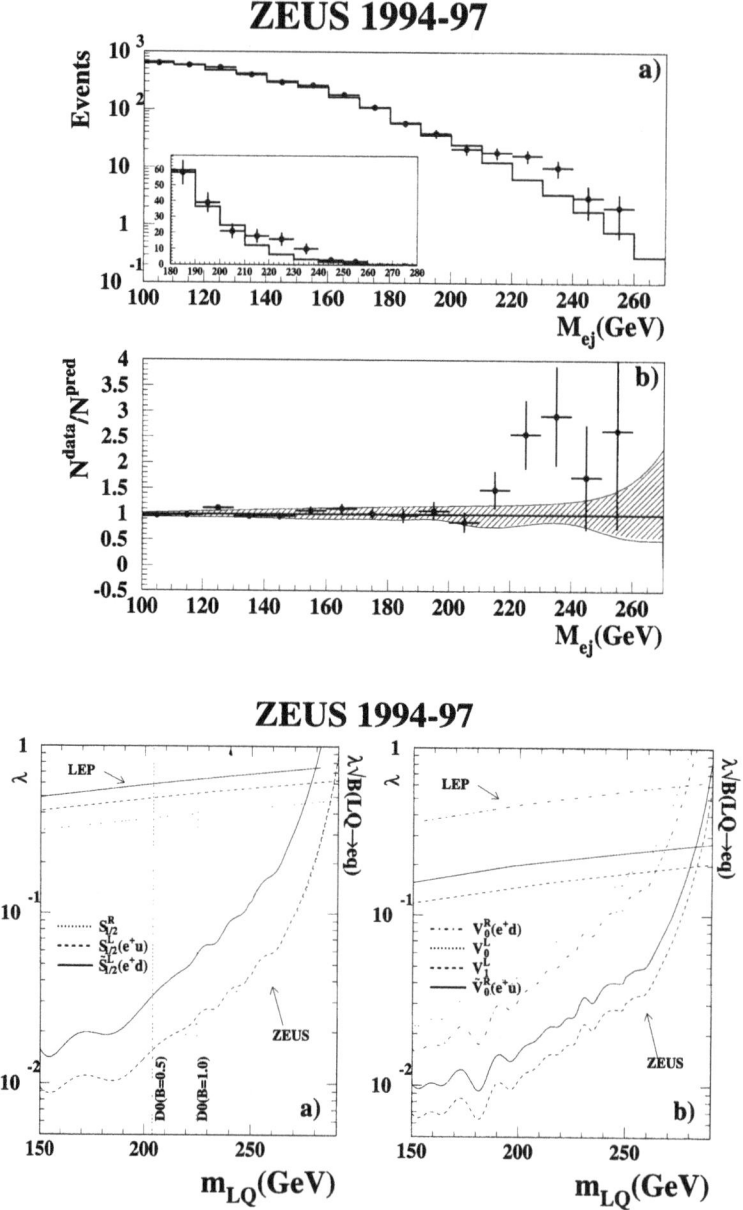

Fig. 12.3 Leptoquark search in the e^+p NC channel. Top two plots. M_{ej} from the ZEUS experiment.(a) Comparison of the SM (histogram) with the reconstructed e-jet mass. The inset shows the region with $M_{ej} > 180\,\text{GeV}$ on a linear scale. (b) Ratio of the number of events observed to the SM expectation, with the shaded band showing the uncertainty in the latter. Bottom plots. Coupling-mass limit plots for $F = 0$ leptoquarks: (a) 95% CL limits for scalar LQs and (b) the same for vector LQs. Limits from LEP and the Tevatron are also shown. (From ZEUS 2000c.)

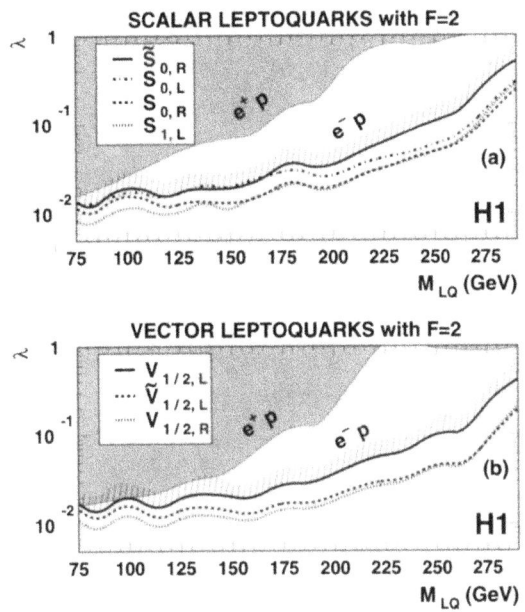

Fig. 12.4 Coupling-mass limit plots for $F = 2$ leptoquarks from a search by H1 in e^-p NC and CC channels: (a) 95% CL limits for scalar LQs and (b) the same for vector LQs. The shaded regions show the limits obtained from searches in the e^+p channels. (From H1 2001d.)

It also shows the ratio of observed events to the SM expectation, with the uncertainly in the latter (dominated by the uncertainty in the parton densities) as the shaded band. The excess of events at mass around 230 GeV contain those that gave rise to the 'HERA anomaly' of 1997 (but at that time there was only 20 pb^{-1} of data). For the full data set shown here, requiring $M_{ej} > 200$ GeV and $\cos\theta^* < 0.5$, seven events are observed where five events are expected. The SM expectation is calculated using the CTEQ4 parton distributions. Since neither the coupling strength λ_{LQ} nor the mass m_{LQ} is known, it is only possible to extract combined limits from the data. The results are often presented as a two-dimensional plot of λ_{LQ} versus m_{LQ}. For the ZEUS e^+p e-jet analysis, examples of limit plots are shown in the bottom two plots of Fig. 12.3, separately for scalar LQs (LH plot) and vector LQs (RH plot). For a coupling strength comparable to QED, $\lambda_{LQ} = \sqrt{4\pi\alpha} \approx 0.3$, LQs with masses in the range 150–280 GeV are excluded by the ZEUS data. Similar results have been found by H1.

The H1 experiment has also performed a search in the e^-p channel for $F = 2$ LQs using both eq and νq final states (H1 2001d). An integrated luminosity of 15 pb^{-1} with 27.5 GeV electrons interacting with 920 GeV protons was recorded during 1998-9 running. The CM energy is $\sqrt{s} = 318$ GeV, somewhat higher than that for the ZEUS analysis above. The scattered electron is required to have at least $E_T = 15$ GeV and the search domain is defined by $Q^2 > 2500$ GeV^2 and $0.1 < y < 0.9$. The SM expectation is calculated using the MRST(98) parton densities. No signal is found and the resulting limit plots for $F = 2$ scalar and vector LQs are shown in Fig. 12.4. The shaded regions in the plots show the excluded regions for $F = 2$ LQs obtained using data in the e^+p channels. The mass limits are lower than those from e^-p scattering for a given coupling strength because the LQ is a u-channel exchange in e^+p scattering and hence its effect is smaller than for direct s-channel formation. For the broader picture on how tighter limits and possible signals for LQs may be obtained by combining data from LEP, HERA, the Tevatron and low energy experiments on parity violation in atomic physics, see the article by Zarnecki (2000).

12.4 Contact interactions

Direct searches for objects like leptoquarks that can couple to the s-channel in ep scattering are clearly limited by the CM energy available at HERA. However it is possible for new interactions beyond the SM with much higher energy or mass scales to have an effect on the cross-sections at presently accessible collider energies. The formalism of contact interactions provides a general way of searching for such effects. The basic idea is not new – the Fermi theory of low energy weak interactions may be considered as a contact interaction derived from the full electroweak Lagrangian. In essence, the chiral form of the weak coupling is retained but propagator effects of the form $g_W^2/(Q^2 + M_W^2)$ are collapsed to a four-point contact interaction with the Fermi coupling $G_F \sim g_W^2/M_W^2$. Note that the dimensions of G_F are [mass]$^{-2}$, this will be a feature of the couplings of the more general contact interaction formalism. It means that it is only possible to extract limits on the ratio of coupling/mass, unless further assumptions are made.

A systematic, almost model independent, approach to contact interactions (CI) for $(e\bar{e})(q\bar{q})$ NC couplings was developed by Haberl et al. (1991) for the 1991 HERA Workshop and references to earlier work in the context of more specific models may be found there. For vector-like contact interactions, respecting chiral symmetry, an extra term $\mathcal{L}_V^{CI}$ is added to the SM Lagrangian,

$$\mathcal{L}_V^{CI} = \sum_{q=u,d} \{ \eta_{LL}^i (\bar{e}_L \gamma_\mu e_L)(\bar{q}_L \gamma^\mu q_L) + \eta_{LR}^i (\bar{e}_L \gamma_\mu e_L)(\bar{q}_R \gamma^\mu q_R) \\ + \eta_{RL}^i (\bar{e}_R \gamma_\mu e_R)(\bar{q}_L \gamma^\mu q_L) + \eta_{RR}^i (\bar{e}_R \gamma_\mu e_R)(\bar{q}_R \gamma^\mu q_R) \},$$

where the indices L and R denote the left-handed and right-handed fermion helicities. In deep inelastic scattering at high Q^2, the contributions from the first generation u- and d-valence quarks dominate as one is also at large x. For simplicity, flavour symmetry is assumed: $\eta_{ab}^d = \eta_{ab}^s = \eta_{ab}^b$ and $\eta_{ab}^u = \eta_{ab}^c$. The small size of the parton densities for second and third generation quarks suppresses their contribution and that of the top quark is ignored because of its very large mass. This leaves eight independent effective coupling coefficients, four for each quark flavour, η_{ab}^i, with a and b labelling the L, R helicities. The formalism can be applied to any new phenomenon, for example, compositeness, leptoquarks or new gauge bosons, by an appropriate choice of the coefficients η_{ab}^i. Scalar and tensor interactions, which could in principle also contribute to $\mathcal{L}^{CI}$, involve strongly suppressed helicity flip couplings at HERA and are therefore not considered.

The SM formalism for high Q^2 DIS at HERA is outlined in the appendix to Chapter 8. The A_i, B_i coefficients contain the electroweak couplings of the lepton and quarks and they may be written (for e^-p scattering with e^- polarization P),

$$2A_i = (1-P)\left[(V_i^L)^2 + (A_i^L)^2\right] + (1+P)\left[(V_i^R)^2 + (A_i^R)^2\right],$$
$$B_i = (1-P)V_i^L A_i^L - (1+P)V_i^R A_i^R, \qquad (12.2)$$

where

$$V_i^m = e_i - (v_e \pm a_e)v_i P_Z,$$
$$A_i^m = -(v_e \pm a_e)a_i P_Z \qquad (12.3)$$

The superscript $m = L$ corresponds to the '$v + a$' combination on the RH side and the superscript $m = R$ corresponds to the '$v - a$' combination. The Z propagator factor P_Z and the fermion electroweak couplings v_i, a_i are given in the appendix to Chapter 8. The contact interaction terms add to the SM expressions for V and A as follows:

$$V_i^m = e_i - (v_e \pm a_e)v_i P_Z + \frac{Q^2}{2\alpha}\left(\eta_{mL}^i + \eta_{mR}^i\right),$$
$$A_i^m = -(v_e \pm a_e)a_i P_Z + \frac{Q^2}{2\alpha}\left(\eta_{mL}^i - \eta_{mR}^i\right), \qquad (12.4)$$

Contact interactions are a non-renormalizable, 'low energy' approximation to a full theory at a much higher mass scale. As such, they are only formulated at leading order, with contributions to F_2 and xF_3 but not to F_L. However, the SM expectation is more reliably calculated using NLO partons. The effect of this mismatch is minimised by using the DIS renormalization scheme in which the NLO and LO expressions for F_2 are the same, by definition, and the NLO contribution to xF_3 is relatively small.

12.4.1 Compositeness

The approach here is somewhat different to that used for the form factor analysis. The mass scale is that of the common constituents assumed in composite models of leptons and quarks. Many possible coupling structures may be investigated in a systematic way. The CI coefficients are assumed to take the form $\eta^i_{ab} = \epsilon\epsilon^i_{ab}(g^2_{CI}/\Lambda^2)$, where g_{CI} is the overall coupling strength, Λ is the mass scale parameter and $\epsilon = \pm 1$ determines the overall sign of the interference term. Having factored out the overall sign and a common coupling/scale factor, the relative signs of the terms are taken to be $\epsilon^i_{ab} = \pm 1, 0$. Unless otherwise stated, it is conventional to quote limits for $g^2_{CI} = 4\pi$ and to assume that the scale Λ is the same for u and d quarks. To avoid violating very tight limits from parity violating transitions in caesium atoms, the following constraint is imposed on all scenarios:

$$\eta^i_{LL} + \eta^i_{LR} - \eta^i_{RL} - \eta^i_{RR} = 0.$$

With these restrictions, a variety of contact interaction scenarios may be explored by varying the ϵ^i_{ab} factors and some examples are listed in Table 12.4.1. Each row of the table represents two CI 'models' corresponding to the two choices $\epsilon = \pm 1$ for the overall sign of the interference term.

Table 12.1 Scenarios for contact interactions. Each row corresponds to two different CI scenarios corresponding to $\epsilon = \pm 1$ for the overall sign of the interference terms. The 95% CL limits for the scales corresponding to the two scenarios from a ZEUS analysis of 47.7 pb^{-1} of e^+p NC data are also given.

Label	ϵ^u_{LL}	ϵ^u_{LR}	ϵ^u_{RL}	ϵ^u_{RR}	ϵ^d_{LL}	ϵ^d_{LR}	ϵ^d_{RL}	ϵ^d_{RR}	Λ_- [TeV]	Λ_+ [TeV]
VV	+1	+1	+1	+1	+1	+1	+1	+1	5.0	4.7
AA	+1	−1	−1	+1	+1	−1	−1	+1	3.7	2.6
VA	+1	−1	+1	−1	+1	−1	+1	−1	2.6	2.5
X1	+1	−1	0	0	+1	−1	0	0	2.8	1.8
X2	+1	0	+1	0	+1	0	+1	0	3.1	3.4
X3	+1	0	0	+1	+1	0	0	+1	2.8	2.9

In general, the CI terms produce two sorts of effect: terms proportional to $(Q^2/\Lambda^2)^2$ which enhance the cross-section at very large Q^2; and terms proportional to $\epsilon(Q^2/\Lambda^2)$ — the interference term with the SM amplitude — which can cause either an increase or decrease. Apart from changing the Q^2 dependence of the cross-section at fixed x, the CI terms may also alter the x dependence at fixed Q^2 because of the different dependence on y of the parity conserving and violating terms in the cross-section. These points are illustrated in Fig. 12.5 which shows log–log plots of the ratio $\sigma(SM + CI)/\sigma(SM)$, at a few values of x, versus Q^2 for a number of scenarios. The ZEUS collaboration has published a detailed study of 30 different contact interaction scenarios using 47.7 pb^{-1} of e^+p NC data at $\sqrt{s} = 300$ GeV (ZEUS 2000b). Two different log likelihood functions were used to find the lower limits on the CI scales, one used a likelihood function calculated from the individual kinematics of each event in the sample and the other from the binned Q^2 distribution. In both cases, the likelihood is a function of $\xi = \epsilon/\Lambda^2$ and is of the form

$$L(\xi) = -\sum_i \log p_i(\xi),$$

where i runs over the data and p_i are appropriately normalized probabilities, constructed from a comparison of measured and simulated event distributions. The value of ξ giving the minimum of L, $L_{\min}$, is then determined and the limits for the mass scales $\Lambda_\pm$ found by integrating $L(\xi) - L_{\min}$ over the appropriate one sided interval as follows

$$\frac{\int_0^{1/\Lambda_+^2} \exp(-L(\xi))\, d\xi}{\int_0^\infty \exp(-L(\xi))\, d\xi} = 0.95, \qquad \frac{\int_{-1/\Lambda_-^2}^0 \exp(-L(\xi))\, d\xi}{\int_{-\infty}^0 \exp(-L(\xi))\, d\xi} = 0.95.$$

The resulting 95% CL limits are compared for the two likelihood functions and the one with greatest sensitivity chosen — though for most scenarios the two methods give results within 15% of each other. The final results are given in Table 12.4.1 and are in the few TeV range. The value of the limit for a particular scenario will depend to some extent on how well that model reproduces any fluctuations in the data at high Q^2. From the table of couplings, it can be seen that for the VV mode all terms contribute with the same sign. This produces the most stringent limit as there are no cancellations between the additional terms. An example of a VV fit from an H1 CI study based on 35.6 pb^{-1} of e^+p NC data at $\sqrt{s} = 300$ GeV (H1 2000) is shown in the LH plot of Fig. 12.6. The plot shows the ratio $\sigma(SM + CI)/\sigma(SM)$, calculated using CTEQ5D partons, as a function of Q^2. The limit for the '+' VV combination is 5.9 TeV, somewhat larger than that found in the equivalent ZEUS fit (4.7 TeV), whereas the '−' limit is only 2.9 TeV compared to 5.0 TeV found by ZEUS. The reason for the differences between experiments and for the larger spread between the two H1 VV limits is a strong downward fluctuation in $\sigma(SM + CI)/\sigma(SM)$ for the H1 data around $Q^2 = 12000$ GeV^2, as may be seen clearly in the figure. For details of a global CI analysis using data from LEP, the Tevatron and low energy experiments, as well as HERA, see Zarnecki 1999.

12.4.2 Leptoquarks

The contact interaction formalism may be used to set limits on leptoquarks at mass scales above those that can be reached using the direct searches already described. The LQ mass and coupling are related to those of general CI formalism by $g_{CI}/\Lambda = \lambda_{LQ}/m_{LQ}$, full details for the η_{ab}^i coefficients are given in Table 12.3 of the Appendix to this chapter. As for the general contact interaction study, leptoquarks giving positive interference effects will give better limits than those with negative interference. The vector LQs coupling to u quarks give the most restrictive results. The table gives

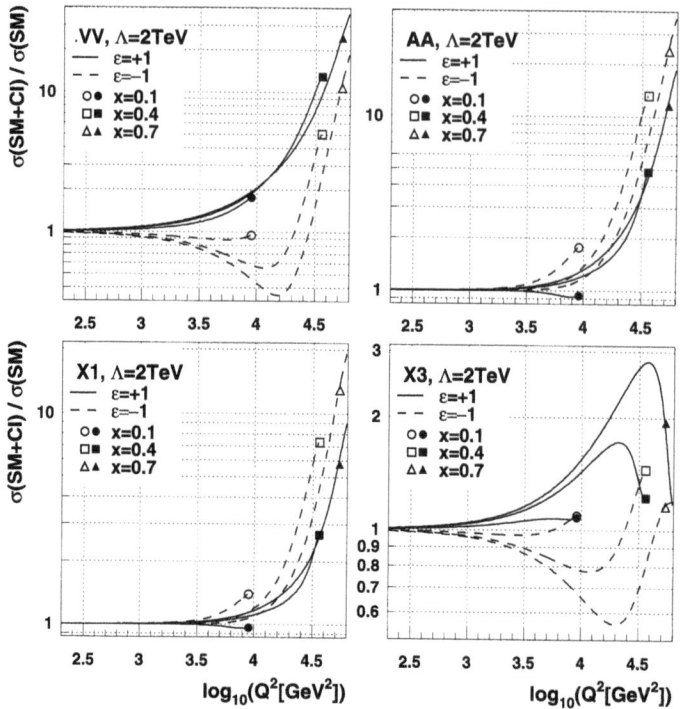

Fig. 12.5 Examples of the effect of contact interactions at a scale of 2 TeV for four scenarios as indicated. The ratio $\sigma(SM+CI)/\sigma(SM)$ is shown for three x values as a function of Q^2. In each case, curves are shown for both signs of the interference term. (From ZEUS 2000b.)

the lower limits on m_{LQ}/λ_{LQ} found by H1 and the results include the effect of full error propagation and variation of parton densities. Examples of LQ CI fits for $S^R_{1/2}$ and $V^R_{1/2}$ states to the H1 NC data are shown in the RH plot of Fig. 12.6. The table shows that both have RL couplings to u and d quarks, but the terms for $V^R_{1/2}$ are both positive and this gives rise to a much tighter limit than the negative couplings of $S^R_{1/2}$.

12.5 W production and anomalous high p_T leptons

Some of best signatures for new particles beyond the standard models are events at high CM energy with large p_T leptons and missing energy. Such events are produced in ep collisions at HERA by W production and decay. Clearly this process has to be understood before any claims for new effects can be made. The leading order Feynman diagrams are shown in Fig. 12.7. Diagrams (a) and (b) correspond to W radiation from the incoming or scattered quark, (c) contains the $WW\gamma$ triple gauge–boson coupling, (f) and (g) are strongly suppressed by the presence of two W propagators.

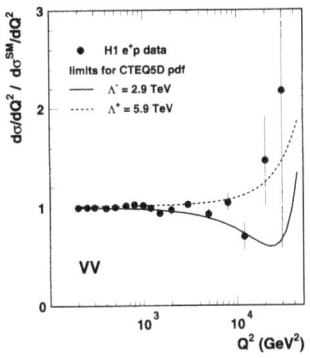 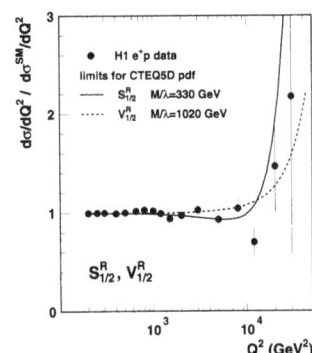

Fig. 12.6 Contact interaction study from H1 using 35.6 pb^{-1} of e^+p NC data. The plots show the same data, the ratio of the measured cross-section to the SM prediction as a function of Q^2 and two model fits: LH plot compositeness VV model; RH plot LQ models. (From H1 2000.)

Diagrams (d) and (e) are required by gauge invariance and contain off mass-shell Ws which tend to give rise to charged leptons with low p_T. The calculation of the cross-section is quite tricky as the u-channel pole of diagram (a) has to be regularised. The Monte Carlo code EPVEC (Baur et al. 1992) is used to handle the problem by splitting the phase-space into two regions

$$\sigma = \sigma(|u| > u_{\text{cut}}) + \int^{u_{\text{cut}}} \frac{d\sigma}{d|u|} d|u|,$$

where $u = (p_q - p_W)^2$, with p_q, p_W the 4-momenta of the incoming quark and final state W respectively. The first part of the above expression is calculated using helicity amplitudes for $e^+q \to e^+Wq'$, $W \to f\bar{f}'$. The second term, for small $|u|$, is calculated by folding the cross-section for $q\bar{q}' \to W \to f\bar{f}'$ with the parton densities in the proton and with the effective parton densities for the resolved photon emitted by the incoming electron. The final result for the total cross-section does not show much sensitivity to the precise value of u_{cut}, which is taken to be 25 GeV^2. Using EPVEC with MRS(G) parton densities at scale M_W^2 for the proton and GRVG-LO partons for the photon at scale $p_W^2/10$, the ZEUS collaboration calculated the combined cross-section for $W^\pm$ to be 0.95 pb (ZEUS 2000d). Different parton densities give a variation in the result of 5–10% and a NLO calculation confirms the EPVEC result. Inevitably there are quite large uncertainties in the choice of hard scale as the calculation is at LO. The combined uncertainty in the W cross-section is estimated to be 20%. The equivalent neutrino cross-section (for $e^+p \to \bar{\nu}W^+X$) is estimated to be much smaller — about 5% of the above. Given that photon exchange is present in diagrams (a)–(e), the dominant $e^+W^\pm X$ final states will be

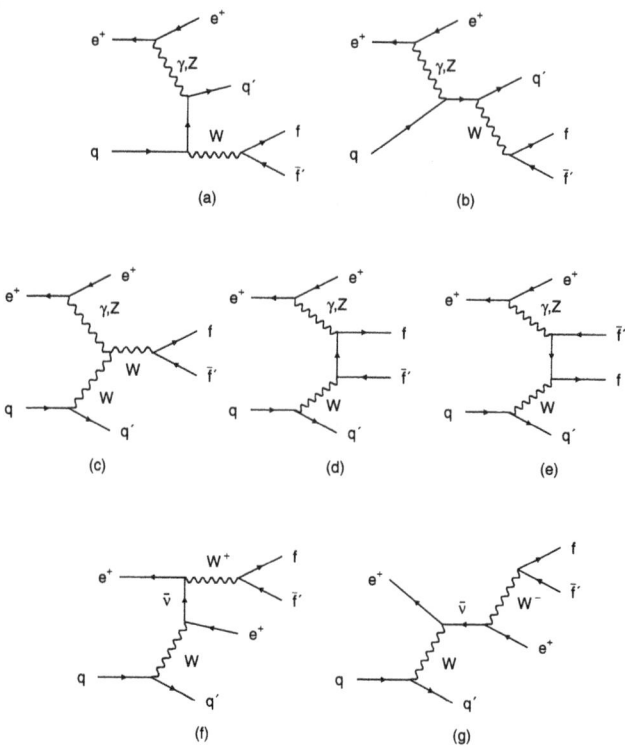

Fig. 12.7 Leading order Feynman diagrams for $e^+p \to e^+W^{\pm}X$, $W \to f\bar{f}'$. (From ZEUS 2000d.)

produced at low Q^2. The scattered positron will remain close to the incoming lepton beam direction and the hadronic system X will typically carry a small amount of transverse momentum. The signature for the presence of W is through its leptonic decay modes $W \to \ell\nu_\ell$, where $\ell = e$ or μ, giving a large p_T lepton and large missing E_T.

Background processes that could produce final states with real or 'fake' missing energy and high p_T electrons are: high Q^2 CC events with a real or fake high p_T lepton; high Q^2 NC events in which the scattered beam positron has not been identified correctly and in which there is missing E_T through a badly or incompletely reconstructed hadronic final state and photoproduction events in which there is missing energy and a fake high p_T $e^{\pm}$ formed from the overlap of a charged particle with an electromagnetic calorimeter cluster from π^0 decay. For high p_T signature muons, a serious source of background is from two-photon processes (one photon from the beam lepton and the other from the proton). Finally the processes $e^+p \to$

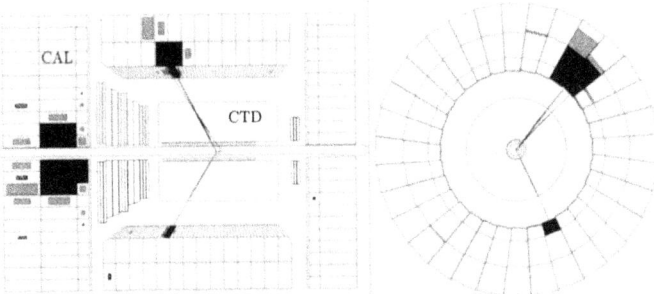

Fig. 12.8 An electron candidate event for $e^+p \to e^+WX, W \to e\nu$ found by ZEUS. The electron is identified as the track pointing towards the lower half of the calorimeter and the transverse mass for the event is $81.4 \pm 3.2\,\text{GeV}$. (From ZEUS 2000d).

e^+Z^0X with a cross-section of $0.3\,\text{pb}$ (using EPVEC) followed by the decays $Z^0 \to \ell^+\ell^-$ or $\nu\bar{\nu}$ have also been studied as a source of background. In all cases the cross-section estimates and event characteristics must be studied very carefully using the best available Monte Carlo simulations.

The crucial attributes of the detector are its hermiticity (to ensure a reliable measurement of missing E_T) and good efficiency for measuring high p_T electrons and muons. To aid discrimination two quantities are calculated from the measured final state momenta and energies: the acoplanarity angle, Φ_{ACOP}, in the plane transverse to the beam (xy-plane) and the transverse mass $M_T^{\ell\nu} = \sqrt{2p_T^\ell p_T^\nu(1 - \cos\Phi^{\ell\nu})}$. Here p_T^ν is the transverse momentum of the neutrino, assuming it to be the only missing final state particle, p_T^ℓ is that of the signature e or μ, $\Phi^{\ell\nu}$ is the angle between the $\ell^\pm$ and ν directions in the xy-plane and Φ_{ACOP} is the angle (again in the xy-plane) between the $\ell^\pm$ and the direction balancing the total hadronic transverse momentum, P_T^X. For well reconstructed high Q^2 NC events Φ_{ACOP} is near to zero, from conservation of transverse momentum. In a study of W production using $47.7\,\text{pb}^{-1}$ e^+p data (ZEUS 2000d), ZEUS required the missing p_T to be at least $20\,\text{GeV}$[4]; $\Phi_{\text{ACOP}} > 0.3\,\text{rad}$; $p_T^\ell > 10\,\text{GeV}$. Three candidate events, all with a final state e^+, are found — one of the events is shown in Fig. 12.8. The Monte Carlo expectations are 2.1 signal events and 1.1 ± 0.3 background events. With an estimated efficiency for identifying the $W \to e\nu$ events of 38%, this gives the measured cross-section of $\sigma(e^+p \to e^+W^\pm X) = 0.9^{+1.0}_{-0.7}(\text{stat.}) \pm 0.2(\text{sys.})\,\text{pb}$ or a 95% CL upper limit of $3.3\,\text{pb}$. For the $\mu\nu$ channel, no candidates are found and none expected — the efficiency for $W \to \mu\nu$ in ZEUS for this analysis is only about 13%.

---

[4] For electron events, the calorimeter alone is used, for muon events tracking information is used in addition.

In principle measurements of W may be used to set limits on anomalous $WW\gamma$ couplings — through the presence of diagram (c) of Fig. 12.7. This has been done as part of the ZEUS analysis, but currently the limits are not very stringent.

In a similar study of $36.5\,\text{pb}^{-1}$ e^+p data, H1 required isolated tracks with $p_T > 10\,\text{GeV}$ and missing $p_T > 25\,\text{GeV}$ from the calorimeter (H1 1998). Six events were found of which five contain muons, with an expectation of 0.53 ± 0.11 events from $W \to \mu\nu$ and 0.25 events from other SM processes (mainly $\gamma\gamma \to \mu^+\mu^-$). The sixth event contained an electron and the expectations are 1.65 ± 0.47 events from $W \to e\nu$ and 0.68 events from other SM sources. Given the discrepancy between the results of the two studies, ZEUS repeated the analysis of its sample following the H1 procedure, but did not find any evidence for an excess of events.

Since these initial studies a total of 120–$130\,\text{pb}^{-1}$ of data have been accumulated by each experiment for the period 1994–2000 of HERA-I operations. The picture has not changed, ZEUS find events with large missing transverse energy and a high p_T lepton in accord with the SM expectation for W production, whereas H1 report an excess, particularly in the muon channel. In the H1 (2003a) study, events are required to have $p_T^{calo} > 12\,\text{GeV}$, where p_T^{calo} is the sum of transverse momentum measured in the calorimeter, and the presence of an isolated e or μ track with $p_T > 10\,\text{GeV}$. The requirement $D = \sqrt{(\Delta\eta)^2 + (\Delta\phi)^2} > 1.0, 0.5$ for nearest jet and track, respectively, is also imposed. In the e^-p sample, $1e$ and 0μ candidates are observed, for which the SM expectations (from all sources) are 1.69 ± 0.22 and 0.37 ± 0.06, respectively. For the e^+p data, 10 electron candidates are found, where 7.20 ± 1.20 W decays and 2.68 ± 0.49 background events are expected, and 8 muon candidates are found, where 2.23 ± 0.43 W decays and 0.33 ± 0.08 other background events are expected. Figure 12.9 shows some of the properties of these events compared with the expectations from W production. The distributions of polar angle of the leptons and the difference between azimuthal angles of the lepton direction and that of the hadronic system ($\Delta\phi_{\ell,X}$) do not show any substantial difference between the data and W production. For the other two plots (the transverse mass $M_T^{\ell\nu}$ and P_T^X), the majority of events follow the SM expectations, but in each case at large M_T and large P_T^X, respectively, there is an excess of events. In M_T, there is an excess of events beyond Jacobian peak expected at $m_W/2$, but the most striking feature is the much flatter distribution of P_T^X for the hadronic system than that expected from W events, with three events having $P_T^X > 40\,\text{GeV}$, where 0.5 is expected.

A comparison has been made between the two experiments, using a common set of cuts to select the events and concentrating on the region with $P_T^X > 25\,\text{GeV}$. Taking e and μ channels together, H1 find a total of 10 events where 2.8 ± 0.7 are expected from all SM sources (2.3 from Ws), ZEUS find 2 events where 2.4 ± 0.2 are expected (2 from Ws). ZEUS do not have any candidates with $P_T^X > 40\,\text{GeV}$. Very detailed discussions have

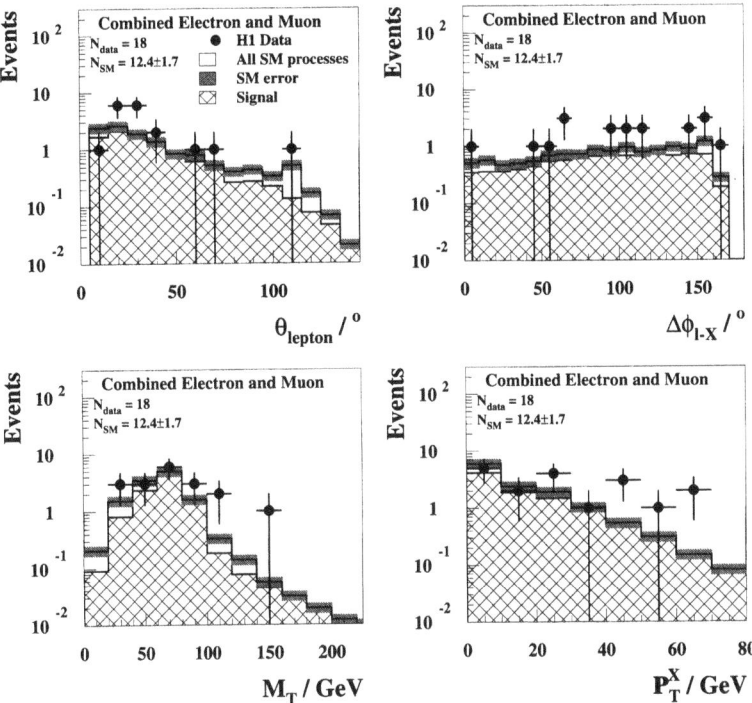

Fig. 12.9 The H1 HERA-I isolated high p_T lepton sample for electrons and muons combined. The plots show the distributions of: the polar angle of the lepton; the difference in azimuthal angles between the directions of the lepton and hadronic system X; the transverse mass of the lepton–neutrino system and the transverse momentum of the hadronic system. The data are shown as points with the SM expectation as the open histogram. The total error on the SM expectation is shown as the shaded band, with the W-production component shown as the hatched histogram. (From H1 2003a.)

taken place between the two groups to eliminate differences that might be caused by differences in detector technology and reconstruction methods. Nothing of significance has been found so the difference remains and is clearly one of the questions to be resolved with data from HERA-II.

12.5.1 Single top production

A possible explanation for the H1 events with large P_T^X is single top production, although SM production at HERA is negligible, there are extensions of the standard model that could give much higher rates. One way to test this idea is to use an effective Lagrangian which has a flavour changing neutral current quantified by anomalous couplings between quarks and gauge bosons corresponding to the diagram shown in Fig. 12.10. The rele-

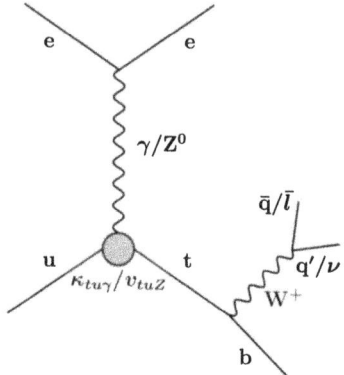

Fig. 12.10 Single top production at HERA via a flavour-changing neutral current transition.

vant coupling here is the magnetic coupling $\kappa_{tu\gamma}$, since a large value of x is needed to produce a top mass and the vector coupling v_{tuZ} contribution is suppressed by M_Z. The measured 95% CL upper bound of BR$(t \to u\gamma) < 3.2\%$ from CDF at the Tevatron gives $\kappa_{tu\gamma} < 0.47$. If the upper limit is saturated, single top events could be produced in ep scattering at HERA with a cross-section of about 1 pb, comparable to that for W production. Once there is a mechanism for its production, the decay of the top, $t \to bW$, $W \to \ell\nu$, $b \to cq\bar{q}'$, produces both the W signature and the large P_T^X from the b decay.

Both H1 (Vallee 2001) and ZEUS (2003c) have searched for this process, using essentially the same 1994–2000 data samples as for their high p_T lepton analyses. H1 add the requirements that: $M_T^{\ell\nu} > 10\,\text{GeV}$; the e or μ has a positive charge and $P_T^X > 25, 35\,\text{GeV}$ depending on jet angle. This leaves 5 events, 3 with e^+ and 2 μ^+ where 1.8 ± 0.5 are expected from SM sources. ZEUS required $P_T^X > 40\,\text{GeV}$ and found no candidate lepton events, where 2 might have been expected from SM processes. In addition to using the leptonic decay modes of the W, one may also look for 3-jet final states with one from the b decay and two from the hadronic decay modes of the W. H1 used $37\,\text{pb}^{-1}$ of data and required 3 jets with minimum p_T's of $25, 15, 10\,\text{GeV}$; ZEUS used $127\,\text{pb}^{-1}$ of data and required 3 jets with minimum p_T's of $40, 25, 14\,\text{GeV}$. Both groups further required the 3-jet and one of the 2-jet masses to be compatible with m_t and m_W respectively. With these selections, H1 find 10 events with 8 expected from SM sources and ZEUS 14 with 18 expected. The absence of an excess in the hadronic modes does not rule out anomalous top production as a possible mechanism for the H1 lepton events, though it does not favour it. The 95% CL upper limit on the single top production cross-section from the H1 hadronic search, corresponds to more than 3 events in the lepton sample.

Taking both the leptonic and hadronic modes together H1 and ZEUS have derived 95% CL upper limits on the coupling $\kappa_{tu\gamma} < 0.174$ for ZEUS and $\kappa_{tu\gamma} < 0.31$ for H1. The tighter limit in the ZEUS case is a reflection of the higher luminosity used for the hadronic W decays and the absence of leptonic decay candidates. Using the ZEUS upper limit for anomalous top production would give an expectation of 2 candidates in the H1 high p_T lepton sample.

12.6 Summary

This chapter has outlined a number of techniques used to probe for effects beyond the standard model in inelastic ep scattering at large Q^2. It is by no means complete in that many possible BSM models and particles have not been covered. The limit results are not necessarily the latest However, the aim has been to give an idea of how certain key techniques may be used to look for BSM signals and set limits. Some of the techniques, such as form factor analysis have been in use almost unchanged since the prehistory of the subject. Others, such as the generalised contact interaction approach, have now been honed to a precision tool able to shed light on many models in a systematic way. No signal for any novel effect has been confirmed, but the ability of the ZEUS and H1 general purpose detectors to provide sensitive measures has been demonstrated. Of particular importance are the hermetic calorimeters and good electron and muon identification. For direct resonance searches, good invariant mass resolution is important and this in turn depends on good resolution for the calorimeter and tracking sub-detectors. Ultimately, though, the most important ingredient is a large sample of high Q^2 events at the highest CM energy.

12.7 Problems

1. Within the context of the BRW classification of leptoquark states, check the fermion numbers and allowed decay channels for the scalar and vector states shown in Table 12.2. Why do the $F = 0, 2$ states get interchanged in the s and u channels for e^+p and e^-p?
2. Using the narrow resonance approximation, Eq. (12.1), with $\lambda_{LQ} \approx 0.3$, estimate the size of the contribution to the high Q^2 NC cross-section at $\sqrt{s} = 318\,\text{GeV}$ of a leptoquark with a mass of $210\,\text{GeV}$. How does this compare to the SM cross-section?
3. Establish that the forms for A_i, B_i given in Eqs (12.2) and (12.3) are consistent with the formalism given in Section 8.7.
4. Using the definitions of the contact interaction coupling structure given in Table 12.4.1, explain qualitatively the behaviour of the curves shown in Fig. 12.5.

12.8 Appendix: LQ classification & CI couplings

Table 12.2 The BRW classification of leptoquark states.

Model	Fermion number F	Charge Q	$BR(LQ \to e^{\pm}q)$ β	Coupling		Squark type
S_0^L	2	$-1/3$	$1/2$	$e_L u$	νd	$\tilde{d}_R$
S_0^R	2	$-1/3$	1	$e_R u$		
$\tilde{S}_0^R$	2	$-4/3$	1	$e_R d$		
$S_{1/2}^L$	0	$-5/3$	1	$e_L \bar{u}$		
		$-2/3$	0		$\nu \bar{u}$	
$S_{1/2}^R$	0	$-5/3$	1	$e_R \bar{u}$		
		$-2/3$	1	$e_R \bar{d}$		
$\tilde{S}_{1/2}^L$	0	$-2/3$	1	$e_L \bar{d}$		$\overline{\tilde{u}_L}$
		$+1/3$	0		$\nu \bar{d}$	$\overline{\tilde{d}_L}$
S_1^L	2	$-4/3$	1	$e_L d$		
		$-1/3$	$1/2$	$e_L u$	νd	
		$+2/3$	0		νd	
V_0^L	0	$-2/3$	$1/2$	$e_L \bar{d}$	$\nu \bar{u}$	
V_0^R	0	$-2/3$	1	$e_R \bar{d}$		
$\tilde{V}_0^R$	0	$-5/3$	1	$e_R \bar{u}$		
$V_{1/2}^L$	2	$-4/3$	1	$e_L d$		
		$-1/3$	0		νd	
$V_{1/2}^R$	2	$-4/3$	1	$e_R d$		
		$-1/3$	1	$e_R u$		
$\tilde{V}_{1/2}^L$	2	$-1/3$	1	$e_L u$		
		$+2/3$	0		νu	
V_1^L	0	$-5/3$	1	$e_L \bar{u}$		
		$-2/3$	$1/2$	$e_L \bar{d}$	$\nu \bar{u}$	
		$+1/3$	0		$\nu \bar{d}$	

In the above table, $F = L + 3B$ is the LQ fermion number, Q the electric charge (in units of the proton's charge), β is the branching ratio for $LQ \to eq$ (with BR $1 - \beta$ for $LQ \to \nu q$). The subscript denotes the weak isospin, the superscript the lepton chirality and a tilde differentiates states with different hypercharge.

In addition, the table also shows the final states to which the LQ may decay and possible squark assignments to the LQ states in the minimal supersymmetric theories with broken R-parity.

Table 12.3 Contact Interaction coupling coefficients η^i_{ab} for leptoquark searches.

LQ	coupling to u quark [GeV^{-2}]	coupling to d quark [GeV^{-2}]	F	m_{LQ}/λ_{LQ} [GeV]
S_0^L	$\eta^u_{LL} = +\frac{1}{2}(\lambda/M_{LQ})^2$		2	620
S_0^R	$\eta^u_{RR} = +\frac{1}{2}(\lambda/M_{LQ})^2$		2	570
$\tilde{S}_0^R$		$\eta^d_{RR} = +\frac{1}{2}(\lambda/M_{LQ})^2$	2	220
$S_{1/2}^L$	$\eta^u_{LR} = -\frac{1}{2}(\lambda/M_{LQ})^2$		0	340
$S_{1/2}^R$	$\eta^u_{RL} = -\frac{1}{2}(\lambda/M_{LQ})^2$	$\eta^d_{RL} = -\frac{1}{2}(\lambda/M_{LQ})^2$	0	320
$\tilde{S}_{1/2}^L$		$\eta^d_{LR} = -\frac{1}{2}(\lambda/M_{LQ})^2$	0	450
S_1^L	$\eta^u_{LL} = +\frac{1}{2}(\lambda/M_{LQ})^2$	$\eta^d_{LL} = +1\,(\lambda/M_{LQ})^2$	2	420
V_0^L		$\eta^d_{LL} = -1\,(\lambda/M_{LQ})^2$	0	670
V_0^R		$\eta^d_{RR} = -1\,(\lambda/M_{LQ})^2$	0	550
$\tilde{V}_0^R$	$\eta^u_{RR} = -1\,(\lambda/M_{LQ})^2$		0	410
$V_{1/2}^L$		$\eta^d_{LR} = +1\,(\lambda/M_{LQ})^2$	2	380
$V_{1/2}^R$	$\eta^u_{RL} = +1\,(\lambda/M_{LQ})^2$	$\eta^d_{RL} = +1\,(\lambda/M_{LQ})^2$	2	960
$\tilde{V}_{1/2}^L$	$\eta^u_{LR} = +1\,(\lambda/M_{LQ})^2$		2	1060
V_1^L	$\eta^u_{LL} = -2\,(\lambda/M_{LQ})^2$	$\eta^d_{LL} = -1\,(\lambda/M_{LQ})^2$	0	450

The above table shows the coupling coefficients and limits on the ratio mass/coupling from an H1 leptoquark CI analysis. The LQ classification and fermion number are listed, together with the 95% CL lower limits on m_{LQ}/λ for scalar (S) and vector (V) LQs. By convention the quantum numbers and helicities are given for e^-q and $e^-\bar{q}$ states. Limits on the coupling λ are only meaningful for leptoquark masses $m_{LQ} > \sqrt{s}$. (Adapted from H1 2000.)

A
Dirac equation and some other conventions

The conventions followed are those of comparable books (Halzen & Martin 1984, Aitchison & Hey 1989, Renton 1990).

Units

Throughout most of this book 'natural units' with $\hbar = c = 1$ are used. This means that energies, momenta and masses have the same units and the same dimension of $[\text{length}]^{-1}$, or equivalently length has dimension $[\text{energy}]^{-1}$. The conversation factor to 'real' units is then $\hbar c = 0.197\,\text{GeV fm}$ or $(\hbar c)^2 = 0.389\,\text{Gev}^2\,\text{mb}$.

The electromagnetic fine structure constant is

$$\alpha \equiv \frac{e^2}{4\pi\epsilon_0 \hbar c} = \frac{e^2}{4\pi} = \frac{1}{137}.$$

Lorentz 4-vectors and metric

The basic (contravariant) space-time and energy-momentum 4-vectors are

$$x^\mu = (t, \mathbf{x}) \quad p^\mu = (E, \mathbf{p}) \quad \text{with scalar product}$$

$$p \cdot x = p^\mu g_{\mu\nu} x^\nu = Et - \mathbf{p} \cdot \mathbf{x}, \quad \text{where}$$

$$g_{\mu\nu} = \begin{pmatrix} 1 & 0 & 0 & 0 \\ 0 & -1 & 0 & 0 \\ 0 & 0 & -1 & 0 \\ 0 & 0 & 0 & -1 \end{pmatrix}$$

is the metric tensor. The covariant 4-vector is $p_\mu = g_{\mu\nu} p^\nu$ so that a scalar product may also be written $p^\mu p_\mu = E^2 - \mathbf{p}^2 = m^2$.

Derivates, note the signs

$$\partial^\mu = \frac{\partial}{\partial_\mu} = \left(\frac{\partial}{\partial t}, -\nabla\right), \quad \partial_\mu = \frac{\partial}{\partial^\mu} = \left(\frac{\partial}{\partial t}, \nabla\right)$$

Dirac equation

The algebra of 4×4 Dirac matrices is defined by

$$\gamma^\mu = (\gamma^0, \boldsymbol{\gamma}), \quad \gamma\gamma^\mu\gamma^\nu + \gamma^\nu\gamma^\mu = 2g^{\mu\nu}, \quad \gamma^{\mu\dagger} = \gamma^0\gamma^\mu\gamma^0$$

$$\gamma^0\gamma^0 = I, \quad \gamma^{0\dagger} = \gamma^0, \quad \gamma^{k\dagger} = -\gamma^k, \quad k = 1, 2, 3$$

$$\gamma^5 = i\gamma^0\gamma^1\gamma^2\gamma^3, \quad \gamma^5\gamma^\mu + \gamma^\mu\gamma^5 = 0, \quad \gamma^{5\dagger} = \gamma^5$$

The slash notation is

$$\not{p} = \gamma \cdot p = \gamma^0 E - \boldsymbol{\gamma} \cdot \mathbf{p}, \quad \text{but note that} \quad \not{\partial} = \gamma \cdot \partial = \gamma^0 \frac{\partial}{\partial t} + \boldsymbol{\gamma} \cdot \nabla.$$

The Dirac equation in coordinate space for a spin 1/2 particle of mass m and momentum p ($E > 0$) is

$$(i\hbar\not{\partial} - mc)\psi(x) = 0, \quad \text{or with} \quad \psi(x) = e^{-ip\cdot x}u(p): \quad (\not{p} - m)u(p) = 0,$$

where $u(p)$ is the 4-component spinor. The conjugate spinor is

$$\bar{u} \equiv u^\dagger\gamma^0, \quad \text{satisfying} \quad \bar{u}(\not{p} - m) = 0.$$

The normalization is

$$u^{(r)\dagger}u^{(s)} = 2E\delta_{rs}, \quad \bar{u}^{(r)}u^{(s)} = 2m\delta_{rs}, \quad \text{with completeness} \quad \sum_{s=1,2} u^{(s)}\bar{u}^{(s)} = \not{p} + m.$$

The negative energy or antiparticle solutions satisfy ($E > 0$):

$$\psi(x) = e^{+ip\cdot x}v(p), \quad (\not{p} + m)v(p) = 0, \quad \bar{v}v = -2m, \quad \sum v\bar{v} = \not{p} - m.$$

The standard representation of the γ matrices is used.

$$\gamma^0 = \begin{pmatrix} I & 0 \\ 0 & -I \end{pmatrix}, \quad \boldsymbol{\gamma} = \begin{pmatrix} 0 & \boldsymbol{\sigma} \\ -\boldsymbol{\sigma} & 0 \end{pmatrix}, \quad \gamma^5 = \begin{pmatrix} 0 & I \\ I & 0 \end{pmatrix}$$

$$\sigma_1 = \begin{pmatrix} 0 & 1 \\ 1 & 0 \end{pmatrix}, \quad \sigma_2 = \begin{pmatrix} 0 & -i \\ i & 0 \end{pmatrix}, \quad \sigma_3 = \begin{pmatrix} 1 & 0 \\ 0 & -1 \end{pmatrix}$$

The standard form of solution for the particle and antiparticle spinors is

$$u(p) = \sqrt{E+m}\begin{pmatrix} \xi_\pm \\ \frac{\boldsymbol{\sigma}\cdot\mathbf{p}}{E+m}\xi_\pm \end{pmatrix}, \quad v(p) = \sqrt{E+m}\begin{pmatrix} \frac{\boldsymbol{\sigma}\cdot\mathbf{p}}{E+m}\eta_\pm \\ \eta_\pm \end{pmatrix}, \quad \text{where}$$

$$\xi_+ = \begin{pmatrix} 1 \\ 0 \end{pmatrix}, \quad \xi_- = \begin{pmatrix} 0 \\ 1 \end{pmatrix}, \quad \eta_+ = \begin{pmatrix} 0 \\ 1 \end{pmatrix}, \quad \eta_- = \begin{pmatrix} 1 \\ 0 \end{pmatrix}.$$

States of definite chirality (equivalent to helicity in the massless limit) are

$$u_R = \frac{1+\gamma^5}{2}u, \quad u_L = \frac{1-\gamma^5}{2}u, \quad v_R = \frac{1-\gamma^5}{2}v, \quad v_L = \frac{1+\gamma^5}{2}v$$

and for the standard solutions

$$u_R(p) = \frac{\sqrt{E+m}}{2}\begin{pmatrix} (1+\frac{\boldsymbol{\sigma}\cdot\mathbf{p}}{E+m})\xi_\pm \\ (1+\frac{\boldsymbol{\sigma}\cdot\mathbf{p}}{E+m})\xi_\pm \end{pmatrix}, \quad u_L(p) = \frac{\sqrt{E+m}}{2}\begin{pmatrix} (1-\frac{\boldsymbol{\sigma}\cdot\mathbf{p}}{E+m})\xi_\pm \\ -(1-\frac{\boldsymbol{\sigma}\cdot\mathbf{p}}{E+m})\xi_\pm \end{pmatrix}$$

$$v_R(p) = \frac{\sqrt{E+m}}{2}\begin{pmatrix} -(1-\frac{\boldsymbol{\sigma}\cdot\mathbf{p}}{E+m})\eta_\pm \\ (1-\frac{\boldsymbol{\sigma}\cdot\mathbf{p}}{E+m})\eta_\pm \end{pmatrix}, \quad v_L(p) = \frac{\sqrt{E+m}}{2}\begin{pmatrix} (1+\frac{\boldsymbol{\sigma}\cdot\mathbf{p}}{E+m})\eta_\pm \\ (1+\frac{\boldsymbol{\sigma}\cdot\mathbf{p}}{E+m})\eta_\pm \end{pmatrix}$$

At high energies $E \gg m$ these solutions tend to

$$u_R \to \sqrt{E}\begin{pmatrix} \xi_+ \\ \xi_+ \end{pmatrix}, \quad u_L \to \sqrt{E}\begin{pmatrix} \xi_- \\ -\xi_- \end{pmatrix}, \quad v_R \to \sqrt{E}\begin{pmatrix} -\eta_+ \\ \eta_+ \end{pmatrix}, \quad v_L \to \sqrt{E}$$

Some trace theorems

$$\text{Tr}(\slashed{a}\slashed{b}) = 4a\cdot b, \quad \text{Tr}(\slashed{a}\slashed{b}\slashed{c}) = 0, \text{ and for any odd number of } \gamma \text{ matrices.}$$

$$\gamma_\mu \slashed{a} \gamma^\mu = -2\slashed{a}, \quad \gamma_\mu \slashed{a}\slashed{b} \gamma^\mu = 4a\cdot b, \quad \gamma_\mu \slashed{a}\slashed{b}\slashed{c} \gamma^\mu = -2\slashed{c}\slashed{b}\slashed{a}$$

$$\text{Tr}(\slashed{a}\slashed{b}\slashed{c}\slashed{d}) = 4((a\cdot b)(c\cdot d) + (a\cdot d)(b\cdot c) - (a\cdot c)(b\cdot d))$$

$$\text{Tr}(\gamma^\mu \slashed{a} \gamma^\nu \slashed{b})\,\text{Tr}(\gamma_\mu \slashed{c} \gamma_\nu \slashed{d}) = 32((a\cdot c)(b\cdot d) + (a\cdot d)(b\cdot c))$$

$$\text{Tr}(\gamma^\mu \slashed{a} \gamma^\nu \gamma^5 \slashed{b})\,\text{Tr}(\gamma_\mu \slashed{c} \gamma_\nu \gamma^5 \slashed{d}) = 32((a\cdot c)(b\cdot d) - (a\cdot d)(b\cdot c))$$

$$\text{Tr}(\gamma^5 \slashed{a}\slashed{b}\slashed{c}\slashed{d}) = 4i\epsilon_{\mu\lambda\nu\sigma} a^\mu b^\lambda c^\nu d^\sigma,$$

where $\epsilon_{\mu\lambda\nu\sigma}$ is the totally antisymmetric tensor of rank 4 with value +1 (−1) according as $\mu\lambda\nu\sigma$ is an even (odd) perm of 0123 and 0 otherwise. A useful contraction of two ϵs is:

$$\epsilon^{\mu\nu\alpha\beta}\epsilon_{\mu\nu\gamma\sigma} = -2(\delta^\alpha_\gamma \delta^\beta_\sigma - \delta^\alpha_\sigma \delta^\beta_\gamma)$$

Rotation matrices

The angular momentum operators in quantum mechanics are the generators of rotations, for example to rotate a state through an angle θ about the y-axis one has

$$R_y(\theta)|\psi\rangle = e^{-iJ_y\theta}|\psi\rangle.$$

If the system is invariant under rotations then the Hamiltonian will commute with the angular momentum operators.

Eigenstates of angular momentum, $|j,m\rangle$, are labelled by the eigenvalues of the total angular momentum operator $J^2 = J_x^2 + J_y^2 + J_z^2$ and J_z, the operator along the axis of quantization. For a fixed value of j, m ranges over the $2j+1$ values $-m,\ldots,+m$.

Under a rotation about the y-axis, the state $|j,m\rangle$ will be transformed into a linear combination of the $2j+1$ states $|j,m'\rangle$ quantized along the rotated axis Oz'. The elements of the *rotation matrices* are the coeffificents of the linear transformations between the two sets of states

$$e^{-iJ_y\theta}|j,m\rangle = \sum_{m'} d^j_{m'm}(\theta)|j,m'\rangle \quad \text{or} \quad \langle j,m'|e^{-iJ_y\theta}|j,m\rangle = d^j_{m'm}(\theta).$$

For $j = 1/2$, one finds

$$d_{++} = d_{--} = \cos(\theta/2); \quad d_{-+} = -d_{+-} = \sin(\theta/2).$$

and for $j = 1$

$$\begin{aligned} d_{11} &= d_{-1-1} = (1+\cos\theta)/2; \\ d_{01} &= -d_{10} = -d_{0-1} = d_{-10} = \sin\theta/\sqrt{2}; \\ d_{1-1} &= d_{-11} = (1-\cos\theta)/2; \\ d_{00} &= \cos\theta. \end{aligned}$$

More details on angular momentum and rotations may be found in the standard texts cited at the start of this appendix.

B
Phase space and cross-sections

Units are such that $\hbar = c = 1$.

General cross-section formula

The state of a particle with 4-momentum $p = (E, \mathbf{p})$ and spin (or helicity) λ has normalization

$$\langle p', \lambda' | p, \lambda \rangle = 2E \delta_{\lambda \lambda'} \delta^3(\mathbf{p} - \mathbf{p}').$$

This corresponds to $2E$ particles per unit volume and a Lorentz invariant phase space factor $d^3p/(2E(2\pi)^3)$. A useful identity is

$$\frac{d^3\mathbf{p}}{2E} = \theta(E)\delta(p^2 - m^2)\, d^4p.$$

The scattering operator S is written as $S = 1 + iT$, and the reduced matrix element, $\mathcal{M}(B:A)$, for a scattering process $A \to B$ is defined by

$$S_{fi} = \langle B | iT | A \rangle = i(2\pi)^4 \delta^4(p_A - p_B) \mathcal{M}(B:A),$$

where p_A, p_B are the total 4-momenta of states A and B respectively.

To calculate the total cross-section for a collider process $a + b \to X$ one starts from the Fermi golden rule

$$\text{flux} \times \sigma(ab \to X) = \int w_{fi} \times [\text{final state phase space}]$$

where $w_{fi} = |S_{fi}|^2/VT$ is the transition rate and VT is the total space–time volume. In a colliding beam experiment the initial state particle flux is $2E_a 2E_b |\mathbf{v}_a - \mathbf{v}_b|$, where $\mathbf{v}_i$ are the particle velocities. Inserting the expression for S_{fi} in terms of the reduced matrix element into the above equation gives rise to the square of the overall 4-momentum conservation delta function. Formally this is handled by using the identity

$$(2\pi)^4 \delta^4(p_f - p_i) = \int e^{ix(p_f - p_i)} d^4x$$

to replace one of the $\delta^4()$, then performing the integration with $p_f = p_i$ on account of the other $\delta^4()$ to give $\int d^4x = VT$. The VT factors then cancel to give

$$\sigma(ab \to X) = \frac{1}{2E_a 2E_b |\mathbf{v}_a - \mathbf{v}_b|} \int \mathrm{dLips}(s:X)|\mathcal{M}(X:ab)|^2$$

where dLips is the Lorentz invariant phase space, which for a final state with n_X particles is

$$\mathrm{dLips}(s:X) = (2\pi)^4 \delta^4(p_X - p_a - p_b) \prod_{i=1}^{n_X} \frac{d^3 \mathbf{k}_i}{(2\pi)^3 2k_i^0},$$

where $s = (p_a + p_b)^2$ and $p_X = \sum_i k_i$. For a total cross-section to all final states X an additional $\sum_X$ is performed. For the unpolarized cross-section for particles with spin, final spin states are summed and initial states averaged. This gives an additional term, $1/[(2S_a+1)(2S_b+1)]$, on the right-hand side, where S_a, S_b are the spins of particles a and b. If a differential cross-section is required, then the relevant variables are excluded from the phase space integral.

Flux factor

The flux factor $2E_a 2E_b|\mathbf{v}_a - \mathbf{v}_b|$ is a Lorentz invariant and may be written in a number of ways. In the fixed-target frame, with b at rest, it is $2E_a 2m_b|\mathbf{v}_a|$. The following relations are also true:

$$2E_a 2E_b|\mathbf{v}_a - \mathbf{v}_b| = 4\sqrt{[(p_1 \cdot p_2)^2 - m_1^2 m_2^2]} = 2\sqrt{\lambda(s, m_a^2, m_b^2)},$$

where $\lambda(s, m_a^2, m_b^2) = (s - m_a^2 - m_b^2)^2 - 4m_a^2 m_b^2.$

Two-body scattering

Consider the special case of two-body scattering $\mathbf{a} + \mathbf{b} \to \mathbf{a'} + \mathbf{b'}$ with 4-momenta $p_a = p_1$, $p_b = p_2$, $p_{a'} = p_3$, $p_{b'} = p_4$. The cross-section in the fixed-target frame (particle b at rest) is given by:

$$d\sigma = \frac{|\mathcal{M}|^2}{2E_1 2m_2 |\mathbf{v}_1|} \mathrm{dLips}(s; p_3, p_4), \qquad (\mathrm{B}.1)$$

where in this case

$$\mathrm{dLips}(s; p_3, p_4) = (2\pi)^4 \delta^4(p_3 + p_4 - p_1 - p_2) \frac{d^3 \mathbf{p}_3}{2E_3 (2\pi)^3} \frac{d^3 \mathbf{p}_4}{2E_4 (2\pi)^3}$$

with $s = (p_3 + p_4)^2 = (p_1 + p_2)^2$. dLips may be evaluated by using the $\delta^4()$ to integrate over 4 of the 6 variables $d^3p_3 d^3p_4$ and doing this in the CM frame gives:

$$\frac{d\sigma}{d\Omega} = \frac{1}{(8\pi w)^2} |\mathcal{M}|^2 \frac{p'}{p} \qquad (\mathrm{B}.2)$$

where $w = \sqrt{s}$ is the total CM energy, p and p' the magnitudes of the initial and final CMS 3-momenta, respectively. The above cross-section has dimensions of [energy]$^{-2}$. To get back to physical units of area one must multiply by $(\hbar c)^2 = 0.389\,\text{mb}\,\text{GeV}^2$.

Optical theorem

The unitarity of the S matrix, $S^\dagger S = 1$, is a consequence of the conservation of the overall probability flux in a scattering process. It may also be written in terms of the T matrix as

$$-i(T - T^\dagger) = T^\dagger T.$$

Taking matrix elements of this expression and inserting a complete set of states on the RHS gives:

$$2\text{Im}\bigl[\langle B|T|A\rangle\bigr] = \sum_X \left(\prod_{i=1}^{n_X} \int \frac{d^3 k_i}{(2\pi)^3 2k_i^0}\right) \langle B|T^\dagger|X\rangle\langle X|T|A\rangle.$$

Using the relationship $\langle B|T|A\rangle = (2\pi)^4 \delta^4(p_A - p_B)\mathcal{M}(B:A)$ and cancelling the overall 4-momentum conservation $(2\pi)^4 \delta^4(p_A - p_B)$ then gives

$$2\text{Im}[\mathcal{M}(B:A)] = \sum_X \left(\prod_{i=1}^{n_X} \int \frac{d^3 k_i}{(2\pi)^3 2k_i^0}\right)$$
$$\times (2\pi)^4 \delta^4(p_A - p_X)\mathcal{M}^\dagger(B:X)\mathcal{M}(X:A).$$

Putting $B = A = a + b$, the RHS of the above expression is essentially that for the total cross-section for $a + b \to X$ apart from the flux factor. Thus one has the *optical theorem*

$$\sigma(ab \to X) = \frac{\text{Im}[\mathcal{M}(ab:ab)]}{\sqrt{\lambda(s, m_a^2, m_b^2)}}, \tag{B.3}$$

where $\text{Im}[M(ab:ab)]$ is the imaginary part of the forward ($\theta_{CM} = 0$) scattering amplitude for elastic $a + b \to a + b$. For the unpolarised total cross-section for particles with spin there is an additional

$$\frac{1}{(2S_a + 1)(2S_b + 1)} \sum_{ss'}$$

on the RHS from the summing and averaging over spin states.

Crossing symmetry

Crossing symmetry states that the matrix element for a process containing an antiparticale of 4-momentum p_μ in the initial (final) state is identical with the matrix element for the *crossed* process which contains the corresponding particle of 4-momentum $-p_\mu$ in the final (initial) state. As crossing symmetry requires the continuation to negative energies, it is intimately connected with the analytic properties of the matrix element and more generally with the theory of dispersion relations. This subsection gives a brief summary of crossing relations for $2 \to 2$ scattering.

Consider the process $a + b \to c + d$ with reduced matrix element $\mathcal{M}_{ab:cd}(s,t,u)$ where

$$s = (p_a + p_b)^2, \quad t = (p_a - p_c)^2, \quad u = (p_a - p_d)^2$$

are the usual Mandelstam kinematic invariants. This defines the 's-channel' with $s > 0$, $t < 0$, $u < 0$. The 't-channel' corresponds to the process $a + \bar{c} \to \bar{b} + d$ with

$$\bar{s} = (p_a + \overline{p_c})^2 = (p_a - p_c)^2 = t, \quad \bar{t} = (p_a - \overline{p_b})^2 = (p_a + p_b)^2 = s, \quad \bar{u} = u,$$

where $\bar{s} = t > 0$, $\bar{t} = s < 0$, $\bar{u} = u < 0$ and

$$\mathcal{M}_{a\bar{c}:\bar{b}d}(\bar{s},\bar{t},\bar{u}) = \mathcal{M}_{ab:cd}(t,s,u).$$

Similarly for the 'u-channel' $a + \bar{d} \to c + \bar{b}$

$$\bar{s} = (p_a + \overline{p_d})^2 = (p_a - p_d)^2 = u, \quad \bar{t} = t, \quad \bar{u} = (p_a - \overline{p_b})^2 = (p_a + p_b)^2 = s,$$

where $\bar{s} = u > 0$, $\bar{t} = t < 0$, $\bar{u} = s < 0$ and

$$\mathcal{M}_{a\bar{d}:c\bar{b}}(\bar{s},\bar{t},\bar{u}) = \mathcal{M}_{ab:cd}(u,t,s).$$

Crossing symmetry may be used to relate the $|\mathcal{M}|^2$ appearing in the cross-section formluae for, say, $a + b \to c + d$ and $a + \bar{c} \to \bar{b} + d$. Note however that crossing symmetry does not apply to the flux and phase–space terms nor to colour factors. One final note of caution about the application of crossing symmetry to processes involving fermions. When the $\sum_{\text{spins}} |\mathcal{M}|^2$ is performed it will involve taking traces over the spinor operators

$$u(p)\bar{u}(p) = \not{p} + m \text{ (fermions)}, \quad v(p)\bar{v}(p) = \not{p} - m \text{ (antifermions)}.$$

When p is crossed to $-p$, on the LHS of the above

$$u(p)\bar{u}(p) \to u(-p)\bar{u}(-p) = v(p)\bar{v}(p),$$

whereas on the RHS $\not{p} + m \to -\not{p} + m$. This is a consequence of the sign conventions chosen for the fermion and antifermion spinors, $\bar{u}u = 2m$ and $\bar{v}v = -2m$. It means that when $\sum_{\text{spins}} |\mathcal{M}|^2$ is crossed it picks up a factor of (-1) for each fermion crossed, thus for

$$e^+e^- \to \mu^+\mu^- \quad \text{and} \quad \mu^-e^- \to e^-\mu^- \quad \text{factor } (-1)^2$$

whereas

$$e^-\gamma \to e^-\gamma \quad \text{and} \quad e^+e^- \to \gamma\gamma \quad \text{factor } (-1).$$

C
DIS cross-sections

General

In the fixed–target frame (proton at rest), the double differential cross-section for inelastic scattering, $e(k) + p(p) \to e(k') + X(p_X)$, mediated by virtual photon exchange only can be written as;

$$\frac{d^2\sigma}{d\Omega\, dE'} = \frac{\alpha^2}{q^4} \frac{E'}{mE} \mathrm{L} \cdot \mathrm{W} \qquad (C.1)$$

where α is the fine structure constant, m the proton mass, $q = k - k'$ the 4-momentum transfer, E and E' the energies of the incident and scattered electron and L·W is the contraction of the leptonic and hadronic tensors.

The leptonic tensor $L_{\mu\nu}$ has the form

$$L_{\mu\nu} = 2[k'_\mu k_\nu + k'_\nu k_\mu - k \cdot k' g_{\mu\nu}].$$

The hadronic tensor $W_{\mu\nu}$ is defined formally in terms of matrix elements of the hadronic EM current j_μ

$$W_{\mu\nu} = \frac{1}{4\pi} \sum_X \left(\prod_{i=1}^{n_X} \int \frac{d^3 k_i}{(2\pi)^3 2 k_i^0} \right) (2\pi)^4 \delta^4(p + q - p_X)$$

$$\times \left(\frac{1}{2} \sum_s \right) \langle p, s | j_\mu^\dagger(0) | X \rangle \langle X | j_\nu(0) | p, s \rangle, \qquad (C.2)$$

but it is more useful to use the form in terms of the structure functions W_i

$$W_{\mu\nu} = W_1 \left(-g_{\mu\nu} + \frac{q_\mu q_\nu}{q^2} \right) + \frac{W_2}{m^2} \left(p_\mu - \frac{p \cdot q}{q^2} q_\mu \right)\left(p_\nu - \frac{p \cdot q}{q^2} q_\nu \right). \qquad (C.3)$$

The W_i are real scalar functions of two variables, E', θ in the lab or any two of the invariant variables, x, y, Q^2 (to be defined below). Note that both $L_{\mu\nu}$ and $W_{\mu\nu}$ are symmetric under the interchange $\mu \leftrightarrow \nu$ and gauge invariance requires

$$q_\mu L^{\mu\nu} = 0, \quad L^{\mu\nu} q_\nu = 0, \quad q_\mu W^{\mu\nu} = 0, \quad W^{\mu\nu} q_\nu = 0.$$

[The hadronic weak current has a similar structure, but there is an additional term $i\varepsilon_{\mu\nu\alpha\beta} p^\alpha p^\beta W_3/(2m^2)$ coming from the parity violating piece.]

Using the above constraints from gauge invariance, the contraction $L \cdot W$ reduces to

$$L \cdot W = 4(k \cdot k')W_1 + 2[(p \cdot k)(p \cdot k') - (k \cdot k')m^2]\frac{W_2}{m^2}.$$

In the fixed-target frame the cross-section then takes the form

$$\frac{d^2\sigma}{d\Omega\, dE'} = \frac{\alpha^2}{4mE^2 \sin^4 \frac{\theta}{2}} (W_2 \cos^2 \frac{\theta}{2} + 2W_1 \sin^2 \frac{\theta}{2}) \qquad (C.4)$$

where θ is the scattering angle of the electron. This form was often used for the early deep-inelastic experiments.

It is convenient to have the cross-section expressed in terms of scaling variables $x = \frac{-q^2}{2p \cdot q}$, $y = \frac{p \cdot q}{p \cdot k}$ and $Q^2 = -q^2$. Also at this point, it is conventional to replace the W_i by their 'scaling limit' equivalents $F_1 = W_1$, $F_2 = \nu W_2/m^2$ where $\nu = (p \cdot q)$. The Jacobian for the transformation $d\Omega\, dE' \to dx\, dy$ is given by $2\pi/J$ where

$$J = \frac{\partial(x,y)}{\partial(u,E')} = \frac{E'}{m(E - E')}, \qquad u = \cos\theta.$$

Then at large $s = (k+p)^2$ when mass terms may be ignored

$$\frac{d^2\sigma}{dx\, dy} = \frac{4\pi\alpha^2 s}{Q^4}\left[xy^2 F_1(x,y) + (1-y)F_2(x,y)\right]. \qquad (C.5)$$

A closely related form, derived from this using $Q^2 = sxy$ and the longitudinal structure function $F_L = F_2 - 2xF_1$, is

$$\frac{d^2\sigma}{dx\, dQ^2} = \frac{2\pi\alpha^2}{xQ^4}\left[Y_+ F_2(x,Q^2) - y^2 F_L(x,Q^2)\right], \qquad (C.6)$$

where $Y_+ = 1 + (1-y)^2$.

For completeness the expression for $d^2\sigma/dxdQ^2$ including the mass terms (which may be necessary at low Q^2 and moderate x) is

$$\frac{d^2\sigma}{dx\, dQ^2} = \frac{2\pi\alpha^2}{xQ^4}\left[1 - y - \frac{m^2x^2y^2}{Q^2} + \frac{y^2}{2}\frac{1 + 4m^2x^2/Q^2}{1+R}\right] F_2(x,Q^2), \quad (C.7)$$

where $R \equiv \sigma_L/\sigma_T = F_L/2xF_1$ and $F_L = F_2(1+4m^2x^2/Q^2) - 2xF_1$ (σ_L, σ_T are the $\gamma^* p$ total cross-sections for longitudinal and transverse polarizations, respectively, see below).

Note that there are a number of different definitions of the W_i structure functions. In this book, following recent conventions (e.g. Yndurátn 1999), the W_i and F_i are both defined to be dimensionless. In older books the factor $1/m$ appearing here as $1/(mE)$ (from the flux factor) in the first equation of this subsection, is absorbed in the definition of $W_{\mu\nu}$. Also here $\nu = (p \cdot q)$ rather than $\nu = (p \cdot q)/m$.

Elastic electron–muon scattering

It is useful to have various forms of the cross-section for the EM elastic scattering of two point-like spin-$\frac{1}{2}$ particles, specifically the process $e(k) + \mu(p) \to e(k') + \mu(p')$, where the electron mass is ignored and m is the muon mass. Start from the formulae in the 'two-body scattering' subsection of Appendix B with $p_3 = k'$, $p_4 = p'$ and

$$|\mathcal{M}|^2 \to \frac{1}{4} \sum_{\text{spins}} |\mathcal{M}(s's : \sigma'\sigma)|^2,$$

$$\mathcal{M}(s's : \sigma'\sigma) = i\frac{ee'}{q^2}[\bar{u}(k',\sigma')\gamma_\mu u(k,\sigma)][\bar{u}(p',s')\gamma^\mu u(p,s)].$$

When squared and summed this reduces to a contraction of two leptonic tensors

$$\frac{1}{4} \sum_{\text{spins}} |\mathcal{M}|^2 = \frac{(4\pi)^2 \alpha^2 e'^2}{q^4} L^{(e)}_{\alpha\beta} L^{(\mu)\alpha\beta}.$$

Ignoring the electron mass

$$L^{(e)} \cdot L^{(\mu)} = 8\Big[(k' \cdot p')(k \cdot p) + (k' \cdot p)(k \cdot p') - m^2(k \cdot k')\Big].$$

Evaluating this in the fixed-target frame gives

$$L^{(e)} \cdot L^{(\mu)} = 16m^2 EE'\left[\cos^2\frac{\theta}{2} + \frac{Q^2}{2m^2}\sin^2\frac{\theta}{2}\right],$$

where E, E' are the energies of the incident and scattered electron. In evaluating the two-body final state phase space, it is convenient not to integrate out all the delta functions. Writing

$$d^3p'/(2p'^0) = d^4p'\,\theta(p'^0)\delta(p'^2 - m^2),$$

the $\int d^4p'$ gives overall 4-momentum conservation leaving $\delta(p'^2 - m^2)$. Now

$$\delta(p'^2 - m^2) = \delta(q^2 + 2\nu) = \frac{1}{2}\delta(\nu - Q^2/2)$$

where $\nu = p \cdot q = m(E - E')$, giving

$$\text{dLips} = \frac{E'dE'd\Omega}{4(2\pi)^2}\delta(\nu - Q^2/2).$$

Taking these results together with the flux factor of $4mE$ gives

$$\frac{d^2\sigma}{d\Omega\,dE'} = \frac{4\alpha^2 e'^2}{Q^4} mE'^2 \left[\cos^2\frac{\theta}{2} + \frac{Q^2}{2m^2}\sin^2\frac{\theta}{2}\right]\delta(\nu - Q^2/2). \quad (C.8)$$

At high energies when all mass terms may be ignored, the summed matrix element squared takes a simple form:

$$\frac{1}{4}\sum_{\text{spins}}|\mathcal{M}|^2 = \frac{e^2 e'^2}{t^2} 2[s^2 + u^2],$$

where $s = 2k \cdot p$, $t = -2k \cdot k'$, $u = -2p \cdot k'$. Using this and Jacobian $J' = Q^2 E'/(my^2 E)$ for the $dx\, dQ^2 \to d\Omega\, dE'$ change of variables gives

$$\frac{d^2\sigma}{dx\, dQ^2} = \frac{2\pi}{J'}\frac{d^2}{d\Omega\, dE'} = \frac{\pi\alpha^2 e'^2}{t^2}\frac{y^2 s^2}{Q^2}\left[1 + \frac{u^2}{s^2}\right]\delta(\nu - Q^2/2).$$

Since $x = Q^2/(2\nu)$, $\delta(\nu - Q^2/2) = \delta(1-x)/\nu$ and using $t^2 = Q^4$, $u/s = -(p \cdot k')/(p \cdot k) = -(1-y)$ and $Q^2 = sxy$, this last results may be written

$$\frac{d^2\sigma}{dx\, dQ^2} = \frac{2\pi\alpha^2 e'^2}{Q^4}[1 + (1-y)^2]\frac{1}{x}\delta(1-x). \tag{C.9}$$

Finally if the $\delta(1-x)$ is integrated out out gets a high energy invariant form of the elastic cross-section

$$\frac{d\sigma}{dQ^2} = \frac{2\pi\alpha^2 e'^2}{Q^4}[1 + (1-y)^2].$$

A closely related form may be deduced easily from the formula for a two-body cross-section in the CM frame (with all delta functions integrated out)

$$\frac{d\sigma}{d\Omega} = \frac{1}{(8\pi)^2 s}\frac{1}{4}\sum_{\text{spins}}|\mathcal{M}|^2,$$

(as in this case the initial and final 3-momenta are identical). Inserting the high energy form for $|\mathcal{M}|^2$ gives

$$\frac{d\sigma}{d\Omega} = \frac{\alpha^2 e'^2}{2st^2}[s^2 + u^2],$$

or since $dt = d(\cos\theta)/2$,

$$\frac{d\sigma}{dt} = \frac{2\pi\alpha^2 e'^2}{t^2}[1 + u^2/s^2].$$

$\gamma^* p$ total cross-section

Both for theoretical and phenomenological studies, it is often convenient to think of EM deep inelastic scattering as $\gamma^* p$ scattering. In this section, the relationship between $\sigma^{\text{tot}}_{\gamma^* p}$ and structure functions will be established.

Using the optical theorem (Eq. B.3) $\sigma^{\text{tot}}_{\gamma^* p}$ is given by $\text{Im}[\mathcal{M}(\gamma^* p : \gamma^* p)]$, the forward $\gamma^* p$ scattering amplitude and this latter is $e^2 T_{\mu\nu}$, the time ordered product of the hadronic EM current operators as shown in Eq. (3.39).

The imaginary part of $T_{\mu\nu}$ is $2\pi W_{\mu\nu}$ and and to form $\text{Im}[\mathcal{M}(\gamma^*p : \gamma^*p)]$ this is contracted with the γ^* polarisation vector ε_μ. Thus, for a γ^* with polarisation $\lambda = \pm 1, 0$

$$\sigma^{\text{tot}}_{\gamma^*p}(\lambda) = \frac{8\pi^2\alpha}{W^2 - m^2}\varepsilon^\dagger_\mu(\lambda)W^{\mu\nu}\varepsilon_\nu(\lambda), \tag{C.10}$$

where W is the γ^*p CM energy. This expression is for unpolarised protons, the spin averaging is included in the definition of $W^{\mu\nu}$. The reader may wonder why $W^2 - m^2$ appears in the denominator rather than $\sqrt{\lambda(W^2, m^2, -Q^2)}$ as a literal application of Eq. (B.3) would imply. The reason is that as the γ^* is a virtual particle there is a certain amount of freedom in the definition of the γ^* flux factor. Here the convention of Hand is followed in which the flux is given by the energy that a real photon would require to create the final state. An alternative is that of Gilman in which the flux is given by the momentum of the γ^*. In this case $W^2 - m^2$ is replaced by $\sqrt{\lambda(W^2, m^2, -Q^2)}$.

To evaluate Eq. (C.10) take the z-axis to be along the direction of $\mathbf{q}$, the γ^* 3-momentum. The γ^* has energy ν/m and mass $-Q^2$. With these choices, the polarization vectors are

$$\varepsilon(\pm 1) = \frac{1}{\sqrt{2}}(0, 1, \pm i, 0), \quad \varepsilon(0) = \frac{1}{\sqrt{Q^2}}\left(\sqrt{(\nu/m)^2 + Q^2}, 0, 0, \nu/m\right),$$

using these and Eq. (C.3) gives[1]

$$\sigma_T \equiv \frac{1}{2}(\sigma(+) + \sigma(-)) = \frac{8\pi^2\alpha}{W^2 - m^2}W_1$$

$$\sigma_L \equiv \sigma(0) = \frac{8\pi^2\alpha}{W^2 - m^2}\left[\frac{(\nu/m)^2 + Q^2}{Q^2}W_2 - W_1\right]$$

From these two a third relation, obtained by adding them, is often used

$$\sigma_T + \sigma_L = \frac{8\pi^2\alpha}{W^2 - m^2}\frac{(\nu/m)^2 + Q^2}{Q^2}W_2.$$

Using $F_1 = W_1$ and $F_2 = \nu W_2/m$, one may also write

$$\sigma_T = \frac{8\pi^2\alpha}{W^2 - m^2}F_1 \tag{C.11}$$

$$\sigma_T + \sigma_L = \frac{8\pi^2\alpha}{W^2 - m^2}\left[\frac{\nu}{Q^2} + \frac{m^2}{\nu}\right]F_2. \tag{C.12}$$

There are two simplified forms of these last two relations:

---

[1] An obvious shorthand is used for γ^*p cross-sections.

- For $W^2 \gg m^2$, ignore the m^2 in common overall factor, replace ν/Q^2 by $1/(2x)$ and drop the 2nd term ($m^2/\nu = 2m^2/(W^2 y)$) in the expression for F_2 to give

$$\sigma_T = \frac{4\pi^2 \alpha}{W^2} 2F_1 \qquad (C.13)$$

$$\sigma_T + \sigma_L = \frac{4\pi^2 \alpha}{W^2} \frac{F_2}{x}. \qquad (C.14)$$

- The second form is appropriate for large W^2 and small Q^2. Use the relation $W^2 - m^2 = Q^2(1-x)/x \approx Q^2/x$ for the overall factor and make the same approximation as before for the F_2 term to give

$$\sigma_T = \frac{4\pi^2 \alpha}{Q^2} 2xF_1 \qquad (C.15)$$

$$\sigma_T + \sigma_L = \frac{4\pi^2 \alpha}{Q^2} F_2. \qquad (C.16)$$

Since real photons have only transverse polarization states, as $Q^2 \to 0$ $\sigma_L \to 0$ and $F_2/Q^2 \to \sigma_T$ (constant independent of Q^2). Using the two relations, one may also write

$$\sigma_L = \frac{4\pi^2 \alpha}{Q^2}\left[F_2 - 2xF1\right] = \frac{4\pi^2 \alpha}{Q^2} F_L, \qquad (C.17)$$

which implies $F_L/Q^2 \to 0$ as $Q^2 \to 0$.

The double diffential cross-section for $e + p \to X$ is related to the $\gamma^* p$ cross-sections by

$$\frac{d^2 \sigma}{dx\, dQ^2} = \Gamma(\sigma_T + \varepsilon \sigma_L), \qquad (C.18)$$

where Γ and ε are the flux and polarization of the virtual photon, respectively. This is easily checked. Start from Eq. (C.6) and using the relations (C.16) and (C.17) replace F_2 and F_L by σ_T and σ_L, finally compare with the above form to give

$$\Gamma = \frac{\alpha[1 + (1-y)^2]}{2\pi x Q^2} \qquad (C.19)$$

$$\varepsilon = \frac{2(1-y)}{1 + (1-y)^2}. \qquad (C.20)$$

These are approximations valid for $W^2 \gg m^2$, the full expressions including the discarded small mass terms may be found in Roberts 1990, p13 or Renton 1990, p305.

D
Feynman rules

The Feynman rules give a systematic way of constructing Lorentz invariant amplitudes corresponding to a given order in a perturbation expansion. The normalization of the amplitudes corresponds to that given in Appendix B, particularly Eq. (B.1). Briefly, the topics covered in this appendix are: rules for QED and QCD; SU(3) λ matrices; electroweak propagators and vertices and spin-1 polarization vectors.

QED

Fermions are assumed point-like and are represented by a straight line (Fig. D.1(a)). Photons correspond to a wavy line (same figure (b)). An external particle has one end unattached, whereas a *propagator* is attached to a *vertex* at both ends.

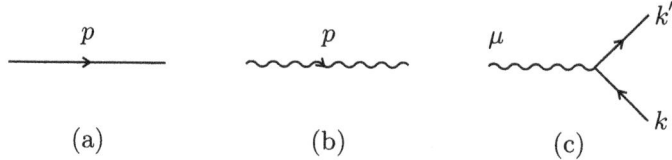

Fig. D.1 QED: (a) fermions; (b) photon; (c) photon-fermion vertex.

External Particle wave-functions

Fermions	Dirac Spinors		
incoming	(initial state)	$f, \bar{f}$ $\;u(p,s),$	$v(p,s);$
outgoing	(final state)	$f, \bar{f}$ $\;\bar{u}(p',s'),$	$\bar{v}(p',s').$

Photons	Polarization vectors	
incoming	(initial state)	$\varepsilon_\mu(k,\lambda);$
outgoing	(final state)	$\varepsilon_\mu^*(k',\lambda').$

Expressions for the Dirac spinors are given in Appendix A and those for the polarization vectors in the last section of this appendix.

Feynman rules

Propagators

fermion $\dfrac{i}{\not{p}-m} = \dfrac{i(\not{p}+m)}{p^2-m^2}$ (Fig. D.1(a))

photon $\dfrac{-ig^{\mu\nu}}{k^2}$ (Lorentz or Feynman gauge)   (Fig. D.1(b))

Vertex

$\gamma ff:$ $-ie\gamma_\mu$ (for charge $+e$)   (Fig. D.1(c)).

QCD

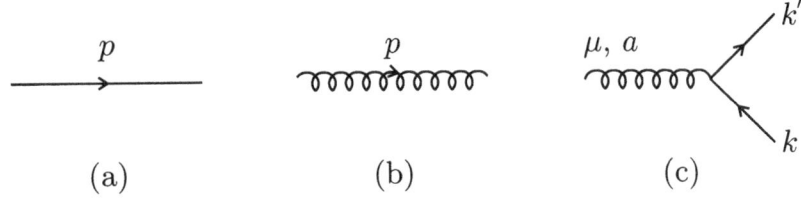

Fig. D.2 QCD: (a) quarks; (b) gluons; (c) quark-gluon vertex.

Refering to Fig. D.2(a), the quarks are assumed to be point-like spin-1/2 particles as in QED, but each external wave-function now has four Dirac spinor components and three colour indices (not shown explicitly). Similarly external gluons are massless spin-1 particles represented by polarization 4-vectors but in addition a colour vector with eight components $(\alpha^a, a = 1,\ldots 8)$.

Gluons	Polarization vectors	
incoming	(initial state)	$\varepsilon_\mu(k,\lambda)\alpha^a$;
outgoing	(final state)	$\varepsilon^*_\mu(k',\lambda')\alpha^{*a}$.

Propagators

quark $\dfrac{i\delta^{ij}}{\not{p}-m}$ (diagonal in colour indices)   (Fig. D.2(a))

gluon $\dfrac{-ig^{\mu\nu}\delta^{ab}}{k^2}$ (Feynman gauge)   (Fig. D.2(b))

Vertices

$gff:$ $-ig_s\dfrac{\lambda^a}{2}\gamma_\mu$ (Fig. D.2(c)).

$ggg:$ $-g_s f_{abc}[g_{\mu\nu}(k_1-k_2)_\lambda + g_{\nu\lambda}(k_2-k_3)_\mu + g_{\lambda\mu}(k_3-k_1)_\nu]$ (Fig. D.3(a)).

$gggg:$ $-ig_s^2[f_{abx}f_{cdx}(g_{\mu\lambda}g_{\nu\sigma} - g_{\mu\sigma}g_{\nu\lambda}) + f_{adx}f_{bcx}(g_{\mu\nu}g_{\lambda\sigma} - g_{\mu\lambda}g_{\nu\sigma})$
$+ f_{acx}f_{dbx}(g_{\mu\sigma}g_{\nu\lambda} - g_{\mu\nu}g_{\lambda\sigma})]$ (Fig. D.3(b)).

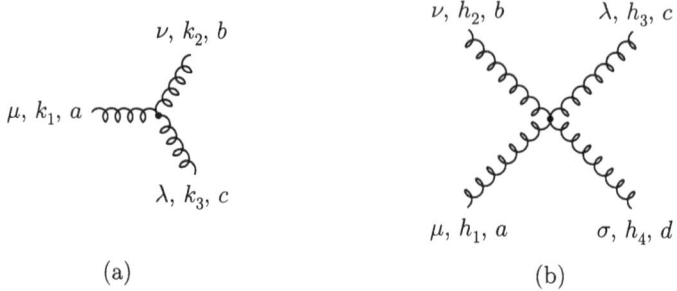

Fig. D.3 QCD gluon self-couplings: (a) trilinear coupling; (b) quadrilinear coupling.

Beyond the tree level, the simple rules given here are not adequate and ghost particles have to be introduced to preserve unitarity. For both QED and QCD, a closed fermion loops requires an additional factor of (-1) — this is the only detail of loop diagrams that is needed in the qualitative discussion of higher order corrections to the propagator in Section 3.4.2.

SU(3)

The SU(3) λ matrices are listed, along with the non-zero elements of the SU(3) structure constants f_{abc}.

$$\lambda_1 = \begin{pmatrix} 0 & 1 & 0 \\ 1 & 0 & 0 \\ 0 & 0 & 0 \end{pmatrix} \quad \lambda_2 = \begin{pmatrix} 0 & -i & 0 \\ i & 0 & 0 \\ 0 & 0 & 0 \end{pmatrix} \quad \lambda_3 = \begin{pmatrix} 1 & 0 & 0 \\ 0 & -1 & 0 \\ 0 & 0 & 0 \end{pmatrix}$$

$$\lambda_4 = \begin{pmatrix} 0 & 0 & 1 \\ 0 & 0 & 0 \\ 1 & 0 & 0 \end{pmatrix} \quad \lambda_5 = \begin{pmatrix} 0 & 0 & -i \\ 0 & 0 & 0 \\ i & 0 & 0 \end{pmatrix} \quad \lambda_6 = \begin{pmatrix} 0 & 0 & 0 \\ 0 & 0 & 1 \\ 0 & 1 & 0 \end{pmatrix}$$

$$\lambda_7 = \begin{pmatrix} 0 & 0 & 0 \\ 0 & 0 & -i \\ 0 & i & 0 \end{pmatrix} \quad \lambda_8 = \tfrac{1}{\sqrt{3}} \begin{pmatrix} 1 & 0 & 0 \\ 0 & 1 & 0 \\ 0 & 0 & -2 \end{pmatrix}$$

Non-zero elements of the totally antisymmetric structure constants are (up to permutations and note that $f_{abc} = f^{abc}$)

$$f_{123} = 1; \quad f_{147} = f_{246} = f_{257} = f_{345} = 1/2;$$

$$f_{156} = f_{367} = -1/2; \quad f_{458} = f_{678} = \sqrt{3}/2.$$

Electroweak

The Feynman rules for tree level graphs in QED and QCD are summarised, together with propagators and fermion–vector–boson couplings for the standard model electroweak theory. The more complicated trilinear and quadrilinear couplings among the vector bosons are omitted as they are not relevant for most of the topics in this book.

External Particle wave-functions

The external particles are fermions (leptons and quarks) and vector bosons (photon, $W^{\pm}$, Z^0). The fermion wave-functions are as for QED and the spin-1 polarization 4-vectors are summarised in the next section.

Propagators

$$\text{fermion} \quad \frac{i}{\not{p}-m} = \frac{i(\not{p}+m)}{p^2-m^2}$$

$$\text{photon} \quad \frac{-ig^{\mu\nu}}{k^2}$$

$$W, Z \quad \frac{i}{k^2-M_V^2}\left(-g^{\mu\nu} + \frac{k^\mu k^\nu}{M_V^2}\right)$$

$$\frac{g_W^2}{M_W^2} = \frac{8G_F}{\sqrt{2}}, \quad g_W \sin\theta_W = e, \quad \text{or } M_W^2 = \frac{e^2}{4\sqrt{2}G_F \sin^2\theta_W}, \quad M_Z = \frac{M_W}{\cos\theta_W},\ldots$$

θ_W is the weak mixing angle, g_W the weak coupling constant and e the electric charge.

Vertices

$$\text{EM} \quad \gamma ff: \quad -ie\gamma_\mu$$

$$\text{CC} \quad W\ell\nu_\ell: \quad \frac{-ig_W}{\sqrt{2}}\gamma_\mu\frac{1-\gamma_5}{2}$$

$$Wqq': \quad \frac{-ig_W}{\sqrt{2}}\gamma_\mu\frac{1-\gamma_5}{2}$$

$$\text{NC} \quad Z^0\nu\nu: \quad \frac{-ie}{2\sin\theta_W \cos\theta_W}\gamma_\mu\frac{1-\gamma_5}{2}$$

$$Z^0qq: \quad \frac{-ie}{\sin\theta_W \cos\theta_W}\gamma_\mu\left[c_L\frac{1-\gamma_5}{2} + c_R\frac{1+\gamma_5}{2}\right]$$

where CKM mixing angles have been ignored in the CC Wqq' couplings and the NC Z^0qq coupling coefficients are given by

Spin-1 polarization vectors

The photon

For a photon with 4-momentum $k^\mu = (k^0, \mathbf{k})$ the 4-vector polarization is $\varepsilon^\mu = (\varepsilon^0, \boldsymbol{\varepsilon})$. To satisfy the Lorentz condition $(\partial_\mu A^\mu = 0)$ requires $k \cdot \varepsilon = 0$

Table D.1 NC couplings of the Z^0.

	c_L	c_R
ν	$+1/2$	0
e^-, μ^-	$-1/2 + \sin^2\theta_W$	$+\sin^2\theta_W$
u, c	$+1/2 - (2/3)\sin^2\theta_W$	$-(2/3)\sin^2\theta_W$
d, s	$-1/2 + (1/3)\sin^2\theta_W$	$+(1/3)\sin^2\theta_W$

and EM gauge invariance ($\varepsilon^\mu \to \varepsilon^\mu + \rho k^\mu$, where ρ is arbitrary) leads to the choice $\varepsilon^0 = 0$. Taking **k** along the z-axis, the two transverse helicity states are given by

$$\varepsilon^\mu(\lambda = \pm 1) = (0, 1/\sqrt{2}, \pm i/\sqrt{2}, 0).$$

The orthogonality and completeness relations are

$$\varepsilon^*(\lambda)\cdot\varepsilon(\lambda') = \delta_{\lambda\lambda'}, \quad \sum_\lambda \varepsilon_\mu(\lambda)\varepsilon_\nu^*(\lambda) = -g_{\mu\nu},$$

where the second relation may also be used for the photon propagator in trace calculations, despite the fact that for a virtual spin-1 state longitudinal and space-like polarization states may contribute (see Aitchison and Hey 1989 pp155-6 for an explanation).

W and Z

For a spin-1 state with mass M_V, $k^2 = M_V^2$ and only the condition $k\cdot\varepsilon = 0$ applies. Taking **k** along the z-axis, the transverse helicity states are as for the photon and the longitudinal state is given by

$$\varepsilon^\mu(\lambda = 0) = \left(\frac{|\mathbf{k}|}{M_V}, 0, 0, \frac{k^0}{M_V}\right).$$

The orthogonality and completeness relations are

$$\varepsilon^*(\lambda)\cdot\varepsilon(\lambda') = \delta_{\lambda\lambda'}, \quad \sum_\lambda \varepsilon_\mu(\lambda)\varepsilon_\nu^*(\lambda) = -g_{\mu\nu} + \frac{k_\mu k_\nu}{M_V^2}.$$

E
Monte Carlo codes

Three sorts of Monte Carlo codes are routinely used in high energy physics experiments: event generators; cross-section calculations and detector simulations. The combination of event generators and detector simulations provide an essential tool for checking detector performance and for calculating acceptance correction functions.

For both cross-section calculation and event generators, the standard reference for HERA ep physics is the DESY Workshop on Monte Carlo Generators for HERA Physics 1998/9[1] with an up to date list of programs at the HERAMC website.[2] Both the proceedings or its online version contain references to the codes mentioned here, unless otherwise stated. A sketch of the ideas underlying the main types of code is given below.

Detector simulation code depends on the detector component technology rather than the physics to be studied. Many groups build their detector simulations using the GEANT package from CERN — a suite of programs providing particle tracking through standard geometries and simulations of many sorts of detectors. For HERA experiments, the FORTRAN based version of GEANT (v3.21),[3] is the standard. For the future, GEANT4[4] has been completely re-organised and rewritten as object-oriented C++ code. Detector simulation will not be considered further here.

Cross-section calculations

Even at leading order and for a fully inclusive process like $ep \to eX$ it is usually necessary to perform a numerical calculation of the cross-section, because the proton structure functions are derived from parton densities available in numerical form. Another reason for needing a numerical model is that detector acceptance cuts in energies and angles are often very difficult, if not impossible to apply analytically. For higher order calculations, whether for radiative corrections (QED or full electroweak) or higher order

---

[1] Proceedings edited by A T Doyle, G Grindhammer, G Ingelman & H Jung, DESY-PROC-99-02, published by DESY. Also www.desy.de/~heramc/

[2] www.desy.de/~heramc/mclist.html

[3] wwwasd.web.cern.ch/wwwasd/geant3.21/

[4] wwwasd.web.cern.ch/wwwasd/geant4/

QCD, the calculations become much more difficult and often require specialized numerical methods. Many of the problems arise from the infrared and collinear singularities associated with massless (or nearly massless) particles. QED offers the paradigm for the basic understanding of these problems as outlined in Section 3.3. First-order radiative corrections for NC and CC ep scattering may be calculated using the HECTOR package.[5] For fully polarized NC ep scattering, the code RADGEN has been developed and is part of the HERMES Monte Carlo package.

Similar problems to those in QED occur in the NLO QCD calculation of cross-sections for jet production in DIS. Three codes are available (DISASTER, DISENT and MEPJET). A brief overview of the problem and how the programs handle the singular integrals has been given in Section 7.4.2 and will not be repeated here.

One other cross-section calculation mentioned in the text (Section 5.6.4) is HVQDIS from Harris and Smith (1998). This is based on the NLO QCD matrix elements for massive $Q\bar{Q}$ pair production in DIS. It is a fixed order, fixed flavour scheme calculation. The code includes the fragmentation of the heavy quark into hadrons (e.g. $c \to D$ or D^*) using the Peterson form of fragmentation function. The infrared and collinear divergences are handled by the subtraction method and Monte Carlo integration is used, thus making it possible to calculate differential distributions for heavy meson production with and without acceptance cuts.

Event generators

An event generator is a computer model of a high energy physics scattering reaction which attempts to simulate the complete process from a given initial state to a proper sampling of possible final states. The Monte Carlo method, that is the sampling of given distribution by random numbers, is at the heart of the procedure. The models described here are based on hard parton–parton interactions, but also have to describe the fragmentation process to observable hadrons. The output of an event generator is a set of 4-vectors describing the 'stable' final state particles (that is particles with lifetimes of order 10^{-8} s and longer). Although event generators are based on physics, there are still many parameters to be tuned. This is particularly true of the hadronization process. One of the advances in understanding that has followed from the quark–parton model and QCD, and which is built in to the event generators, is the extent to which the hadronization of quarks and gluons should be independent of production process. Since a consistent framework is used for many different types of interaction (e^+e^- annihilation, ep DIS and hard $p\bar{p}$ processes), common hadronization parameters have often been fixed by the detailed studies of jet physics at LEP. However there are still parameters specific to ep processes to be fixed by tuning to HERA data.

[5] Arbuzov A et al. (1996), Comp. Phys. Com., **94**, 128.

Broadly the structure of Monte Carlo codes of interest in deep inelastic scattering follow the structure dictated by the factorization theorems and shown diagramatically in Fig. 10.1. Since one is dealing with lepton–hadron or hadron–hadron interactions much of the work concerns modelling aspects of QCD. The codes split logically into three parts: the exact leading order calculation of the hard parton–parton matrix elements; the treatment of the initial state including gluon radiation before the hard scatter; the treatment of the final state with more gluon radiation and hadronization. There are two 'families' of codes that have been developed around different ways of approximating soft gluon emission and fragmentation: those associated with the so-called 'Lund string' approach and those based on angular ordering in gluon emission. An important aspect of QCD that both approaches address, but in different ways, is coherent emission of gluon radiation. One of the most difficult aspects of the event generators is terminating the parton shower and matching the partons to observed particles.

The Lund codes

The Lund codes (PYTHIA, JETSET) and associated programs (LEPTO, ARIADNE, AROMA) use the Lund string approach to parton fragmentation. The basic idea is that as, say a $q\bar{q}$ pair, move apart following an interaction the colour electric field is compressed to a linear vortex line — the Lund string. When sufficient energy is stored in the string it can break to form a new $q'\bar{q}'$ pair and so on. The dynamics of and radiation from colour dipoles are also important ingredients in the Lund approach. The ideas underlying this approach and much more are described in detail in the book by Andersson (1998). JETSET is the package that realises the string fragmentation and the final hadronization to known particles. PYTHIA is a general purpose generator for e^+e^-, ep and p$\bar{\text{p}}$ interactions based on LO matrix elements and parton showers. It is interfaced to JETSET to provide the complete event. Parton showers are a way of simulating gluon emission by parton branching based on QCD splitting functions. LEPTO is a package for deep inelastic lepton–nucleon scattering for NC and CC processes and is based on the full electroweak cross-sections, first order QCD matrix elements and parton showers. AROMA is a similar program which realises heavy quark production via boson–gluon fusion with non-zero quark masses. ARIADNE is an extension of the Lund string idea to take account of coherence in gluon emission. This is done by considering the string as a series of linked colour dipole sources. As one now has an extended source, the coherence condition reduces emission when the antenna size is larger than the wavelength. ARIADNE provides one of the best overall descriptions of the hadronic final state in deep inelastic scattering at HERA.

HERWIG and related codes

The alternative approach to the Lund string are codes based on the idea of angular ordering — which is described in Section 9.5. HERWIG is a general purpose package, similar in scope to PYTHIA, but with angular ordering imposed to give colour coherence both within and between jets. This is done in the parton shower evolution as follows. In the parton splitting $i \to jk$, the variable $\xi_{jk} = (p_j \cdot p_k)/(E_j E_K)$ is introduced in addition to the energy fraction $z_j = E_j/E_i$ which controls the distribution of energy and longitudinal momentum. For massless partons at small angles, $\xi_{jk} \approx \theta_{jk}^2/2$. The value of z_j is given by the DGLAP splitting functions and the value of ξ_{jk} chosen so that it is smaller than the ξ for the previous branching of the parent parton, i. HERWIG also has an alternative approach to hadronization, the cluster model. References for the underlying physics are to be found via the HERAMC pages. The code (v6.5[6]) has now been extended to include supersymmetric processes. CASCADE is a code designed for low-x physics and based on the CCFM equation. It is also applicable to photoproduction and has been extended to include heavy quark production. For heavy quarks, the quark mass modifies the angular-ordered phase space such that soft radiation in cone with angle $\sim m_Q/E$ around the direction of motion of the quark is suppressed.

Other codes

Many other, more specialised codes, are available from the HERAMC sources. Of relevance to high Q^2 DIS at HERA and beyond are those for exotic or beyond the standard model physics. Some BSM physics processes, particularly aspects of supersymmetry are now included in the 'standard' general purpose codes as described above. In addition, there is the code SUSYGEN. W production must be modelled accurately as a potential standard model background to new physics signals in high energy ep interactions. The code EPVEC (Baur et al. 1992) simulates the production of on-mass-shell W's at HERA, but in its original form does not simulate hadronic decays and QCD effects. Some work has been undertaken to rectify this by individual authors in H1 and ZEUS.

A very important aspect of event simulation for ep interactions is the proper description of initial state radiative processes. This is provided by the HERACLES code, which may be interfaced to LEPTO or ARIADNE by use of the DJANGO interface. The advantage of including radiative processes in the simulation is that the radiative correction is included along with other corrections in a single calculation, rather than having to be calculated separately.

---

[6]Corcella G et al. (2001) JHEP01, 010.

F
Data sources

The single most comprehensive archive of high energy physics data is the HEPDATA archive at Durham University UK. It is most easily accessed through the website http://durpdg.durham.ac.uk/HEPDATA.

Unpolarized structure function data

Most of the data sets described in Chapter 5 (Table 5.1) and Chapter 8, together with full references to the original publications, are available from HEPDATA.

Note that some of the data sets included on this website have now been superseded. In the late 1980s there were significant disagreements between various data sets. The muo-production experiments EMC and BCDMS disagreed in both normalization and shape and the neutrino production data of the CDHSW and CCFR collaborations differed in shape at low x. A complete re-analysis of the old SLAC experiments (SLAC-OLD) was undertaken incorporating modern understanding of relevant corrections, in order to help to resolve these discrepancies. The EMC data have now been superseded by the increased precision of the NMC data, and the CDHSW data have been superseded by the increased precision of the CCFR data. A long standing disagreement in the shape of F_2 as measured in muon (NMC) and ν scattering experiments (CCFR) has been resolved by an improved heavy quark treatment in the CCFR experiment (see Section 5.6.2). The data of the WA25, WA59 and IHEP-JINR NDC experiments cannot compete statistically with the modern high precision data sets.

The DIS data sets which are used in current global analyses are firstly, the 1997 NMC F_2 data on muon scattering from p and d targets, including the special extraction of the ratio $F_2^{\mu d}/F_2^{\mu p}$ with minimal systematic error. These data are supplied with full details on point-to-point correlated systematic errors. Secondly, the E665 and BCDMS 1989 F_2 data on muon scattering from p and d targets. Information on correlated systematic errors is available for these experiments, but not in electronic form. Thirdly, 1997 CCFR F_2 and xF_3 data on ν, $\bar{\nu}$ scattering on an Fe target. These data are supplied with full details on point-to-point correlated systematic errors. Fourthly, the reanalysed SLAC data on electron scattering on a proton target. Correlated systematic errors are not available for these data sets.

Finally, HERA data from ZEUS and H1 on electron/positron scattering on a proton target. The high statistics data sets from 1996/1997 onwards have replaced the earlier data, with the exception of the 1994–1997 combined high Q^2 ($> 200\,\text{GeV}^2$) analyses. In due course the high Q^2 NC and CC cross-section measurements from HERA and the structure function xF_3 will also be available from the HEPDATA archive. All the HERA data sets are supplied with full details on point-to-point correlated systematic errors.

Other data for global fits

The HEPDATA archive contains results from on hadronic production of lepton pairs by the Drell–Yan process: $pp \to \ell^+\ell^- X$, $\bar{p}p \to \ell^+\ell^- X$, $pA \to \mu^+\mu^- X$, $\bar{p}A \to \mu^+\mu^- X$ and $\pi^{\pm}A \to \mu^+\mu^- X$. It also contains data on direct photon production in hadronic interactions: $pp \to \gamma X$, $\bar{p}p \to \gamma X$, $\pi^{\pm}p \to \gamma X$, $pA \to \gamma X$ and $\pi^{\pm}A \to \gamma X$ For both Drell–Yan and direct photon production, the experiments were performed at CERN or Fermilab.

For the processes discussed so far, structure functions, D–Y and direct photons, the data in the archive has been the subject of a published 'data review'. Apart from a critical evaluation of the data, this means that the data has been gathered together under one heading in the archive. For other data, such as that for high E_T jet production at hadron colliders, this has not been done yet. The data from individual experiments is available, but it requires a little more work to collect it together. A review on jet data is planned by the HEPDATA team.

Polarized structure function data

There is an almost complete collection of data on polarized structure functions at HEPDATA. The most recent experiments tend to be the ones with highest precision. The data referred to in Chapter 11, Table 11.1 and used in most of the NLO QCD fits to determine the polarized parton densities, are from SMC (CERN), E143, E154/5 (SLAC) and HERMES (DESY). The targets are polarized protons, deuterons and ^{3}He.

G
Parton Parameterizations

Historically, many different parameterizations by various authors were used and these were collected together in the CERNLIB program PDFLIB. However, PDFLIB is no longer maintained at CERN and it is now recommended that the parameterizations and code to be used with them be obtained from the HEPDATA website.[1] This website contains links to the pages of all the individual groups and gives a very comprehensive summary of all PDF sets available world wide. It also includes a link to details of the Les Houches Accord on parton distribution functions (LHA–PDF) which will be important for future applications at the LHC.

Obviously the most up to date PDF sets will give the best description of the data, but some of the older PDF sets are still of interest for investigating the effect of varying input assumptions. A brief summary of the PDF sets available from the main theoretical groups and experimental collaborations is given below.

PDF sets from the theoretical groups
MRST PDF sets

MRST PDFs are extracted to NLO in the $\overline{MS}$ renormalisation scheme using the general mass variable flavour number scheme of Thorne and Roberts to treat heavy quark production (see Section 4.5). In Table G.1, the features of the more recent MRST PDF sets are summarised.

The MRST2002 NLO and NNLO sets represent only a minor update of the 2001 sets with a slightly enhanced high-x gluon. The MRST 2001 sets come in variants which account for a slightly higher or lower value of $\alpha_s(M_Z^2)$ and a variant which gives a better fit to the Tevatron jet data. The MRST2001E sets are the eigenvector PDF sets corresponding to the default MRST2001 PDFs. These can be used to calculate the experimental uncertainties on the PDFs and quantities derived from them (see Section 6.7). The MRST2001 PDFs are also available at LO and at NNLO. The NNLO PDFs come in variations which span the range of our current knowledge of the NNLO splitting functions from 'faster' evolution to 'slower' evolution.

[1] http://durpdg.durham.ac.uk/HEPDATA.

The MRST99 PDFs have more than just historical interest because of the variations on input assumptions.

- There are variants with lower and higher $\alpha_s(M_Z^2)$ values
- There are variants with a harder or softer high-x gluon, deriving from fits to direct photon data with less or more intrinsic-k_T.
- There are the variants labelled 'quarks up and down', which give larger or smaller amounts of u and d quarks at small x, accordingly as the normalizations of the HERA data are varied by $\pm 2.5\%$.
- There are the 'strange up and down' variants, in which the default 50% strangeness suppression factor at Q_0^2 are changed by $\pm 10\%$,
- There are the 'charm up and down' variants, which correspond to varying the charm quark mass from 1.28 to 1.58 GeV around its default value of 1.43 GeV.
- There is a variant with the d/u ratio non zero as $x \to 1$.

All these variants are available in the DIS scheme as well as the $\overline{\text{MS}}$ scheme. The MRST98 PDFs are available in a subset of these variants and they are also available in LO as well as NLO. An interesting additional variant in the MRST98 PDF sets is one which included higher twist terms in the fit. Earlier sets of PDFs from the MRS group are now of historical interest only.

MRST supply the PDFs in files of values on x, Q^2 grids and they supply an interface routine which returns the values of the PDFs at any input x, Q^2 value. Beyond leading order in pQCD, the structure functions are related to parton distributions by convolutions with coefficient functions. As these must match the Thorne-Roberts scheme for heavy quark production, it is easiest to use the code supplied by the authors to calculate structure functions. Such routines are available from the HEPDATA website for processes mediated by virtual photon or $W^\pm$ exchange.

CTEQ PDF sets

In Table G.2, the features of the more recent CTEQ PDF sets are summarised. CTEQ PDFs are extracted to NLO in the $\overline{\text{MS}}$ renormalisation scheme using the zero mass variable flavour number scheme to treat heavy quark production. The CTEQ group have always supplied a LO version of each of their fits as well as NLO versions in both the $\overline{\text{MS}}$ scheme and the DIS scheme. More recently, they have supplied special versions (CTEQXHJ) with more gluon at high-x to give better fits to the Tevatron jet data.

In the CTEQ6 series of PDF fits, two LO versions are supplied. The CTEQ6L version has α_s running at NLO, although the PDF evolution and hard cross-sections are calculated at LO. The CTEQ6L1 version has α_s running at LO and a correspondingly higher value of $\alpha_s(M_Z^2)$. Eigenvector PDF sets, which can be used for estimates of experimental uncertainties, are also supplied corresponding to the CTEQ6M default PDF set. There is also a CTEQ6.1M update of these, which has more symmetrical and reliable errors for some of the eigenvectors. The CTEQ6.1 PDF sets gives as good a fit to the Tevatron jet data as the older special CTEQXHJ fits, so that no special CTEQ6HJ set is supplied.

In the CTEQ5 series of PDF fits, there are additional variants for different treatments of heavy quarks: CTEQ5F3/4 for the fixed flavour number scheme for three or four flavours and CTEQ5HQ for the general mass variable flavour number scheme of ACOT(see Section 4.5). The CTEQ4 PDF sets have the same variants as CTEQ5 and additionally there is a series of variants CTEQ4A1-5 determined with different fixed values of $\alpha_s(M_Z^2)$.

The CTEQ PDFs are generally supplied on grids with an interface routine which returns the values of the PDFs at any input x, Q^2 value.

GRV PDF sets

The GRV98 PDF sets are available at NLO in both $\overline{MS}$ and DIS scheme versions, with $\alpha_s(M_Z^2) = 0.114$, and at LO with $\alpha_s(M_Z^2) = 0.125$. An interface routine returns the light parton densities at any input x, Q^2, and the contributions of the light quarks, charm and bottom quarks to F_2 (for γ^* exchange). Since GRV use the fixed flavour number scheme with three flavours, there are no heavy quark parton densities (see Section 4.5).

PDF sets from experimental collaborations

The ZEUS PDF fit results, described in Chapter 6, are available from the HEPDATA website as the ZEUS 2002 PDFs. These are supplied both as grids and as eigenvector PDF sets from which the experimental uncertainties can be calculated (see Section 6.7). There is an interface routine which calculates the parton densities, structure functions and reduced cross-sections for any input x, Q^2. Variants of these PDF sets for the zero mass variable flavour number scheme, fixed 3-flavour number scheme and the Thorne-Roberts general mass variable flavour number scheme are also available.

The ZEUS PDF sets are the successors to the Botje PDFs which were extracted using the fixed target data and earlier HERA data. The Botje PDFs were the first to take into account experimental uncertainties. They are available from the QCDNUM website
http://www.nikhef.nl/ h24/qcdnum.

The H1 collaboration have not issued publically available PDF sets.

Table G.1 A summary of MRST PDF.

PDFset	$\alpha_s(M_Z^2)$	Features
MRST2002 NLO	0.119	default
MRST2002 NNLO	0.1155	default
MRST2001 NLO	0.119	default
	0.117	lower α_s
	0.121	higher α_s
	0.121	Tevatron Jets
MRST2001E	0.119	eigenvector PDF sets
MRST2001 NNLO	0.1155	default
	0.1155	'faster' evolution
	0.1155	'slower' evolution
	0.118	Tevatron Jets
MRST2001 LO	0.130	
MRST99 NLO	0.1175	default
	0.1175	higher g
	0.1175	lower g
	0.1225	higher α_s
	0.1125	lower α_s
	0.1178	quarks up
	0.1171	quarks down
	0.1175	strange up
	0.1175	strange down
	0.1175	charm up
	0.1175	charm down
	0.1175	larger d/u
MRST99 DIS	0.1175	available in all the same variants as NLO $\overline{\text{MS}}$
MRST98 NLO	0.1175	default
	0.1175	higher g
	0.1175	lower g
	0.1225	higher α_s
	0.1125	lower α_s
MRST98 DIS	0.1175	available in all the same variants as NLO $\overline{\text{MS}}$
MRST98 LO	0.125	default
	0.125	higher g
	0.125	lower g
	0.130	higher α_s
	0.120	lower α_s
MRST 1998 HT	0.117	higher twist

Table G.2 A summary of CTEQ PDF sets.

PDFset	$\alpha_s(M_Z^2)$	Features
CTEQ6M	0.118	NLO $\overline{MS}$
CTEQ6D	0.118	NLO DIS
CTEQ6L	0.118	LO
CTEQ6L1	0.130	LO
CTEQ6M	0.118	eigenvector PDF sets
CTEQ6.1M	0.118	eigenvector PDF sets
CTEQ5M	0.118	NLO $\overline{MS}$
CTEQ5D	0.118	NLO DIS
CTEQ5L	0.127	LO
CTEQ5HJ	0.118	Tevatron Jets
CTEQ5HQ	0.118	Heavy Quarks
CTEQ5F3	0.108	FFN-3
CTEQ5F4	0.112	FFN-4
CTEQ4M	0.116	NLO $\overline{MS}$
CTEQ4D	0.116	NLO DIS
CTEQ4L	0.132	LO
CTEQ4HJ	0.116	Tevatron Jets
CTEQ4HQ	0.116	Heavy Quarks
CTEQ4F3	0.106	FFN-3
CTEQ4F4	0.111	FFN-4
CTEQ4A1	0.110	varying α_s
CTEQ4A2	0.113	varying α_s
CTEQ4A3	0.116	varying α_s
CTEQ4A4	0.119	varying α_s
CTEQ4A5	0.122	varying α_s

Bibliography

Abramowicz, H & Caldwell, A (1999). *Rev. Mod. Phys.* **71**, 1275.
Adams, M R et al. (1990). *Nucl. Instr.. Meth.* **A 291**, 533.
Aitchison, I J R & Hey, A J G (1989). *Gauge Theories in Particle Physics*, 2nd ed, Adam Hilger.
Altarelli, G & Martinelli, G (1978). *Phys. Lett.* **B 76**, 89.
Altarelli, G (1982). *Phys. Rep.* **81**, 1.
Altarelli, G, Ball, R D & Forte, S (2000). *Nucl. Phys.* **B 575**, 313.
Anandam, P & Soper, D E (2000). *Phys. Rev.* **D 61**, 094003.
Andersson, B. (1998). *The Lund Model*, Cambridge University Press.
Anselmino, M, Efremov, A & Leader, E (1995). *Phys. Rep.* **261**, 1, ibid (1997) **281** 399, Erratum.
Arneodo, M (1994). *Phys. Rep.* **240**, 301.
Arneodo, M et al. (1996). *Proceedings of the Workshop on Future Physics at HERA*, Ingelman G, De Roeck A & Klanner R, Eds, DESY Hamburg, Vol.2, p887.
Aurenche, P et al. (1999). *Eur. Phys. J.* **C 9**, 107.
Ball, R D & Forte, F (1994). *Phys. Lett.* **B 335**, 77.
Barone, V, Pascaud, C and Zomer, F (2000). *Phys. Rev. Lett.* **82**, 2451.
Barreiro, F et al. (1996). *Z. Phys.* **C 72**, 561.
Bartels, J, Golec-Biernat, K & Kowalski, H (2002). *Phys. Rev.* **D 66**, 014001.
Bassler, U and Bernardi, G (1999). *Nucl. Instr.. Meth.* **A 426**, 583.
Baur, U, Vermaseren, J A M & Zeppenfeld, D (1992). *Nucl. Phys.* **B 375**, 3.
Bernreuther, W (1983). *Annals Phys.*, **151**, 127.
Bethke, S (2000). *J. Phys.* **G 26**, R27.
Beyer, R et al. (1996). *Proceedings of the Workshop on Future Physics at HERA*, Ingelman G, De Roeck A & Klanner R, Eds, DESY Hamburg, Vol.1, p140.
Bjorken, J D (1969). *Phys. Rev.* **179**, 1547.
Blümlein, J et al. (1996). *Proceedings of the Workshop on Future Physics at HERA*, Ingelman G, De Roeck A & Klanner R, Eds, DESY Hamburg, Vol.1, p23.
Blümlein, J & Böttcher, H (2002). *Nucl. Phys.* **B 636**, 225.
Bodek, A & Yang, U K (2000). *Phys. Rev. Lett.* **84**, 5456.
Botje, M (1997). *QCDNUM16, a fast QCD evolution code*, ZEUS Note 97-066 (unpublished).

Bowler, M (1990). *Femtophysics*, Pergamon Press.
Bravar, A, von Harrach, D & Kotzinian, A (1998). *Phys. Lett.* **B 421**, 349.
Brosky, S J & Farrar, G (1973). *Phys. Rev. Lett.* **31**, 1153.
Buchmüller, W, Rückl, R & Wyler, D (1987). *Phys. Lett.* **B 191**, 442; erratum (1999) ibid **448**, 320.
Bunce, G et al. (2000). *Ann. Rev. Nuc. Part.* **50**, 525.
Buza, M et al. (1996). *Nucl. Phys.* **B 472**, 611.
Cashmore, R J et al. (1996). *Proceedings of the Workshop on Future Physics at HERA*, Ingelman G, De Roeck A & Klanner R, Eds, DESY Hamburg, Vol.1, p163.
Catani, S & Seymour, M H (1997). *Nucl. Phys.* **B 485**, 291, Erratum ibid **510**, 503.
CCFR Collab., Bazarko, A.O. et al. (1995). *Z. Phys.* **C 65**, 189.
CCFR Collab., Seligman, W G et al., CCFR (1997). *Phys. Rev. Lett.* **79**, 1213.
CDF Collab., Abe, F et al. (1996). *Phys. Rev. Lett.* **77**, 5336.
CDF Collab., Abe, F et al. (1998). *Phys. Rev. Lett.* **81**, 5754.
CDF Collab., Affolder, T et al. (2001a). *Phys. Rev.* **D 64**, 032001.
CDF Collab., Affolder, T et al. (2001b). *Phys. Rev. Lett.* **87**, 131802.
CDF Collab., Affolder, T et al. (2001c). *Phys. Rev.* **D 64**, 012001.
Ciafaloni, M, Colferai, D & Salam, G P (1999). *Phys. Rev.* **D 60**, 114036.
Collins, J C et al. (1989). *Factorisation of Hard Processes in QCD in Perturbative Quantum Chromodynamics*, A H Mueller, Ed, World Scientific.
Commins, E D & Bucksbaum, P H (1983). *Weak Interactions of Leptons and Quarks*, Cambridge University Press.
Cooper-Sarkar, A M, De Roeck, A & Devenish, R C E (1998). IJMPA 13 (3385).
Cooper-Sarkar, A M et al. (1999). *J. Phys.* **G 25**, 1387.
Cowan, G (1998). *Statistical Data Analysis*, Oxford University Press.
Cudell, J R et al. (1997). Preprint hep-ph/9712235.
DØ Collab., Abbott, B et al. (1999a). *Phys. Rev. Lett.* **82**, 2451.
DØ Collab., Abbott, B et al. (1999b). *Phys. Rev. Lett.* **82**, 2457.
DØ Collab., Abbott, B et al. (2000). *Phys. Rev. Lett.* **84**, 2786.
DØ Collab., Abazov, V M et al. (2002). *Phys. Lett.* **B 525**, 211.
Dasgupta, M & Webber, B R (1996). *Phys. Lett.* **B 382**, 273.
de Groot et al. (1979). *Z. Phys.* **C 1**, 143.
De Rujula, A et al. (1974). *Phys. Rev.* **D 10**, 1649.
Dittmar, M, Pauss, F & Zürcher, D (1997). *Phys. Rev.* **D 56**, 7284.
Dokshitzer, Yu. L et al. (1991). *Basics of Perturbative QCD*, Editions Frontiers.
Donnachie, A & Landshoff, P V (1992). *Phys. Lett.* **B 296**, 227.
Donnachie, A & Landshoff, P V (1993). *Z. Phys.* **C 61**, 139.
Donnachie, A & Landshoff, P V (2001). *Phys. Lett.* **B 518**, 63.
Donnachie, A et al. (2002a). *Pomeron Physics and QCD*,

Cambridge University Press.
Donnachie, A & Landshoff, P V (2002b). *Phys. Lett.* **B 550**, 160.
Drell, S D & Yan, T M (1970). *Phys. Rev. Lett.* **25**, 316.
Durham (2002). *Advanced Statistical Techniques in Particle Physics*, Whalley M R & Lyons L, Eds, Durham IPPP/02/39.
E155 Collab., Anthony, P L et al. (2000). *Phys. Lett.* **B 493**, 19.
E155 Collab., Anthony, P L et al. (2003). *Phys. Rev. Lett.* **553**, 18.
E605 Collab., Moreno, G et al. (1991). *Phys. Rev.* **D 43**, 2815.
E866 Collab., Hawker, E H et al. (1998). *Phys. Rev. Lett.* **80**, 3715.
E866/NuSea Collab., Webb, J C et al. (2003). Preprint hep-ex/0302019, submitted to PRL.
Ellis, R K, Stirling, W J & Webber, B R (1996). *QCD and collider physics*, Cambridge University Press.
Feynman, R P (1969). *Phys. Rev. Lett.* **23**, 1415.
Feynman, R P (1972). *Photon-Hadron Interactions*, Benjamin.
Forshaw, J R & Ross, D A (1997). *Quantum Chromodynamics and the Pomeron*, Cambridge University Press.
Forshaw, J R, Kerley, G & Shaw, G (1999). *Phys. Rev.* **D 60**, 074012.
Friedman, J I, Kendall, H W & Taylor, R E (1991). *Rev. Mod. Phys.* **63**, 573,597,615.
Gayler, J (2002). Presented at the XXXII Symposium on Multiparticle Dynamics, Alushta, Crimea, Sept. 2002 and to be published in the proceedings, hep-ex/0211051.
Gehrmann, T & Stirling, W J (1996). *Phys. Rev.* **D 53**, 6100.
Giele, W T, Glover, E W N & Kosower, D A (1994). *Phys. Rev. Lett.* **73**, 2019.
Glover, E W N (2002). *Nucl. Phys. Proc. Suppl* **116** 3.
Glück, M, Reya, E & Vogt, A (1992). GRV92, *Z. Phys.* **C 53**, 127.
Glück M, Reya, E & Vogt, A (1994). GRV94, *Z. Phys.* **C 67**, 433.
Glück M, Reya, E & Vogt, A (1998). GRV98, *Eur. Phys. J.* **C 5**, 461.
Glück, M et al. (2001). GRSV, *Phys. Rev.* **D 63**, 094005.
Golec-Biernat, K & Wüsthoff, M (1999). *Phys. Rev.* **D 59**, 014017, *Phys. Rev.* **D 60**, 114023.
Golec-Biernat, K (2002). *Acta Phys. Pol.* **33**, 2771.
Goto, Y etal (2000). AAC Collab., *Phys. Rev.* **D 62**, 034017.
Graudenz, D (1997). DISASTER++ Version 1.0, hep-ph/9710244 (unpublished).
Gribov, L V, Levin, E M & Ryskin, M G (1983). *Phys. Rep.* **100**, 1.
Gruppen, K (1998). *Particle Detectors*, Cambridge University Press.
H1 Collab., Aid, S et al. (1995). *Phys. Lett.* **B 354**, 494.
H1 Collab., Adloff, C. et al. (1997a). *Phys. Lett.* **B 393**, 452.
H1 Collab., Adloff, C. et al. (1997b). *Nucl. Phys.* **B 485**, 3.
H1 Collab., Adloff, C et al. (1998). *Eur. Phys. J.* **C 5**, 575.
H1 Collab., Adloff, C et al. (2000). *Phys. Lett.* **B 479**, 358.
H1 Collab., Adloff, C et al. (2001a). *Eur. Phys. J.* **C 19**, 269.

H1 Collab., Adloff, C et al. (2001b). Eur. Phys. J. C **19**, 289.
H1 Collab., Adloff, C. et al. (2001c). Eur. Phys. J. C **21**, 33.
H1 Collab., Adloff, C et al. (2001d), Phys. Lett. B **523**, 234.
H1 Collab., Adloff, C. et al. (2002). Phys. Lett. B **528**, 199.
H1 Collab., Andreev, V et al. (2003a). Phys. Lett. B **561**, 241.
H1 Collab., Adloff, C et al. (2003b). DESY-03-038, accepted by Eur. Phys.J. C
Haberl, P, Martyn, H-U & Schrempp, F (1991). *Proc. Workshop on Physics at HERA*, Buchmüller W & Ingelman G, DESY, Hamburg. Vol.2, p1133.
Halzen, F & Martin, A D (1984). *Quarks and Leptons*, Wiley.
Hamberg, R et al. (1991). Nucl. Phys. B **359**, 343; Van Neerven, W L & Zijlsrtra, E B (1992), Nucl. Phys. B **382**, 11.
Harris, B W and Smith, J (1998). Phys. Rev. D **57**, 2806.
HERMES Collab., Ackerstaff, K et al. , (1999). Phys. Lett. B **464**, 123.
HERMES Collab., Airapetian, A et al. , (2000). Phys. Rev. Lett. **84**, 2584.
Iancu, E (2002). *Plenary talk at Quark Matter 2002, to be published in the proceedings*, hep-ph/0210236.
Iancu, E, Leonidov, A & McLerran, L (2002). *Lectures at the 2001 NATO Advanced Study Institute on QCD perspectives on hot and dense matter*, Cargèse, Corsica, published in the proceedings, hep-ph/0202270.
Ingelman, G & Rückl, R (1989). Z. Phys. C **44**, 291.
Jackson, J D (1975). *Classical Electrodynamics* 2nd ed, Wiley.
Jung, H & Salam, G P (2001). Eur. Phys. J. C **19**, 351.
Kalinowski, J et al. (1997). Z. Phys. C **74**, 595.
Khoze, V A et al. (2001). Eur. Phys. J. C **19**, 313.
Klein, M & Riemann, T (1984). Z. Phys. C **24**, 151.
Köpp, G et al. (1995). Z. Phys. C **65**, 545.
King, B.J. et al. (1991). Nucl. Instr.. Meth. A **302**, 254.
Krämer, M et al. (2000). Phys. Rev. D **62**, 096007.
Kwiecinski, J, Martin, A D & Stasto, A M (1997). Phys. Rev. D **56**, 3991.
Martin, A D et al. (1998). MRST98, Eur. Phys. J. C **4**, 463.
Martin, A D et al. (2000). MRST, Eur. Phys. J. C **18**, 117.
Martin, A D et al. (2001). MRST2001, Eur. Phys. J. C **23**, 73.
Martin, A D et al. (2002). MRST, preprint hep-ph/0211080.
Mirkes, E & Zeppenfeld, D (1996). Phys. Lett. B **380**, 205; and MEPJET 2.0 Program Manual 1997.
Mueller, A H & Qiu, J (1986). Nucl. Phys. B **260**, 427.
Muta, T (1998). *Foundations of QCD*, 2nd Ed, World Scientific.
NMC Collab., Arneodo, M et al. (1997). Nucl. Phys. B **487**, 3.
PDG, Groom, D E et al. (2002). Phys. Rev. D **66**, 010001.
Pascaud, C & Zomer, F (1995). preprint LAL-95-05 (unpublished).
Peskin, M E & Schroeder, D V (1995). *An introduction to quantum field theory*, Addison Wesley.
Prytz, K (1993). Phys. Lett. B **311**, 286.
Renton, P (1990). *Electroweak Interactions*, Cambridge University Press.

Roberts, R G (1990). *The structure of the proton*, Cambridge University Press.
Roberts, R G & Thorne, R (1998). *Phys. Rev.* **D 57**, 6871.
Roberts, R G & Thorne, R (2001). *Eur. Phys. J.* **C 19**, 339.
Sakumoto, W.K. et al. (1990). *Nucl. Instr.. Meth.* **A 294**, 179.
Salam, G P (1997). *Proc. Ringberg Workshop on New Trends in HERA Physics. Tegernsee, May 1997*, Kniehl B A, Kramer G & Wagner A, Eds, World Scientific, p71.
Salam, G P (1999). *Acta Phys. Pol.* **B30**, 3679.
SMC Collab., Adeva, B et al. (1998). *Phys. Rev.* **D 58**, 112001.
Smith, S R et al. (1981). *Phys. Rev. Lett.* **46**, 1607.
Spiesberger, H et al. (1991). *Proc. Workshop on Physics at HERA*, Buchmüller W & Ingelman G, DESY, Hamburg. Vol.2, p798.
Stasto, A M, Golec-Biernat, K & Kwiecinski, J (2001). *Phys. Rev. Lett.* **86**, 596.
Sterman, G (1993), *An Introduction to Quantum Field Theory*, Cambridge University Press.
Stösslein, U (2002). *Intl. J. Mod. Phys.* **A 17**, 3220.
Stump, D et al. (2003). CTEQ Collab., Preprint hep-ph/0303013.
Thorne, R S (2000). *Phys. Lett.* **B 474**, 372.
Thorne, R S (2002). private communication.
Thorne, R S (2003). private communication.
Tung, W et al. (2002). CTEQ6, JHEP **0207**, 012.
Tymieniecka, T & Zarnecki, A F (1992). DESY 92-137 (unpublished).
Vallee, C (2001). *Proc. EPS HEP Conf. Budapest*, Horvath D, Levai P & Patkos A, Eds, JHEP Proceedings hep/2001/161.
van Neerven, W L & Vogt, A (2000). *Phys. Lett.* **B 490**, 111.
Veltman, M (1994). *Diagrammatica*, Cambridge University Press.
Virchaux, M & Milsztajn, A (1992). *Phys. Lett.* **B 274**, 221.
Webber, B R (1993). *J. Phys.* **G 19**, 1567.
Wüsthoff, M & Martin, A D (1999). *J. Phys.* **G 25**, R309.
Yndurâin, F J (1999). *The theory of quark and gluon interactions* 3rd ed, Springer Verlag.
Zarnecki, A F (1999). *Eur. Phys. J.* **C 11**, 539.
Zarnecki, A F (2000). *Eur. Phys. J.* **C 17**, 695.
ZEUS collab., Derrick, M et al. (1996). *Z. Phys.* **C 72**, 399.
ZEUS Collab., Breitweg, J et al. (2000a). *Phys. Lett.* **B 487**, 53.
ZEUS Collab., Breitweg, J et al. (2000b). *Eur. Phys. J.* **C 14**, 239.
ZEUS Collab., Breitweg, J et al. (2000c). *Eur. Phys. J.* **C 16**, 253.
ZEUS Collab., Breitweg, J et al. (2000d). *Phys. Lett.* **B 471**, 411.
ZEUS Collab., Breitweg, J et al. (2001a). *Eur. Phys. J.* **C 21**, 443.
ZEUS Collab., Chekanov, S et al. ZEUS (2001b). *Phys. Lett.* **B 511**, 19.
ZEUS Collab., Chekanov, S et al. (2002a). *Phys. Lett.* **B 539**, 197.
ZEUS Collab., Chekanov, S et al. (2002b). *Eur. Phys. J.* **C 23**, 13.
ZEUS Collab., Chekanov, S et al. (2002c). *Phys. Lett.* **B 547**, 164.

ZEUS Collab., Chekanov, S et al. (2003a). Eur. Phys. J. **C 28**, 175.
ZEUS Collab., Chekanov, S et al. (2003b). Phys. Rev. **D 67**, 012007.
ZEUS Collab., Chekanov, S et al. (2003c). DESY-03-012, to be published.
ZEUS Collab., Chekanov, S et al. (2003d). DESY-03-093, submitted to Eur. Phys. J. C.
ZEUS Collab., Chekanov, S et al. (2003e). In preparation.

Index

A bold entry indicates the primary source.

Abelian, **9**,36
Acceptance, 117,**119**
Adler sum rule 31,**32**
  Bell Jackiw anomaly, 325
Aligned jet configuration, 265
Altarelli and Parisi, 69
Angular ordering, 256
Anomalous dimension, 54,**55**,78,100
  high p_T leptons, 354
Asymptotic freedom, 9,**51**
Axial anomaly, 320,**325**
  vector couplings, 27,210,221,**225**

BFKL, 232,**274**
  amplitude, 246
  equation, 244
BGF, **42**,74,189
BK equation, **263**,271
Balitsky Kovchegov equation, **263**,271
Balitsky, Fadin, Kuraev
       and Lipatov, 232
Beam flux, 138
Bernstein moments, 179
Bjorken x, **5**,15
  scaling **7**,21,23
  sum rule, **324**,330
Bloch–Nordsieck theorem, 43
Boson gluon fusion, **42**,74,189
Breit frame, **190**,207
Brick wall frame, 191
Burkhardt–Cottingham
       sum rule, 324

CCFM equation, 256
CCFR, 110
CGC, **262**,271
CTEQ, **141**,156,186,392
Callan–Gross relationship, **21**,23,82
Callan and Symanzik, 53
Casimir operators, 40
Charge renormalization, 47,**49**
Chew–Frautschi diagram, 240
Chi-squared (χ^2) tolerance, 171
Chirality, **18**,33
Chudakov effect, 256
Ciafaloni, Catani, Fiorini
       and Marchesini, 257
Coefficient functions, 81
Collinear Singularities, 43
Colour, 36
  factor, 38
  dipole models, 263
Collider, 104
Collinear approximation, 23
Colour glass condensate, **262**,271
Compositeness, 344,**351**
Cone algorithm, 195
Contact interactions, **350**,362,363
Correlated systematic
       uncertainties, 165
Counter terms, 47
Counting rules, 144
Cross-sections, 372
  formula, 368
Crossing symmetry, 370

DA method, 106
DAS, 234
DGLAP, **69**,75,77,232
  equations, 79

evolution, 84
DIS, 6
DLLA, **232**,234
Deep inelastic scattering, 6
Detectors, 107
Diagonalization, 169
Diffraction, **11**, 266
Dijet production, 197,299
Dimensional regularisation, 46
Dipole models, **263**,266,269,271
Dirac equation, 364
Direct photon production, 158,**304**
Dokshitzer, Gribov, Lipatov,
 Altarelli and Parisi, 10,**69**
Donnachie and Landshoff, 242
Double Angle method, 106
Double asymptotic scaling, 232,**234**
Drell–Yan process, 281
Dynamically generated partons, 175

E665, 108
EMC effect, 120
Efficiency, 119
Efremov–Leader–Teryaev
 sum rule, 324
Eigenvector PDF set, 141,**169**
Electroweak
 couplings, 27,210,221,**225**,380
 parameters, 218
 radiative corrections,220
Ellis–Jaffe sum rule, **324**,331
Event generators, 384
Exclusive process, 5

FFNS, **91**,161
Factorization, 10,**65**,280,281
 scale **66**,163
Feynman diagrams, 2
 rules, 378
 evolution, 85
 ladder, 65,245
Fixed flavour number scheme, 91
Flavour number schemes, 89

GLS sum rule, **31**,183

Gluon
 distribution, 148–151,154,158,203,205
 evolution, 85
 ladder, **65**,245
Golec-Biernat–Wüsthoff model, 267
Gottfried sum rule, **34**,146
Gribov, Levin, Ryskin,
 Mueller and Qiu, 261
Gross Llewellyn-Smith
 sum rule, **31**,183

H1 detector, 111
HERA collider, 104
 detectors, 111
Hadronic current, 58
Hand convention, 42,**376**
Handedness, **18**,33
Hard pomeron, 242
Heavy flavour, 89
Heavy quark, **89**,161
Hessian method, 168
High density QCD, 261
High p_T jet production, 159,**294**
Higher twist, 64,**87**,162,183,188,275

Impact factors, **249**,264
Inclusive process, 5
Inelastic scattering, 4
Infinite momentum frame, 23
Infrared singularities, 43

Jacobi polynomials, 179
Jacquet–Blondel estimator, 106
Jet algorithms, 195
 measures, 194
 production, 181,**189**,294

k_T algorithm, 196
 broadening, 158
Kinematic peak, 105
Kinoshita, Lee, Nauenberg
 theorem, 45

LL(Q^2), 232
LL($1/x$), 232

LLA, **79**,232
LO, 52,**82**
Lambda-QCD (Λ_{QCD}), **52**,181
Leading logs, **57**,78
  approximation, **79**,232
Leading order, 50,52,56,64,**82**
Leading twist, **63**,64
Leptonic current, 58
Leptoquarks, **345**,353,362,363
Light cone dominance, 59
  cone variables, 59
Limiting fragmentation, 263
Lipatov pomeron, 242
Longitudinal structure
    function, **82**,153,165,274
  momentum ordering, 245
Loops, **45**,82
Low x, 228
Luminosity, **138**,311
  monitor, 292
Lund string, 385

MRST, **141**,153,154,156,186,392
$\overline{MS}$, 47
$\overline{MS}$, 47
Mandelstam variables, 295
Massive quark, 89
Matching prescriptions, 89
Mellin moments, 179,**232**
Minimal subtraction scheme, 47
Model assumptions, 159,**160**
  uncertainties, 159,183
Modified minimial
    subtraction scheme, 47
Moments, **63**,64,78,179
Moment space, 178
Momentum sum rule, **9**,32
Monte-Carlo data, 118
  simulation, 117,**383**

n-space, 178
NLL(Q^2), 232
NLL($1/x$), 232
NLO, 52,56,**82**
NNLO, **82**,95,163,188

Next-to-leading logs, 57
  approximation, 232
Next-to-leading order, 52
Non-linear effects, 260,**261**
Non-Abelian, **9**,37
Non-singlet, **80**,143
Normalization, 170
Nuclear effects, **120**,141
Number sum rules, 31

OPE, **60**,61
Offset method, 167
Operator product expansion, **60**,61
Optical theorem, 370

PDF, 19
  set, 141
PT method, 106
Parton densities, 10
  density function, 65
  distribution function, **19**,140,147
  model, 7,22
  parametrization, **143**,144,389
  parton luminosity, 312
Pauli Villars regularisation, 46
Phase-space, 368
  slicing, 193
Physical region, 102
Pipeline, 113
Polarization parameter, 130
Polarized asymmetries, 318,331,338
  lepton beams, 217
  parton densities, 327
  PDFs, 327
  structure functions, 319
Pomeron, 241
  intercept, **241**,251
Pomeranchuk ad Okun, 241
Prompt photon production, 158,**304**
Propagator, 378
Protojet, 196
Prytz method, 84
Purity, 119

Q^2, 5

QCD, **9**,36
  Compton process, **40**,71,189
  improved parton model, **10**,66,69
QCDC, **40**,189
QPM, 14
Quantum chromodynamics, **9**,36
Quark distributions, 157
Quark–parton model, 14,**15**

RGE, 53
Radiative corrections, 43,**115**,220
Rapidity, 278
Regge intercept, 241
  pole, 239
  theory, 236
  trajectory, 239
Regularization, 46
Renormalization, 45
  group method, 52
  group equation, 53
  scale, 163
Renormalon, 87
Resolution, **103**,106,107
Resummation at low-x, 274
Rotation matrices, 366
Running coupling constant, 46,**50**
Rutherford scattering, 1

SU(3) λ matrices, 380
Saturation, 261
  scale, **262**,268
Scale uncertainty, **163**,297
Sea distribution, 150,151,157
  quarks, **26**,30
Shadowing, 261
Sigma (Σ) method, 107
Single scale process, 250
Singlet, **80**,143
  evolution, 85
Singularities, 43
Slavnov and Taylor, 49
Soft gluon emission, 95,**158**
  resummation, 95,**158**
Soft pomeron, 241
Splitting functions, 10,**72**,75,78,100

Spin crisis, 319
Strong fine structure constant, 11
  coupling constant, 11,**50–52**, 181,200
Structure functions, 6
  constants, 37,**380**
Stuckelberg, Peterman,
        Gell-Mann and Low, 53
Subtraction techniques, 193
Sum rules, **31**,**32**,34,324
Summation schemes, 230

Theoretical uncertainties, 159,183
t'Hooft and Veltman
        regularisation, 46
Tolerance, 171
Top production, 359
Transition region, 266
Transverse momentum
        ordering, 79
Trigger, **102**,109,113
Truncated moemnts, 179
Twist, 62
Two scale process, 250

Unitarity, 260

VFNS, 91
Valence distribution, 150,154,216
  quarks, **26**,30
  evolution,84
Variable flavour number
        schemes, 91
Vector couplings, 27,210,221,**225**
Vertex, 378

W decay asymmetry, 292
  production, **291**,354
Wandura–Wilczek relation, 324
Ward–Takahashi identities, 49
Weak neutral couplings, 27,210,221,**225**

Z production, 289
ZEUS detector, 111
ZMVFNS, **91**,161
Zero–mass flavour schemes, 91

The manufacturer's authorised representative in the EU for product safety is
Oxford University Press España S.A. of el Parque Empresarial San Fernando de
Henares, Avenida de Castilla, 2 – 28830 Madrid (www.oup.es/en or product.
safety@oup.com). OUP España S.A. also acts as importer into Spain of products
made by the manufacturer.

www.ingramcontent.com/pod-product-compliance
Lightning Source LLC
LaVergne TN
LVHW011000250326
834688LV00003B/43